FUNDAMENTALS OF
HAZARDOUS
MATERIALS
INCIDENTS

REGINALD L. CAMPBELL
ROLAND E. LANGFORD

Technical Editor
TERRY L. McARTHUR

 LEWIS PUBLISHERS

Library of Congress Cataloging-in-Publication Data

Campbell, Reginald L.
  Fundamentals of hazardous materials incidents / authors, Reginald L.
Campbell, Roland E. Langford.
          p.   cm.
  Includes bibliographical references and index.
  ISBN 0-87371-362-1
  1. Hazardous substances — Accidents — Management.   I. Langford,
Roland E.   II. Title.
T55.3H3C353   1991
604.7 — dc20   90-13477
                                                                        CIP

Second Printing, 1991

LEWIS PUBLISHERS, INC.
121 South Main Street, Chelsea, Michigan 48118

PRINTED IN THE UNITED STATES OF AMERICA

Mr. Campbell was born in Hartford, Connecticut. His undergraduate degree was from Fairmont State College, Fairmont, West Virginia, and Yale-New Haven Hospital, New Haven, Connecticut. He received his Masters of Science from Marshall University, Huntington, West Virginia. He was an Assistant Professor of Medicine, Hahnemann Medical College of Allied Health Sciences, and instructor of Occupational Health at the U.S. Department of Labor's National Mine Health and Safety Academy, Beckley, West Virginia. He is presently Chief Executive Officer of Campbell Associates, Inc., 533 Suffolk Drive, Sierra Vista, Arizona, a consulting firm specializing in occupational safety and health issues.

Mr. Campbell is noted in several biographical references, including *Who's Who in the Southwest*, *Who's Who in the World*, and *Personalities of the Americas*.

Dr. Langford was born in Owensboro, Kentucky and grew up in Savannah, Georgia. He received an A.A. from Armstrong State College in 1965, a B.S. in chemistry and physics from Georgia Southern College in 1967, and an M.S. in solution thermodynamics in 1971 and a Ph.D. in physical geochemistry in 1974, both from the University of Georgia under a National Defense Education Act Fellowship.

He taught chemistry, geology, environmental chemistry, and physical science at the Savannah Science Museum, Bainbridge College, Georgia Military College, and Georgia Southern College before becoming the Chief of Clinical Chemistry at the U.S. Army Academy of Health Sciences. Following that assignment, Dr. Langford served as a consultant in hazardous waste management at the U.S. Army Environmental Hygiene Agency and then as Commander of the Environmental Sanitation Detachment in Taegu, Korea for the U.S. Army's 5th Preventive Medicine Unit. In Korea, he served as an adjunct faculty member to the University of Maryland. Upon his return to the U.S., he became Chief of Environmental Health at Fort Huachuca, Arizona and Preventive Medicine Officer for the U.S. Army Information Systems Command.

Dr. Langford has served as a Judge at the Westinghouse International Science Fair, a member of the review panel for the National Science Foundation's Comprehensive Assistance to Undergraduate Science Education, a member of the scientific review panel for the Hazardous Substances Databank of the National Library of Medicine, and a consultant to the International Union of Operating Engineers. He is a Certified Industrial Hygienist (CIH), Certified Hazardous Materials Specialist (CHMS), Registered Sanitarian (RS), Fellow of the American Institute of Chemists (FAIC), Diplomate of the American Academy of Sanitarians, and Engineer-in-Training (EIT). He is author or coauthor of some twenty papers in thermodynamics, geochemistry, and hazardous materials and is listed in thirteen biographical references, including *Who's Who in the World* and *Who's Who of Emerging Leaders.*

In 1988, he enrolled at the University of North Carolina - Chapel Hill to work toward a Ph.D. in radiological hygiene.

PREFACE

This book grew out of notes and lesson plans used by R.L. Campbell and R.E. Langford of Campbell Associates, Inc. in training hazardous sites workers and first responders. Given the requirements for the training of all employees working with hazardous materials and in the spirit of workers' right-to-know, there has been a need for a simple, yet fairly complete, text for use by trainers. It is hoped that this book will help fill this need.

Our appreciation is expressed to many people, including Dr. C.E. Melton, Dr. James E. Watson, Mr. Terry McArthur, Mr. Jon Lewis of Lewis Publishers, and Miss Janice M. Morey of CRC Press, as well as all the students who have been through our training classes, for the helpfulness of their suggestions and comments in improving the content of this book.

Both of us would like to express deep appreciation for the understanding from the ladies in our lives during the many 2:00 AM typing sessions and the hours spent away from them in front of the word processor.

To all others whom we may have overlooked, thank you.

R.L. Campbell
R.E. Langford

Sierra Vista, Arizona, 1990

CONTENTS

1   INTRODUCTION TO HAZARDOUS MATERIALS INCIDENTS

1.1. PURPOSE

This instructional manual has been developed from various training programs conducted by Campbell Associates, Inc. to fill the need for a text that can be used in the 8-, 24-, or 40-hour Occupational Safety and Health Administration (OSHA) mandated courses and in general training for hazardous waste site workers, first responders to emergency incidents, handlers of hazardous materials, and general employees working around potentially hazardous substances and situations. The purpose of the book is to present the information needed by these groups to adequately prepare them for most of the hazards that may occur.

The book has been produced through the joint efforts of Reginald L. Campbell and R. Everett Langford. Both authors have extensive experience in hazardous materials/wastes, emergency response, and training/education.

1.2. DEVELOPMENT

The course and book have been developed in part from materials Mr. Campbell accumulated while an instructor of Occupational Health at the National Mine Health and Safety Academy of the U.S. Department of Labor's Mine Safety and Health Administration, Beckley, West Virginia. Other portions were developed by Dr. Langford for use in training programs for hazardous materials workers.

Portions of these materials have been presented in-house and externally to various groups, including labor unions, government employees, emergency response personnel, and the general public.

While the content is generally all-encompassing, other courses, seminars, and texts may be used to supplement this book or to delve more deeply into some of its topics. Some of these supplementary materials are presented in Appendix A.

The instructor is free to add or delete material to better fit the particular training course. For example, the 8-hour introductory course would only briefly touch upon many of the topics discussed in these pages. For the 40-hour course, the entire volume would possibly be covered. Specific chapters might be skipped; for example, if radioactivity is not present, Chapter 14 may be left out.

1.3. TO THE STUDENT

Much of this material is presented in greater detail than you are expected to know; however, the book will serve as a permanent reference for review later if you become interested in or need more specific information.

Don't be frightened by the amount of material; it is intended to contain just about anything you might need in order to respond or function safely and healthfully in your work environment. Much may not apply to your job or situation; trust your instructor to let you know how much does and what you need to know. Good luck — and safe working!

2 INTRODUCTION TO ANATOMY AND PHYSIOLOGY

2.1. INTRODUCTION

The human body is composed of many specialized organ systems, including the digestive, respiratory, circulatory, lymphatic, endocrine, nervous, and reproductive systems. Each of these can be affected by chemicals.

It is not the purpose of this chapter to make you a biologist; only those technical terms that are necessary will be used, and every attempt will be made to keep the topics as simple as possible.

In order to understand why laws and regulations have been made to limit human and environmental exposures to chemicals, it is necessary to have a basic understanding of human anatomy and physiology.

2.2. DEFINITIONS

Some of the definitions needed in this chapter are as follows:

- *Anatomy* is the study of the structure of plants and animals.
- *Physiology* is the study of how plants and animals function.
- *Metabolism* is the sum total of all the chemical reactions taking place in the body.
- *Exposure* refers to proximity to or contact with something.

3

2.3. INTRODUCTION TO ANATOMY AND PHYSIOLOGY

Most chemical exposures experienced by workers and others are by means of only a few organ systems, chiefly the respiratory system, skin, and digestive system. However, once chemicals enter the body, they can be moved about by other systems, such as the circulatory and lymphatic, and affect even others, such as the nervous and endocrine. Therefore, we need to consider all these human systems.

2.4. THE RESPIRATORY SYSTEM

Humans can exist for many days without food and for hours without water, but only for minutes without oxygen. In addition, metabolism produces carbon dioxide, which is toxic in sufficient concentrations. The respiratory system brings the needed oxygen to the body and removes the carbon dioxide.

Just as energy is given off by burning a log in the fireplace, so do humans get their energy from the "burning" of food (oxidation) in their cells. In this process, oxygen combines with food to produce water and carbon dioxide. The oxygen and carbon dioxide are carried by the blood throughout the body. Therefore, the respiratory and circulatory systems are closely related. It might be said that these two automatic systems work together so that all the cells that make up your body have an acceptable environment.

The earth's atmosphere, which normally contains about 21% oxygen, is breathed into the lungs by the force of the *diaphragm*, a sheet of muscles located in the body cavity, which pulls on the lungs to force air in and then pushes on the lungs to force air out. The entry air contains 21% oxygen and very little carbon dioxide (0.04%) and water (0.50%), and the exit air contains about 4% carbon dioxide and 6% water with only about 16% oxygen.

Figure 2-1 is a diagram of the human respiratory system.

Air entering the nose is warmed and filtered by a series of bony shelves that are covered with mucous membranes. Larger foreign particles are filtered out by the nasal hairs and mucous plates. The air then passes through the *pharynx* and past the *glottis*, a flap that prevents food from the mouth from entering the respiratory system. The air then passes through the *larynx*, which contains the vocal cords, into the largest of the respiratory passages, the *trachea*. The trachea is a large tube, about 2 cm (1 inch) in diameter, ringed with bands of cartilage that prevent it from collapsing. All these passages are lined with *cilia*, which are small hairlike projections that beat like small brooms to sweep foreign materials up and out of the respiratory tract.

The trachea branches into two major passages called *bronchi* (singular: bronchus), one going to each lung. These branch in turn into ever smaller passages called *bronchioles*. All these contain cartilage to prevent collapse.

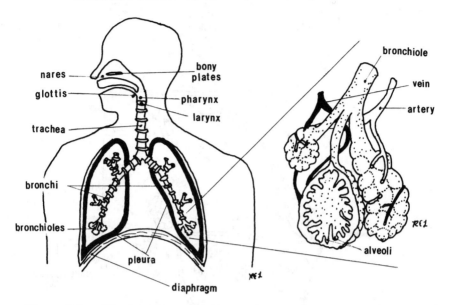

Figure 2-1.   The human respiratory system.

The very smallest bronchioles, however, do not have cartilage rings and simply end in sacs called *atria* (singular: atrium), each of which contains a number of chambers called *alveoli* (singular: alveolus). Alveoli are shown in Figure 2-1. It is at the alveoli that the exchange of oxygen to the blood's hemoglobin and carbon dioxide from the blood takes place. The alveoli on the end of the bronchiole appear like bunches of grapes on a vine. The cells of the alveoli are very sensitive to many chemicals, and the blood lies only one cell thickness away so that transfer of these gases and other substances may occur.

In the normal, healthy person, the lungs appear like pink, spongy masses, but they can become black and rigid for smokers and miners. The lungs are surrounded by tissues called *pleura*, which are two sheets separated by a fluid allowing motion of the lungs as they expand and contract. The disease pleurisy results when this fluid dries up and breathing becomes painful.

2.5. THE SKIN

The skin is the largest organ of the body and is constantly exposed to the environment, so quite often it suffers from injuries due to chemicals.   In humans, the skin forms a sheath to protect the body from the environment, so it is a physical barrier; but it is much more, since it is the location of hair follicles, sweat glands, and oil glands. A cross-section diagram of human skin is shown in Figure 2-2. It is a layered system with an outer layer called the

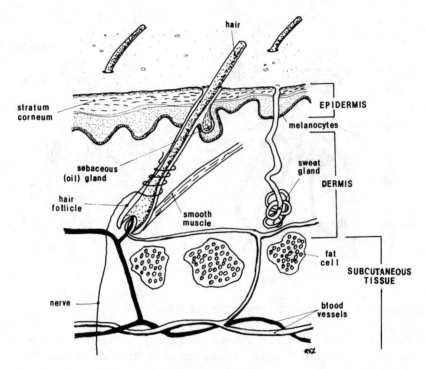

Figure 2-2.   Cross section of the human skin.

epidermis, itself made of two parts: the outermost *corneum,* which is a tough mixture of flattened dead skin cells, oil, free fatty acids, and dirt, and the *germinativum* just below made of living skin cells. Skin cells are made here and move upward with time, taking from 14 to 28 days to reach the surface, where they live for up to 14 days and eventually are removed by rubbing and washing. Thus, an injury to the skin can take up to a month to heal, since it takes that long for cells to reach the surface.

The next layer is the *dermis,* consisting of the corium (which is connective tissue), hair roots, sweat glands, oil glands, and fat deposits. The oil glands produce oils that are oxidized to free fatty acids. These fatty acids tend to make the skin a hostile environment for most bacteria and viruses. In addition, the oils and acids tend to act as a thin chemical barrier for exposure to some chemicals. Unfortunately, many common organic solvents remove the oils and fatty acids, causing the skin to become dry and inflamed. Then bacteria and viruses can more easily attack it.

The *sweat glands* are important in the regulation of body temperature. When the body becomes too hot, sweat is produced. Sweat then evaporates off the skin, cooling it and the blood in the vessels below. The skin also contains nerve receptors for heat and cold, touch, and pain. It is also the anchor for the hair shafts, which grow from the papilla in the dermis and pass through the *follicles* to the outside.

2.6. THE DIGESTIVE SYSTEM

The digestive, or gastrointestinal, system is a long tube from the mouth to the anus. Its function is to pass food through the body and wastes out. A diagram of the digestive system is shown in Figure 2-3. Food enters the mouth, passes over nerve sensors in the tongue called taste buds, and passes the *glottis*, which acts as a flap to separate the mouth from either the digestive or respiratory systems, as necessary. The food is then moved by ripplelike motion down the *esophagus*, a tube leading to the stomach. Muscles at each end of the stomach act to close the contents off from the esophagus at one end and the intestines at the other. Although some digestion of food takes place for starch in the mouth, the first real digestion occurs in the stomach, where the food is mixed with hydrochloric acid and stomach enzymes. The *stomach* is a saclike muscle; it contracts and expands to pulverize and mix food with the acid.

From the stomach, food passes into the upper portion of the small intestine called the *duodenum*, where various other enzymes break down other food constituents. The duodenum is the primary location where fatty foods are broken down into simpler compounds. The duodenum is only about 25 centimeters (10 inches) long. The next three-meter (ten-foot) portion is called the *jejunum*, which is very rich in enzymes that further digest the food. The remaining 4 meters (12 feet) of small intestine is called the *ileum*, where the food products, broken down into simpler compounds, pass to the blood through hairlike projections into the intestines. These hairlike objects are called *villi* (singular: villus) and are shown in Figure 2-3. They are very small, so they greatly increase the area of absorption.

The small intestine then joins to the large intestines, collectively called the *colon*. The colon is chiefly a storage location, although there is a small amount of water transfer into and out of the colon. The food is mixed with mucus and bacteria and becomes waste matter called feces. When the feces reach the *rectum*, which is normally empty, the desire to eliminate becomes great (usually after about 36 hours). When the muscles of the anus open, the wastes are passed to the outside of the body, and the process of digestion and elimination is complete.

2.7. THE NERVOUS SYSTEM

The human nervous system is very complex, and we shall only briefly

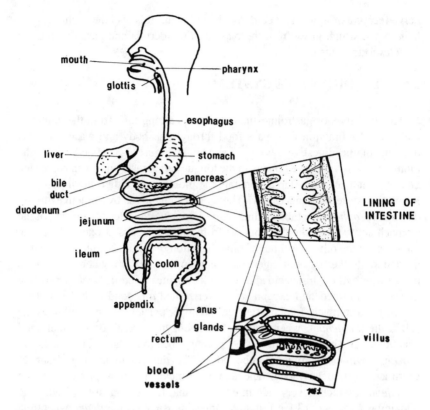

Figure 2-3.   The human digestive system.

discuss it. The *brain* is the center of the system. It functions as a processing center, with impulses coming from receptor nerves throughout the body and going out to control all the body's functions. The brain is connected with all parts of the body by nerves. These nerves are not continuous — they have junctions where they almost, but not quite, touch. Electrical impulses can jump these spaces, called *synapses*, by chemical processes. Some chemicals can block the transfer of these electrical pulses. Other chemicals can cause, in effect, a short circuit. In both cases, normal nerve activity cannot take place.

The nervous system is usually divided into two parts: the *central nervous system* (often abbreviated CNS) consisting of the brain and spinal cord, which is a bundle of nerves fibers running through holes in the protective vertebrae of the backbone; and the *peripheral nervous system* consisting of nerves that connect the spinal cord with sensory receptors and glands and muscles of the body.

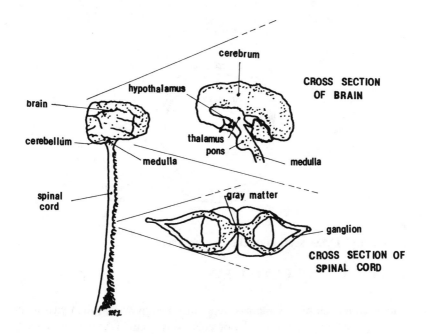

Figure 2-4.   The human central nervous system, showing cross sections of the brain and spinal cord.

The central nervous system is shown in Figure 2-4. The brain consists of several parts, each with a different function. The chief parts are the *cerebrum*, *cerebellum*, and the *medulla oblongata*. The cerebrum is the seat of thought, memory, and reasoning. The cerebellum is the seat of muscular coordination. The medulla oblongata controls automatic functions of the body such as breathing, heartbeat, etc.

2.8. THE EYES

A cross-section diagram of the eye is shown in Figure 2-5. The eye is fundamentally a sac filled with fluid. Light enters through the transparent cornea and is reduced in intensity by the *iris*. The resulting circular opening is called the *pupil*. The light passes the pupil and travels through the *lens*, which adjusts in thickness to focus the image at the rear of the eye on the *retina*. Receptor cells in the retina convert light impulses into nerve impulses, which are transmitted to the brain by way of the optic nerve. The intensity of light is detected by cells called rods, and colors are detected by cells called cones.

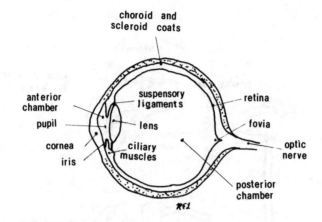

Figure 2-5.   Cross section of the human eye.

2.9. THE REPRODUCTIVE SYSTEM

Humans, like all higher animals, reproduce by a process called *meiosis* or sexual reproduction. All the cells of your body contain within the nucleus certain chains of chemicals called *chromosomes*. The chemical that composes these chromosomes is *deoxyribonucleic acid* (DNA). The ways that this chemical can arrange itself determine all the characteristics of the offspring (hair color, eye color, height, etc.). DNA is like a template that transfers information from one generation to the next. All the cells of the human body, except for the sex cells, contain the same number of chromosomes, 46 in 23 pairs. The sex cells contain only one half this number or 23 chromosomes. This is so because a union of the two sex cells leads to the original 46. This process is the basis of heredity.

Male sex cells, called *sperm*, are formed in the male gonads, the *testes*. The process of sperm formation is called spermatogenesis. These cells begin their development from precursor cells, which are formed at the edges of the testes. These precursor cells contain 46 chromosomes. They divide by a process called mitosis or cell division whereby identical daughter cells result from a splitting of the parent cell. As the population of cells increases, the cells move to the center of the testes where they become mature sperm following the last two stages of meiosis and end with only 23 chromosomes. These sperm look like small tadpoles, having a rounded or arrow-shaped head and a long tail.

Female sex cells are called eggs or *ova* (singular: ovum) and are formed in the female gonads, the *ovaries*. The eggs are formed by a process initially quite similar to that producing sperm. The immature cell divides by mitosis but, unlike the sperm, into two unequal-sized cells, with one receiving almost all the cellular material. A second division of the larger cells produces another unequal division, with the larger body becoming the egg and the other, like the first smaller cell, not involved in reproduction.

We now have an egg cell, which is much larger, and a sperm cell, both with 23 chromosomes carrying the hereditary coding in their sequence of genes. The union of the two cells is called *fertilization*, with the gene sequence of the new DNA decided by chance from the two parents' DNA. The details of coding and heredity are too complex to cover in this text.

The single fertilized cell then divides by mitosis repeatedly in the process of making a child. Usually this process continues in a normal fashion until birth; however, sometimes there are changes as a result of chance, a physical cause, or a chemical cause.

The process of formation and development of an embryo is actually a programmed sequence of cell proliferation followed by differentiation (or specialization) and cell migration leading to organ formation.

The most critical period of human embryo formation is during the first three months (trimester) of pregnancy, since it is at this time that the most rapid cell proliferation takes place, upon which all subsequent development is based. During the very first two weeks, however, chemicals cannot generally cause birth defects, since prenatal death followed by natural abortion usually occurs if damage to the embryo takes place at this time.

2.10. THE ENDOCRINE SYSTEM

The endocrine system consists of those glands of the body that do not have ducts. While this might seem a strange definition, it really is not. Many glands in the human body produce chemicals, called *enzymes*, that promote chemical changes. Most of these glands pass these enzymes to other parts of the body by ducts. For example, the salivary glands of the mouth pass saliva directly into the mouth to digest starches, and the liver passes bile to digest fats into the duodenum through a duct. Other glands, however, pass their chemicals, called *hormones*, to other parts of the body through the blood. A few glands are both ducted and ductless, passing different chemicals for different functions.

There are seven ductless glands in the body: the *gonads*, which produce hormones to promote the development of the sex cells; the *pancreas*, which produces a hormone used in the control of the amount of blood sugar used as food by the cells; the *thyroid*, which produces a hormone used in control of metabolism; the *parathyroids*, which produce a hormone that maintains calcium and phosphorus levels; the *pituitary*, which is the most complex endocrine gland and produces growth hormones and also plays a role in control of

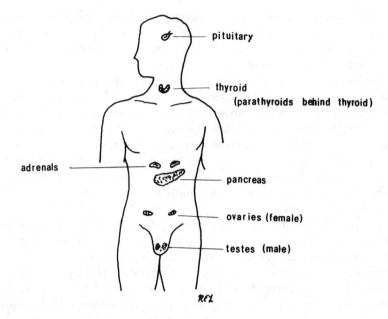

Figure 2-6.   Locations of the endocrine glands.

the other endocrine glands; and the *adrenals*, which produce several hormones that function in the conversion of proteins to carbohydrates and also to stimulate the body to respond in emergencies. The approximate location of these glands is shown in Figure 2-6.

2.11. THE BLOOD CIRCULATORY SYSTEM

It is believed that life began in ancient seas, and even today life needs to be surrounded by a mild salt solution. Every cell of your body needs to live in such an environment. Since the cells of the human body may be far removed from sources of food, some mechanism is needed for food materials to be brought to the cells and for wastes to be removed. The blood performs this function, among others.

The circulatory system is shown in Figure 2-7. The *heart* is the center of the circulatory system. It is basically a muscular pump, pulling blood from parts of the body and pumping it under pressure to other parts. The heart is almost totally cardiac muscle. The parts of the heart are such that they act as if all the cells were really one. Since the blood flows in a continuous circuit, we can start our description at any point; however, let's begin at the point in the heart where blood is pumped to the lungs.

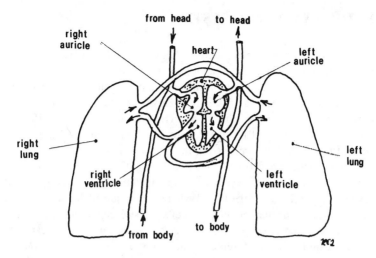

Figure 2-7.   The human circulatory system.

Blood is pumped to the lungs by the lower right chamber (as viewed from the back), called the right *ventricle*. The blood is pumped through the lungs, where oxygen is absorbed by hemoglobin and waste gases are expelled. The now-oxygen-rich blood is pumped back to the heart's left *atrium* (sometimes also called auricle). The bicuspid value between the atrium and ventricle opens, and the blood is pumped into the left ventricle. From here, it is pumped to all the parts of the body. The first vessel that blood passes through is the *aorta*, the largest artery of the body. From the aorta, the blood is pumped through ever-narrowing vessels until at last the very tiny *capillaries* are reached. It is at the capillaries that the gaseous exchange occurs in which oxygen is given to every cell and the cell's wastes are transferred to the blood. From the heart to the capillaries, all the vessels are called *arteries*. As the blood leaves the capillaries, the return system is used. All the vessels from the capillaries back to the heart are called *veins*. Thus, the blood in the arteries is rich in oxygen and low in wastes, but the blood in the veins is low in oxygen and rich in wastes. The blood moves into ever-enlarging vessels until it returns to the right atrium through the *vena cava*, the largest vein of the body, from which it is pumped to the right ventricle. The cycle then repeats itself.

Many factors influence the rate of the heart's pumping and the pressure of the blood in the vessels. There are chemicals that can affect both.

In addition to carrying oxygen and carbon dioxide, the blood also contains white blood cells (*leukocytes*), which attack foreign bodies such as bacteria and viruses; chemicals that cause clotting to seal off breaks in the vessels; hormones from the various endocrine glands; and electrolytes involved with nervous system functioning and general chemical balance in the body.

2.12. THE LYMPHATIC SYSTEM

Another system, sometimes called the body's second circulatory system, is the lymphatic system. It represents another route by which fluids can flow from the cells into the blood. Larger molecules, such as proteins, cannot pass from the cells into the blood in the capillaries. However, there must be a way for these molecules to return to the blood. The lymphatic system fulfills this function. The fluid carried by this system is called *lymph* and is produced throughout the body, but chiefly from the liver and intestines.

Protein molecules are carried by the lymph from the cells back to the blood. The vessels of the lymphatic system are similar to those of the blood circulatory system, being larger at the points where they enter the blood stream and tiny capillaries where they contact the cells. These capillaries are different from the blood capillaries, since they have minute spaces that can allow molecules such as proteins to enter. The lymph is returned to the blood at several locations. One-way valves allow the flow to pass into the blood but not for the blood to enter the lymphatic system.

Another important role of this system is to carry fats and some other nutrients from the small intestines to the blood, rather than having this happen by way of the villi.

Finally, very large particles such as bacteria can enter the system. As the lymph passes through the lymph nodes, these particles are removed and destroyed. Sometimes, though, the lymph nodes can be temporarily overwhelmed, as during a cold. The lymph nodes can become swollen and painful. Locations of lymph nodes under the neck and armpits are commonly swollen during infections.

2.13. THE URINARY SYSTEM

Liquid wastes are removed from the body by means of the urinary system, as shown in Figure 2-8. As blood passes through the *kidneys*, waste substances are removed. These wastes are carried to the *bladder*, which is a storage sac, until discharged from the body. The liquid waste is called *urine* and is generally a light yellow color from the breakdown products of protein metabolism.

The two kidneys are able to remove wastes from the blood and concentrate them for removal. The kidneys are very rich in blood vessels from which they remove the products of metabolism, which would be quickly fatal if not removed. The wastes pass from the kidneys through tubes called *ureters* to the urinary bladder. Muscles at the bottom of the bladder hold the urine until it is ready to be passed. When this happens, the urine goes through the tube called the *urethra* to the outside of the body. In the female, this system is separate

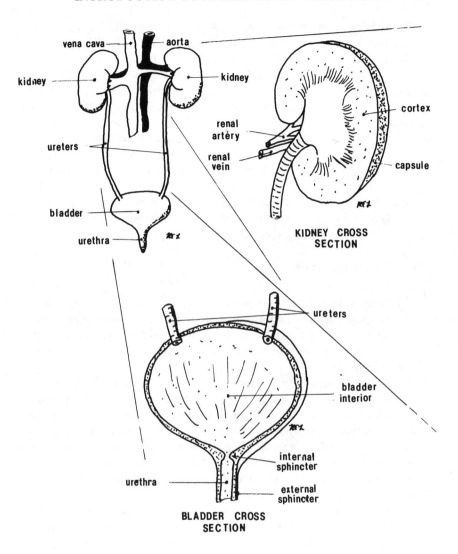

Figure 2-8. The human urinary system, showing cross sections of the kidney and bladder.

from the reproductive system; however, in the male, the urine is routed through the penis, with constricting muscles blocking flow to the reproductive system.

2.14. THE LIVER

One of the most amazing organs of the human body is the liver. It acts as a chemical filter to remove toxic materials from the body and then detoxifies the chemicals into less harmful substances. In addition, the liver also produces an enzyme, called *bile*, that aids in the digestion of fats.

Removing toxics from the blood is a very large order. The various types of chemicals that may enter the bloodstream can each require a different reaction to be rendered less harmful. The liver is able to detoxify many different types of chemicals; however, it can be harmed itself by some substances, either by their direct action on the liver or by their ability to overwhelm the liver's capacity to detoxify. Several types of cancerous tumors can be promoted in the liver by chemicals. Also, long-term use of certain chemicals, such as alcohol, can destroy the liver.

The liver can usually continue its function, filtering hundreds of gallons of blood daily and detoxifying harmful chemicals, throughout a person's life. This organ is one of the most critical for life.

2.15. CONCLUSION

The study of the anatomy and physiology of the human body is important if the effects of chemicals upon people are to be understood. Many organ systems of the body can be affected by the toxic effects of chemicals; however, the organ systems most commonly affected are the respiratory system, the skin, and the digestive system.

The skin is the largest organ of the body and acts to keep foreign substances out. The respiratory system brings oxygen to the cells of the body and removes gaseous waste products. The digestive system brings food to the cells of the body and removes solid wastes from the body. The eyes form images by conversion of light into nerve impulses to the nervous system. The nervous system acts as the communication network to coordinate the functions of the body. The purpose of the reproductive system is to produce offspring. The endocrine system assists other organ systems and controls enzyme reactions in the body. The urinary system removes toxic products of metabolism from the body. The lymphatic system carries protein molecules back to the blood, carries fats and other nutrients from the intestines to the blood, and removes foreign bodies like bacteria from the body. The liver produces enzymes to help digest fats, and also detoxifies chemicals that enter the body.

The study of anatomy and physiology will help us better understand the effects of chemicals upon the systems of the body as discussed in the next chapter.

2.16. CHAPTER REVIEW QUESTIONS

1. The brain and spinal cord make up which system?
   (a) Central nervous system (b) Peripheral nervous system
   (c) Gastrointestinal system (d) None of the above

2. Food (and swallowed chemical) absorption via the digestive tract (ingestion) occurs primarily in which part?
   (a) Stomach   (b) Small intestines
   (c) Large intestines   (d) Throat

3. Oxygen and carbon dioxide transfer takes place in what part of the
   respiratory system?
   (a) Trachea   (b) Bronchioles
   (c) Alveoli   (d) Glottis

4. About how long does it take for new skin cells to reach the surface?
   (a) 1 day   (b) 1 week
   (c) 1 month   (d) 1 year

5. The outer layer of the skin is known as which of the following?
   (a) Dermis   (b) Epidermis
   (c) Follicle   (d) Vascular

6. The part of the eye that converts light into electrical impulses is which
   of the following?
   (a) Retina   (b) Cornea
   (c) Lens   (d) Iris

7. The primary function of the liver is which of the following?
   (a) To produce a hormone for growth
   (b) To produce red blood cells
   (c) To digest food
   (d) To detoxify harmful substances in the blood

8. The term *endocrine* refers to which of the following?
   (a) Hard or tough   (b) Ductless
   (c) Bacteria   (d) Ducted

9. The junction where nerves almost but not quite touch is which of the following?
   (a) Lapse   (b) Synapse
   (c) Functionality   (d) Partum

10. Chromosomes are important in what process?
    (a) Digestion   (b) Respiration
    (c) Vision   (d) Reproduction

3   INTRODUCTION TO TOXICOLOGY

3.1. INTRODUCTION

In any text on hazardous materials, it is important to consider the effects of chemicals on people, on animals, and on the environment. This chapter will discuss these effects, both from short-term and long-term exposure.

It is not the intent of this chapter to make you a toxicologist, but only to give you the basic information needed to understand how chemicals can affect living systems.

3.2. DEFINITIONS

Some of the terms in this chapter may be new to you; here are a few definitions.

- *Toxicology* is the study of the effects of chemicals on living organisms.
- The *threshold dose* is the minimum amount of a chemical needed to cause a stated response.
- *Exposure* occurs whenever an individual is in the same environment or in contact with something.
- *Chemical agents* are substances that can have an effect on the body.
- *Physical agents* are those conditions that can affect the body due to characteristics such as heat, cold, pressure, noise, or radiation.
- *Chronic* refers to longtime exposure, repeated exposures, or long periods of time between exposure and injury.
- *Acute* refers to brief exposures or a short time between exposure and injury.

- The LD_{50} is the median lethal dose of a chemical, that is, the amount that will cause death to 50% of a population exposed to it. For gases and some liquids, this is modified to LC_{50} or median lethal concentration.
- *Teratogens* are substances that can cause birth defects. *Terato-* means *monster.*
- *Mutagens* are substances that can cause genetic changes in future generations. *Muta-* means *change.*
- *Carcinogens* are substances that can cause abnormal cell growth or development. *Carcino-* means *cancer* from the Latin word for *crab* since such growths often appear crablike.

3.3. HEALTH HAZARDS

Some substances can affect living things so as to harm them. Some act slowly (chronic); others act rapidly (acute). Many human body systems can be affected by chemicals, but most occupational exposures occur by way of the skin, respiratory tract, eye, and digestive tract. The ways in which chemicals can affect these organ systems are varied and often depend upon characteristics of the individual, such as size, weight, age, etc. In the previous chapter we learned about the various organ systems of the body. Let's now consider how chemicals can affect these systems.

3.4. ACUTE EXPOSURES

In toxicology, acute effects are generally defined as those resulting from a single exposure, or from multiple exposures within 24 hours or less. Quite often, for a given agent, the acute effects can be quite different from the chronic effects. For example, acute exposure to excessive amounts of the metal lead can cause a coliclike condition; however, chronic exposure to this same substance can lead to wrist drop and possibly nervous-system damage, depending upon the form of the metal.

One important factor in acute exposures is the speed at which the chemical is absorbed. If a substance is rapidly absorbed, the effects are likely to occur almost immediately.

Another factor that must be considered is the frequency of exposure. Two exposures of one half the amount of a substance usually produce less effect than one larger exposure. This may be due to the body's ability to change or alter some of the chemical with time. Also, the body may be able to repair some of the damage if sufficient time passes between exposures.

3.5. CHRONIC EXPOSURES

Chronic effects from a chemical may include some immediate effects in addition to the long-term effects.

The range of effects is wide; some effects are immediate and others may take dozens of years to appear. Therefore, terms for immediate exposures have been introduced.

Short-term refers to exposures of a week or so. *Subchronic* refers to exposures of about 3 months.

Chronic toxic effects take place whenever the agent accumulates in a biological system, that is, whenever the absorption is greater than elimination or metabolism. Metabolic processes involve the breakdown of certain chemicals and the formation of others. Sometimes toxic chemicals are either destroyed or weakened by metabolism.

Chronic exposures are much more difficult to study than acute exposures, since long time periods may be involved and different routes of entry of the chemical to the body may have been used.

Several common organic solvents can lead to a skin exposure while also being present in the air, leading to respiratory or ocular (eye) exposure.

The liver is often affected by the entry of toxic substances, since its primary function is to detoxify foreign substances in the blood. It can be overwhelmed by amount or by toxicity. Chlorinated solvents and alcohols in particular can damage the liver.

Other examples of chronic exposure include exposure to iodine, which seeks out the thyroid gland, and to heavy metals, which affect the nervous system and bones.

3.6. TOXIC CLASSES

Toxic chemicals can be divided into several classes, generally depending upon the effects produced. These broad classes are *irritants*, *asphyxiants*, *hepatotoxins*, *nephrotoxins*, *neurotoxins*, *anesthetics*, *hematopoietic toxins*, and *lung-damaging toxins*.

Irritants cause irritation of the organs that they contact. They may be further divided into primary irritants, which simply cause inflammation, and secondary irritants, which cause inflammation plus additional effects.

Asphyxiants deprive tissues of oxygen. They, too, may be subdivided into simple asphyxiants, which are physiologically inert gases and only displace oxygen, and chemical asphyxiants, which render the body incapable of utilizing the oxygen in the blood.

Hepatotoxins affect the liver, often rendering it incapable of detoxifying poisons.

Nephrotoxins affect the kidneys, altering the removal of liquid wastes from the body.

Neurotoxins affect the entire nervous system.

Anesthetics are central nervous system (CNS) depressants.

Hematopoietic toxins affect the blood-forming organs.

Lung-damaging toxins can damage the lungs but are not irritants.

These classes depend upon the target organ, that is, the organ or system directly affected by the chemical. Toxic substances are sometimes alternatively described by the use of the chemical (e.g., pesticides, solvents, food/water additives, etc.), by the source of the chemical (e.g., animal, plant, organic, inorganic, etc.) or by the effects of the chemical (like carcinogen, mutagen, teratogen, etc.).

Whenever toxic classes are discussed , you must remember that a chemical might be described by different terms depending on the situation. For example, a chlorinated compound might be called a carcinogen, hepatotoxin, synthetic organic chemical, or degreasing solvent, depending upon the situation being considered.

3.7. POISONING POTENTIAL

Often, in addition, toxic substances are divided into classes according to their poisoning potential, or how much it takes to cause harm.

Most toxicologists have agreed to certain terms that relate to ranges of quantity of substances. These are usually based upon lethal dose, usually taken orally, for humans. The usual values are for milligrams of toxin per kilogram of body weight of the exposed individual. This is often abbreviated mg/kg.

In this system, shown in Table 3-1, practically nontoxic substances would require more than 15,000 milligrams (15 grams, or about one half ounce) per kilogram (2.2 pounds) of weight to cause death. Since most adult humans weigh about 150 pounds (approximately 70 kilograms), one would have to ingest about 2.5 pounds of such a substance to cause death.

The next group of substances is those considered slightly toxic, requiring 5000 to 15,000 mg/kg to cause death.

Next are those considered moderately toxic, requiring 500 to 5000 mg/kg to cause death..

Toxic substances require between 5 and 500 mg/kg to be fatal.

Very toxic chemicals need only 0.5 to 5 mg/kg to be fatal.

The next group is those substances considered extremely toxic, requiring only 0.05 to 0.5 mg/kg.

Table 3-1. Poisoning Potential

Term	Oral Dose (mg/kg)	Median Amount for Average Human
Practically nontoxic	>15,000	2.3 pounds
Slightly toxic	5,000–15,000	1.5 pounds
Moderately toxic	500–5,000	7 ounces
Toxic	5–500	2/3 ounce
Very toxic	0.05–5	2 drops
Extremely toxic	0.05–0.5	2/10 drop
Supertoxic	≤0.05	trace

Chemicals that can cause death at amounts less than 0.05 mg/kg are called supertoxic. These would require less than five drops of liquid to kill the average human adult.

Bear in mind that these classes are broad and the numbers are not exact. However, these groups can be used for comparison of two or more substances to give an indication of the poisoning potential.

3.8. DOSE RESPONSE

We all know someone who seems to be able to drink alcohol all day and night without seeming to become very drunk. On the other hand, we probably know people who get tipsy after one small drink. What's going on here?

Over the years, toxicologists have come to the correct conclusion, first spoken in the early 1500s by Paracelsus, that "the dose makes the poison." Almost any substance can cause injury or death in sufficient amount.

Therefore, it is important that we consider response to different doses of chemicals, not just the nature of the chemical.

Since *dose* refers to an amount, we must first consider how we measure amounts. One of the most valuable ways to do this is to consider the amount of chemical needed to cause a certain response (such as death, burns, irritation, etc.) in terms of the number of milligrams of the chemical (or ounces, etc.) per number of kilograms (or pounds) of body weight of the person exposed to the chemical. This gives the poisoning potential. However, this is not an individual characteristic; it is based on thousands of trials and represents an average value. An individual might vary widely from this statistical value.

Response means whatever effect is sought. Response can range from slight, temporary impairment to death; however, the specific response must be defined in order to make sense of dose-response relationships.

Most values for occupational exposures, or indeed any exposures, have been arrived at on the basis of animal studies.

3.9. ANIMAL MODELS

Experimental results for chemical exposures have been based upon animal models and adjusted for human characteristics. Use of human subjects is considered unethical, so most toxicity testing uses animals. There are problems with this approach. Aside from the concerns of animal rights groups, it has been demonstrated that different species respond differently to the same chemical. Even closely related animals such as rats and mice often exhibit widely different effects from some chemicals. In addition, there is no animal (or other) model that exactly duplicates human response.

Today, animals continue to be used for testing but with greater concern for their well-being and comfort, since there are no better models available. New techniques, for example those using bacteria or plants, are being investigated; one of the standard tests for carcinogens uses bacteria. Rest assured, research toxicologists would delight in finding nonanimal models that duplicate human response, if only for the fact of not having to deal with housing and feeding the animals. We shall probably see much more research in the area of substitute models in the coming years.

As a dose is slowly increased from zero to a maximum, the response does not appear suddenly, but is graded. For animal testing, the endpoint most commonly chosen is death. A dose is selected and the response, as defined, is observed. Other doses, more and less, will be given until a range is found. This range is bounded on one side by the dose that causes the death of all the animals and on the other by the dose at which all the animals survive. The data are usually presented as a response curve relating the amount of the dose to the percentage of the animals responding.

3.10. RESPONSE CURVE

The graph of dose versus response is generally an S-shaped curve. At the upper end, it approaches but never quite reaches 100%. At the lower end, it approaches but never quite reaches 0%. It is at the low-dose end that most of

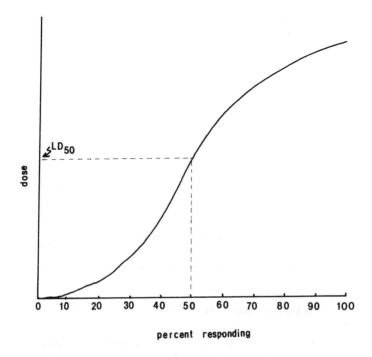

Figure 3-1.   A typical dose-response curve.

the experimental problems result. Responses are usually very small and might be missed. For most chemicals, it is not known whether the curve really goes to zero or not. Likewise, for most chemicals, we do not know whether there exists a threshold below which very small doses do not lead to a response. Since we are unsure, we assume that there is a linear (straight-line) approach to zero and there is no threshold; that is, any dose will cause some response. A typical response curve is shown in Figure 3-1. The dose that causes the death of one half the animals is called the LD_{50}, lethal dose 50%.

3.11. DISTRIBUTION CURVE

Now that we have data from toxicity testing using the dose-response curve, we can extend the results in terms of a population exposed to that chemical.

Individual members of a group will respond differently to the same dose; these differences are referred to as biologic variances or individual susceptibility.

Some of the differences can be caused by age, sex, heredity, physical condition, diet, etc.

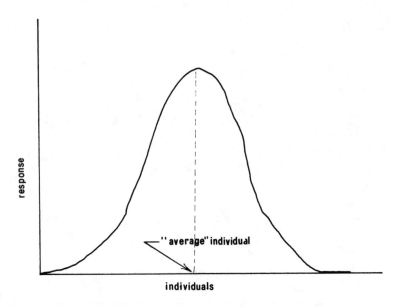

Figure 3-2.   A typical distribution curve.

A graph of reaction versus the number of individuals responding is called a distribution graph or curve and is usually bell-shaped. An example is given in Figure 3-2. A few individuals will respond strongly and a few hardly at all. These are the tapering ends of the curve. Most individuals fall somewhere between the extremes. The members of the groups will thus show a range of responses from slight to great. We consider that the middle of the curve represents the "average" member of the group.

3.12. ANALYTICAL LIMITS

Another difficulty of small doses, apart from the minute responses for very low doses, is the problem of measuring very small amounts of anything. Scientific instrumentation has limits on how small a value can be determined. This limit depends upon the equipment, the procedures, the ability of the laboratory workers, and even random errors, which occur in any human endeavor. The smallest amount that can be detected is called the limit of detection. For quantities up to that value, it is not possible to know whether the amount is zero or slight.

Since we assume that any dose causes some response, we are limited in not being able to measure very small doses. Even what we consider a "pure" chemical might actually contain a small quantity of a substance that is harmful.

The dioxin (TCDD) situation reflects this: at the levels tested for by the manufacturers, no dioxin was observed in the herbicides being produced. Only with more sensitive equipment was it realized that dioxin was present in these compounds.

3.13. CLASSES OF CHRONIC TOXINS

The chronic toxins are usually divided into three groups, depending upon their effects: *teratogens*, *mutagens*, and *carcinogens*. It is also possible for some substances to affect reproduction by reducing sex drive; however, the effects of these agents are usually temporary, and conditions improve upon removal of the exposure.

The next few sections will discuss each of these three groups of chronic toxins in more detail.

3.14 TERATOGENS

Many of you may be familiar with the sedative thalidomide, which, although never approved by the U.S. Food and Drug Administration, was used in several nations during the late 1950s and early 1960s. Over 10,000 deformed babies were born to mothers taking this drug. Most of the babies were born without arms or legs or both. Some babies had limbs that were very short in length.

You may also be aware of the occurrence of hundreds of malformed babies in Japan as a result of their mothers eating fish contaminated by organic mercury compounds from industrial contamination of a coastal bay. The children suffer from cerebral palsy as a result.

Another teratogen formerly used to induce abortions is aminopterin. If the planned abortion did not occur, the infant was usually born with skull and brain damage, defective palate, and abnormalities of the limbs.

Another proven teratogen might surprise you: ethyl alcohol. This common chemical in beer, wine, and liquor can cause growth failure and impaired brain development in the baby.

Iodine has been implicated as a teratogen, as have other common drugs taken during pregnancy, such as aspirin, insulin, several antibiotics, amphetamines, barbiturates, and tobacco.

Heroin, morphine, and methadone can also be teratogens, as can several tranquilizers and sedatives. Lysergic acid diethylamide (LSD) was once thought to be a teratogen; however, recent research has indicated it may not be.

Apart from pharmaceuticals, several chlorinated solvents have been indicted as causing newborn depression. Likewise, TCDD (dioxin) is a teratogen in addition to being a mutagen and carcinogen. The metal lead may possibly also cause birth defects, as can the hospital sterilizing gas ethylene oxide.

Even excessive amounts of some vitamins can lead to defects in the offspring if taken by a pregnant mother. Too much vitamin A, in addition to potentially harming the mother's liver and skin, can also cause abnormalities in the baby's spine.

The greatest threat to the development of the baby is during the early weeks of development; however, birth defects caused by chemicals should not be ruled out at any time during pregnancy.

3.15. MUTAGENS

You may recall from the previous chapter that the basis of heredity is the genetic code carried by the gene sequence of the chromosomes. There are chemicals that can alter this code, producing mutations.

Teratogens usually only produce their change in the next generation, but mutagens can continue to cause defects for untold numbers of generations.

Human genetic alterations may be caused in several ways by which the genetic code fails to be transmitted accurately. Chromosomes are composed of genes made of DNA in a particular sequence. If the sequence is changed, for example, by a part being left out or extra parts put in, the message will be altered and a mutation may result.

It should be mentioned that not all changes in the DNA will automatically lead to a mutation and, if they do, the change will not necessarily be harmful. However, in general, we must consider that chemical exposures are abnormal and therefore the results may be harmful.

Mutations may also be caused by the breaking of a chromosome, resulting in rearrangement of the pieces or even just fragments of the original. Another way that mutations are caused is by the gain or loss of complete chromosomes.

Any of these three methods can be caused by a chemical acting upon the sex cells of either parent; however, such changes can also take place naturally or as a result of exposure to other factors such as ionizing radiation.

One scientific study has indicated that about 0.6% of all newborn babies have some detectable chromosomal abnormality and some 0.3% vary in the total number of chromosomes.

However, among spontaneously aborted fetuses, the rates are as much as 30% with defects in chromosomes. It is believed that this is one reason for such abortions.

Some of the substances thought to cause mutations include caffeine, beryllium metal, LSD, marijuana, diethyl sulfate, alkyl mustards, and nitrous oxide. Most mutagens appear to cause their damage by chemical reactions with the amino acids of the DNA.

Ethylene dibromide (EDB), several now-banned pesticides, vinyl chloride, and some polyaromatic hydrocarbons (PAHs) have been linked to mutagenic ability; however, they have not been proven as yet to be mutagens.

Ionizing radiation from X-rays, cosmic rays, and exposure to radioactive materials have also been proven to cause mutations.

3.16. CARCINOGENS

The exposures discussed above affect future generations; however, there are chemicals that can harm the exposed individual. Some of these chemicals can cause changes in cell growth or cell metabolism. These substances are called carcinogens.

Remember, though, that there can be wide spectrum of harm caused by these substances. In other words, cancer is not one disease like polio or smallpox but is many different diseases all grouped together, since all of them involve changes to cells. A chemical might "cause" one kind of cancer but not any others.

How chemicals can lead to cancers is not clearly understood; however, it is thought that at least some cancers result from chemical changes to the DNA of cells of the exposed individual. Since these cells are not necessarily involved with reproduction, genetic changes may not occur. Other chemicals appear to weaken the body enough to allow other things to cause the cancer.

3.17. HISTORY OF CANCER

Cancers have apparently been with us since the beginning of time — ancient Egyptian mummies have been found with cancer tumors. Whether these were caused by exposure to chemicals can never be known, although it is unlikely, considering the small number of synthetic chemicals available in those days. However, in the late eighteenth century, it was noted that chimney sweeps in England often suffered from an otherwise rare form of scrotal cancer. Years of exposure to soot rubbed into the skin had caused cancer.

Shortly thereafter, it was discovered that coal tar could cause cancer. Recall that coal was the primary fuel in those days, so millions of people were being exposed. Later research found that a group of chemicals called polyaromatic hydrocarbons (PAHs) were to blame.

In the 1930s, it was observed that cancers of the urinary tract were common among German workers in the dye industry. Aromatic amines were found to be the cause.

About the same time, other dyes called azo (nitrogen-containing) compounds were found that could both directly cause cancer and also predispose an individual to cancer.

3.18. PROVEN CARCINOGENS

There are a number of chemicals that have been shown to cause various types of cancer.

One of these is benzene, an aromatic hydrocarbon, which has been shown to lead to leukemia. However, it is now suspected that the action of benzene weakens the immune system to the point that leukemia-causing viruses can cause the disease, rather than benzene being the direct cause.

Dioxin is really the chemical 2,3,7,8-tetrachloro-dibenzo-p-dioxin (TCDD). It is also an acute toxin and a teratogen. As a carcinogen, TCDD is suspected of causing liver and kidney tumors and possibly injury to the CNS. TCDD has experimentally induced tumors in test animals at concentrations as low as 5 parts per billion (ppb). This compound brings up an interesting aspect of chemistry and toxicology; TCDD was never made on purpose. There are no known uses for this material and no need for its production. It is only found as an unintentional byproduct of the manufacture of two herbicides (plant killers). Here is a case where a very minute contaminant of a mixture has been shown to be the harmful component.

Vinyl chloride was shown to be a carcinogen after workers in plastics manufacture who were exposed to it were found to be suffering from angiosarcoma (cancer of blood vessels). Central nervous system and liver damage due to vinyl chloride had been known since the 1940s; this is one reason it was never used as a surgical anesthetic, as had once been proposed.

3.19. CLASSES OF CARCINOGENS

Eight classes of carcinogens are usually recognized: direct-acting, secondary, inorganic, solid-state, hormonal, immunosuppressives, cocarcinogens, and tumor promoters.

Direct-acting carcinogens interact directly with the DNA of the cell. Examples are ethylene imine and bis(chloromethyl)ether.

Secondary carcinogens require chemical conversion by metabolism to cause cancer. Examples are vinyl chloride, benzo(a)pyrene, and several nitrosoamines. Nitrosoamines can be formed in very small amounts from cooking cured bacon and ham.

Inorganic carcinogens apparently cause changes in the DNA by disrupting the normal duplication process. Examples are nickel and chromium.

Solid-state carcinogens seem to affect only certain cells and tissues. The physical size seems to be very important. Examples are fibers of polymers, metals, and asbestos.

Hormonal carcinogens affect the organism by disruption of the normal balance of the endocrine system. Examples are estradiol and diethylstilbestrol.

Immunosuppressive carcinogens weaken the immune system so that viruses or other factors can cause cancer. Examples are benzene and azathioprine.

Cocarcinogens are not true cancer-causing agents but act to enhance the effects of direct-acting and secondary carcinogens. Examples are ethyl alcohol and sulfur dioxide.

Tumor promoters are also not true carcinogens but enhance the effects of other carcinogens when exposure to the promoter follows the earlier exposure. Examples are phenol (carbolic acid) and saccharin.

3.20. FEARS VERSUS REAL CONCERNS

As just seen, numerous chemicals can cause or promote cancer; however, we must not fall into the trap of thinking that "everything causes cancer," which we sometimes hear. There are hundreds of thousands of chemicals, but only a few hundred have been proven to cause cancer or even suspected of doing so.

Remember that there are chemicals that can slow or even stop the growth of cancer cells. Not all chemicals are villains.

Since we generally know which chemicals are suspected of causing cancer, there is no reason that exposures to these substances should not be reduced to the lowest possible level. However, there is a problem: we don't know what "safe" levels are (if indeed they exist at all). Since not everyone develops cancer — including workers and others exposed to these chemicals — there is probably some level below which they are "safe." But this level would be different for every person. Since we cannot conduct testing on every person, and since we do not know if there is an acceptable level, we make the choice to assume that any exposure can potentially lead to cancer.

In some cases, it is clear that a given chemical causes cancer — for example, scrotal cancer in chimney sweeps and leukemia in benzene workers. In other cases, for example, cancers in military personnel exposed to Agents Orange and Pink, the evidence is more difficult to obtain and subject to question.

In a typical occupational setting, there might be dozens of chemicals to which workers are exposed. Which one caused the cancer, and at what concentration?

And while we legislate and control most of these harmful substances, there are others just as dangerous that are not controlled. Tobacco and alcohol are both acute and chronic toxins causing cancer, birth defects, and possibly genetic damage.

We need rational concern for the effects of chemicals on our bodies. We should not automatically assume that all chemicals are harmful or that all chemicals are harmless.

Not everything causes cancer or we would all have cancer! Do not let irrational fears cloud your learning about how to protect yourself from those harmful substances that do pose a threat to health.

3.21. CONCLUSION

Just as a basic knowledge of anatomy and physiology is important in learning how to deal with hazardous chemicals, so is a basic knowledge of toxicology.

There are chemicals that can cause harm quickly (acute) or slowly (chronic). Among the chronic toxins are mutagens, teratogens, and carcinogens.

Toxicologists use experimentation to arrive at dose-response relationships. From these data, values are determined that can be used to estimate relatively safe levels of human exposure.

Using the information from toxicology, occupational health professions are able to recognize, evaluate, and attempt to control exposures in workplaces, whether these be factories, hazardous waste sites, spill sites, or elsewhere. The next chapter will discuss how industrial hygiene and occupational medicine use toxicological data to help workers.

3.22. CHAPTER REVIEW QUESTIONS

1. Toxicology is the study of which of the following?
   (a) Structures of plants and animals
   (b) Effects of chemicals on living organisms
   (c) Effects of radiation on living organisms
   (d) None of the above

2. A substance that causes sudden or onetime effects is said to be which of the following?
   (a) Chronic toxin   (b) Acute toxin
   (c) Mutagen   (d) Carcinogen

3. Substances that can cause birth defects are called which of the following?
   (a) Carcinogens   (b) Mutagens
   (c) Teratogens   (d) None of the above

4. Substances that can cause abnormal cell growth are called which of the following?
   (a) Carcinogens   (b) Mutagens
   (c) Teratogens   (d) None of the above

5. Substances that cause harm by damaging the nervous system are called which of the following?
   (a) Asphyxiant
   (b) Neurotoxins
   (c) Hepatotoxins
   (d) Irritants

6. Chemical A is toxic at 75 mg/kg and Chemical B is toxic at 500 mg/kg. Which has the greater poisoning potential?
   (a) Chemical A
   (b) Chemical B
   (c) Both have the same
   (d) Depends only upon victim's weight

7. The S-shaped dose-response curve indicates which of the following?
   (a) All chemicals are carcinogens
   (b) Increasing the dose increases the response
   (c) No chemicals are really poisonous
   (d) Decreasing the dose increases the response

8. For the majority of chemical exposures, which of the following is usually the most important individual factor?
   (a) Weight
   (b) Sex
   (c) Age
   (d) Height

9. LD_{50} refers to the amount of a substance needed to do which of the following?
   (a) Cause death to 50 individuals of a given species
   (b) Cause death to one half the members of a given group
   (c) Have the smallest observable response
   (d) None of the above

10. A substance that has a poisoning potential of 10 mg/kg would be in what class?
    (a) Nontoxic
    (b) Moderately toxic
    (c) Highly toxic
    (d) Extremely toxic

4   HAZARDOUS MATERIALS

4.1. INTRODUCTION

This chapter will build upon the information learned in the preceding chapters on anatomy, physiology, and toxicology, with emphasis on an introduction to chemistry, incompatible chemicals, and industrial hygiene.

4.2. DEFINITIONS

A few terms need defining:

- *Industrial hygiene* is the recognition, evaluation, and control of health hazards in the work environment.
- *Incompatible chemicals* are those substances that, when combined or placed in proximity to each other, may produce harmful results.
- *Hazardous chemicals* are those substances that, by themselves or when mixed with others, can produce harmful results.
- *Chemistry* is the study of matter and the changes that take place in matter.

4.3. INTRODUCTION TO CHEMISTRY

We need to briefly discuss some basic chemistry, since we are going to be talking about chemicals for the rest of this chapter.

The study of chemistry is usually divided into two major areas: *organic chemistry*, which studies chemicals made of carbon, and *inorganic chemistry*, which studies all other substances. While this might seem like an unfair division, in reality there are many more organic (carbon-containing) chemicals than all others. In fact, of the some three million known chemicals, over two million are organic.

While chemistry has been traditionally the study of matter (things), today much of chemistry is more interested in the changes and energy relationships of changes of matter. These energy relationships are important in our study of hazardous materials.

To begin, the world around us is made of chemicals, some simple and many more complex. The simplest are those substances that cannot be broken down into any simpler materials. These are known as *elements*. There are some 105 elements that have been discovered, although only about 90 of these are found in nature. Examples of elements are iron, hydrogen, oxygen, and sulfur. Everything is made of the elements — the air, earth, stars, and you. The differences lie in how the elements are joined together.

The smallest unit of an element is called an *atom*. For most of history, it was thought that atoms were like little solid balls and could not be subdivided. But scientific studies beginning at the end of the nineteenth century proved that atoms themselves are made of smaller parts — but these parts cannot exist by themselves for long. Representations of an atom are shown in Figure 4-1. Atoms are very small — there are some 600 million million million of them in one third of an ounce of coal.

Atoms can join together to form what are called *molecules*. Representations of a molecule are shown in Figure 4-2. Molecules make up *compounds*, which are combinations of elements. Examples of compounds are water, made of hydrogen and oxygen; iron sulfide, made of iron and sulfur; rust, made of iron and oxygen; and iron sulfate, made of iron, sulfur, and oxygen.

Compounds are formed from combinations of elements, but just putting two or more elements together will not necessarily make a compound. Something else is needed. That "something" is *energy*. In order to be made spontaneously, the resulting compound (molecule) must contain less energy than the elements (atoms) making it up. Just as rocks only roll downhill, where they have less energy, compounds can only be made if they have less energy (are more stable) than the elements. Now, this energy has to go somewhere. All spontaneous chemical reactions give off energy.

For example, an iron nail placed in water will rust. A chemist would describe this as a chemical reaction between iron (in the nail) and oxygen (from the water) forming the more stable iron oxide (rust). A rusty nail will never become pure iron again by itself, since iron plus oxygen has more energy than the rust. An example of a typical energy diagram for a chemical reaction is shown in Figure 4-3.

How, then, can the reverse reaction take place? In other words, how do we get iron? Iron ore is really a form of rust. In order to change it into iron, energy must be put into it so that the resulting iron and oxygen can have more energy

2-D   MODEL

3 - D   MODEL

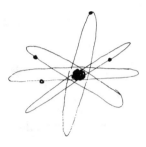

"FUZZY BALL" MODEL

Figure 4-1.   Representations of the atom.

than the ore. This is exactly what happens in an iron foundry: heat is a form
of energy and is put into the iron ore in a blast furnace to separate the iron and
oxygen into the two elements. The iron sinks to the bottom because of its
density and the oxygen is captured by other materials in the furnace.

In summary, spontaneous reactions give off energy, and to reverse the
reaction, energy must be put in. The rock will roll down the hill, but energy
is needed to push it back up. Just as the rock might need a small shove to get
it moving, some chemical reactions also need a small shove (energy) to get
them going. For example, lighting a piece of wood requires a match to get the
wood burning, but the energy given off by the fire that results is much more

2-D MODEL

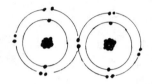

3-D MODEL

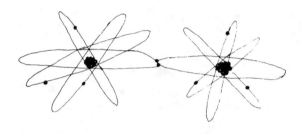

"FUZZY BALL" MODEL

Figure 4-2.   Representations of a molecule.

than the amount given off by the match. Some chemicals need some energy to get a reaction started. For example, some chemicals need a high temperature or a source of ignition in order to spontaneously react.

4.4. INCOMPATIBLE CHEMICALS

When two or more chemicals do spontaneously react, energy is given off. This may appear in the form of heat, fire, or even explosion. Some of these chemicals must be kept separated — they are incompatible. Also, some chemicals that are relatively harmless can react to form very hazardous substances. These, too, are incompatible chemicals.

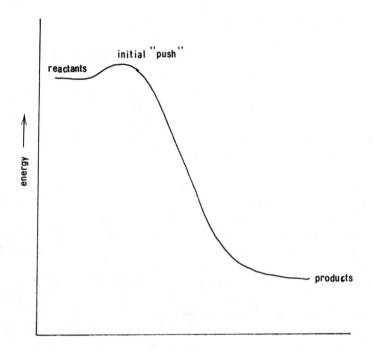

Figure 4-3.   A typical chemical-reaction energy diagram.

To give an example, chlorine is a very poisonous green gas, and sodium is a dangerous silvery metal that can explode into flame if placed in water — yet when chlorine and sodium combine, they form harmless table salt. The resulting compound is harmless, but a very large amount of energy is given off — in fact, an explosion results. These two chemicals would be considered incompatible, since the reaction is violent even though the product is harmless.

On the other hand, household bleach, which is a compound containing chlorine, is only slightly toxic, and vinegar can be eaten and so is fairly harmless, but reacting these two can result in the liberation of chlorine, which is very toxic. These two, bleach and acids, are incompatible, since a reaction would give off a harmful substance.

There are groups of chemicals that are incompatible with other groups. For example, acids that contain hydrogen and bases that contain hydroxide react to form water and another compound. The reaction is often quite violent, with heat being given off. Therefore, acids and bases are considered incompatible, since they react to give off heat. Other groups of incompatible chemicals include reactive metals and nonmetals; reactive metals and water or steam; flammable substances and oxygen; and many others.

4.5. CHEMICAL SAFETY

Care must always be exercised in the storage and handling of chemicals, since so many are incompatible. Many chemicals even need to be separated by a fireproof wall. Acids and bases should never be stored near or above each other. Flammables should be stored in explosion-proof containers and separate from all oxidizers, including many common acids. Some chemicals must be stored and used while cooled. Other chemicals, such as ethers, can form explosive products simply upon standing. Any substances that could react should be separated.

There are thousands of chemicals in common use today and no person could remember all the hazards of each. Therefore, laws have been enacted to require that all workers be taught the hazards of the chemicals with which they work. This training, plus Material Safety Data Sheets (MSDSs), which are discussed in Chapter 5, and warning signs and labels, allow workers to better protect themselves from the harmful effects of hazardous materials.

Remember, just about any substance can be harmful if used or stored improperly. All spontaneous chemical reactions liberate energy, either a little or a lot. Never mix unknown or unfamiliar chemicals. Never trust any chemical even if you have worked with it for years. Never intermix different chemicals in storage. Never assume any chemical is harmless. These cautions are especially true for workers in hazardous waste sites, since they really do not know what chemicals there are or how those chemicals could react.

4.6. ROUTES OF ENTRY

Now that we know about some of the hazards, how do chemicals do their harm?

Just about any part of the body can be the route by which chemicals can enter the body. However, there are a few routes of entry that are seen most common. These are the skin, respiratory tract, digestive tract, and eye.

To be toxic, that is, to have a negative effect on the body, a chemical first has to come into contact with and then interact with the body. The place of contact is called the site of entry, and the mechanism by which the substance interacts is the route of entry.

The most common route of entry is the skin, but exposures here are not necessarily the most severe. The second most common route is the respiratory tract. The third is the digestive tract. The eye is not commonly seen as a route of entry to other parts of the body, but is commonly injured itself.

4.7. DERMAL (SKIN) EXPOSURE

The skin is the most common site of occupational injuries; however, most of these are simply irritation of the skin, called *dermatitis*. Exposure to chemicals can dry the skin and remove the protective fatty acids. The dry skin can become cracked so that bacteria and viruses can enter into the tissue and blood vessels below.

The cells of the skin can also be killed by caustics and corrosives, with repair taking a month or more due to the slow growth of skin cells. Other damage to the skin can include removal of the fatty layer of the dermis, with consequent destruction of the developing cells.

Toxic chemicals can also enter the body through the skin and enter the circulatory system, leading to poisoning of remote parts of the body. This is called *systemic poisoning*, because the entire system can be affected.

4.8. RESPIRATORY EXPOSURE

In the normal individual, the respiratory system effectively removes particles larger than about 10 microns (1/2500th of an inch). Also, very small particles, like gas molecules, are simply breathed out as easily as they are breathed in. It is particles between about 1 and 10 microns that can move all the way to the alveoli and become lodged there, leading to disease. Unfortunately, smoke particles and many others, such as asbestos fibers, lie in this size range.

Molecules of other chemicals can also be harmful if they pass through the alveoli and into the blood, leading to systemic poisoning.

Inhalation exposure of chemicals occurs quite frequently, although less so than dermal exposures. Inhalation exposures can be much more severe, since many dusts, fumes, fibers, and gases can penetrate to the deepest regions of the lungs, where the blood lies only one cell away.

Some of the common results of exposures to the respiratory system include asbestosis, emphysema, mesothelioma, miner's black lung disease, and lung cancer.

4.9. DIGESTIVE EXPOSURE

Ingestion of chemicals occurs fairly often, but less often than the exposures discussed above, since the mouth is usually closed and swallowing is required for chemicals to enter the digestive system. Most poisonings by the digestive system occur from toxic substances transmitted either from the hands to the mouth or by mixing with food.

Once swallowed, chemicals can be passed to the blood by way of the villi, just like food molecules. The chemical processes of the digestive system can affect the toxicity of the chemical ingested, either neutralizing it or enhancing it.

One entry of toxins to the digestive system not often considered is by swallowing materials breathed into the nose. While this event is rare, it should be noted.

Some of the effects of chemical exposures to the digestive system include stomach and intestinal cancer, ulcers, and systemic poisoning.

4.10. NERVOUS SYSTEM EXPOSURE

Many chemicals can affect either the central nervous system (CNS) or the peripheral nervous system. This effect is the basis of many drugs and recreational chemicals.

Some chemicals serve to anesthetize the nervous system, reducing the intensity of nerve impulses. Others can block the electrical impulses at the synapses, acting like an open circuit. Still others can cause a short circuit at the synapse. In either of these last two cases, the normal nervous-system response cannot occur. The basis of many military war gases and commercial pesticides is interference with synaptic transmission.

Chemicals can also affect only parts of the nervous system; for example, ethyl alcohol and other substances affect the higher centers of the brain, causing a reduction in thinking ability. Only with large doses are the automatic functions of the medulla and cerebellum affected.

Pain relievers and tranquilizers function to affect the cerebrum and to some extent the other parts of the brain.

Some of the effects seen in the nervous system from these exposures include dizziness, confusion, breathing irregularities, changes in heart rhythm, sweating, and even death.

4.11. OCULAR (EYE) EXPOSURE

The eye is not very often the location of entry of chemicals to other parts of the body, although it is possible for the brain to be affected through the eye. However, the thin outer lining of the eye, the cornea, is very sensitive to many chemicals, especially acids and bases. Temporary or permanent blindness can result from ocular exposure to many chemicals. The cornea, or outer layer of tissue, can become completely opaque from even brief chemical splashes.

Some of the effects of chemicals on the eye include clouding of the cornea and even damage to the interior of the eye, including the lens and fluids.

4.12. EFFECTS ON REPRODUCTION FROM EXPOSURE

During weeks 3 through 7 of pregnancy, chemical substances can cause birth defects of major organ systems. It is interesting to note that the damage occurs at the time of formation of the organs. Thus, exposures at specific times cause specific damage. For example, damage to the CNS of the embryo usually takes place during weeks 3 to 6; to the heart during weeks 4 to 7; to arms and legs during weeks 4 to 8; to the teeth and palate during weeks 7 to 12; and so on. After about the 12th week, such effects are generally more minor, since all the major organ systems are in place by about week 10. After this period of $2^{1}/_{2}$ months, a miniature human being exists; the rest of the time to week 38 (full term) is a period of fetal growth.

4.13. EFFECTS ON THE ENDOCRINE SYSTEM FROM EXPOSURE

The seven glands of the endocrine system, being chemical producers themselves, can be greatly affected by chemicals. There are substances that can prevent the particular glands from functioning normally, with attendant effects upon the system controlled by the hormone from that gland.

An excellent example of this is birth control pills. These chemicals function either to prevent fertilization of the egg by the sperm or to prevent the attachment of the fertilized egg to the wall of the uterus.

Other chemicals can affect each of the endocrine glands. A person with a defective pancreas can take injections of insulin in order to stabilize the metabolism of sugars. Likewise, other chemicals can negatively affect the endocrine system in many complex ways.

If exposure occurs while one is young, some chemicals can retard or stop growth due to effects on the pituitary; others can affect balance of blood chemicals.

4.14. EFFECTS ON THE CIRCULATORY SYSTEM FROM EXPOSURE

Some chemicals can alter the formation of blood cells. Red blood cells, which contain hemoglobin and transfer oxygen, are formed in the marrow of the long bones of the body. Various chemicals can reduce or destroy their formation or development, leading to anemia.

Other chemicals can affect the white blood cells, the leukocytes, which have a variety of functions, but are important in fighting infection and other foreign bodies in the blood. If these cells are damaged or destroyed, the body loses its ability to fight off infection.

4.15. HEPATIC (LIVER) EFFECTS FROM EXPOSURE

Since one of its primary missions is to detoxify harmful chemicals in the blood, the liver is quite often affected by foreign substances. Many chemicals can overwhelm the liver's capacity to detoxify. This result may arise from the toxicity of the particular chemical or by its concentration in the blood.

Alcohols and chlorinated compounds can damage the liver. If the liver ceases to function, death ensues. This is the usual cause of death for alcoholics and some other drug abusers. In addition, workers around some chlorinated solvents can also be harmed.

The chemicals that result from liver detoxification can be tested for in the blood and can be a clue to the kinds and amounts of chemicals introduced into the body. In addition, the liver produces chemicals on its own; these can be a clue to how well the liver is working. Such clues are the basis of the liver function tests often used in medical surveillance.

Some of the effects seen on the liver include altered liver enzyme levels, cancers, lack of detoxification capacity, and even death.

4.16. URINARY EFFECTS FROM EXPOSURE

The urinary system can also be damaged by exposure to toxic chemicals. The kidneys, which remove toxic liquid and soluble materials from the blood, can be damaged so that they no longer are able to remove these toxins. Systemic poisoning can result. Many of the products of body metabolism are toxic to the body; if these are not removed, symptoms of poisoning appear. Damage to the kidneys can even be fatal.

However, as in the case with the liver, the chemicals found in the urine can be used in medical surveillance as a clue to the functioning of the urinary system, as well as to measure exposure to chemicals. Urine testing for metabolic products of marijuana, for example, is the basis of drug screening programs.

Some of the harmful effects of chemicals on the urinary system include bladder cancers, kidney malfunction, and systemic toxemia.

4.17. INDUSTRIAL HYGIENE

The protection of those who work with hazardous materials from the chemicals they deal with is the job of the industrial hygienist. The study of industrial hygiene involves testing for the presence of harmful chemicals, determining the extent of the hazard, protecting workers if there is a hazard, and making changes necessary to remove the potential for contact with the hazard.

The basic philosophy of industrial hygiene is that all workers can and should be protected from harm as much as possible where they work. Many occupations are hazardous, but even workers in those fields should be protected. For example, car racing is thrilling and dangerous, but a driver should not die, as have some drivers, from breathing asbestos fibers found in their flameproof suits. Mining is also dangerous, but workers should not be exposed to poisonous gases. Industrial hygiene is the science of protecting workers.

4.18. HISTORY OF INDUSTRIAL HYGIENE

Industrial hygiene, while thought of as a new science, really began a long time ago. About 400 B.C., the Greek physician Hippocrates noticed that lead miners commonly suffered from mental diseases. Hippocrates could even be considered the first industrial hygienist as well as the Father of Medicine. Unfortunately, nothing was done for the miners to prevent the problems caused by chronic lead poisoning.

Some 500 years later, the Roman scholar Pliny the Elder noted that miners and other workers of zinc and sulfur suffered from various occupational diseases. About 200 A.D., the Greek physician Galen recognized that acid mists were harmful to copper miners; however, once again, nothing was done to solve the problem.

The first real attempts at industrial hygiene were made by Ulrich Ellenborg in the late 1400s. His book on occupational diseases included suggestions for control of the hazards.

About a century later, the German scholar best known by his Latin name, Georgius Agricola, recognized and evaluated the health hazards in the mining industry. He suggested controls by use of protective masks to lessen the chance of lung diseases and the use of ventilation within the mine shafts.

In the last years of the seventeenth century, an Italian physician named Bernado Ramazzini clearly saw that occupations and diseases were related. He suggested that all physicians, when taking a patient's history, first ask the patient's occupation.

The next century saw rapid advances in industrial hygiene. It was found, for example, that workers in the cider industry often suffered from a coliclike disease that was caused by leaching of lead from metal pipes and containers. Chimney sweeps were seen to suffer from scrotal cancer; then in 1788, laws were passed in England to protect chimney sweeps. From this beginning, industrial hygiene has become a vital part of worker education and protection.

However, it is not always just workers who are exposed to hazardous chemicals. Some historians feel that part of the decline in power of the ancient Roman Empire was due to the increase in mental illness, especially of the ruling class, as a result of drinking water containing lead leached from the supply pipes. Today, persons living near industrial plants may actually be exposed to greater concentrations of chemicals than the workers in the plant, due to exhaust ventilation systems. The general public may also be exposed to chemicals from former and present hazardous waste sites and from chemical spills or accidents.

4.19. CONCLUSION

There are many hazardous chemicals found in today's world. Everyone, from workers to the general public, needs to be protected from the potentially harmful effects of such chemicals.

Everything is made of elements, joined together into compounds. All spontaneous chemical reactions give off energy, sometimes violently. Many chemicals need to be separated during storage, handling, and use. These incompatible chemicals are potentially hazardous, and care must be exercised with them.

Most of the organ systems of the body can be affected by chemicals. The skin, being the most exposed to the environment, is most often the site of occupational injuries, although not necessarily the most severe. Inhalation of chemicals in the respiratory tract can be very serious due to the potential for damage at the level of the alveoli, where oxygen-carbon dioxide transfer takes place and where fibers can cause scarring. The eye can be damaged by splashes, leading to cloudiness or opacity of the cornea. The nervous system can be blocked by some chemicals or overworked by others. The circulatory system can transport chemicals throughout the body and can itself be damaged by chemicals. The liver, whose function is to detoxify foreign substances, can be overwhelmed by some chemicals. The urinary tract can be damaged, especially the kidneys and bladder, by chemicals. The endocrine system can be prevented from functioning with following effects on the organs controlled by the hormones produced by the glands. Since so many systems can be harmed in so many ways, the protection of people exposed to chemicals is very important.

Industrial hygiene attempts to protect all workers from harm in their workplaces.

As a result of the potential for harm from chemicals, many organizations have been developed, some by industries, some by interest groups, and some by governments. The next chapter will discuss some of these organizations.

4.20.   CHAPTER REVIEW QUESTIONS

1. Spontaneous chemical reactions always proceed in which direction?
   (a) To the smallest number of substances
   (b) To a state of lower energy
   (c) To the formation of gases
   (d) To a state of higher energy

2. A mutagen causes which condition?
   (a) Genetic changes in future generations
   (b) Tumors in susceptible individuals
   (c) Birth defects
   (d) None of the above

3. The term *terato-* has what meaning?
   (a) Monster-forming   (b) Cancer-forming
   (c) Change-producing   (d) None of the above

4. Most occupational exposures to chemicals occur by means of which system?
   (a) The skin   (b) The ears
   (c) The eyes   (d) The digestive tract

5. Airborne particles of what size can be trapped in the alveoli?
   (a) Greater than 10 microns   (b) Less than 1 micron
   (c) Between 1 and 10 microns   (d) About the same size as atoms

6. A class of chronic toxins is which of the following?
   (a) Carcinogens   (b) Teratogens
   (c) Mutagens   (d) All of the above

7. Which of the following chemical classes are always incompatible?
   (a) Salts and water   (b) Bases and water
   (c) Oxidizers and acids   (d) Oxidizers and combustibles

8. Industrial hygiene involves which of the following with respect to occupational hazards?
   (a) Recognition   (b) Evaluation
   (c) Control   (d) All of the above

9. An uncharged group of atoms, chemically combined, is called which of the following?
    (a) Molecule   (b) Superatom
    (c) Element   (d) Ion

10. Which of the following is a TRUE statement?
    (a) Industrial hygiene is a new science (since the 1930s).
    (b) Industrial hygiene is only interested in finding problems, not solving them.
    (c) Industrial hygiene is concerned with both finding and solving problems in workplaces.
    (d) Safety is the same as industrial hygiene.

5   WORKER PROTECTION

5.1. INTRODUCTION

We have seen how chemicals can be harmful if not stored, used, and disposed of correctly. In this chapter, we shall discuss the organizations and systems that have been designed to protect the health of workers.

5.2. DEFINITIONS

Some of the definitions we need to consider are contained in regulations and laws. Thus, some may seem complex or strangely worded; however, in order to meet certain legal requirements under federal regulations, they have to be defined as they are.

- An *employee* is any person who works in a business affecting commerce.
- *Commerce* means trade, commerce, transportation, or communication among the states, or between a state and any place outside thereof, or within the District of Columbia, or a possession of the United States, or between points in the same state but through a point outside the state.
- *Standard* (OSHA) is defined as a condition reasonably necessary or appropriate to provide safe or healthful employment and a place of employment.

5.3. OCCUPATIONAL SAFETY AND HEALTH ACT

In 1970, the U.S. Congress passed and President Nixon signed into law the most comprehensive worker protection package ever developed. This law, Public Law 91-596, the Williams-Steiger Occupational Safety and Health Act of 1970, required that the Secretary of Labor promulgate "as soon as possible" as a federal standard any national consensus standard for the protection of the safety and health of employees unless such a standard would not result in improved safety or health. The Act was entered as 84 U.S. Statute 1590.

Virtually every employee in the United States was thus assured a safe workplace. The Department of Labor (DOL) was directed to establish and enforce standards to protect employees.

Federal, state, and local governments are exempt from the law; however, a separate program was set up by executive order for federal employees, and most of the states have likewise set up similar programs for public employees.

5.4. OCCUPATIONAL SAFETY AND HEALTH ADMINISTRATION

In order to implement the Act, the DOL established the Occupational Safety and Health Administration (OSHA) to decide upon safety and health standards and to enforce those standards. OSHA has enforcement authority in any state unless the state has its own program with equal or more stringent standards.

OSHA, in order to implement this nationwide authority, has established ten regional offices located in Boston, New York, Philadelphia, Atlanta, Chicago, Dallas, Kansas City, Denver, San Francisco, and Seattle. Addresses of these offices are given in Appendix B.

OSHA issued the standards as Title 29, Code of Federal Regulations (CFR), Part 1910 for general industry and as Title 29 CFR, Part 1926 for the construction industry. Incidently, construction according to the law includes even alterations to a homeowner's garage. OSHA promulgated chemical exposure standards in 1972 as Title 29 CFR, Part 1910.1000; new standards were released in 1989.

5.5. NATIONAL INSTITUTE OF OCCUPATIONAL SAFETY AND HEALTH

While OSHA is an agency of the Department of Labor, the Act required that OSHA seek technical assistance and consider the recommendations of the National Institute of Occupational Safety and Health (NIOSH), today an agency of the Department of Health and Human Services (DHHS).

NIOSH is the scientific and research arm of the occupational safety and health program. NIOSH does not promulgate or enforce standards; only OSHA has the power to issue codified standards and to punish violators.

NIOSH does, however, conduct the research in order to arrive at standards that are valid and sound.

5.6. OSHA STANDARDS

In order to arrive at a standard for comparison of worker exposure to whatever amount might be harmful, OSHA assembles a committee of interested parties representing employers, employees, NIOSH, and state officials. This committee makes a recommendation to OSHA, which publishes it as a proposed rule in the Federal Register (FR) within 60 days of receiving it. All interested parties then have 30 days after publication to submit comments. During this period, a request for a public hearing may be submitted.

Whether a hearing is called or not, OSHA then has 60 days after the hearing or close of time for comments to issue the standard or explain why one was not issued. Generally, most standards become effective 90 days after publication of the final standard in the FR. Standards for chemicals are called Permissible Exposure Levels (PELs) and may be found in Title 29 CFR, Part 1910.1000.

If OSHA feels there is a potential for grave danger from a work practice or substance, it may issue an emergency standard. At the same time, OSHA must begin the same formal procedure discussed above. A permanent standard must be in effect with six months to replace any emergency standard.

5.7. NIOSH RECOMMENDED STANDARDS

One of the important functions of NIOSH is the development of standards. NIOSH conducts the research necessary to validate the exposure values proposed for worker safety and health. NIOSH also develops and issues testing methods to determine if the standards are being met. In order to differentiate these from OSHA enforceable standards, NIOSH calls their values Recommended Exposure Levels (RELs).

Another important function of NIOSH is the evaluation of protective equipment. All respirators, for example, must be NIOSH-certified in order to be acceptable to OSHA for use by workers. As a result of a joint NIOSH-MESA (Mining Enforcement and Safety Administration) (today, MSHA, Mining Safety and Health Administration) agreement, a new term was issued in Title 30 CFR, Part 11 for approval of respirators: Immediately Dangerous to Life and Health (IDLH), that is, "conditions that pose an immediate threat of severe exposure to contaminants ... which are likely to have adverse cumulative or delayed effects on health". Included are substances that are carcinogens, for example. The IDLH is the limiting concentration for which air-purifying respirators may be used.

NIOSH, as an agency of DHHS, is responsible for scientific studies to ensure that standards are safe. To this end, NIOSH publishes criteria documents based upon formal research into health effects of exposure. Criteria documents each address a specific chemical or class of chemicals. They usually consider previous standards, biological effects of exposure, environmental data, engineering controls, work practices, research needs, and the actual development of the recommended standard. The NIOSH recommended standards are the result of this work.

OSHA weighs these recommendations with other considerations such as feasibility and means of implementation in the issued standards. Thus, the OSHA standard (PEL) might be different from the NIOSH recommended standard (REL).

5.8. OSHA GENERAL INDUSTRY STANDARDS

OSHA standards for general industry are contained in 29 CFR, Part 1910. These standards cover medical records, walking, working surfaces, escape, platforms, ventilation, noise, radiation, compressed gases, personal protective equipment, environmental controls, first aid, fire protection and fire fighting, material handling, machinery, tools, welding, electricity, diving, and toxic and hazardous substances.

These standards cover most of the situations found in industry. However, there may be cases when an operation is not specifically covered by a standard. In this case, the general duty clause still requires the protection of worker safety and health.

5.9. OSHA STANDARDS FOR CONSTRUCTION

OSHA construction standards, as contained in Title 29 CFR, Part 1926, apply to all aspects of construction, including painting. Some of the Part 1926 standards include equipment, signs, tools, ladders, welding, vehicles, trenching, shoring, concrete, steel, tunnels, caissons, compressed breathing air, explosives, electricity, and others.

Hazardous waste sites must meet the requirements of Part 1926 for any permanent or temporary construction. Such sites present much more complex problems than simple construction due to the unknown nature of most sites. There is always a potential for chemical exposure, fire or explosion, oxygen deficiency, biological hazards, heat stress, cold exposure, noise, and even ionizing radiation. Carcinogens, explosives, pesticides, and radioisotopes have been found in hazardous waste sites. For these reasons, training, including formal classes and field exercises, is required by law for safe operation at hazardous waste sites. This training must include, as a minimum, aspects of the site safety plan; safe work practices; anticipated hazards; vehicle use; field equipment; emergencies and escape, plus self-rescue; handling, storage, and transportation of materials; use and care of personal protective clothing and equipment; sampling techniques; and general safety and health procedures. Documentation of such training is a requirement under the law.

5.10. LEGAL REQUIREMENTS OF EMPLOYERS

Employers are responsible for assuring compliance with OSHA requirements for medical records, exposure records and posting warning signs; ensuring that all applicable standards, as a minimum, are met; that proper protective equipment is available and that employee procedures are taught and enforced; informing employees when rules are not being followed; retraining employees as needed; terminating employees who willfully fail to follow regulations; and ensuring that there are no safety or health hazards in the workplace.

5.11. LEGAL REQUIREMENTS OF EMPLOYEES

Employees, too, have certain requirements. They must follow safety and health rules, warning signs, and standard operating procedures that are in compliance with OSHA regulations. An employee can be forced to wear protective equipment. Employees are also required to notify their employer if they see any safety or health hazards or violations. If the employer fails to address these violations, the employee is required to notify OSHA without fear of retaliation.

Employees must be trained by the employer; however, this does not mean that the employees do not have to do any studying on their own. Any worker who deals with hazardous materials or situations should learn as much as possible about the hazards.

5.12. LIMITATIONS OF OSHA STANDARDS

Unfortunately, there is a problem with OSHA standards — many are years out of date. Since these standards are law, they must be written into the law very specifically. Newer values of exposure standards can take years to be finally entered into the law. The NIOSH recommended standards might be quite different from the OSHA legal ones.

In addition, the OSHA standards are the very minimum to meet the letter of the law. Prudent employers and employees should use the OSHA standards as the definite minimum but should also consider other values from NIOSH or other organizations, as long as those values are at least as strict as the OSHA standards.

In this matter of the spirit versus the letter of the law, all workers and employers need to be aware at all times that the reason for the law is to protect life and health as much as possible, not just to meet the absolute wording of the regulations.

OSHA standards should be approached as the minimum, not necessarily the best, value or procedure for the protection of workers.

5.13. HAZARDOUS MATERIALS WARNING AND IDENTIFICATION SYSTEMS

Since there are many chemicals that can affect the organ systems of the human body, a way is needed to alert workers, and even the general public, to the hazards presented by these chemicals.

The simplest warning signs are simply words calling attention to some hazard, such as "watch your step" or "low ceiling." This approach is acceptable if all the potential victims can read the language of the sign and understand its meaning. Because often they cannot, more and more we are seeing pictures or pictographs being used to warn of hazards. These pictures may be used alone or with words. Even in countries with high literacy rates and a common language, it has been found that people respond more quickly to a picture than to words. Thus, such signs are becoming more popular worldwide even for highway and traffic warnings.

For most physical hazards, these signs are very valuable because the danger can be visualized; however, for chemical exposures, there are problems. For example, how does a sign indicate the amount of a substance that is harmful? In such cases, more is expected from the worker to understand the hazards and the concentrations that might be harmful.

There are other governmental organizations that issue standards and warning signs. Of these, the Environmental Protection Agency (EPA), which will be discussed in the next chapter, the Department of Transportation (DOT), and Nuclear Regulatory Commission (NRC) are of most interest for first responders and other hazardous materials workers.

Many private (nongovernmental) organizations have developed standards for their industries or areas of interest; these so-called consensus organizations issue recommendations and standards that can be more quickly changed as new discoveries are made than can laws. In fact, the original OSHA standards were simply standards in use by various consensus organizations.

The next paragraphs will discuss some of these organizations and their standards and warning signs.

5.14. DEPARTMENT OF TRANSPORTATION

The U.S. Department of Transportation (DOT) is a federal regulatory organization and is responsible for rules and regulations involving transport of all materials, with only a few exceptions. The basic law by which the DOT regulates chemicals is the Hazardous Materials Transportation Act of 1974 (HMTA), as amended. Those chemicals declared hazardous by the DOT are contained in Title 49 CFR, Parts 100-199.

DOT issues standards for labeling and marking of individual packages containing hazardous substances and for placards on vehicles carrying hazardous cargo. Examples of container labels and vehicle placards are given in Figure 5-1.

Some of the materials regulated by the DOT are also regulated by the EPA under a joint agreement that requires shippers to comply with both DOT and EPA standards but that attempts to make the regulations as compatible as possible.

5.15. NUCLEAR REGULATORY COMMISSION

The NRC issues standards for protection from manmade radioisotopes and ionizing radiation other than X-ray machines. These standards are similar to those of several national and international consensus organizations dealing with radioactivity. Like the chemical exposure standards, these are based upon animal models and cases in which humans have received exposures to radiation. Many of the data are from victims of the Hiroshima and Nagasaki atomic bomb explosions plus newer information from other human experiences like the Chernobyl accident and medical exposures.

The NRC standards for radioisotopes must be considered in addition to any chemical toxicity of those substances. Standards for area warning signs, shipping labels and markings, and vehicle warning signs are contained in the NRC regulations in 10 CFR, Part 20 and in DOT regulations of 49 CFR, Part 172 (see Figure 5-2).

5.16. AMERICAN CONFERENCE OF GOVERNMENTAL INDUSTRIAL HYGIENISTS

The American Conference of Governmental Industrial Hygienists (ACGIH)

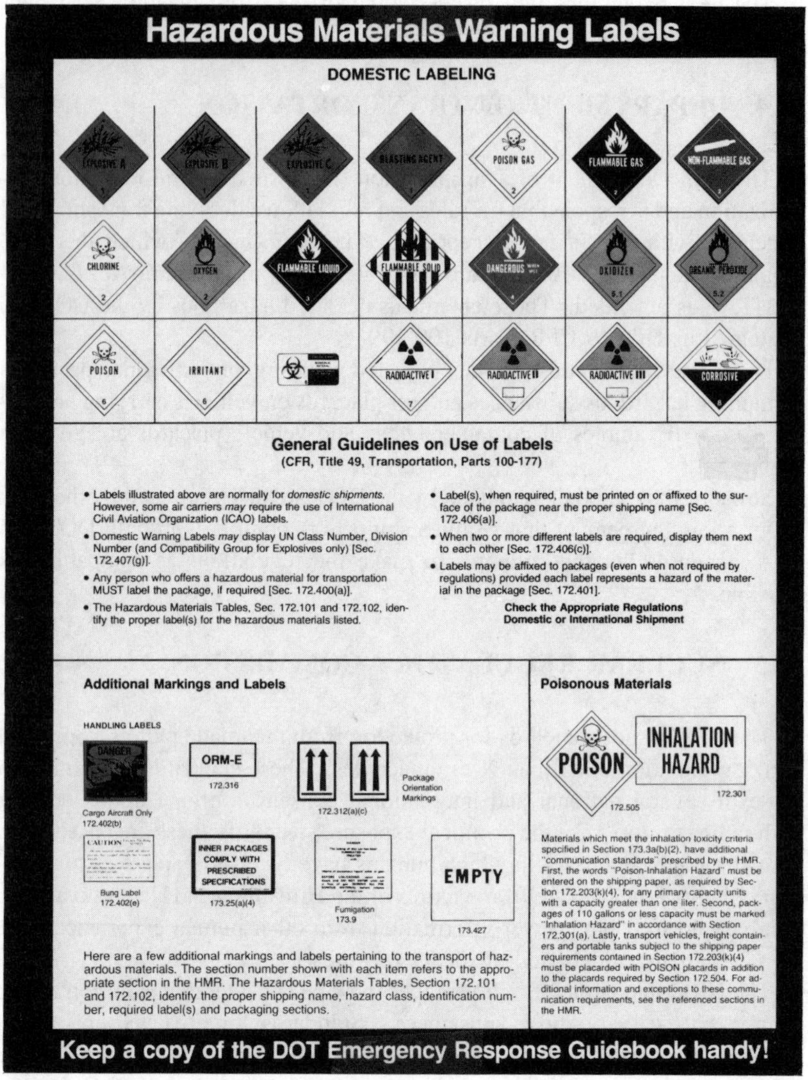

Figure 5-1.   Department of Transportation warning labels and placards.

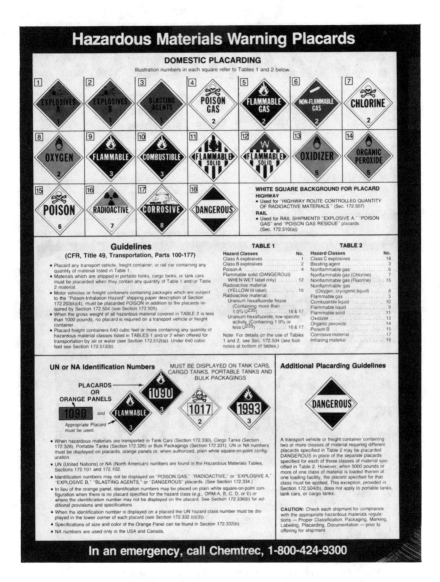

Figure 5-1.   Continued.

HAZARDOUS WASTE

HANDLE WITH CARE!

ACCUMULATION
START DATE_____

CONTENTS _____

CARLTON INDUSTRIES INC. 409-242-5055 STOCK NO. ACL-10

NON-REGULATED WASTE

OPTIONAL INFORMATION:

SHIPPER_____

ADDRESS_____

CITY, STATE, ZIP_____

CONTENTS _____

_____

NON-REGULATED WASTE

CARLTON INDUSTRIES INC. LA GRANGE, TEXAS

Figure 5-1.   Continued.

NON-HAZARDOUS WASTE

OPTIONAL INFORMATION:

SHIPPER _____

ADDRESS _____

CITY, STATE, ZIP _____

CONTENTS _____

NON-HAZARDOUS WASTE
CARLTON INDUSTRIES INC. LA GRANGE, TX STOCK NO. 10402

Figure 5-1.   Continued.

was founded in 1938 to exchange ideas and experiences and for the promotion of standards and techniques in industrial health. Over the years, the ACGIH has conducted research to arrive at concentrations of chemicals and physical agents that appear to be safe for most workers. After several changes in terminology, these values have been named Threshold Limit Values® (TLVs®) and refer to airborne concentrations of chemicals to which it is believed that most healthy workers may be repeatedly exposed day after day without adverse effects.

The TLVs are carefully documented using state-of-the-art techniques and are generally considered to be the best values commonly available for worker protection from airborne concentrations of chemicals.

ORM-E

HAZARDOUS WASTE

FEDERAL LAW PROHIBITS IMPROPER DISPOSAL

IF FOUND, CONTACT THE NEAREST POLICE, OR
PUBLIC SAFETY AUTHORITY, OR THE
U.S. ENVIRONMENTAL PROTECTION AGENCY

PROPER D.O.T.
SHIPPING NAME_____ UN OR NA#_____
GENERATOR INFORMATION:

NAME_____

ADDRESS_____

CITY_____ STATE_____ ZIP_____
EPA   EPA
ID NO._____   WASTE NO._____

ACCUMULATION   MANIFEST
START DATE_____   DOCUMENT NO._____

HANDLE WITH CARE!
CONTAINS HAZARDOUS OR TOXIC WASTES

CARLTON INDUSTRIES INC. 1-800-231-5988 LA GRANGE, TX STOCK NO. 1401PF

HAZARDOUS WASTE

FEDERAL LAW PROHIBITS IMPROPER DISPOSAL

IF FOUND, CONTACT THE NEAREST POLICE, OR
PUBLIC SAFETY AUTHORITY, OR THE
U.S. ENVIRONMENTAL PROTECTION AGENCY

PROPER D.O.T.
SHIPPING NAME_____ UN OR NA#_____
GENERATOR INFORMATION:

NAME_____

ADDRESS_____

CITY_____ STATE_____ ZIP_____
EPA   EPA
ID NO._____   WASTE NO._____

ACCUMULATION   MANIFEST
START DATE_____   DOCUMENT NO._____

HANDLE WITH CARE!
CONTAINS HAZARDOUS OR TOXIC WASTES

CARLTON INDUSTRIES INC. LA GRANGE, TEXAS STOCK NO. 1402

Figure 5-1.   Continued.

RADIATION SYMBOL

1. Cross-hatched area is to be magenta or purple.
2. Background is to be yellow.

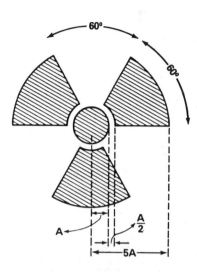

In addition to the contents of signs and labels prescribed in this section, licensees may provide on or near such signs and labels any additional information which may be appropriate in aiding individuals to minimize exposure to radiation or to radioactive material.

Radiation areas. Each radiation area shall be conspicuously posted with a sign or signs bearing the radiation caution symbol and the words:

CAUTION

RADIATION AREA

High radiation areas. Each high radiation area shall be conspicuously posted with a sign or signs bearing the radiation caution symbol and the words:

CAUTION

HIGH RADIATION AREA

Each airborne radioactivity area shall be conspicuously posted with a sign or signs bearing the radiation caution symbol and the words:

CAUTION

AIRBORNE RADIOACTIVITY AREA

Additional requirements. (1) Each area or room in which licensed material is used or stored and which contains any radioactive material (other than natural uranium or thorium) in an amount exceeding 10 times the quantity of such material specified in Appendix C of this part shall be conspicuously posted with a sign or signs bearing the radiation caution symbol and the words:

CAUTION

RADIOACTIVE MATERIAL(S)

Each area or room in which natural uranium or thorium is used or stored in any amount exceeding one hundred times the quantity specified in Appendix C of this part shall be conspicuously posted with a sign or signs bearing the radiation caution symbol and the words:

CAUTION

RADIOACTIVE MATERIAL(S)

Containers. (1) Except as provided in paragraph (f)(3) of this section, each container of licensed material shall bear a durable, clearly visible label identifying the radioactive contents.

A label required pursuant to paragraph (f)(1) of this section shall bear the radiation caution symbol and the words "CAUTION, RADIOACTIVE MATERIAL" or "DANGER, RADIOACTIVE MATERIAL". It shall also provide sufficient information ⇒ to permit individuals handling or using the containers, or working in the vicinity thereof, to take precautions to avoid or minimize exposures.

Figure 5-2.   Nuclear Regulatory Commission Warning Sign.

Substance	[CAS #]	ADOPTED VALUES TWA ppm	TWA mg/m³	STEL ppm	STEL mg/m³
L.P.G. (Liquified petroleum gas) [68476-85-7] (1987) ...		1000	1800	—	—
Magnesite [546-93-0] (1986) .		—	10(e)	—	—
Magnesium oxide fume [1309-48-4] (1977)		—	10	—	—
◄Malathion [121-75-5] — Skin (1977)		—	10	—	—
Maleic anhydride [108-31-6] (1977)		0.25	1.0	—	—
Manganese [7439-96-5], as Mn					
• Dust & compounds (1988)		—	5	—	—
Fume (1979)		—	1	—	3
Manganese cyclopentadienyl tricarbonyl [12079-65-1], as Mn — Skin (1986)		—	0.1	—	—
Marble, see Calcium carbonate					
Mercury [7439-97-6], as Hg — Skin					
Alkyl compounds (1980) .		—	0.01	—	0.03
All forms except alkyl Vapor (1982)		—	0.05	—	—
• Aryl & inorganic compounds (1982) ..		—	0.1	—	—
•Mesityl oxide [141-79-7] (1981)		15	60	25	100
Methacrylic acid [79-41-4] (1981)		20	70	—	—
Methane [74-82-8] (1981)....		—(d)	—	—	—
Methanethiol, see Methyl mercaptan					
Methanol, see Methyl alcohol					
◄Methomyl [16752-77-5] (1977)		—	2.5	—	—
Methoxychlor [72-43-5] (1977)		—	10	—	—
•2-Methoxyethanol [109-86-4] — Skin (1984) ...		5	16	—	—
2-Methoxyethyl acetate [110-49-6] — Skin (1984) ...		5	24	—	—
4-Methoxyphenol [150-76-5] (1982)		—	5	—	—
Methyl acetate [79-20-9] (1976)		200	606	250	760
‡Methyl acetylene [74-99-7]..		1000	1640	(1250)	(2050)
Methyl acetylene-propadiene mixture (MAPP) (1976) ..		1000	1640	1250	2050
Methyl acrylate [96-33-3] — Skin (1977)		10	35	—	—
Methylacrylonitrile [126-98-7] — Skin (1986) ...		1	2.7	—	—

28

Substance	[CAS #]	ADOPTED VALUES TWA ppm	TWA mg/m³	STEL ppm	STEL mg/m³
Methylal [109-87-5] (1987) ...		1000	3110	—	—
Methyl alcohol [67-56-1] — Skin (1976)		200	262	250	328
Methylamine [74-89-5] (1977)		10	13	—	—
Methyl amyl alcohol, see Methyl isobutyl carbinol					
Methyl n-amyl ketone [110-43-0] (1987).........		50	233	—	—
◄N-Methyl aniline [100-61-8] — Skin (1986)		0.5	2.2	—	—
••Methyl bromide [74-83-9] — Skin (1986)		5	19	—	—
•Methyl n-butyl ketone [591-78-6] — Skin (1981) ...		5	20	—	—
••Methyl chloride [74-87-3] — Skin (1976)		50	103	100	207
• -Methyl chloroform [71-55-6] (1976)		350	1910	450	2460
Methyl 2-cyanoacrylate [137-05-3] (1976).........		2	9.1	4	18
Methylcyclohexane [108-87-2] (1987).........		400	1610	—	—
Methylcyclohexanol [25639-42-3] (1987)		50	234	—	—
o-Methylcyclohexanone [583-60-8] — Skin (1976) ...		50	229	75	344
2-Methylcyclopentadienyl manganese tricarbonyl [12108-13-3], as Mn — Skin (1986)		—	0.2	—	—
◄Methyl demeton [8022-00-2] — Skin (1986) ..		—	0.5	—	—

(d) Simple asphyxiant; see definition in the "Introduction to the Chemical Substances."
(e) The value is for total dust containing no asbestos and < 1% crystalline silica.
( ) Adopted values enclosed are on the Notice of Intended Changes list. Consult it for current proposal for change.
◄ Identifies substances for which there are also BEIs (see BEI section). Substances identified in the BEI documentations for methemoglobin inducers (for which methemoglobin is the principle toxicity) and organophosphorus cholinesterase inhibitors are part of this notation.
■ Substance identified by other sources as a suspected or confirmed human carcinogen. See the compilation in the Appendix to the Documentation of TLVs, pp. A-5(86)—A-9(86).
• Substance for which OSHA and/or NIOSH has a Permissible Exposure Limit (PEL) or a Recommended Exposure Limit (REL) lower than the TLV.
‡ See Notice of Intended Changes.

29

Figure 5-3.   Sample page from the Threshold Limit Values booklet. (By permission of the ACGIH.)

It is important to recall, however, that these values are based on healthy workers — there may be persons who could be harmed at these levels — and that the limits are not fine lines separating "safe" from "harmful," since many factors relating to dose-response are involved.

The TLVs are really several different standards: the TLV-Time Weighted Average (TLV-TWA), which is the average maximum concentration for a normal 8-hour workday and 40-hour work week; the TLV-Short Term Exposure Limit (TLV-STEL), which is the concentration to which workers may be exposed for short periods of time repeatedly; and the TLV-Ceiling (TLV-C), which is the concentration that should never be exceeded at any time.

TLVs are constantly being evaluated and updated, and therefore the values are published annually. A sample page is shown in Figure 5-3.

The original OSHA chemical exposure standards issued in 1972 were generally the ACGIH TLVs for 1968.

5.17. NATIONAL FIRE PROTECTION ASSOCIATION

One of the early organizations to develop standards and to suggest warning signs was the National Fire Protection Association (NFPA). This organization has issued a number of construction design standards to lessen or prevent the effects of fires. These standards are found in the *NFPA Life Safety Code 101*, and many have been codified into law by cities and counties.

For the purpose of this book, we are most interested in the NFPA warning diamond. This is presented in *NFPA Code 704A*. The system is based upon a diamond-shaped symbol that is divided into four squares. An example of the sign is presented in Figure 5-4.

The upper square indicates the fire hazard. The background color should be red with overprinted white numerals from 0 for materials that will not burn to 4 for very flammable materials.

The left-hand square indicates the health hazards. The background color should be blue with overprinted white numerals from 0 for materials that present no special health hazards to 4 for highly toxic substances. Remember that this rating is generally for acute effects, since the system is designed to be an emergency warning system; chronic hazards are of lesser importance.

The right-hand square indicates the chemical reactivity. The background in this case should be yellow with the white numerals from 0 for normally stable substances to 4 for materials capable of detonation or explosion.

The lower square indicates the presence of any special or specific hazard, especially reactivity with water. In that case, the capital letter W with a line through it alerts fire personnel not to use water. Other warnings placed in this white square with overprinted black letters include OXY for oxidizers, ACID for strong acids, ALK for strong alkaline materials, the radioactive warning trefoil for radioisotopes, and COR for corrosives.

The code allows the use of the colors to be optional; however, most of the time these are used. If colors are not used, all the squares are white with overprinted black numerals or words.

This system is an alerting system for emergency response personnel, chiefly fire fighters, so chronic health hazards are not emphasized.

5.18. AMERICAN NATIONAL STANDARDS INSTITUTE

Another organization that sets standards is the American National Standards Institute (ANSI), which has representatives of various professional organizations meet to decide, by committee, on standards ranging from characteristics

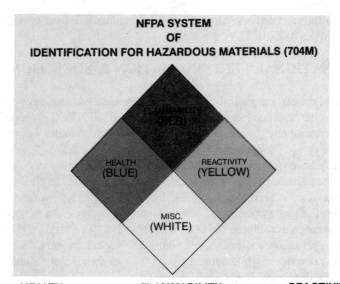

NFPA SYSTEM OF IDENTIFICATION FOR HAZARDOUS MATERIALS (704M)

HEALTH	FLAMMABILITY	REACTIVITY
4 - TOO DANGEROUS TO ENTER VAPOR	4 - EXTREMELY FLAMMABLE	4 - MAY DETONATE. VACATE AREA IF IN ADVANCED STAGES OF FIRE.
3 - EXTREMELY DANGEROUS. USE FULL PROTECTIVE CLOTHING	3 - IGNITES AT NORMAL TEMPERATURES.	3 - STRONG SHOCK OR HEAT MAY DETONATE. USE MONITORS FROM BEHIND EXPLOSION RESISTANT BARRIERS.
2 - HAZARDOUS - USE BREATHING APPARATUS	2 - IGNITES WHEN MODERATELY HEATED	2 - VIOLENT CHEMICAL CHANGE POSSIBLE - USE HOSE STREAMS FROM DISTANCE.
1 - SLIGHTLY HAZARDOUS	1 - MUST BE PREHEATED TO BURN	1 - UNSTABLE IF HEATED. USE NORMAL PRECAUTION
0 - LIKE ORDINARY MATERIAL	0 - WILL NOT BURN	0 - NORMALLY STABLE

MISCELLANEOUS: Used for special situations; e.g.

W̶ DO NOT USE WATER.

P- Material may self-react or polymerize

☢ Radioactivity

OXY- Oxidizing agents

Figure 5-4.   The National Fire Protection Association warning diamond.*

of safety glasses to composition of compressed breathing gases to the size of movie-film sprocket holes to attenuation characteristics of hearing-protective equipment. For almost anything used in industry, there is an ANSI standard. ANSI standards have generally been the basis of standards for protective equipment. Always be sure that the standard you are using is the most current, since these standards are often updated but not on an annual basis.

5.19. INTERNATIONAL WARNING SYSTEMS

There are international organizations having standards involving exposures to chemicals. We shall consider the United Nations (UN) Shipping Number, the International Air Transport Association (IATA), and the International Maritime Organization (IMO).

The UN classification system for chemicals requires that all chemicals in international transport must be labeled using signs agreed to by member nations. In this system, an individual number is given to almost every one of the chemicals used in commerce. This number must be placed on the shipping documents, on the container, and on the placard on the vehicle or vessel transporting it. In addition, there are general hazard-class numbers assigned to cargos that must be posted.

The UN hazard-class numbers are 1 — explosives; 2 — compressed gases; 3 — flammable liquids; 4 — flammable solids; 5 — oxidizers; 6 — poisonous or infectious; 7 — radioactive; 8 — corrosive; and 9 — miscellaneous dangerous substances.

5.20. INTERNATIONAL AIR TRANSPORT ASSOCIATION

The International Air Transport Association (IATA), composed of the airlines of the world, has developed standards for types and quantities of materials to be carried aboard both passenger and cargo aircraft. Most passenger aircraft are very limited in both the types of chemicals and the amounts that can be transported. In fact, a glance at your airline ticket the next time you fly will show that even a passenger is prohibited from carrying matches, oxidizers, corrosives, compressed gases except for small spray cans of personal cosmetics, and electromagnetic or radioactive materials.

IATA has one label not seen in other modes of transport; this is the warning sign "not to be carried on passenger aircraft."

5.21. INTERNATIONAL MARITIME ORGANIZATION

The International Maritime Organization (IMO) issues standards for ocean shipping of chemicals. This organization has given numbers to most of the materials commonly transported by sea. Generally, these are the same as the UN shipping numbers.

5.22. LIMITATIONS OF WARNING SYSTEMS

There are some difficulties in warning systems, especially for marking of individual containers of chemicals. At this time, there is no standard label on small containers to warn of the hazards. Different manufacturers have devised different systems, often using color codes, pictographs, or specific formats. As long as the worker understands the particular system in use, there is no problem; but the new employee may be used to a different system or to none at all.

The U.S. Department of Defense is a major user of chemicals and has proposed a standardized label for chemicals. This may become the model for a nationwide system, but until such time as warning labels are uniform, it is up to the worker to know the system in use for the chemicals in his or her workplace.

Another problem lies in the fact that different people respond differently to the same amount of a chemical. Most people may be hurt by a bee sting, but usually the pain and swelling go away in a few days; however, other people are so sensitive to the toxin that one bee sting can cause death. Rating and warning systems cannot take individual characteristics into account.

Mixtures present a third problem, since very few mixtures have been studied. One chemical may cause another to be more or less hazardous in the mixture than when alone. Most warning systems address only the most hazardous component of a mixture.

In order for any warning system to work, it must be understood. This means that a rating system, warning sign, label, placard, etc. needs more than simply to be in place. The proper warning must be conveyed to the user. Therefore, users must be educated in the meanings of the system and the reasons behind the warning. Simply tacking up a sign does not protect workers or the public unless they are educated in the reasons for the signs. One purpose of this text is to fulfill that need.

5.23. VALUES OF WARNING SYSTEMS

While the preceding paragraph might make you think that all such systems and warnings are not important, it must be stressed that these systems are indeed valuable. Almost everything in life has limitations, and these systems are no exception.

Poisoning potentials allow us to categorize substances into broad groups so as to warn us of the relative degree of hazard and the amount of protection needed. Shipping labels alert us to hazards in the event of an accident or spill. Standards for shipping containers lessen the possibility of accidents. Warning signs caution us to be especially careful or to wear protective equipment. International agreements mean that no matter where we are or where a package has come from, we can understand the hazards within.

While there are many different systems, each has a purpose and can be of great value.

5.24. CONCLUSION

Many organizations are working to protect the worker from harm in the workplace. There are regulatory organizations such as OSHA, NRC, and DOT; and there are consensus organizations such as ACGIH, ANSI, NFPA, and many others. Most of these have issued standards and conducted studies to arrive at values considered safe for most workers. Using additional data from NIOSH and other laboratories, standards have been agreed to by OSHA for the protection of workers' safety and health.

The next chapter will continue this discussion into protection of the environment and the general public.

5.25. CHAPTER REVIEW QUESTIONS

1. PELs are published by which of these organizations?
   (a) OSHA   (b) NFPA
   (c) ACGIH   (d) NIOSH

2. Which of the following is a FALSE statement?
   (a) ACGIH, OSHA, and NRC are regulatory organizations
   (b) NFPA, ACGIH, and ANSI are consensus organizations
   (c) NRC, EPA, and OSHA are regulatory organizations
   (d) Regulatory organizations have the force of law behind them

3. If OSHA, ACGIH, and NIOSH standards differ, the prudent person will follow what course?
   (a) Ignore all of them
   (b) Follow the OSHA standard
   (c) Develop a fourth, less strict standard
   (d) Follow the most stringent standard

4. OSHA standards are found in which title of the Code of Federal Regulations (CFR)?
   (a) Title 20   (b) Title 29
   (c) Title 40   (d) Title 10

5. PEL is an abbreviation for which of the following?
   (a) Practical Evaluation Limit
   (b) Permissible Exposure Limit
   (c) Permissible Extended Limit
   (d) Possible Excursion Limit

6. TLV® is an abbreviation for which of the following?
   (a) Threshold Limit Value   (b) Theoretical Limiting Value
   (c) Total Loss of Vision   (d) Temporary Loss of Viability

7. If the PEL is 10 mg/m^3 and the TLV® is 1 mg/m^3, which is the OSHA legally enforceable limit?
   (a) 10 mg/m^3 (b) 1 mg/m^3
   (c) 5.5 mg/m^3 (d) None of the above

8. OSHA was created by which of the following?
   (a) Gramm-Rudman Act of 1986
   (b) The Roosevelt Administration
   (c) The U.S. Constitution
   (d) Williams-Steiger Act of 1970

9. Employee protection standards required of a hazardous waste facility are overseen by which of the following organizations?
   (a) MSHA   (b) NIEHS
   (c) NIOSH   (d) OSHA

10. The organization that issues TLVs® is which of the following?
    (a) NIEHS   (b) NIOSH
    (c) ACGIH   (d) OSHA

6 ENVIRONMENTAL PROTECTION

6.1. INTRODUCTION

In 1976, the U.S. Congress decided that earlier laws on disposal of solid wastes were inadequate to address the variety and quantity of hazardous waste substances being placed in the environment. Therefore, Congress passed the Resource Conservation and Recovery Act, commonly called RCRA. This Act laid the foundation for a comprehensive national program of hazardous waste management.

However, by 1980, it was realized that simply requiring present and future proper management of hazardous wastes was not sufficient. Many highly publicized former disposal sites such as Love Canal at Niagara Falls, New York; the so-called Valley of the Drums in Kentucky; and a multitude of chemical dumps in New Jersey clearly indicated that there were hundreds, if not thousands, of former disposal sites posing threats to groundwater and human health. Because of the massive costs potentially involved with cleanup of these sites, as well as questions of legal liability, Congress enacted the Comprehensive Environmental Response, Compensation, and Liability Act, often called CERCLA.

These two laws, with amendments, form the basis of protection of the environment from hazardous materials.

6.2. DEFINITIONS

RCRA, as the first law to regulate the management of hazardous wastes, required several new definitions not covered under earlier solid waste acts. One of these was *hazardous waste* itself:

... a solid waste (which by law includes gases and liquids), or combination of solid wastes, which because of its quantity, concentration, or physical, chemical, or infectious characteristics may cause, significantly contribute to an increase in mortality or an increase in serious irreversible, or incapacitating reversible, illness; or pose a substantial present or potential hazard to human health or the environment...

For the purpose of regulations under RCRA, several groups of handlers of hazardous waste were also defined:

- A *generator* is any person (owner, operator, company, facility, etc.) who produces hazardous waste through a manufacturing process or who accumulates hazardous wastes (such as a laboratory or service station).
- A *transporter* is any person engaged in the offsite transportation of hazardous waste by air, rail, highway, or water.
- *Treatment* is any method, technique, or process designed to change the characteristics of any hazardous waste to neutralize such waste, or to recover energy or material resources from the waste, or to render such waste nonhazardous.
- *Disposal* is the discharge, deposit, injection, dumping, spilling, leaking, or placing of any waste into or on any land or water so that the waste may enter the environment or be emitted into the air or discharged into any waters.

6.3. INTENT OF THE RESOURCE CONSERVATION AND RECOVERY ACT OF 1976

The Resource Conservation and Recovery Act of 1976, Public Law 94-580, codified as 90 U.S. Statute 2795, was passed by Congress and signed by the President in order to control hazardous wastes from their generation to their ultimate disposal — so-called cradle-to-grave management. However, another important aspect of the Act that has not received as much attention was the intent of Congress that important natural resources be conserved by reuse or recycling of materials. Millions of tons of recoverable material have been needlessly buried because separation methods and recovery techniques have not been developed. The philosophy behind RCRA is thus twofold: to protect health and the environment, and to recover and conserve materials and energy.

Thus, recovery should be the first consideration of any generator of waste. Only after concluding that there are no practical recycling or recovery methods should disposal be considered. And even then, disposal must conform to all laws and regulations so as to prevent hazards to human health or the environment. The regulations promulgated by the U.S. Environmental Protection Agency (EPA) to implement RCRA are deliberately so stringent as to encourage reuse and to discourage damage to the environment. The EPA has authority unless a state has a more stringent program; in that case, the state may have primary authority.

6.4. ELEMENTS OF THE RESOURCE CONSERVATION AND RECOVERY ACT

RCRA and the implementing regulations in Title 40, Code of Federal Regulations, Parts 122–265 contain the elements of a complete cradle-to-grave program for hazardous wastes.

These elements are identification of hazardous waste; standards applicable to generators, transporters, treaters, storers, and disposers of hazardous wastes; inspections; enforcement; and research, training, and demonstrations in the field of hazardous waste handling.

6.5. WHAT IS A HAZARDOUS WASTE?

What seems like a trivial question can really be quite complex. The definition presented earlier is the legal terminology; however, the determination of whether or how much of a substance may "cause, or significantly contribute to an increase in mortality or serious illness or pose a hazard to health and the environment" is not as clear as the definition would seem to indicate.

To this end, the EPA has developed criteria for determining if a waste is hazardous or not. First, it has to be a waste, that is, something to be discarded. Anything that can be recycled or reused is not a waste, although it may still be a hazardous material governed by other regulations such as Department of Transportation (DOT) Hazardous Materials regulations, the Clean Air Act, the Clean Water Act, the National Pollution Discharge Elimination System (NPDES), the Toxic Substance Control Act (TSCA), or others.

Once it is decided that a material is indeed a waste, then it is necessary to find out if it is hazardous under RCRA.

6.6. HAZARDOUS WASTE IDENTIFICATION

The generator, based upon knowledge of the waste, may simply declare it to be hazardous. In this case, all the provisions of RCRA come into play without the expense of laboratory analysis, etc. However, since the cost of transportation and disposal of hazardous wastes is many times that of nonhazardous wastes, most generators try to prove the waste does not come under

RCRA rather than saying it does. The criteria for general hazardous wastes are given in Title 40 CFR, Part 261.20; however, the following gives the general information.

Waste can be hazardous based upon laboratory testing for one or more of four characteristics: ignitability, corrosivity, reactivity, or toxicity. The testing must be performed by a laboratory acceptable to and certified by the EPA.

The waste may be, in addition, one of the listed wastes in RCRA regulations. These include wastes from specific as well as nonspecific sources, chiefly industrial waste streams. Wastes from nonspecific sources are found in Title 40 CFR, Part 261.31 and are prefixed by the letter D. These wastes are listed in Appendix C. Wastes from specific sources are contained in 40 CFR, Part 261.32 and have an identification number prefixed by the letter F. Such chemicals are listed in Appendix D.

There are some 120 or so chemicals listed in 40 CFR, Part 261.33e that are called acute hazardous wastes. These are, generally, the most toxic or most dangerous chemicals commonly found in industry. Such wastes include not only the chemicals themselves but even off-specification rejects, containers, and spills. These wastes carry an identification number with the prefix letter P. These chemicals are listed in Appendix E.

In addition, there are another 240 chemicals or so that are called toxic wastes and are, in general, slightly less hazardous than the P wastes. This list also includes the chemicals themselves, off-specification rejects, and spills. Such wastes are found in 40 CFR, Part 261.33f. These wastes carry an identification number with a prefix U and are listed in Appendix F.

If the waste falls into any of these categories, it must be managed under the hazardous waste regulations. There are, however, some exclusions based upon quantities from small operations, laboratories, etc.

Figure 6-1 gives a decision matrix for determination of what a hazardous waste is and how to handle the waste to conform to RCRA.

6.7. STANDARDS APPLICABLE TO GENERATORS

Once a waste is categorized as hazardous, there are certain requirements forced upon the generator of the waste. While the standards may be found in detail in Title 40 CFR, Part 262, in general the requirements include a manifest system, pretransportation procedures, recordkeeping, and reporting.

The generator must prepare a manifest for all offsite shipments. This document must contain such information as the identification number of the generator as furnished by the EPA, the complete address of the generator, a description of the nature and quantity of the waste, the intended treatment or disposal facility, and one emergency alternate treatment or disposal facility. The generator must sign the manifest and obtain the signature of the initial transporter. There must be sufficient copies to provide one for the generator's

HAZARDOUS WASTE DECISION MODEL

WASTE STREAMS

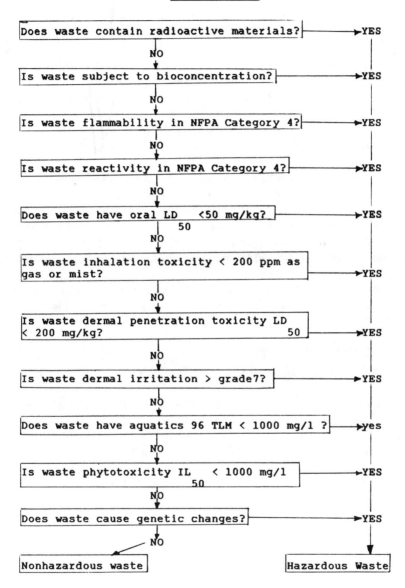

Figure 6-1.   Waste decision matrix.

file, one for each transporter, one for the treatment/disposal facility, and one copy to be returned from the treatment/disposal facility back to the generator to close the loop.

The generator's pretransport requirements include proper packaging, labeling, marking, and placarding in accordance with DOT regulations.

The generator must maintain complete records of each original manifest for three years or until the copy from the receiving facility is returned. This copy must then be retained for three years. Generators must also keep records of annual reports filed with the EPA, any exception reports for shipments experiencing a problem, and results of tests and analyses for three years.

An annual report must be submitted to the EPA (or more often if requested by the agency) or to the state organization if it has primacy.

6.8. STANDARDS APPLICABLE TO TRANSPORTERS

The legal requirements for transporters include obtaining a permit from the EPA, refusal of any shipment not accompanied by a proper manifest or any shipment not properly packaged, appropriate actions to protect health and the environment in the event of a spill or other discharge during transportation, and maintenance of records for three years.

In the case of a spill or other discharge, the transporter is legally responsible for cleanup. The complete regulations for transporters are found in Title 40 CFR, Part 263.

6.9. STANDARDS APPLICABLE TO TREATMENT, STORAGE, AND DISPOSAL FACILITIES

Owners and operators of hazardous waste storage, treatment, and disposal facilities are required to obtain a permit from the EPA for their operation. Requirements for these facilities may be found in Title 40 CFR, Part 264. These facilities are also required to sign the manifest and to return a copy to the generator. Facilities must maintain an operating record containing descriptions and quantities of the wastes received; the method and date of treatment, storage, or disposal; the location within the facility of the waste, and the quantity at each location; test and analysis reports; records of inspections and all unusual incidents; and summary reports of all incidents. This record must be maintained until closure of the facility.

There are regulations covering receipt of unmanifested wastes; wastes at variance with the manifest; monitoring; closure and postclosure standards; financial considerations of bonding, stability, etc.; contingency plans; and operational considerations for all the possible types of treatment and disposal facilities.

6.10. BENEFITS FROM RCRA

Since the enactment of RCRA and the promulgation of standards by the EPA, the amount of unlawful dumping has diminished considerably. The manifest system with its tracking requirements, especially in the event of "lost" or delayed shipments, has meant that it is almost impossible for a hazardous waste generator to get away with improper offsite disposal. Since civil and criminal penalties for improper storage, treatment, or disposal can be considerable, most generators today are pretty much in compliance with the law.

RCRA has meant that hazardous wastes are tracked from the point of generation to the ultimate acceptable disposal. "Midnight dumpers" have been largely removed from the hazardous waste business.

6.11. COMPREHENSIVE ENVIRONMENTAL RESPONSE, COMPENSATION, AND LIABILITY ACT OF 1980

Although RCRA has meant that most hazardous wastes are stored, transported, treated, and disposed of in accordance with standards considered acceptable for the protection of human health and the environment, there are many former dumps and other sites posing potential damage. RCRA only concerns hazardous wastes now being generated.

By 1980, Congress realized that many old sites, estimated to number as many as 50,000, were posing a hazard to public health. Of these, perhaps 2000 were considered immediate hazards.

The costs of cleanup of many of these sites are astronomical — often beyond the financial capabilities of even very large companies. Litigation was slow and complicated even in those cases where a defendant could be found. In addition, a chemical dump might contain drums of chemicals from dozens of companies unrelated to the owner of the dump, due to disposal by midnight dumpers.

This situation forced Congress into passing the Comprehensive Environmental Response, Compensation, and Liability Act of 1980 (Public Law 96-510) which was filed as 94 U.S. Statute 2767; however, it is commonly known as Superfund because of the large amounts of money involved in site cleanup of those past operations and because of the scope of the problem of such cleanups nationwide. We shall use the acronym CERCLA or Superfund to mean this law, although the term Superfund sometimes is applied only to the trust fund established by Section 9507 of the Internal Revenue Code.

CERCLA defined a new term, *Reportable Quantity* (RQ), as that single quantity that, when released into the environment, may present substantial danger to the public health or welfare or the environment. RQs have been established for many chemicals. Release of an RQ requires notification to the National Response Center. For most substances, the RQs are any of 1, 10, 100, or 1000 pounds. There are currently some 700 substances so regulated.

CERCLA and its reauthorization provide for federal matching funds to states that establish programs for remedial action to bring past hazardous waste sites up to standards established by the EPA.

CERCLA contains provisions for determination of liability for costs associated with cleanup. In addition, CERCLA provides funding for research, demonstrations, and training activities in the field of hazardous wastes.

Provisions for operators of service stations are included for recycling of lubricating oil. If handled in accordance with applicable laws and regulations and if not comingled with any hazardous substances, such oil is not considered as a hazardous material by the EPA.

CERCLA is chiefly a law to clearly define liability and cost recovery for cleanup of hazardous waste sites, especially when the site is an abandoned facility.

CERCLA has been amended by the Superfund Amendments and Reauthorization Act of 1986 (Public Law 99-799), discussed in the next section.

6.12. SUPERFUND AMENDMENTS AND REAUTHORIZATION ACT OF 1986

After two years of sometimes heated discussion, the U.S. Congress, in October 1986, passed the reauthorization of CERCLA. President Reagan signed it into law. There were many changes to CERCLA in the new law, which was called the Superfund Amendments and Reauthorization Act of 1986 (SARA). This law provided changes in the program through increased funding and taxes and in the schedule for cleanups. It also created provisions totally separate from the old CERCLA: the underground storage tanks (UST) program, a new response program for correcting leaking underground storage tanks (LUSTs), and requirements for financial assurance for corrective action and compensation to third parties for injury and property damage. The law also prohibits remedial actions in any state that has not demonstrated a hazardous waste treatment/disposal facility capable of handling 20 years of waste in compliance with RCRA.

SARA requires that a hazard ranking system be developed to accurately assess the relative risk to human health and the environment posed by sites and facilities on the National Priorities List (NPL), which is a listing of those sites determined by the EPA as most in need of immediate remediation. This list is periodically updated as sites are cleaned up and as new sites are discovered.

A separate part, Title III, of SARA entitled *The Emergency Planning and Community Right-to-Know Act* provides that communities be alerted to dangerous chemicals stored, manufactured, transported, or used in their locale. This Act also requires development of emergency response plans and organizations; in addition, it creates reporting forms and procedures to inform communities of the existence and release of specific chemicals. Appendix A of the implementing regulations, issued first in the January 27, 1987 Federal Register, lists the approximately 400 "extremely hazardous substances" covered by the law and gives a Threshold Planning Quantity (TPQ) for each. This is the quantity above which a facility must meet the reporting requirements of the Act.

Examples of the chemicals and quantities involved for both TPQs and RQs are as follows: allyl alcohol has a TPQ of 1000 pounds and an RQ of 100 pounds; ammonia has a TPQ of 500 pounds and an RQ of 100 pounds; chlorine has a TPQ of 100 pounds and an RQ of 10 pounds; etc.

SARA creates stiff fines for violations and also adds provisions for citizen suits. With the passage of Title III, Congress clearly showed its intent to regulate the chemical industry in the United States and hopefully to prevent serious accidents such as occurred at Bhopal, India. This law will be expensive and time-consuming for the industry and for the EPA, but it is hoped that because of such laws, such accidents will never occur in the U.S.

6.13. NATIONAL PRIORITIES LIST

How, then, does a site get on the NPL? Since it is expected that some 2000 sites will be added within the next few years, we need to consider how a location makes the list.

The EPA decides upon a site and adds it to the list when it is believed that there have been releases of any designated hazardous substance that may present a hazard to the public health or welfare.

The NPL is updated regularly as sites are cleaned up and as other sites are discovered. It should be stressed that the sites on the NPL are not the only potentially hazardous locations; there are many more awaiting discovery or completion of studies to determine the risk.

To date, only a very few sites have been completely cleaned up. The process is slow and expensive, since it involves monitoring groundwater by drilling and maintaining sampling wells, often removal of large quantities of wastes and packing material, formation of an impervious barrier in many cases, repacking, capping with impervious material, and contour restoration of the area. The $8.5 billion authorized through 1991 under SARA will, of itself, not be sufficient to clean up all the locations potentially harming health. However, the regulations now clearly indicate liability and force the parties who created the hazard in the first place to bear financial responsibility. While the dollar costs are high, the costs in human suffering and environmental damage of not cleaning up these sites would be considerably greater.

6.14. SITES ENCOUNTERED IN CLEANUP OPERATIONS

Although almost any site can meet the criterion of being a risk to health or to the environment, some general statements can be made about these dumps.

Most commonly, these are industrial dumping locations that may or may not be near industrial plants. While some bulk chemicals are found in these sites, most of the chemicals were usually buried in drums or other containers. One problem often seen is that the drums used were not full and may have been in poor condition to begin with. As these drums rust, they collapse and release their contents. In addition, partially filled containers of many volatile flammable substances are more hazardous than full drums due to flammability. Mixed wastes, chiefly organic solvents but potentially any chemical, present another hazard in that selection of proper protective clothing becomes more complicated.

Other chemicals commonly found in these sites include paints, pesticides, polychlorinated biphenyls (PCBs), and heavy metals (lead, arsenic, chromium, etc.). Less commonly discovered are inorganic chemicals and radioactive materials. In the majority of cases, there are no records to indicate the types or quantities of the chemicals disposed of, the locations of the drums, the quality of the bottom liner, if any, or even who was responsible for the site. There are usually no monitoring wells, and the discovery of the pollution by leachate is often from drinking-water wells in the area.

The worst possible case is the unlawful dump, where chemicals have been placed, often for years, without any records at all. So-called "midnight dumpers" would collect chemical wastes for a fee and simply unload the truck, even a bulk liquid tanker, wherever no one was looking. These sites may contain just about any chemical ever used. In fact, sometimes laboratory and research chemicals never manufactured can even be found in such dumps.

Former landfill sites are often only slightly better. In these cases, someone is usually responsible, although that person or firm may have claimed bankruptcy; there may even be some records as to the identity and quantity of wastes buried.

Any former site can present many hazards, not just to the public and the environment, but to the workers attempting to clean up the site. Extreme care must be taken at all steps of the operation. Preliminary work must be undertaken very carefully until the extent of the site is known.

6.15. LEAKING UNDERGROUND TANKS

The addition of leaking underground storage tanks, sometimes called LUSTs, increases the cleanup problem, since there may be thousands of such tanks in the U.S. Also, their location underground makes them more difficult to find and more hazardous to work with. SARA divides such tanks into two categories: petroleum product tanks and all others. The standards for these classes are somewhat different. It is expected that simply finding all the leaking tanks will take years, with full remediation taking even longer.

6.16. CONCLUSION

All workers with hazardous materials need to be aware of the provisions under RCRA and CERCLA. RCRA governs present and future operations involving hazardous wastes. CERCLA and SARA govern cleanup of past operations, including underground tanks, and require emergency operations plans at all locations using or storing chemicals.

RCRA will hopefully mean that past practices that have made many locations serious health hazards due to improper management of hazardous wastes will never occur again. We have a means to account for and track hazardous wastes from the cradle to the grave.

CERCLA will allow us to clean up our past mistakes so that we no longer need fear what might be lurking in old dumps, mines, lagoons, and industrial sites.

SARA will allow all of us to know what chemicals are being used in and transported through our communities and to have in place an emergency plan in the event of accidents.

These EPA-managed programs will make life safer and better for all of us, both hazardous materials workers and the general public.

6.17. CHAPTER REVIEW QUESTIONS

1. RCRA empowered the EPA to issue regulations for which?
   (a) Cleanup of past hazardous waste sites
   (b) Management of hazardous waste now and in the future
   (c) Protection of workers from exposures to chemicals
   (d) Notifying communities of chemicals used and stored therein

2. CERCLA empowered the EPA to issue regulations for which?
   (a) Cleanup of past hazardous waste sites
   (b) Management of hazardous waste now and in the future
   (c) Protection of workers from exposures to chemicals
   (d) Notifying communities of chemicals used and stored therein

3. SARA empowered the EPA to issue regulations for which?
   (a) Cleanup of past hazardous waste sites
   (b) Management of hazardous waste now and in the future
   (c) Protection of workers from exposures to chemicals
   (d) Notifying communities of chemicals used and stored therein

4. Title III of SARA is also called what?
   (a) Clean Air Act   (b) Superfund Act
   (c) Solid Waste Act   (d) Community Right-to-Know Act

5. Hazardous wastes are so designated because of which?
   (a) Ignitability   (b) Reactivity
   (c) Toxicity   (d) All of the above

6. RQ is an abbreviation for which of the following?
   (a) Required Quantity   (b) Registered Quality
   (c) Requested Quotation   (d) Reportable Quantity

7. RCRA regulations may be found in which Code of Federal Regulations (CFR)?
   (a) 40 CFR, Parts 260-265   (b) 40 CFR, Part 761
   (c) 29 CFR, Part 1910   (d) 29 CFR, Part 1926

8. SARA changed CERCLA by, among other things, which of the following?
   (a) Eliminating many Superfund sites from the NPL
   (b) Adding the availability of citizen suits
   (c) Reducing the fines and penalties for environmental damage
   (d) Removing liability from companies for past burial sites

9. The intent of RCRA was which of the following?
   (a) To force proper disposal of hazardous chemicals
   (b) To force proper treatment and storage of hazardous chemicals
   (c) To force recycle and reuse where possible
   (d) All of the above

10. Underground storage tanks are covered by which law?
    (a) RCRA   (b) CERCLA
    (c) OSHA   (d) SARA

7   HAZARD COMMUNICATION AND MEDICAL SURVEILLANCE

7.1. INTRODUCTION

In earlier chapters, we have learned about general human anatomy, introductory physiology, basic toxicology, and organizations and laws to protect
health and the environment. In this chapter, we shall discuss how information
on health hazards is transmitted to workers and the public.

7.2. OSHA HAZARD COMMUNICATION STANDARD

On November 25, 1983, the Occupational Safety and Health Administration
(OSHA) issued its hazard communication standard as Title 29, Code of Federal
Regulations (CFR), Part 1910.1200. This standard requires that chemical
manufacturers and importers make a comprehensive determination of hazards
for chemical products they sell. In addition, they and any distributors must
furnish information to anyone buying or using those products. The hazard
information must be conveyed in labeling on the product and in Material Safety
Data Sheets (MSDSs).
Anyone using chemicals must protect their employees by

- preparing a written hazard communication program that includes an
  inventory of all hazardous chemicals used in their facility
- obtaining an MSDS for each hazardous chemical they use
- displaying appropriate warning signs and placards
- preparing a hazard communication training plan
- providing training to employees potentially exposed to hazardous
  chemicals

By having such a plan and fulfilling the requirements of the law, employers will be able to make their workplaces safer and healthier. Earlier OSHA regulations, while valuable in providing standards for workplaces, neglected the important aspect of education of employees to the hazards with which they are confronted. Title 29 CFR, Part 1910.1200 means that all employees will be notified and aware of the hazardous chemicals with which they must work.

7.3. REASONS FOR THE STANDARD

In spite of OSHA regulations governing concentrations and other exposures to hazardous chemicals, some workers were not aware of the hazards. There have been cases where workers died of exposure to toxic substances because they did not know what they were dealing with. A classic example was the case of a worker who died of a cyanide overdose; he and his fellow workers were never told that the chemical was cyanide or to avoid its contact with acids.

This standard seems so logical that no one can really object to it. The small investment in time and money to train and notify workers will reap rewards in fewer accidents and injuries. Now that every chemical must include an MSDS down to the ultimate user and that new labeling has been required, every person dealing with hazardous chemicals will know what the hazards are.

7.4 RELATION OF MSDSs TO SARA

The Superfund Amendments and Reauthorization Act of 1986 (SARA) contains Title III, which is commonly called the Community Right-to-Know Law. SARA requires that facilities subject to OSHA workplace regulations must submit MSDSs or lists of chemicals needing an MSDS to the local emergency planning committee, state emergency response commissions, and local fire departments.

For the first time, occupational safety and health and community emergency response have been linked. Extending the facility's hazard communications program to the local emergency responders will allow these people to be better prepared to assist in the event of an accident.

Thus, MSDSs for hazardous chemicals will be used to inform workers at a facility and also the local response personnel of the type and degree of hazard present in any facility using chemicals.

7.5. MSDSs

What, then, are these MSDSs and what information must they contain?

MSDSs are required by OSHA on every chemical produced, imported, or used in the United States, as required by 29 CFR, Part 1910.1200. When a person purchases a chemical other than an over-the-counter product, they are to receive an MSDS. Figure 7-1 shows an example of an MSDS.

An MSDS must contain the following information:

- Chemical Name — the chemical or common name of the substance must be given. Under the name may be listed the OSHA Permissible Exposure Limit (PEL) or the American Conference of Governmental Industrial Hygienists (ACGIH) Threshold Limit Value (TLV), which was discussed in the previous chapter.
- Physical and Chemical Characteristics — these may include such items as boiling point, melting point, specific gravity (density), evaporation rate, solubility in water, color, and form. This information can be very important to an industrial hygienist or chemist and also to the worker, since it can give clues as what to look for, where the chemical might be, and what some of the hazards might be in the event of a spill or other release.
- Fire and Explosion Hazard Data — includes the flash point of flammables, what types of extinguishing equipment to be used, any special techniques or equipment needed to fight a fire, and if there are any unusual fire or explosion hazards.
- Reactivity Data — contains information of the stability of the chemical. If it is unstable, what conditions to be avoided are included. Warnings against mixing with other chemicals will also be contained in this section.
- Health Hazard Data — gives information on routes of entry, acute and chronic effects, signs, and symptoms of exposure. Also noted are any medical conditions that might be aggravated by an exposure as well as emergency and first aid procedures to be used.
- Precautions for Safe Handling and Usage — describes such things as whether the chemical can be flushed to a sewer or not, whether it can be disposed of in a landfill or not, what special steps are needed in the event of a spill or other release, and if there are any special precautions in handling, storage, or usage of the chemical.
- Control Measures — deals with personal protective equipment needed when working with the chemical plus any work practices involved in the handling of drums, barrels, etc. This section will note whether or not respiratory protection or gloves or overalls or chemical suits are needed in the event of an accident involving that chemical.

Material Safety Data Sheet

May be used to comply with
OSHA's Hazard Communication Standard,
29 CFR 1910.1200. Standard must be
consulted for specific requirements.

U.S. Department of Labor
Occupational Safety and Health Administration
(Non-Mandatory Form)
Form Approved
OMB No. 1218-0072

IDENTITY *(As Used on Label and List)*

Note: Blank spaces are not permitted. If any item is not applicable, or no
information is available, the space must be marked to indicate that.

Section I

Manufacturer's Name	Emergency Telephone Number
Address *(Number, Street, City, State, and ZIP Code)*	Telephone Number for Information
	Date Prepared
	Signature of Preparer *(optional)*

Section II — Hazardous Ingredients/Identity Information

Hazardous Components (Specific Chemical Identity; Common Name(s))	OSHA PEL	ACGIH TLV	Other Limits Recommended	% *(optional)*

Section III — Physical/Chemical Characteristics

Boiling Point		Specific Gravity (H$_2$O = 1)	
Vapor Pressure (mm Hg.)		Melting Point	
Vapor Density (AIR = 1)		Evaporation Rate (Butyl Acetate = 1)	

Solubility in Water

Appearance and Odor

Section IV — Fire and Explosion Hazard Data

Flash Point (Method Used)		Flammable Limits	LEL	UEL

Extinguishing Media

Special Fire Fighting Procedures

Unusual Fire and Explosion Hazards

(Reproduce locally)

OSHA 174, Sept. 1985

Figure 7-1.   Example of a Material Safety Data Sheet.

Section V — Reactivity Data

Stability	Unstable		Conditions to Avoid
	Stable		

Incompatibility (*Materials to Avoid*)

Hazardous Decomposition or Byproducts

Hazardous Polymerization	May Occur		Conditions to Avoid
	Will Not Occur		

Section VI — Health Hazard Data

Route(s) of Entry:	Inhalation?	Skin?	Ingestion?

Health Hazards (*Acute and Chronic*)

Carcinogenicity:	NTP?	IARC Monographs?	OSHA Regulated?

Signs and Symptoms of Exposure

Medical Conditions
Generally Aggravated by Exposure

Emergency and First Aid Procedures

Section VII — Precautions for Safe Handling and Use

Steps to Be Taken in Case Material Is Released or Spilled

Waste Disposal Method

Precautions to Be Taken in Handling and Storing

Other Precautions

Section VIII — Control Measures

Respiratory Protection (*Specify Type*)

Ventilation	Local Exhaust		Special
	Mechanical (*General*)		Other

Protective Gloves	Eye Protection

Other Protective Clothing or Equipment

Work/Hygienic Practices

Page 2   ∗ U S G P O 1986-491-529/45775

Figure 7-1.   Continued.

7.6. PROBLEMS WITH THE MSDS

While the preceding might imply that all the information needed on a chemical is contained in the MSDSs, unfortunately sometimes sections of the sheet are blank. In these cases, it is not possible to know if there is a hazard or not. In addition, most commercial products are mixtures; MSDSs address the hazardous components of the mixture as individual substances, whereas the combined effects might actually be greater than those of the isolated chemical.

Sometimes general statements appear on MSDSs such as "chemical sinks in water", or "chemical has vapors", or "chemical may give off toxic products during fires". While some information is better than none, it is limited and may even confuse the user or responder.

7.7. VALUE OF THE MSDS

While there are shortcomings to the system, many valuable data are contained on MSDSs. The information needed for first responders and other workers with hazardous materials may usually be found on the MSDS. MSDSs are not intended to be long texts of chemical data; they are intended to briefly describe the hazards associated with the material.

7.8. POSTING OF MSDSs

A requirement of the law is that the MSDSs for any chemicals being used be available for all shifts in a readily accessible place. Most employers have chosen to place the MSDSs in looseleaf binders, often bright yellow or another highly visible color, chained to a wall in the work area. Placing the MSDSs in a file in the office does not meet the requirement of the law, especially if the office is locked at night and on weekends. Since the MSDSs are printed on paper, often employers will either laminate the pages between sheets of plastic or place them in plastic sleeves, especially in humid or dirty work areas.

Many employers also post a listing of emergency telephone numbers next to the book containing the MSDSs. Ideally, both should be near a telephone so that alerts may be called out and information obtained beyond that contained in the MSDS.

7.9. ELEMENTS OF A HAZARD COMMUNICATION PROGRAM

There are some general components of any hazard communication program. Each facility manager should address the following items:

- Prepare a written hazard communication plan.
- Identify and evaluate all chemical hazards in the workplace.
- Prepare a complete hazardous substance inventory.
- Develop a file of MSDSs and arrange for copies to be in all work areas on all shifts.
- Ensure that incoming products have proper labels and warnings.
- Develop a local system for labeling in the event that the commercial products are not properly labeled.
- Develop a training program, calling in outside professionals if necessary.
- Identify and train employees potentially exposed to hazardous chemicals.
- Evaluate the program and update it as necessary.

If all of the parts of the program are properly addressed, there should be a good hazard communication program for the protection of the safety and health of employees.

All employees should see to it that all the above elements of the program are present in their work areas. If not, suggest to the employer that you would like additional training or help. Offer your expertise to help train and warn others.

7.10. TRAINING

Probably the most vital part of hazard communication is the training of all employees in the hazards they face and in the proper techniques and equipment needed to protect themselves.

This training, if properly conducted, will serve not only to meet the requirements of the law, but may even improve efficiency and productivity, since the workers are better informed as to what they are working with.

Training should include how to understand the MSDSs, types of personal protective equipment available, limitations of equipment, what to do in the event of an emergency, whom to contact should help be needed, general safety and health procedures, decontamination if needed, and specific data on the particular chemicals or classes of chemicals in use.

This text may be used as the basis of such a training program.

7.11. MEDICAL SURVEILLANCE

Sampling of workplace air, determination of ventilation flow rates, and estimation of quantities to reduce the amount of hazardous chemicals to which a worker or responder may be exposed are all valuable in making occupations safer; however, for an individual worker, personal medical surveillance is the only method to know accurately if or how much of a chemical is affecting the worker.

Medical surveillance consists of preemployment, routine, and postemployment assessments of the worker's health. The next sections will discuss the parts of a good medical surveillance program.

7.12. PREEMPLOYMENT PHYSICAL ASSESSMENT

Prior to beginning work at any hazardous materials site or entry into any hazardous environment, a complete physical examination should be conducted. This preemployment physical should include

- a medical history—to determine if there are any preexisting condi tions or problems
- an occupational history—to discover where the worker has been employed and if he or she was exposed to any hazardous conditions in those previous jobs
- a thorough examination—to include specific organs or systems that could be affected by expected exposures
- blood and urine studies—to determine the level of any harmful chemicals in the worker's body
- a pulmonary function test, if needed— to measure the worker's baseline respiratory capacity, especially if respiratory-protective equipment is called for by the job
- a stress test, if needed—to determine if the worker can handle the stress of wearing the level of personal protective equipment needed on the job
- other tests—as determined by the physician to discover the worker's general and specific health

7.13. PERIODIC PHYSICAL EXAMINATIONS

Routine and periodic physical examinations are usually needed by all hazardous materials workers. Depending upon the types and concentrations of chemicals being worked with, these examinations might be scheduled annually or more often. Often samples of blood and urine are obtained for testing to determine if the worker has received a harmful amount of a chemical into the body.

Also, some physical agents require testing: laser and microwave workers should have routine eye examinations, and radiation workers should have routine blood counts. Many other occupations should have specific testing: pesticide applicators should have routine cholinesterase tests; painters should have lead tests; benzene workers should have routine blood studies; and nurses and other health care workers should have routine blood tests for hepatitis, AIDS, and other diseases.

The frequency of these tests will depend upon the type of employment and the results of previous examinations.

7.14. BIOLOGICAL EXPOSURE INDICES®

The American Conference of Governmental Industrial Hygienists (ACGIH) has developed tests for exposure to specific chemicals; these are called Biological Exposure Indices® or BEIs. A sample of a listing of those chemicals for which a BEI exists is given in Figure 7-2.

7.15. POSTEMPLOYMENT PHYSICAL EXAMINATIONS

When a worker finishes working at a hazardous materials site, the physical examinations do not necessarily end. At the conclusion of the job, a thorough examination should be performed to ensure that harmful exposures did not take place. In addition, since many chemicals exhibit chronic effects evident only after many years, anyone who has worked with hazardous chemicals should continue receiving examinations on an annual or more frequent basis.

In addition to the records maintained by employers, workers should maintain a duplicate file of all medical results in the event that the other files are lost or destroyed.

7.16. OCCUPATIONS REQUIRING MEDICAL SURVEILLANCE

Many occupations should receive medical examinations. You should seek advice from a physician, preferably one trained in occupational medicine, as to the types and frequency of physicals you should take.

7.17. CONCLUSION

Title 29 CFR, Part 1910.1200 contains the regulations for informing workers and the public as to the chemical hazards in workplaces. Since many companies may be unaware of the chemical hazards in their areas, the require-

ADOPTED BIOLOGICAL EXPOSURE DETERMINANTS

Airborne Chemical [CAS #] Indices	Sampling Time	BEI	Notation
BENZENE [71-43-2]			
Total phenol in urine	End of shift	50 mg/L	B, Ns
Benzene in exhaled air:			
mixed-exhaled	Prior to next shift	0.08 ppm	Cf
end-exhaled		0.12 ppm	Cf
CADMIUM [7440-43-9]			
Cadmium in urine	Not critical	10 μg/g creatinine	B
Cadmium in blood	Not critical	10 μg/L	B
CARBON DISULFIDE [75-15-0]			
2-Thiothiazolidine-4-carboxylic acid (=TTCA) in urine	End of shift	5 mg/g creatinine	Sc
CARBON MONOXIDE [630-08-0]			
Carboxyhemoglobin in blood	End of shift	less than 8%	B, Ns
CO in end-exhaled air	End of shift	less than 40 ppm	B, Ns
DIMETHYLFORMAMIDE [68-12-2]			
N-Methylformamide in urine	End of shift	40 mg/g creatine	

Figure 7-2.   Selected substances for which a BEI exists. (By permission of the ACGIH.)

ment for MSDSs to accompany shipments means that the workers who were not aware of any danger will now be informed.

SARA, Title III, requires that MSDSs be also sent to local emergency planning agencies so that they may be aware of the chemicals being used, stored, and transported through their community.

Communication of the hazards presented by certain chemicals to workers and to the public will mean that we all are better informed so as to lessen the effects of any spill, release, or other accident.

The information in MSDSs and in training classes will better protect us from those substances that could potentially cause harm to us or our environment.

Medical surveillance is an important part of any program for the protection of workers. Hazard communications informs the worker of the potential exposures; medical surveillance ensures that harmful amounts of the substances are not entering the worker's body. Preemployment, routine, and postemployment physical examinations are needed by just about any person working with hazardous materials.

Medical examinations will prove if protective measures are sufficient. In fact, one individual's exposure to harmful levels should be a clue that the overall worker protection program may have a flaw; thus, others may benefit from one worker's physical examination.

If the parts of a hazard communications program or required physical assessments are not being performed where you work, notify your employer or OSHA.

7.18. CHAPTER REVIEW QUESTIONS

1. The OSHA Hazard Communication Standard is found in which of the following?
    (a) 29 CFR, Part 1910.1000 (b) 29 CFR, Part 1910.1200
    (c) 40 CFR, Part 263.33b   (d) 29 CFR, Part 1926.100

2. Which of the following must be contained on an MSDS?
    (a) Chemical name   (b) Health hazard data
    (c) Reactivity data   (d) All of the above

3. Which of the following is a FALSE statement?
    (a) MSDSs do not cover consumer items
    (b) Either the formal or common chemical name may be used on MSDSs
    (c) MSDSs do not have to be posted as long as someone knows where they are
    (d) MSDSs are only one part of the hazard communication program

4. Which of the following are parts of a hazard communication plan?
   (a) Preparing a written plan
   (b) Preparing a complete chemical inventory
   (c) Developing and conducting a formal training program
   (d) All of the above

5. A preemployment physical examination should include which of the following?
   (a) Medical history   (b) Occupational history
   (c) Laboratory tests   (d) All of the above

6. Which of the following medical surveillance elements should be conducted on workers wearing respirators?
   (a) Stress test   (b) Pulmonary function test
   (c) Physical examination   (d) All of the above

7. When should postemployment physical examinations take place?
   (a) Immediately upon ending the job and then annually
   (b) Every fifth year
   (c) After about 20 years
   (d) Only when you feel sick

8. BEI® is an abbreviation for which of the following?
   (a) Background Effectiveness Index
   (b) Basic Elements of Information
   (c) Breathing — Exhale and Inhale
   (d) Biological Exposure Index

9. How often should routine physicals be taken?
   (a) Every other year
   (b) As often as the physician feels necessary
   (c) Upon retirement
   (d) After working for ten years

10. How are MSDSs and SARA related?
    (a) SARA requires that MSDSs be filed with community emergency groups
    (b) MSDSs include a section on SARA
    (c) MSDSs have always been required by OSHA and SARA
    (d) None of the above

8 FACTORS AFFECTING RESPONSE/SITE PERSONNEL

8.1. INTRODUCTION

There are many factors that may put stress upon personnel responding to a hazardous materials incident or working at a hazardous materials or waste site. It is important for all personnel from the Incident Commander or Site Superintendent to the incident responder and site worker to be aware of the limitations of personal protective equipment, limitations on personnel when wearing chemical-resistant suits, and physical and mental stressors placed directly upon the individual response-team member or site worker.

Personal protective equipment can cause much physical and mental stress upon personnel. In the chapter on personal protective equipment, it is emphasized that personal protective equipment and clothing pose inherent problems. There are difficulties in walking, reaching, picking up small objects, grasping and retaining objects, moving items, hearing, and most importantly, lack of good visual acuity or the ability to see all around (peripheral vision).

With vision being markedly reduced, the risk of having an accident or injury increases. Each of these problems can become burdens that add extra stress to the situation.

Not only will you have problems inherent to the personal protective clothing and equipment, but also the natural environment may add its own types of hazards. There are many questions you must ask yourself: Are you working in a wooded area, a ravine, along side of a busy interstate highway, in a residential area, near a lake, stream, or river, etc.? Will the contaminants be able to enter a local watershed? Are there inhabitants near the incident or waste site? Are you in a rural area or a city or a suburban situation? Each of these situations will present unique problems. For instance, we would be more concerned if we were working around a body of water, since this could pose myriad problems; for example, if we fell into the water, contamination of the water would be possible, and protective clothing is not made to float!

And what about the contaminant getting into the body of water? Are you able to control the runoff? Will the contaminant leach underground into an artesian system or a community water supply?

What kind of weather conditions are present? Is it a very nice day, with no wind, or is it a windy, dry day? Is it raining, snowing, etc.? Each of these situations will present different hazards and complications.

The three most important naturally occurring hazards are heat, cold, and wind. Each of these factors can cause problems for equipment and personnel.

8.2. HEAT

Heat is the most significant naturally occurring hazard you will encounter. Heat stress is the most dangerous condition that you can expect to encounter (other than the problems of chemical exposures), not only from the environment but also within confined, nonbreathing, chemically resistant suits. Heat stress can occur even during cool weather because of the design of the suits. Depending upon the physical conditions around you and the type of work demands placed upon you, heat stress can occur very quickly, within 15 minutes in some cases. Rash, dizziness, cramps, and drowsiness are some of the first signs and symptoms of heat stress that you should be aware of.

Heat stress cannot be taken lightly, for it can lead to heat stroke or even death.

Your fatigue will increase markedly during hot weather or when you have to stay in the chemically resistant suits for long periods of time. You will exhibit a loss of body fluids and during this situation, you must have fluid replacement. You should not drink soda, coffee, or tea, but rather a drink that will give you a chemical replacement for the chemicals you have lost through perspiration. These replacements could be commercial drinks like Gatorade®, Quickick®, or Squincher® or just plain water.

8.3. COLD

The next most significant natural hazard that can cause health and safety concerns is cold. In some geographic areas of the country, you may have to respond during cold weather. The basic problem caused by cold is hypothermia or a cooling of the body more quickly than the body can produce heat. Hypothermia can sneak up on you just like its opposite, heat stroke (hyperthermia). When you feel yourself shivering, your body is trying to warm itself. In the chapter on physiology, the central nervous system was discussed; if the fluid surrounding the brain is not kept warm, serious injury or death can occur. Frost nip or a tingling or numbing of the skin are two of the most common early symptoms of cold injury. Other symptoms of hypothermia are slowing down of muscle coordination, slurred speech, slow reaction time, and drowsiness. These can occur even on only a relatively cool day. For instance, if the ambient temperature (the temperature of the air) is 40 degrees Fahrenheit and the wind velocity is 30 mph, then the temperature that you feel is 13 degrees Fahrenheit. The air temperature is not that cold, yet the wind's velocity causes the air to feel like the lower temperature. In any wind-chill situation, it is possible to suffer a case of hypothermia, depending possibly upon the amount of physical exertion that you are performing.

The effect of the wind speed on your body is referred to as the *wind-chill index* or *wind-chill factor*. Outdoor occupations run great risks of acquiring a case of hypothermia. It is advisable when you are going to have to work in a situation in which there is potential for hypothermia to wear multiple layers of light clothing, with an outer layer of windproof material.

During cold weather you will also be burning up extra calories. You will need to replace these calories if you are going to have to work in chemically resistant clothing or suits. You may even have to limit your use of chemically resistant suits if you are not able to maintain a high level of energy.

With cold, your protective clothing will have some problems, especially in the areas of zippers, snaps, catches, etc. since these can become brittle and allow contaminants to enter into the suit. Constant checking to see if the equipment maintains its integrity is important. This may be time-consuming but is necessary to safeguard the wearer.

Each of these natural hazards will require special training for acclimatization. Again, as part of the training, it will be necessary to work with your personnel to bring each individual up to his or her maximum tolerance of the hazards.

8.4. OXYGEN DEFICIENCY

Oxygen deficiency is not usually recognized as a hazard in emergency response or in working at a hazardous materials waste site. Oxygen deficiency means that the air around you does not contain enough oxygen to support life. Although the human body can survive in ambient air of 16% oxygen, OSHA has established regulations that the ambient air be 19.5% or more before you can enter and work without a self-contained breathing apparatus (SCBA). This regulation exists to protect you. Any concentration less than 19.5% oxygen can impair functions and, in some cases, cause nausea, brain damage, heart damage, or even death.

Oxygen deficiency can occur at almost any hazardous materials incident, especially if there is a chemical present that displaces air or is heavier than air. The decomposition of wood or wood products and of other organic materials will produce carbon dioxide, which can displace oxygen. The problem of oxygen deficiency can also occur in low-lying areas such as gullies, culverts, drain pipes, etc., where chemicals have had a chance to settle.

Symptoms of an oxygen-deficient atmosphere include impaired vision, reduced attention span, irrational judgment, coordination changes, and increased heart and breathing rates.

8.5. UNUSUAL FACTORS THAT MIGHT BE ENCOUNTERED

During a hazardous materials incident or when handling waste at a hazardous waste site, a small possibility exists for encountering radioactive materials or other unexpected conditions. These could be parts of abandoned medical equipment for example. In Brazil, members of the local population pried open a sealed container, exposing themselves to radioactive cesium; consequently, a number of them died. At a hazardous incident site, there may be radioisotopes that are used in hospital therapeutic work or other types of radioactive materials. For instance, some types of household smoke detectors contain a minute amount of radioactive material. In this particular case, a large number of the detectors would be necessary to create a hazard, but the potential does exist. Radioactive materials and radiation are discussed in Chapter 14.

Biological or etiological agents having disease-causing organisms or cultures are another unusual group that may be found, for example, in the waste from a hospital or medical research center.

In addition to these materials, there may be other naturally occurring materials such as poison ivy, poison oak, and poison sumac. These items can affect the skin or other systems in the body.

Animals can pose problems, especially if your area has a animal population that may be infected with diseases such as rabies. Disease or disease-causing materials found in a certain area are said to be indigenous pathogens. For example, portions of some of the western states have indigenous plague. Approximately 5 to 10 people are infected with plague annually in the New Mexico area.

Not only may you find problems with animals, but also there may be other biological hazards at an incident or waste site that will have to be dealt with. It should be a standing rule that individuals at the site who cut, scratch, or puncture themselves get immediate first aid, wash out the wound, and, if necessary, go to the hospital or clinic.

8.6. SAFETY CONSIDERATIONS

Once you are aware of the physical and natural hazards that you have encountered and the potential problems of temperature, radiation, and biological agents, as well as the lay of the land, you will have to concern yourself with safety factors needed to work the incident or site.

It is appropriate for the Incident Commander to appoint a Safety Officer to carry out the program.

The Safety Officer can look for hazards such as holes, ditches, sharp objects, slippery surfaces, or any other hazard that may be present.

It will be his or her responsibility to prevent bystanders from getting in the way of responding equipment, to see that the personnel entering the Incident Zone have the proper equipment and are wearing it correctly, and to see that none of the responding or cleanup personnel perform their duties in a way that will endanger themselves or their fellow workmates.

There may be power lines, underground gas lines, or other utilities that should be pointed out to the Safety Officer. It will be the Safety Officer's responsibility to contact the individual utility companies to have crews dispatched to deal with the specific hazard.

A noise hazard may be involved either in usual work or in a rescue attempt, especially if special rescue cutting saws are employed. In addition to the noise, these saws are not intrinsically safe and should not be used in a hazardous atmosphere. High noise levels may be generated from heavy equipment being used. The easiest hearing protection is ear plugs, which can easily be worn under the chemically resistant clothing or suits.

As was mentioned earlier, the environment inside of chemically protective suits and equipment, the rescue of personnel trapped in vehicles, the setting up of berms to prevent runoff, or the actual cleanup will greatly increase physical and mental demands as well as increasing the risks of accidents by impairing the worker's senses and reflexes.

There is always the danger of fires and explosions, which can spontaneously erupt at an incident or waste site. Often, fires at a waste site will start due to investigative or remedial activities, such as moving drums, accidentally mixing incompatible chemicals, or introducing an ignition source (e.g., sparks from a front-end loader or dozer hitting a rock) into a flammable or explosive mixture.

Fires at a hazardous incident are a major concern because in addition to intense heat and smoke, open flame, or flying objects such as barrels exploding, a fire may also cause the release of toxic chemicals into the environment. However, fire, under certain conditions, may also be a friend of the responder, since it may consume some of the hazardous materials. In some pesticide incidents, for example, since it is best to burn off the material instead of allowing it to contaminate the soil or groundwater. It may even be best to actually ignite some of the material to burn it off. Technical advice is necessary to determine which chemicals should be burned and under what conditions.

8.7. PERSONAL HYGIENE

Personal hygiene is another area to which the Safety Officer and everyone concerned must pay attention. Personal hygiene can involve everything from taking a shower to the way you eat and drink around the incident or waste site.

In the chapter on decontamination you will be introduced to how to take a series of showers to remove contamination. But even after you finally go through the decontamination process, you will have to take a personal shower. You should go immediately to the designated area to remove your outer garments and then take the shower. Try not to waste time before taking this shower after you come out of the decontamination line. This will reduce the risk of your exposure to any contaminants that may have not been removed from your equipment and will also reduce the risk of taking home any contamination to your family.

No eating, drinking, smoking, chewing tobacco products, or use of chewing gum is permitted in the Incident or Exclusion Zones or other designated areas. The potential for contamination is very real and can easily happen if you eat, drink, smoke, etc. at the site. For instance, if you chew tobacco, you can't expel the juice in protective gear, and you shouldn't swallow it! If you are wearing a face mask and raise the mask to spit, you expose yourself to the dangers of the contaminant. Even the act of chewing gum when wearing a mask can cause a break in the seal around the mask. There should be an area far enough from the site to allow you to eat or drink. This is important during both an immediate incident and the long-term cleanup operations at a waste site.

8.8. ALCOHOL AND MEDICATION

One of the leading problems confronting our society, even in the workplace, is the use of legal and illegal chemical substances. In industry, one in five workers are dependent on some type of chemical substance over and above nicotine and caffeine.

Alcohol is the number-one abused chemical in American society. Marijuana is the second most abused substance. Smoking tobacco cigarettes is third, and the abuse of over-the-counter and prescription drugs is fourth.

Many studies have been conducted by various health agencies as to the effects of alcohol, smoking, drugs, etc. upon the human body. But there have not been as many studies into what happens when, say, you are on a prescription medication from your doctor and you work a hazardous materials incident. What interactions might there be between the medication and the various chemicals with which you come in contact? No one really knows, and some physicians are becoming very concerned, especially if the worker is also using illegal drugs.

Since medical concern is so strong, no medication (legal or illegal) or alcohol should be used while working a hazardous materials site.

Illicit or illegal drugs should never be allowed or tolerated at a hazardous materials site. Many problems may arise from their usage: senses are dulled, and hearing, speech, eye coordination, and, most markedly, body movement and reflexes are diminished. This physical and mental impairment can cause serious accidents to you or to others.

When working in protective equipment, many of your senses are already impaired and visibility is cut down — so do not use any drugs or medications without approval of your or the company doctor. If you are under the care of a physician, check with him to see if he has any concerns about your being exposed to chemicals either at a site or incident — let the doctor know what kind of work you do. Chemical impairment may cause you to trip and fall into a heavily contaminated area.

8.9. THE BUDDY SYSTEM

The buddy system is a safety system. It allows you and your workmate to keep an eye on each other in case one of you should run into a situation where help is needed. The buddy system exists because of the inherent dangers, seen or unseen, at the incident or waste site. If you run into trouble, you want to know that there is someone nearby who can come to your aid or get assistance. Therefore, always stay in visual or radio contact with your partner. Request frequent radio checks to maintain contact. If you can see your partner, good, but if you cannot and have lost radio contact, respond to him or her immediately.

8.10. RESPONSE TEAM SAFETY PLAN

Throughout this chapter we have emphasized safety. Every response team should have a written safety plan and procedures manual. All personnel should be briefed and, if possible, given their own individual copy. The response team safety plan should contain standard operating procedures for the following:

- incident or site entry procedures
- emergency evacuation and routes to evacuate the surrounding areas
- types of personal protective equipment to be worn or used at the incident or waste site
- decontamination procedures, general and specific
- first aid procedures and emergency equipment needs
- fire-fighting procedures and equipment for specific situations
- containment procedures
- procedures for removal or disposal of hazardous wastes

In keeping with the safety manual, the Safety Officer will assist the training officer in educating all personnel about the response team safety plan. Training will be the key factor in how safely you respond and how you work the incident or waste site. This material gives you information about the many problems encountered and the recognition of the potential or actual hazards. Ongoing training is necessary to point out weaknesses in your unit or specific problems that can be or have been encountered.

We cannot overemphasize the importance of training with the equipment and clothing, developing various scenarios that would fit your geographic area or locale, and most importantly, educating your response team with regard to their equipment, their training level, their limitations, and the technical expertise available to assist you. Each of these can be a critical factor during an actual response. Safety for your personnel will be the most important factor, but you will also have to protect any inhabitants of the area and the surrounding environment.

8.11. COMMUNICATIONS

Almost every fire fighter, police officer, or emergency medical response person has experienced difficulty in communications at some time. Communications must be maintained by radio, visual contact, hand signals, horns/sirens, or other method. Horns and sirens are usually reserved for emergency signals. As was mentioned earlier, the buddy system should be maintained so that you have visual or radio contact with your workmate. There should always be other response personnel or site safety personnel observing as many of the responders or workers as possible in case there is an emergency. The safety personnel will also need to monitor any radio communications. Everyone at an incident or waste site has to act as a safety person.

8.12. DECONTAMINATION

In Chapter 13, we will discuss decontamination of both personnel and equipment. These procedures have to be identified prior to your entry into the site. Emergency decontamination has to be addressed as a separate area of concern, since this may necessitate a quicker procedure to get you safely out of the danger area. Emergency showers or hoses should be set up and identified as separate from the normal decontamination area.

8.13. CONCLUSION

Many unusual factors will face the response team member and the hazardous waste site worker. These may range from naturally occurring situations such as cold, heat, and wind to manmade objects such as personal protective clothing and equipment, power lines, gas lines, or other barriers that personnel have to use or work around.

In any response or hazardous waste site operation, safety must be a very important factor so that everyone may come out of the operation safely and healthfully. Never underestimate the risk of working in this hazardous environment, since many of the dangers are insidious and you have to become aware of many varied potential hazards that can be encountered. So think *safety* — this will protect you, your workmates, your community, and your family.

8.14. CHAPTER REVIEW QUESTIONS

1. Which of the following is not a physical hazard encountered at a hazardous materials incident?
   (a) Stress   (b) Noise
   (c) Wind   (d) Radiation
   (e) Cold/heat

2. Fully encapsulated chemical-resistant suits pose which of the following?
   (a) Difficulty in walking   (b) No problem in reaching an item
   (c) Impaired vision   (d) Difficulty in communicating
   (e) a, c, and d   (f) All of the above

3. Stress can put a lot of undue mental and physical fatigue on the responder/worker because:
   (a) The equipment that is being used was not designed to minimize stress.
   (b) Physical factors such as noise, wind, rain, and radiation are considered to be stressors.

4. Radiation is a known hazard. Which of the following poses the greatest external danger?
   (a) Alpha particles   (b) Smoke detectors
   (c) Gamma rays   (d) Beta particles
   (e) All of the above

5. Define etiologic agents:
   (a) Radioactive waste   (b) Blood, serum, or viruses
   (c) Waste from hospitals   (d) Waste from a chemical plant
   (e) b and c   (f) None of the above

6. One method to reduce exposure to etiologic agents is:
   (a) Take a shower every time you come in contact with etiological materials.
   (b) Wash your hands after handling materials.
   (c) Wear the proper level of protection when handling materials.
   (d) None of the above.
   (e) All of the above.

7. Why is it important to use the "buddy system" when wearing fully encapsulated chemical-resistant suits?
   (a) You do not need to have someone with you.
   (b) Wearing a fully encapsulated suit impairs your senses.
   (c) It is a good safety practice to have someone working with you during an incident in case you have an accident.
   (d) None of the above.
   (e) All of the above.
   (f) a and c

8. Communication equipment used during an incident or hazardous materials site requires what typing or rating?

9. Rescue personnel should be in the hot zone in order to respond quickly.
   (a) True   (b) False

10. Airline respirators can be used in the hot zone, as long as the air compressor is in the cold zone. Explain your answer.

    (a) True   (b) False

9 CLASSIFICATION OF HAZARDOUS MATERIALS

9.1. INTRODUCTION

It is necessary to have a basic understanding of the various types of hazards that may be encountered during a response to a hazardous materials incident or in working at a hazardous waste site. You should be able to understand what the different signs, symbols, and placards mean on the barrels and vehicles.

We will list the classifications of some hazardous materials as they are found in the Code of Federal Regulations (CFR).

9.2. HAZARDOUS MATERIALS

A hazardous material is any substance that has been determined by the Secretary of Transportation to be capable of posing an unreasonable risk to health, safety, and property, when transported in commerce, and which has been so designated (Title 49 CFR, Parts 186.8 and 172.101) or declared a hazardous waste under the Resource Conservation and Recovery Act (RCRA) (Title 40 CFR, Part 261.2) or included in definitions in the Comprehensive Environmental Response, Conservation, and Liability Act (CERCLA) (Sections 101(14) and 104).

9.3. HAZARDOUS WASTE

A hazardous waste is any waste or combination of wastes as defined in Title 40 CFR, Parts 261.2 or 261.3, or Title 49 CFR, Part 171.8.

9.4. MULTIPLE HAZARDS

Any material meeting the definition of more than one hazard class is then classified according to the sequence given in Title 49 CFR, Part 173.2.

9.5. CLASSIFICATIONS OF HAZARDS

When you begin to deal with response situations or begin work at a hazardous waste site, it is imperative that you learn the different categories of hazardous materials as well as the specifics of each hazard classification. Some basic categories of hazardous materials include

- explosives (UN Class 1, Explosives)
- flammables/combustibles (UN Class 3, Flammable Liquids, or Class 4, Flammable Solids)
- poisons (UN Class 6, Poisonous)
- radioactive materials (UN Class 7, Radioactive)
- corrosive materials (UN Class 8, Corrosives)
- etiological agents (UN Class 6, Infectious)
- oxidizers and organic peroxides (UN Class 5, Oxidizers)
- asphyxiants (UN Class 2, Gases).

You should also familiarize yourself with major subgroups within each of the different classifications, since they each have characteristics that make them unique. Some have a higher toxicity or a lower flashpoint or are capable of replacing oxygen.

9.6. EXPLOSIVES

There are several subgroups of explosives as follows:

1. *Class A.* Capable of detonating or otherwise of *maximum hazard.*
   - *Type 1.* Solid explosive that can be caused to defaulter by contact with sparks or flames.

     Examples: black powder and low-level explosives.
   - *Type 2.* Solid explosives containing a liquid explosive ingredient, which, when unconfirmed, can be detonated by means of a blasting cap.

     Examples: high explosives and commercial dynamite containing a liquid explosive ingredient.

- *Type 3.* Solid explosives that contain no liquid explosive ingredient and that can be detonated, when unconfirmed, by means of a blasting cap.

  Examples: high explosives and commercial dynamite containing no liquid explosive ingredient and trinitrotoluene, amatol, tetryl, picric acid, urea nitrates, pentolite, and commercial boosters.

- *Type 4.* Solid explosives that can be caused to detonate, when unconfirmed, by contact with sparks or flame such as produced by safety fuses or an electrical squib.

  Examples: initiating and priming explosives, lead azide, fulminate of mercury, and some high explosives.

- *Type 5.* Desensitized liquid explosives that may be detonated separately or, when absorbed in absorbent cotton, by a blasting cap.

  Examples: liquid high explosives and desensitized nitroglycerine.

- *Type 6.* Liquid explosives that can explode by impact in a drop of less than 10 inches.

  Example: nitroglycerine.

- *Type 7.* Blasting caps.

  a. Detonating primers are devices for commercial use that contain a detonator and an additional charge of explosives, all assembled in a suitable envelope.

  b. Detonating fuses, Class A, are used by the military services to detonate high explosives.

- *Type 8.* Any solid or liquid compound, mixture, or device that is not specifically included in any of the above types.

  Examples: shaped charges, commercially produced.

- *Ammunition for cannon.* Ammunition for cannon is fixed, semifixed, or separate loading ammunition that may be fired from a cannon, mortar, gun, howitzer, or recoilless rifle.

- *Ammunition for cannon with projectiles.* Ammunition for cannon with explosive projectiles includes smoke projectiles, incendiary projectiles, and illumination projectiles.

- *Explosive projectiles.* Explosive projectiles are shells, projectiles, warheads, or rocket heads, loaded with explosives or bursting charges, with or without other materials, for use in cannon, guns, tubes, mortars, or other firing and launching devices.

- *Grenades.* Grenades are small metal or other containers designed to be thrown by hand or projected by a rifle.

- *Explosive bombs.* Explosive bombs are metal or other containers filled with explosives. This group includes airplane bombs and depth charges.

- *Explosive mines.* Explosive mines are metal or composition containers filled with high explosives.
- *Explosive torpedoes.* Explosive torpedoes are metal devices containing a means of propulsion and a high explosive.
- *Rocket ammunition.* Rocket ammunition, including guided missiles, is ammunition designed for launching from a tube, launcher, rails, trough, or other launching device in which the propellant material is a solid propellant explosive.
- *Ammunition for small arms with explosive projectiles or incendiary projectiles.* Generally a 20-millimeter round that consists of a metallic cartridge case, the primer, and the propelling charge, with an explosive or incendiary projectile with or without a detonating fuse.
- *Chemical ammunition.* Chemical ammunition is all kinds of explosive chemical projectiles, shells, bombs, grenades, etc., loaded with toxic, tear, or other gases, smoke or incendiary agents, and also such miscellaneous apparatus as cloud-gas cylinders, smoke generators, etc., that may be utilized to project chemicals.
- *Boosters, bursters, and supplementary charges.* Boosters and supplementary charges consist of a casing containing a high explosive and are used to increase the intensity of the explosion of the detonator of a detonating fuse. Bursters consist of a casing containing a high explosive and are used to rupture a projectile or bomb to permit release of its contents.

2. *Class B.* These items are rapidly combustible, but are less hazardous than Class A explosives.

    Examples: propellants and flares.

3. *Class C.* These items are of a low hazard but may explode under high heat when many are tightly packed together.

    Examples: common fireworks (Roman candles, cherry bombs, firecrackers, etc.).

9.7. FLAMMABLES AND COMBUSTIBLES

Flammables and combustibles are the two most common categories of hazardous materials you may encounter. You can look under your kitchen cabinet and find bottles marked either flammable or combustible.

1. *Flammables.* Flammable means any liquid with a *flash point less than 100 degrees Fahrenheit* as measured by the test specified in Title 49 CFR, Part 173.115, with the following exceptions:
   a. A flammable liquid with a vapor pressure greater than 40 psi at 100 degrees Fahrenheit, as defined in section 173.300.
   b. Any mixture having one or more components with a flash point of 100 degrees Fahrenheit or higher that makes up at least 99% of the total volume of the mixture.
   c. A water-alcohol solution containing 24% or less alcohol by volume if the remainder of the solution does not meet the definition of hazardous material.

2. *Combustibles.* Combustible means any liquid with a *flash point from 100 degrees Fahrenheit to 200 degrees Fahrenheit* as measured by the test specified in Title 49 CFR, Part 173.115, except any mixture having one or more components with a flash point of 200 degrees Fahrenheit or higher, that makes up at least 99% of the total volume of the mixture.

9.8. POISONS

The various categories of poisons include the following:

1. *Poison A.* Extremely dangerous poisons: poisonous gases or liquids of such nature that a very small amount of the gas, or vapor of the liquid, mixed with air is *dangerous to life.*
   Examples: hydrocyanic acid, arsine, and phosgene.

2. *Poison B.* Less dangerous poisons: substances, liquids, or solids, including pastes and semisolids, other than Class A or irritating materials, that are known to be so toxic to man as to afford a hazard to health during transportation; or that in the absence of adequate data on human toxicity, are *presumed to be toxic to man.*
   Examples: aniline and arsenic.

9.9. RADIOACTIVE MATERIALS

Radioactive materials are any materials, or combination of materials, that spontaneously emit ionizing radiation and have a specific activity greater than

0.002 microcuries per gram for shipping purposes. Some of these materials are regulated by the Nuclear Regulatory Commission (NRC) and others by the individual states. Types of such radioactive materials are too extensive to cover in this general manual.

Examples: uranium, radium, cesium, and medical isotopes.

9.10. CORROSIVE MATERIALS

Corrosive materials are either any solid or liquid that cause destruction of human skin tissue or a liquid that has a severe corrosion rate on steel.

Examples: hydrochloric acid, sulfuric acid, sodium hydroxide solutions, etc.

9.11. OXIDIZERS

Oxidizers are substances that yield oxygen readily to stimulate the combustion of organic matter.

Examples: chlorates, permanganates, inorganic peroxides, nitrocarbo nitrate, or any nitrate.

9.12. ETIOLOGICAL AGENTS

Etiological agents are viable microorganisms or their toxins that cause or can cause human disease. For additional information on etiological agents, please refer to Health and Human Services, Title 42 CFR, Part 72.3(c).

Examples: tuberculosis bacterium and polio virus.

9.13. IRRITATING AGENTS

Irritating materials are any liquids or solids that, upon contact with fire or when exposed to air, give off dangerous or intensely irritating fumes, but not including any Class A poisons.

Examples: tear gas and other riot control gases.

9.14. ORGANIC PEROXIDES

An organic peroxide is a compound containing the bivalent O–O structure and that may be considered a derivative of hydrogen peroxide.

Examples: benzoyl peroxide and methyl ethyl ketone peroxide.

9.15. ASPHYXIANTS

Asphyxiants are materials that exert their effects by interfering with oxygenation of the tissue, with the result that the individual may literally suffocate. This class is generally divided into *simple asphyxiants* and *chemical asphyxiants*.

1. *Simple asphyxiants.* Simple asphyxiants are physiologically inert gases that act by diluting the air and the atmospheric oxygen to a level below that required to maintain blood oxygen levels sufficient for normal tissue respiration.

   Examples: carbon dioxide, methane, helium, propane, and hydrogen.

2. *Chemical asphyxiants.* Chemical asphyxiants act through direct chemical action, either preventing the uptake of oxygen by the blood, or by interfering with the transportation of oxygen from the lungs to the tissues, or by preventing normal oxygenation of tissues if the blood is oxygenated.

   Examples: carbon monoxide, hydrogen cyanide, and hydrogen sulfide.

9.16. IDENTIFICATION OF HAZARDOUS MATERIALS

When a response unit arrives at the scene of an incident or accident involving hazardous materials, the first thing that must be done is to identify the substances. This may not always be an easy job: the containers may be broken open or burned; the truck or carrier may have lost its placarding or may not have had any at all; the driver may be injured or unable to give you the bill of lading or shipping papers; or these papers may be in the cab of a vehicle that is on fire. Pick the worse scenario and it will probably happen. In such cases, what other means are available to you to identify the materials?

The basic means available to personnel to assist in the identification of hazardous materials are as follows:

- color coding of barrels or cylinders
- labels
- bill of lading
- shipping containers
- identification numbers (DOT, UN, or NA)
- placards

9.17. UNITED NATIONS HAZARD CLASSES

These classes may be displayed at the bottom of placards or described on shipping papers.

1. Class 1, Explosives
   - Division 1.1. Explosives with mass explosion hazard
   - Division 1.2. Explosives with projection hazard
   - Division 1.3. Explosives with fire hazard
   - Division 1.4. Explosives without blast hazard
   - Division 1.5. Very insensitive explosives
2. Class 2, Gases
   - Division 2.1. Flammable gases
   - Division 2.2. Nonflammable gases
   - Division 2.3. Poison gases
   - Division 2.4. Corrosive gases (Canadian category)
3. Class 3, Flammable Liquids
   - Division 3.1. Flash point below -18°C
   - Division 3.2. Flash point from -18 to 23°C
   - Division 3.3. Flash point from 23 to 61°C
4. Class 4, Flammable Solids; Spontaneously Combustible Materials; and Materials Dangerous when Wet
   - Division 4.1. Flammable solids
   - Division 4.2. Spontaneously combustible materials
   - Division 4.3. Materials dangerous when wet
5. Class 5, Oxidizers and Organic Peroxides
   - Division 5.1. Oxidizers
   - Division 5.2. Organic peroxides
6. Class 6, Poisonous and Etiologic Materials
   - Division 6.1. Poisonous materials
   - Division 6.2. Etiologic materials
7. Class 7, Radioactive Materials
8. Class 8, Corrosives
9. Class 9, Miscellaneous Hazardous Materials

In addition to these broad classes, the U.S. Department of Transportation (DOT) has specified a particular hazard number to many substances. For example, vinyl chloride has been assigned the ID number 1086. It should be noted that some chemicals have more than one ID number, often depending upon the physical state or degree of hazard of different forms of the chemical. A listing of these, together with guides as how to handle a transportation accident, is included in the Emergency Response Guidebook published by the DOT.

9.18. COLOR CODING

Color coding is used primarily on compressed gas cylinders, both medical and commercial, although color coding has sometimes been noted on the 55-gallon drums of some companies.

9.19. COMPRESSED GASES—MEDICAL

Color	Gas Content
• green	• oxygen
• brown	• helium
• gray	• carbon dioxide
• green and gray	• oxygen/carbon dioxide
• blue	• nitrous oxide
• black	• hydrogen
• red and orange	• nitrogen
• silver and orange	• prepared nitrogen
• silver	• cyclopropane

9.20. COMPRESSED GASES — COMMERCIAL

Color	Gas Content
• green and orange and silver	• 75% argon, 25% carbon dioxide
• silver and orange	• 100% argon
• black or yellow	• acetylene
• orange	• 100% Oxygen

9.21. COLOR CODING OF BARRELS

There are no existing regulations concerning color coding of 55-gallon drums or other containers.

9.22. LABELS

Labels are required by DOT Regulations in Title 49 CFR, Parts 172-400 dealing with the transportation of hazardous materials by land, sea, or air. Title

49 CFR, Part 172.402 requires multiple labeling when the materials meet the definition of more than one hazard. Labels are to be placed conspicuously on all containers and are to be approximately four inches square.

9.23. BILL OF LADING

There are two federal agencies that have requirements for having a bill of lading when transporting hazardous materials. The Environmental Protection Agency (EPA) requires a Hazardous Waste Manifest for any shipment of hazardous wastes. It is very similar to the DOT bill of lading.

The DOT requires a Hazardous Materials Bill of Lading. Under Title 49 CFR, Part 172.205(h), the bill of lading must include the following information:

- company's preprinted name and address
- identification number
- number of hazardous materials units
- container type
- EPA hazardous waste identification number (if waste)
- description and classification
- flash point, if applicable
- units of weight or volume
- total quantity of materials
- special handling instructions

9.24. PLACARDING

Placarding is the means of identifying hazardous materials by the use of pictographs. Pictographs are signs that are made up of pictures. Each placard has a symbol that identifies what the hazard is. Below are listed the type of substance and the specific placard.

Substance and Placard Name

1. Explosives
   - Class A explosive
   - Class B explosive
   - Class C explosive

2. Compressed gases
   - flammable gas
   - nonflammable gas
   - oxygen
   - chlorine
     - poison gas
     - cryogenic
     - liquefied
3. Flammable liquids
4. Combustible liquids
5. Pyrophoric liquids
6. Flammable solids
7. Oxidizers
8. Organic peroxides
9. Poisons A
10. Poisons B
11. Radioactive materials:
    - Radioactive Class I (white label)
    - Radioactive Class II (yellow label)
    - Radioactive Class III (yellow label)
12. Corrosives
13. Dangerous materials

There are additional placards for special cases, such as for rail or air transportation. These should be learned if they might be used in your area.

9.25. NATIONAL RESPONSE CENTER

Federal law requires that anyone who releases any amount exceeding the Reportable Quantity (RQ) of a chemical into the environment must immediately report to the National Response Center (NRC) by calling 1-800-424-8802 (in Washington DC, the number is 267-2675).

9.26. CHEMTREC

CHEMTREC is an acronym for the Chemical Transportation Emergency Center, which is a public service of the Chemical Manufacturer's Association. This center may be reached at 1-800-424-9300 (in Washington DC, the number is 483-7616 and in Alaska, the number to call is collect at 0-202-483-7616).

The center is operated 24 hours a day to provide advice and assistance in the event of a transportation accident.

For addition information about CHEMTREC, the nonemergency number is 202-887-1255.

9.27. CONCLUSION

In order for first responders or workers at a hazardous waste site to be able to work in a safe and healthy environment, they must know the different classifications and identification methods, as well as any other means by which they can protect themselves from mishandling or identifying a hazardous material.

9.28. CHAPTER REVIEW QUESTIONS

1. Which of the following is not a classification of hazardous materials?
   (a) Radiation   (b) Corrosivity
   (c) Noise   (d) Etiologic agents
   (e) Blasting agents

2. Hazardous materials are:
   (a) Chemicals which cause harm
   (b) Any substance which can cause unreasonable risk to health, safety, and the environment
   (c) Based on a level of toxicity
   (d) None of the above
   (e) All of the above

3. Explosives A do not need to be placarded.
   (a) True   (b) False

4. Poisons need to be placarded. Describe the DOT placard.
   (a) Yellow with warning
   (b) White with skull and crossbones
   (c) White with black, with diagonal lines
   (d) None of the above

5. A chemical with a flash point from 100 to 200°F is defined as:
   (a) Flammable   (b) Combustible
   (c) Explosive   (d) Toxic
   (e) Corrosive

6. Oxidizers are chemicals which release oxygen to help stimulate combustion of organic materials.
   (a) True   (b) False

7. Irritating materials are materials that cause burns or are an irritation to the eyes and skin, and when exposed to air or fire will give off an irritating fume or smoke. Give three examples.
   (a)
   (b)
   (c)

8. Which of the following is considered a simple asphyxiant?
   (a) Oxygen   (b) Nitrogen
   (c) Methane   (d) Propane
   (e) Nitrous oxide   (f) b, c, and d
   (g) None of the above

9. Which of the following are chemical asphyxiants?
   (a) Carbon monoxide   (b) Hydrogen sulfide
   (c) Carbon dioxide   (d) Hydrogen cyanide
   (e) None of the above   (f) All of the above

10 INCIDENT COMMAND SYSTEM

10.1. INTRODUCTION

During an emergency response situation, the senior official, generally a fire service officer, shall establish a command post and become the officer-in-charge. In some areas of the country, law enforcement officials may assume command. Whoever takes command and establishes leadership of the situation is referred to as the Incident Commander (IC) and initiates the Incident Command System (ICS).

By utilizing the ICS, the positioning of response units and personnel is expedited. All units need to direct their communications through the command post. This system will allow for coordination of communications by the Incident Commander.

The difference between hazardous waste site operation and this emergency response is that, at a hazardous waste site, the senior official will usually be the site-designated individual having responsibility for the site. The expertise of that individual can thereby be utilized. Since he or she has been working the site, this individual knows the chemicals, their hazards, and the limitations of what can be done to deal with the incident.

However, in an emergency response situation, the first arriving fire service officer or police officer assumes command, although this responsibility may be passed up to higher authorities, e.g., fire chief, battalion chief, site coordinator, etc., as the need may arise.

The senior official is directed under 29 CFR 1910.120 to establish a command post and become the official-in-charge. This individual will assume responsibility for the actions of the response personnel. It is his or her responsibility to oversee the total operation, evaluate the site, determine engi-

119

neering controls, have a working knowledge of maximum chemical exposure limits, know hazardous-chemicals handling procedures, and know how to use any special technologies.

10.2.   RESPONSIBILITIES OF THE INCIDENT COMMANDER

The Incident Commander should have a thorough knowledge of personal protective equipment and clothing; why it is important in reducing exposure of personnel; what limitations there are on personnel wearing the chemical-resistant suits; and the physical and mental stressors that can be placed upon the response team.

The individual that has been designated Incident Commander will have responsibility for controlling the operation at the site. This responsibility will consist of:

- utilization of technical staff to determine the level of protection and the type of protective equipment necessary to safely work the incident.
- designation of personnel to
    1. acquire proper equipment to safely work the site if the response team does not have such equipment available
    2. coordinate with state and federal agencies regarding any special equipment or personnel that are needed
    3. organize a communications system or net to meet the needs of the emergency so as to be able to communicate with any responding unit, whether local, regional, or federal
    4. act as liaison to coordinate operations with responding local, regional, or federal units
    5. coordinate with local emergency medical units, hospitals, or clinics regarding their capability to receive casualties
    6. coordinate possible evacuation of people from the immediate area surrounding the incident site
    7. ensure that personnel from all units are meeting the criteria for minimum protective equipment contained in 29 CFR 1910.156(e)
    8. serve as a Safety Officer, who has an in-depth knowledge of the incident, specific responsibility to identify and evaluate the hazards, and the following authority:

    If the Safety Officer deems the situation an imminent danger, the Safety Officer shall alter, suspend, or terminate all activities involving response personnel. This action will be taken immediately following notification of the Incident Commander or his designate, and will include corrective action necessary to reduce or eliminate the hazards at the emergency scene

10.3. INCIDENT COMMAND CHECKLIST

The first official on the scene immediately initiates the Incident Command System. This first official, who may eventually pass command up to a more senior official, will determine and initiate the following, as discussed above:

- establish a command post or area
- determine the presence of hazardous material
- begin identification of hazardous materials
- begin evacuation of people living in the immediate area of the incident site, if deemed necessary
- identify boundaries of the incident and isolate the incident
- prepare entry and exit areas for personnel entering incident site. Identify specific areas of activity
- prepare a decontamination station
- prepare a medical aid station
- contain the incident until specialized teams arrive, not attempting to perform rescue or initial entry, in turnout equipment
- request additional specialized teams when necessary
- appoint a Safety Officer
- appoint personnel to direct communications and coordinate response-unit placement
- appoint a Coordinator for emergency personnel
- appoint a Coordinator to handle equipment needs
- appoint a Coordinator for state and federal response teams, if necessary
- appoint a Public Relations Officer to brief the media

10.4. CESSATION OF RESPONSE

Following cessation of response activities, the Incident Commander will implement appropriate decontamination of personnel, equipment, and site-scene activities. The Incident Commander will coordinate with law enforcement personnel to safeguard the incident site so as to prevent unauthorized entry and to protect people living in the area as well as the environment from unnecessary contamination.

The Incident Commander or his designate will coordinate with environmental units involved in the cleanup and removal phases of the incident, which may require that fire or medical units be available.

The Incident Commander should not permit the removal of personnel and equipment until the incident has been stabilized, removed, or disposed of. A hazardous materials incident is not a typical fire-fighting situation, where a return to the scene may take place sometime later. There might be no second chance in a hazardous materials incident.

10.5. CONCLUSION

The ICS allows for one person to assume overall command and control of the incident. This person can then delegate authority to other individuals having specific expertise for special tasks. The Safety Officer appointed by the Incident Commander is a key individual in safeguarding the health and welfare of response personnel at the scene.

10.6. CHAPTER REVIEW QUESTIONS

1. The incident command system allows for:
   (a) Organized control of the incident site
   (b) Various agencies to assist at the incident
   (c) Giving the Fire Chief or on-scene commander full control of the incident
   (d) Giving federal personnel control of the incident
   (e) a, b, and c
   (f) All of the above
   (g) None of the above

2. The incident commander (IC) must, according to OSHA 29 CFR 1910.120, be able to:
   (a) Set up a command post
   (b) Appoint a safety officer
   (c) Delegate the responsibility to the first senior federal official that arrives on the scene
   (d) Control the incident until the problem is resolved
   (e) None of the above
   (f) All of the above
   (g) a, b, and d

3. The IC has the authority to request state and federal assistance during a hazardous materials incident. Explain your answer.
   (a) True   (b) False

4. In some states, law enforcement personnel are the ICs, not the senior fire official. Explain your answer.
    (a) True   (b) False

5. At a hazardous waste site the IC is the senior official who is on site. Explain your answer.
    (a) True   (b) False

11 PERSONAL PROTECTIVE CLOTHING AND EQUIPMENT

11.1. INTRODUCTION

As a first responder or waste site worker, you must have a thorough knowledge of personal protective clothing and equipment available for use at a hazardous materials incident or waste site. You have to know what can and cannot be used, since some items should not be used in particular situations. In this chapter, we will attempt to identify some of the protective clothing that is available. Although the selection of the specific type of clothing is the responsibility of technical personnel, you should have enough working knowledge to realize that there is a mistake if someone hands you a wrong or improper piece of equipment.

In this chapter, we will discuss the ways protective clothing is selected to allow you to work in the area safely. We will also discuss the types of chemically resistant suits that are available, the *breakthrough* or permeation rate, and the limitations of the various levels of protection that each suit provides. No single combination of protective clothing or equipment will protect the worker from all hazards. Therefore, we have to seek the best combination of protective suits and equipment to safeguard the workers.

It should be noted that personal protective equipment can in itself cause significant hazards for personnel working around hazardous materials. Heat stress, psychological and physical stress, impaired movement and vision, and reduction in communications ability can all result from use of personal protective equipment.

11.2. TYPES OF PERSONAL PROTECTIVE CLOTHING

The use of personal protective equipment and clothing is required by the Occupational Safety and Health Administration (OSHA) in their regulations of 29 CFR 1920. In addition, the Environmental Protection Agency (EPA) has regulations requiring the use of personal protective equipment in 40 CFR 300. All responders, whether fire, police, emergency medical personnel, or site cleanup workers, are required to obey these regulations.

Many types of personal protective clothing are available, each addressing a unique and distinct need. We will discuss the various types of suits that may be encountered or needed at a hazardous materials incident or site. Listed below are some of the specialized clothing available to response teams:

- fire-fighting protective clothing
- proximity or approach garments
- flotation gear
- blast and fragmentation suits
- antiradiation suits
- cooling garments
- chemical-protective suits

We shall discuss each of these suits and their limitations as they apply to a hazardous materials incident. There are times that you may need to wear a specialized suit other than a chemical-resistant suit at the incident.

11.3. FIRE FIGHTING

A fire fighter will use *turn-out* gear. This will consist of a helmet, bunker coat, bunker pants, boots, canvas or fire-resistant gloves, and self-contained breathing apparatus (SCBA, positive pressure, units). All of this equipment has been designed for one specific purpose — to fight fires — and must meet certain standards set forth by the National Fire Protection Association (NFPA) in NFPA 1500-1501, 1987.

Fire-fighting personal protective equipment is proper for the purpose of fighting fires, but it is not designed to provide protection against chemical exposure. The bunker gear will afford limited protection against solids or liquids but not against gases or vapors. The first responding units must go in with bunker gear in order to size up the situation, but they should not attempt to perform recovery or rescue because of the limitations of their equipment. Your personal safety could be compromised if you attempt to enter the Incident Zone for a prolonged period of time in such gear.

11.4. PROXIMITY OR APPROACH GARMENTS

Proximity or approach garments are another type of equipment available to the fire fighter. These garments are used in situations where it is necessary to enter a site where a high amount of heat being generated, such as an overturned tanker truck carrying gasoline or an oil-refinery fire. These suits would be employed to allow the response team entry into the incident zone to assist in extinguishing a large liquid fire.

The major limitation of these suits is similar to the bunker gear: they offer little, if any, protection against penetration of chemicals or the degradation of the suit itself. They offer some protection against minor chemical splashes but not against larger exposures.

11.5. FLOTATION GEAR

Flotation gear is another type of gear that has one specific purpose: to keep you afloat if you happen to fall into water. Flotation gear is cumbersome, and if it is worn in conjunction with other protective clothing and equipment, it can become a hazard in itself.

This equipment can weigh between 15 and 25 pounds, which can lead to a decrease in your mobility when you are wearing a fully encapsulated suit. The flotation gear should not be worn outside of your personal protective clothing since the gear is not chemical resistant and could degrade very quickly.

Special precautions should be observed when working around water during any response situation. You should wear the flotation gear, the chemical-resistant ensemble, and the SCBA; therefore, your visibility and movement can be greatly impaired, so you run a risk of having an accident with you being the victim. Great caution and employing the buddy system will reduce this risk.

11.6. BLAST AND FRAGMENTATION SUITS

As the names imply, where there may be a potential for a blast or explosion, trained explosive or ordnance disposal personnel will wear blast or fragmentation suits to reduce the effects of the explosion. There are no suits available that will provide complete protection against all blast or fragmentation. Most blast and fragmentation suits will stop a small amount of explosive materials at a distance of less than 20 feet. If there is a large amount of explosive materials, the suits are of limited value, and ordnance disposal personnel will probably not try to remove the explosive materials but rather will attempt to explode them in place, so as not to jeopardize the lives of the personnel in the area.

Blast and fragmentation suits are not designed for chemical exposure.

11.7. ANTIRADIATION SUITS

Antiradiation suits are not designed to stop all external radiation, but will reduce the exposure of the wearer from skin and other contact with radioactive particles. By themselves, these suits will not prevent the inhalation or ingestion of radioactive materials, since respiratory protection is also needed.

Specialized teams will be needed if there is a radiation incident. These teams can come from the federal agencies or commercial nuclear facilities that have such personnel at their sites. In Appendix G, we have listed the regional offices of the Nuclear Regulatory Commission (NRC), which would have information on the nearest radiation response team.

11.8. COOLING GARMENTS

As the name implies, cooling garments have one purpose — to cool the wearer. In warm climates, one of the greatest stresses you can face is heat. It may be necessary to incorporate a cooling garment in your response equipment. Three types of cooling garments are available commercially. The first relies on the evaporative principle of circulating cool dry air back through the suit or portions of the suit. Cooling can be enhanced by the addition of a vortex cooler or a refrigeration unit. These units are normally uneconomical and do not lend themselves to an incident situation because of their bulk and the need for 10 to 20 cubic feet per minute of respirable air. This creates the need for a large compressor to draw in uncontaminated air. The need to have a supply hose from the compressor to the individual using the system can inhibit the movements of the user.

An innovation based on the common shooting vest is a garment consisting of multiple pockets in which you insert ice packets. This allows for frequent changing of the ice packets when they lose their usefulness. This vest is very effective, but you have to keep changing the ice packets, and you may have to maintain a large supply of them to meet the needs of this system. This system works well under protective clothing.

The third device consists of a water ice reservoir, with tubes circulating a cooling solution through the suit by means of a battery-powered pump. The unit is very similar to the units that the American astronauts wore on their moon missions. The ease of changing the battery and not needing to replenish the cooling solution makes this unit acceptable during a hazardous materials incident.

11.9. CHEMICAL-PROTECTIVE CLOTHING

Chemical-protective clothing is the type of clothing that you will wear at most hazardous materials incidents or waste sites. There are two basic types of suits available: fully encapsulating and nonencapsulating.

11.10. FULLY ENCAPSULATING SUITS

The fully encapsulating suits are just that: they create a chemical-free environment inside the suit with a separate air supply. There are one-piece garments that provide protection for the whole body. In some types of suits, the boots and the gloves are an integral part of the suit. In others, the boots and gloves can be removed. There are advantages to both types. In the first instance, the fact that the boots and gloves are a part of the suit prevents contamination from entry at the junction of the suit with the boots and gloves. However, a problem is that the boots and gloves can become torn or punctured, and cannot be easily replaced. In the case of the removable boots and gloves, the opposite situation exsists: contaminants can enter through the union of the boots or gloves to the suit, although the design of the union helps reduce this possibility. However, gloves and boots can easily be replaced if they become torn or punctured.

Air supply to the suits can be provided by two methods: by an SCBA or by an airhose that can be attached to a one-way valve assembly on the suit. The use of self-contained apparatus has a disadvantage in that the amount of air in the cylinder is limited. The airhose allows unlimited time in the Incident Zone, since it is attached to an air compressor located away from the contamination. However, the suit permeation rate is more often the limiting factor than the air supply.

The fully encapsulating suit is very restrictive in that it greatly reduces mobility and vision. Mobility can be a critical factor, especially if the alarm rings on your SCBA and you have to exit the site in a hurry. You will find you will not be able to easily run in this type of suit; you will have to learn to judge time and distance when wearing it. Vision restriction is caused by the mask of the SCBA and the window of your suit. You do not have the peripheral vision that you normally would have, and you have to keep looking down to see any hazards that might be on the ground.

One of the greatest problems when wearing a fully encapsulating suit is heat stress. We will discuss heat stress later in this chapter, but because of this problem, it is recommended that you have a complete medical examination including a stress test to determine if you are capable of wearing this type of equipment and functioning in it. Not only are you going to experience heat stress, but some people are claustrophobic and cannot work in such confined spaces.

11.11. NONENCAPSULATING SUITS

Nonencapsulating suits commonly consist of jacket, trousers, and hood. In some units the suit is a complete ensemble, but in others the hood and boot covers are separate. This depends on the manufacturer. Nonencapsulating suits do not afford the same level of protection as fully encapsulating suits.

However, nonencapsulating suits are like the fully encapsulating suits in that they do not "breathe" and allow for cooling. Evaporation of body fluids cannot be accomplished in either suit. You will find that you will perspire profusely in both. It may be a way to lose weight, but can be very dangerous.

Some workers feel that the two-piece is more comfortable than the fully enclosed suit; however, the two-piece suits do not have the gas-tight protection that the fully encapsulating suit will give.

11.12. SELECTION OF PROTECTIVE CLOTHING

The types of hazards and conditions of the hazardous materials incident or waste site dictate the level and type of protective clothing. These factors can be

- the types and toxicity of the chemicals and hazardous materials encountered
- the type of work to be done, e.g., rescue, removal, or transfer of the hazardous material to barrels or to another vehicle, and/or disposal of containers, etc. — in each of these situations, there may be a requirement for different equipment or a placement of different physical demands upon you
- permeation or breakthrough of the suit, which can greatly affect the level of protection and type of suit that will be required

11.13. PERMEATION

Permeation is the process by which a liquid chemical moves across or through the barrier of your protective clothing on the molecular level. You cannot see this happen, since the process is submicroscopic. Sometimes you will hear the term *breakthrough*, which is the time it takes for the chemical to penetrate through the material of the suit. Protective suits have different breakthrough times for different chemicals. You should know the limits of your equipment, because there is no universal list of breakthrough times for all chemicals. Manufacturers have guides that list the limitations of their product. Always attempt to determine the breakthrough time before entering the Incident Zone; however, this may not always be possible, so you may have to make a judgment call and hope it is correct. In such cases, it is best to be very conservative.

There are other factors that have to be considered when determining the type of chemical-resistant suit as well as the level or type of protection that is needed. You might be in need of a suit that is impervious to acids, or you could need a suit that is good against solvents. Once this specific determination is made, then there are some other factors that have to be identified. Some of these follow.

11.14. DURABILITY

Will the suit be durable enough to withstand the chemical with which it is coming in contact? Can you bend and reach without tearing or damaging the suit? Will the suit be able to withstand punctures, abrasions, or tears? All these questions need to be answered in order to safeguard the worker or responder.

11.15. FLEXIBILITY

As with durability, will the suit impair the individual's ability to pick up small objects? This is very critical in chemical-resistant suits, where there may be a need to turn a small valve to shut off the flow of a spilling chemical. When selecting gloves this is especially true, since you want to be able to remove the lids or caps of various containers.

11.16. TEMPERATURE

Temperature will affect the durability and protection of some of the garments used in hazardous material incidents. Temperature will also affect the integrity of the clothing. This is especially true in colder weather, where the suit's zippers and snaps might become brittle. In very warm situations, the permeation rate can increase markedly.

11.17. EASE OF DECONTAMINATION

Is it possible to decontaminate the clothing? Do you have to throw the clothing away after using it or can it be reused? What are the costs of each garment — can you replace the more expensive suits with disposable suits or will the costs of the disposable clothing itself be prohibitive? What types of decontamination solutions will be employed? Will they cause damage to the suit? Again, the series of questions needs to be answered before one is sent into the Incident Zone.

11.18. COMPATIBILITY WITH OTHER EQUIPMENT

Can you use the protective clothing safely with other equipment? Are there restrictions on the usage of other equipment? Is the suit compatible with the other suits and is it compatible with the decontamination solution or any chemical with which it will come in contact?

11.19. DURATION

Will the wearer be able to determine if breakthrough occurs? Can the worker perform his necessary tasks before degradation or the breakthrough occurs? How long will the wearer be able to work in the suit without the problems of breakthrough or degradation? Keep in mind that heat stress can limit the time in the protective suits.

11.20. LEVELS OF PROTECTION

Personnel must wear protective clothing and equipment when there are known or potential exposures to airborne contaminants including gases, vapors, and particulates or when there is potential contact with chemicals that can affect the skin or that use the skin as a means of entering the body. Chemical-resistant suits are designed to afford the highest level of skin protection.

Every incident will be different, requiring different levels of protection for the responder. These different levels are designated as: A, B, C, or D. We will now describe each level in detail. It is the responsibility of the responder to be able to differentiate between the levels and to know what level of protection is required.

11.21. LEVEL A

The highest degree of protection that can be provided is level A. This level of protection should be worn when there is a need for the maximum degree of respiratory, skin, and eye protection.

11.22. LEVEL B

Level B is chosen when the highest level of respiratory protection is needed, but a lesser level of skin protection is indicated. During a hazardous materials incident, level B is the minimum level of protection recommended on initial incident assessment You should not use a lower level of protection until the hazards have been identified and it has been determined that a lower level is safe to utilize.

11.23. LEVEL C

In order to use level C, either the airborne contaminants must have been identified and found to be relatively harmless so that there can be a lesser level of respiratory protection or the concentration of the contaminants must have been measured and a determination made that disposable protective suits are suitable for use at the site. The testing must have determined that given the level of contamination, it is safe to breathe ambient air with respiratory protection (filter respirators) instead of SCBA.

11.24. LEVEL D

Level D is basically nothing more than ordinary work clothes; therefore this level should not be worn at any hazardous materials incident that has a skin or respiratory protection requirement. This level is normally seen in the safe zone.

11.25. DETERMINATION OF LEVEL OF PROTECTION

Determination of the level of protection that is needed is based upon the chemical characteristics, the concentration, and the type of identified contaminant at the incident or site.   Airborne and ground samples are taken and analyzed to determine the contaminant.   If the substance cannot be identified, an assumption has to be made that you are dealing with very toxic materials. Then you would go to the highest level of protection until the substance has been identified.

What are the specific requirements or criteria for the selection of the various levels of protective clothing and equipment? We mentioned earlier that there are skin and respiratory protection levels that have to be met, but there are other requirements dealing with the toxicity, concentration, amount, and type of materials encountered.

11.26. COMPONENTS AND CRITERIA FOR SELECTING LEVEL A PERSONAL PROTECTIVE EQUIPMENT

Level A personal protective equipment requires the following:

- SCBA, positive pressure
- chemical-resistant gloves, outer shell
- chemical-resistant gloves, inner shell
- boots, chemical-resistant with a steel toe and shank, and, depending upon the suit's construction, worn either inside or outside of the chemical suit body
- communications capability

The parameters that are necessary for the selection of level A follow:

1. The chemical substance has been analyzed and identified and requires the highest level of skin and respiratory protection. This is determined by the following information:
    - known or potential high concentrations of gases, vapors, or particles in the incident area
    - the chance that you can be splashed, fall into the contaminant, or be unexpectedly exposed to gases, vapors, or particulate matter at the the site
    - identification of the contaminants as being very dangerous or toxic, for example: dioxin, cyanide compounds, concentrated pesticides, poison, suspected carcinogens, radioactive materials, or infectious substances
    - the substances have a potential for destroying the skin, such as concentrated acid or alkaline solutions

- if you might have to work in a confined, poorly ventilated space, until the area has been declared less hazardous
2. The fully encapsulating suit provides the highest level of protection to skin, eyes, and respiratory system only if the suit material is resistant to the chemicals of concern during the time the suit is being utilized. While level A is designed to provide the maximum protection technologically available, the chemicals may rapidly permeate through the suit and the suit can be penetrated by certain chemicals from extremely high concentrations, splashes, or immersion of boots or gloves in concentrated liquids or sludges. It is important for the response team to be able to quickly identify the chemicals at the incident site and choose the appropriate suit.

11.27. COMPONENTS AND CRITERIA FOR SELECTING LEVEL B PERSONAL PROTECTIVE EQUIPMENT

Equipment that is required under Level B is very similar to Level A:

- SCBA, positive pressure
- chemical-resistant clothing overalls and long-sleeved jackets; coveralls; hooded one- or two-piece chemical-splash suit; or disposable chemical-resistant coveralls
- chemical-resistant gloves, outer shell
- chemical-resistant gloves, inner shell
- boots, outer, chemical-resistant with steel toe and shank
- hard hat, sometimes with a full face shield
- communications capability

Criteria for selecting clothing for level B is determined by the following information:

1. The level and concentration of the toxic substances have been identified and are of such nature that they require the highest level of respiratory protection, but a *lower* level of skin and eye protection.
2. The concentration of the chemicals or substances can be classified as Immediately Dangerous to Life and Health (IDLH).
3. Concentrations exceed the level of protection afforded by a full-face, air purifying mask.
4. The site contains substances requiring air-supplied protection, but these do not represent a serious skin hazard.

5. The atmosphere contains less than 19.5% oxygen. You need an oxygen concentration greater than 19.5% to use a respirator; otherwise you must wear an SCBA.

11.28. COMPONENTS AND CRITERIA FOR SELECTING LEVEL C PERSONAL PROTECTIVE EQUIPMENT

The major difference between level B and level C is in the equipment used to protect the respiratory tract, assuming that you are wearing the same chemical-resistant ensemble. The primary difference is the use of air-purifying respirators, not SCBA.

The recommended NIOSH/MSHA-approved air-purifying device is a full face mask equipped with suitable canister for the particular chemical or airborne hazard. The cartridge should be specific to remove the hazards encountered or expected. Air-purifying respirators can only be used if the oxygen content is above 19.5%, the hazards have been identified and tested, and the concentrations of the hazards are known.

The substances encountered must have an adequate warning property so when there is a breakthrough you will be able to recognize it. Many chemicals do not have a smell or warning property that will allow you to tell when your canisters are saturated and breakthrough has occurred. Appropriate cartridges will be used for known substances, and the service limit must not be exceeded.

You must have been qualitatively and quantitatively fit tested for use of the full-face respirator.

Personal protective equipment under level C is as follows:

- full-face, air-purifying, canister-equipped respirator
- chemical-resistant clothing (coveralls, hooded two-piece chemical-resistant splash suit, chemical-resistant hood and apron, or disposable chemical-resistant coveralls)
- chemical-resistant gloves, outer shell
- chemical-resistant gloves, inner shell
- boots, chemical-resistant, steel-toed and shank
- hard hat with face shield
- communications capability

In order to utilize level C, *all* points of the criteria selection have to be met; in other words, if one point cannot be met, then you have to go to a higher level of protection.

Measured concentrations of the identifiable substances must be below the substance's exposure limits, and the concentrations must not exceed the capacity of the air-purifying respirator canister. The atmospheric contaminants must not exceed the NIOSH IDLH levels.

Atmospheric contaminants and liquid splashes must not contact any skin, such as areas of the face or neck not covered by protective clothing. Also, your duties or work requirements at the incident site must not require the use of SCBAs.

11.29. COMPONENTS AND CRITERIA FOR SELECTING LEVEL D PERSONAL PROTECTIVE EQUIPMENT

The lowest level of protection is level D. It is basically your work clothes and is utilized when there are no measurable hazardous air pollutants. Your work requirements at the incident must preclude your being exposed to splashes, immersion, or unexpected inhalation of any of the hazards. Level D can be worn in areas where only boots need to be decontaminated or no toxic inhalation substances exist.

11.30. CONCLUSION

Your protective clothing and equipment is all that is between you and the hazards at the incident site. The proper selection of clothing and equipment will give you the greatest protection factor. Protective clothing is determined by the hazards encountered, whether they are radioactive, chemical, explosive, or infectious.

11.31. CHAPTER REVIEW QUESTIONS

1. Which of the following is not level A protection?
   (a) Work clothes with a canister respirator
   (b) Fully encapsulating chemical-resistant suit with self-contained breathing apparatus
   (c) Chemical-resistant suit with self-contained breathing apparatus
   (d) Chemical-resistant suit with canister respirator

2. Of the following personal protective equipment, which can safely be used at a hazardous waste site/incident involving corrosive chemicals?
   (a) Fire fighters' protective clothing
   (b) Chemical-resistant clothing
   (c) Antiradiation suit
   (d) Proximity or approach garment

3. Fire fighters' protective clothing must meet which of the following standards/regulations?
   (a) OSHA 29 CFR 1910.120   (b) NFPA 1500-1501
   (c) EPA 40 CFR 300   (d) None of the above

4. When considering the level of protection needed, what factor(s) should you be considering?
   (a) What type of chemical(s) is (are) involved
   (b) The type of chemical-resistant suits that you have available
   (c) None of the above
   (d) All of the above

5. When choosing a chemical-resistant suit for entry, which of the following is not a consideration?
   (a) Durability   (b) Ease of contamination
   (c) Flexibility   (d) Permeation rates
   (e) None of the above   (f) All of the above
   (g) a, c, and d

6. Define the following:
   (a) Durability   (b) Breakthrough times
   (c) Decontamination   (d) Flexibility
   (e) Permeation   (f) Penetration
   (g) Degradation

7. Level of protection is classified as the following. Of these, which is the highest skin and respiratory level?
   (a) Level A   (b) Level B
   (c) Level C   (d) Level D
   (e) Level E

8. In determining the level of protection, which of the following is not a consideration?
   (a) Chemical considerations
   (b) Concentration of the chemical
   (c) Time of the day
   (d) Time spent in the hazardous material zone
   (e) All of the above
   (f) None of the above

9. Fully encapsulating chemical-resistant suits are designed to resist etio-logic agents.
   (a) True   (b) False

10. Selecting the proper protecting equipment is the number one priority item for a first responder when responding to a hazardous materials incident. Explain your answer.
    (a) True   (b) False

12 PESTICIDE INCIDENTS

12.1. INTRODUCTION

One of the more serious responses that can be made is to a pesticide accident. Pesticides are made from a myriad of chemicals; some of which are very toxic, can cause cancer or birth defects and even disrupt the genetic code, or can cause damage to the central nervous system.

The term *pesticide* covers a large group of chemicals:

- insecticides, to kill insects
- rodenticides, to eradicate mice, rats, woodchucks, gophers, hares, etc.
- herbicides, to kill weeds
- bactericides, to kill bacteria
- acaricides, to kill mites
- algacides, to kill algae
- nematocides, to kill small worms
- viricides, to kill viruses
- fungicides, to kill fungus
- (botanically) natural insecticides that are toxic

12.2. ROUTES OF ENTRY

These are the ways that pesticides may enter your body. Most commonly, these routes are absorption or ingestion.

Absorption. Skin absorption is the most common route of entry and poisoning by pesticides. You can be poisoned by the smoke, fumes, or contaminated runoff from a pesticide incident. Many pesticides can be absorbed by the skin. Since this can happen without any sensation of pain or irritation to indicate

141

exposure, you may not realize it is happening. Some pesticides are so toxic that only a few drops of the concentrated substance on the skin can cause sickness or death.

Smoke from burning pesticides can enter your body not only through the lungs but also through the eyes. Absorption of a pesticide through the eyes is one of the quickest ways to become poisoned because the eyelid has a high absorbency and the eyes are linked to the central nervous system. Poisoning can occur when the eyes are splashed with the pesticide itself or with pesticide-contaminated water from fire-fighting operations. One or two breaths of some of the concentrated fumigants can also cause rapid poisoning and death.

Ingestion. Accidentally eating a pesticide is easy to avoid. However, poisoning can also result from smoking, eating, or drinking near pesticides. After handling them there have been instances of accidental contamination of foodstuffs that have resulted in mass casualties. Good personal hygiene around pesticides, especially frequent hand washing, is highly encouraged.

On average there are 100 nonfatal acute pesticide poisonings for each fatality. Pesticide poisonings can be acute or chronic, depending on the length of exposure, the concentration, and especially the toxicity of the chemical. Acute poisonings usually result from a massive single exposure to a highly toxic chemical. Chronic effects are caused by repeated exposures to small amounts of some pesticides

12.3. PERSONAL PROTECTIVE EQUIPMENT

Hazardous materials response teams will usually carry chemical-resistant clothing. In a pesticide spill this type of clothing is required for level A and/or level B protection as discussed earlier.

Level A is recommended for the initial entry until the identity and concentration of the pesticide have been determined and a lower level of protection is indicated. Always use a higher level of protection at first to reduce possible exposures.

Fire fighters will wear their helmet, turnout coat, bunker pants, rubber gloves, and self-contained breathing apparatus (SCBA). When wearing this equipment, no skin should be exposed. Tape down all areas that could potentially allow entry, especially sleeves and pants cuffs.

12.4. SYMPTOMS OF PESTICIDE POISONING

If any response personnel come in contact with the pesticide at a fire, waste site, or spill, they should seek immediate medical attention. Any unusual appearance or feeling of discomfort or illness can be a potential sign or symptom of pesticide poisoning. These symptoms may occur immediately or may be delayed for up to 12 hours or longer following exposure.

Symptoms of pesticide poisoning include pinpoint pupils, rapid pulse rate, and rapid breathing. The victim may also experience headache, pain, or dizziness. Increased temperature and blood pressure may also be observed. Pesticide poisoning can easily be overlooked, since the signs and symptoms can be very similar to the flu, smoke inhalation, heat prostration, and other illnesses.

Severe poisoning can produce staggering, drooling from the mouth, slurred speech, profuse sweating, convulsions, and coma. The victim may initially experience a feeling of discomfort such as muscle cramps, stomach pains or cramps, dizziness, skin or eye irritation, or labored breathing. If the victim starts feeling very ill or becomes nauseated and vomits, immediate medical attention is warranted.

12.5. REACTIONS FROM SPECIFIC PESTICIDES

Organophosphates and carbamates. This group of pesticides affects the nerve-impulse transmissions within the body. Interference with nerve transmissions can result in a disruption of bodily functions. For example, the victim may lose ability to walk, stand, move, etc.

Signs and symptoms of this kind of poisoning include pinpoint pupils, blurred vision, increased gland secretions (resulting in massive tearing, salivation, and sweating), decreased pulse rate, increased breathing rate with labored breathing, discharge of intestine and bladder contents, weak and uncontrollable muscles.

Other signs and symptoms of organophosphate poisoning are headache, muscle twitching, tremor, and nausea.

Organochlorines. These insecticides are powerful nervous system stimulators and interfere with the transmission of nerve impulses. They particularly disrupt the function of the central nervous system. This results in behavioral changes, dizziness, and depression of the body's vital centers.

Signs and symptoms of organochlorine poisoning include apprehension, dizziness, disorientation, weakness, muscle twitching, tremor, respiratory depression, paranoia, convulsions, and coma.

Nitrophenols and chlorophenols. Herbicides, wood preservatives, disinfectants, and fungicides that are easily absorbed are in this group. This group particularly affects the kidneys, liver, and the central nervous system. Poisoning is more severe if the individual is exposed to heat at the same time the pesticide exposure occurs.

Signs and symptoms of nitrophenol and chlorophenol poisoning include profuse sweating, headache, thirst, fever, nausea, vomiting, pain in the chest and abdomen, and signs of declining alertness.

12.6. FIRST AID

1. Establish a clear airway and administer oxygen if available. Removal of secretions by suction may be necessary.
2. Remove victim from further contact with the chemical, toxic smoke, or runoff water.
3. Decontaminate the victim with detergent (household) and copious amounts of water. This is necessary if there is concern that the skin and hair have been contaminated. All runoff water must be collected and treated as hazardous waste.
4. Monitor the victim's vital signs. Cardiopulmonary resuscitation (CPR) may be indicated in acute poisoning cases.
5. Oraganochlorine poisonings: Keep the victim from excessive sensory stimulation, such as lights or sirens. Keep the victim calm; do not become aggravated, even if the victim becomes belligerent!
6. Nitro- and chlorophenol poisoning: Reduce body temperature by physical means such as wet sheets or sponge baths. Decreasing the body temperature is a critical factor in this type of poisoning.
7. Eye contact: In all cases, flush with copious amounts of water. The water does not have to be sterile. Delay of a few seconds in flushing the eyes will greatly increase the extent of injury.
8. Skin contact: Wash with detergent and water. If you have the container, check the label for additional decontamination instructions. If the victim is being transported to the hospital, take the container into the emergency room for the physician to see.

In pesticide poisonings, antidote therapy may be necessary, and immediate hospitalization is indicated. First aid measures are required to be listed on the label of the pesticide container. Since signs and symptoms may be very similar, but first aid measures differ for organophosphates, organochlorines, and nitro- and chlorophenols, the product information is very important in obtaining the correct procedures.

Protective suits, rubber gloves, and a respirator or SCBA are indicated when working with decontaminating a heavily contaminated individual. Rescue personnel need to be concerned with their own decontamination as well. If any personal clothing or shoes become contaminated, they should be washed with a detergent and water. If the contaminant cannot be removed, discard the article of clothing or equipment. In some cases it probably would be safest to discard the clothing as hazardous waste.

12.7. CONCLUSION

Pesticide poisonings and exposures pose a very serious health and safety hazard to the responder, fire fighter, and site worker. Fire fighters are also faced with an inhalation hazard. Emergency medical personnel and site workers would more likely be exposed to a contact absorption hazard.

Proper decontamination of personnel and equipment must be performed with great care when dealing with a pesticide incident. All contaminated water and clothing must be disposed of in accordance with hazardous waste regulations.

12.8. CHAPTER REVIEW QUESTIONS

1. Pesticides are a group of chemicals that can:
   (a) Cause cancer
   (b) Disrupt the genetic code
   (c) Be considered very toxic
   (d) Cause damage to the central nervous system
   (e) None of the above
   (f) All of the above

2. Define the following terms:
   (a) Insecticide
   (b) Rodenticide
   (c) Herbicide
   (d) Bactericide
   (e) Germicide
   (f) Algacide

3. Name the most common way for pesticides to enter the body.

4. When dealing with a pesticide emergency, which of the following is the first procedure that should be initiated?
   (a) Suppress the fire with water
   (b) Evacuate the immediate area
   (c) Call for more personnel
   (d) Ignite the pesticide
   (e) All of the above
   (f) None of the above

5. What level of personal protective equipment is needed when dealing with an initial pesticides emergency? Explain your answer.

6. Define the following terms:
   (a) Organophosphates and   (b) Organochlorines
       carbamates
   (c) Nitrophenols and   (d) Antidote therapy
       chlorophenols

7. Emergency response to a pesticide emergency exposes the response personnel to acute contamination to both their person and personal equipment. Explain your answer.
   (a) True   (b) False

13   EQUIPMENT DECONTAMINATION

13.1. INTRODUCTION

The decontamination of equipment is a critical element of both personal and environmental protection. If you do not clean or decontaminate the equipment properly, you may expose yourself to unnecessary risks. Chemicals may stick or adhere to the various parts of your equipment. From the gear you are wearing to the inside of the engine of your response vehicle, everything may need decontamination.

You cannot expose your workmates and yourself to these chemicals, so you have to remove the contaminants thoroughly.

13.2. EQUIPMENT MAINTENANCE

When you deal with the routine maintenance of equipment used at a hazardous materials incident, you may run into some restrictions. Normal maintenance will not restrict the movement of the equipment *unless* it is in an exclusion zone or has been contaminated. How then do you perform emergency maintenance when the equipment *is* located in an incident area or a contaminated area? In this chapter we will be answering these questions on the basis of the technology that is currently available.

13.3. PERSONAL HYGIENE

Personal hygiene, or cleanliness, is very important in working around potentially exposed equipment and personnel. Abrasive hand soap can be used to remove grease, oil, or dirt from the hands. Degreasers can normally be used to remove debris from equipment, but these can be used without gloves only in the safe zone. Make every attempt to remove contaminants from your hands immediately following contact. Wash your hands frequently.

Following an incident or response, you should be decontaminated at the scene, but it is good practice to take a thorough shower following your return to your station. Clean all your equipment, and when you take a shower, clean under your nails, between your toes, and any other body orifices that can allow for residual chemical placement.

13.4. EMERGENCY REPAIRS

Some times you may have to perform basic or emergency maintenance on a vehicle that has been in a contaminated area. The maintenance could be anything from changing a tire or a battery to checking a broken hydraulic line.

Battery charging and maintenance will probably be performed in the maintenance shop at a hazardous waste site, but if you do not have that capability at an incident site, you may not be able to get the equipment moving out of the contaminated area.

At a hazardous waste site, the battery-charging area should be clearly identified and well ventilated to reduce the buildup of explosive hydrogen gas. Sulfuric acid is used to activate the batteries and must be stored in an approved metal cabinet designated for chemicals. The batteries must be stored in an NFPA-approved, well-ventilated cabinet. The offsite maintenance shop should receive only clean or decontaminated equipment, components, or replacement parts.

The fully equipped maintenance shop will be a rare thing during a hazardous waste site cleanup effort. The storage will depend upon the type and number of heavy equipment pieces needed on the site. Most shops will not have replacement tires readily available and will have to go off site to a local distributor; support may also be needed from local hydraulic shops and air-conditioning or air-filtration unit repair shops. If specialized heavy equipment is necessary and is not owned by the contractor, then that equipment will have to be rented or purchased. Most maintenance shops will only be able to make minor repairs or find out what is wrong with the equipment.

Fuel supplies can be established near the supply point, either from 55-gallon drums or a tank truck. These supplies may be necessary either at a hazardous materials waste site or at an incident where equipment has to be kept running for extended periods of time.

At a hazardous waste site, there will usually be a maintenance shop or shed, and the fuel supply should be no less than 50 feet from this structure.

A traffic pattern, preferably one-way, should be established in this area around the fuel depot. Plans should be made for drainage of runoff, such as from rain or a gasoline spill.

13.5. MAINTENANCE ON CONTAMINATED EQUIPMENT

The maintenance and cleaning of contaminated equipment coming out of the Incident Zone depends upon the type and amount of contaminant. The industrial hygienist, chemist, or manufacturer's technical representative at the site will be able to determine the type of decontamination procedure required.

Will there be a need for a chemical to neutralize the contaminant, or will only steam cleaning be needed? Does the equipment need to be scrubbed with a brush, or will rinsing with a liquid detergent remove the contamination? What specific type of decontamination will be necessary?

Most hazardous waste, when mixed with soil, can be removed from the equipment by water, steam cleaning, or a brush and elbow grease. The exact removal method will depend upon the concentration of the contaminants— whether they are soluble in water, whether chemical neutralization is needed, or whether they can just be removed with a gentle brushing followed by a pressure rinsing. The various techniques will also require different levels of personal protective equipment.

Heavy amounts of contamination may require that a plastic-lined pit be prepared and the equipment decontaminated over it, with the toxic sludges being pumped up into barrels for disposal. If submersible or portable pit pumps or sump pumps are used to remove this runoff, the pumps need to be cleaned and decontaminated after every use, since residue could spread contamination.

Any large part of equipment, such as a bucket, a tire, or a boom, etc., must be decontaminated before it is removed from the Incident Zone to the Safe Zone. This can be done in the decontamination corridor, or a specific decontamination area may be set up for heavy equipment.

Electrical parts or equipment will have to be decontaminated in a way that will protect them from water or chemical sprays. Information about how to do this will have to come from the manufacturer, since there is no universal way to decontaminate electrical equipment. Most of the time, electrical equipment is hand wiped, vacuumed, and then covered with plastic sheeting.

13.6. FIELD DECONTAMINATION

It is ideal to follow a set of decontamination procedures, but sometimes you will not be able to decontaminate equipment in the ways just discussed. For instance, during the site investigation phase, when the site is being test drilled and mapped, time is usually too short to set up elaborate decontamination procedures and wash-down locations.

Hoses can be decontaminated if they come in contact with the contaminants by using water and a liquid detergent. Most of the time, this will be sufficient, but there may be times that the hose will have to be destroyed or disposed of as hazardous waste.

13.7. AIR FILTERS/OIL FILTERS/ENGINE OIL

The amount of airborne dust and chemicals that enter the engine will determine the amount of chemically detectable contamination. Oil, which has many additives, may break down after short exposures to contaminants. Any oil that is removed from the response vehicles or heavy equipment will have to be put into containers and considered as hazardous waste. Filters, whether air or oil, will have to be periodically removed and replaced. The filters may have to be considered as hazardous waste.

When removing the filters or changing the oil, it is necessary to wear protective clothing and gloves. Depending upon the incident site contaminants, the level of respiratory protection may be used to determine when it is necessary to change vehicle filters.

Clogged or loaded air filters may serve as a source of contamination, collecting and actually spreading toxic materials into the clean air. As we mentioned, the filters may have to be treated as contaminated hazardous waste, so they should not be put into the local landfill.

Fan blades, guards, and housings will accumulate dirt containing contamination. The fan and its components need to have periodic maintenance to remove the dust and accumulation on the blades.

Contamination of the cab of heavy equipment can be reduced by simply entering and exiting over disposable paper or plastic mats or sheeting, by the use of disposable seat covers, and by vacuuming the interior of the cab before and after usage. The vacuum cleaning system should have an absolute filter on the exhaust side to catch any microscopic dust. This type of filter is found in asbestos removal systems.

Changing the air-conditioning or air-filtration filters routinely will reduce exposure if you are working at a hazardous waste site. There is a method of measuring the pressure drop across the filter to determine whether it is still working, but the best way is just to set up a schedule to replace the filters.

To decontaminate the inside of the cab, the easiest way is to take a rag dipped in decontamination solution and wipe down the surfaces, hand controls, knobs, and contact surfaces of the doors or other surfaces. The seat covers and floormats should be removed, and the cleaning rag should be passed over these surfaces. Cleaning of the inside of the cab should be performed following every exposure. This will lessen the hazard exposure to the operator by reducing the amount of contamination coming in contact with his or her hands.

Another item in the cab that has to be taken into consideration is the heater. It will dry any mud you may have on your boots; this mud can then become airborne dust. If you have walked through the contaminated area, there is a possibility that you will contaminate the inside of the cab in this way. You will have to keep the floor of your vehicle clean. You will also have to wear respiratory protection until the cab has been thoroughly decontaminated.

13.8. FIELD REPAIRS

The most common repairs that you will encounter either at a hazardous materials incident or working on a site will be the changing of a tire and/or the repair of hydraulic line from a puncture or break. When a tire fails or becomes disabled in the Incident Zone, the method of choice is to try to get that piece of equipment farther away from the contaminant.

Try to pull the disabled equipment out with another rig to an area where full protective equipment is not necessary to perform repairs. Decontamination will then render the equipment safe to work on without full protective equipment. If the unit cannot be pulled out, and repairs have to be made on the spot, as in the case of changing a tire, use a crane or tripod to raise the rig. Practice tire changing while in protective gear. This will be a common chore at hazardous waste sites, since there are many sharp-edged barrels. If it is necessary to bring a service truck into the area to assist in changing the tire, then the truck will have to be decontaminated afterwards.

Leaks and breaks in hydraulic lines, if easily accessible, can be handled on the spot, provided the leaking fluid does not form a reactive mixture with any oxidizers on the ground, such as nitrates or chlorates. A catch basin made from plastic sheeting can be placed under the equipment to catch any leaking fluid. This is desirable and may be required locally.

13.9. WELDING

Welding in the Incident Zone can be very dangerous depending on the type of contaminants present. If the vapors are flammable or explosive, it is advisable to remove the equipment to a safe area. Field testing can determine the explosiveness or flammability of the atmosphere. Make sure there is no leaking fuel in the area if you have to weld. Follow all safety regulations concerning welding.

If welding has to be performed while wearing personal protective equipment, a unique set of problems arises. First, the protective equipment is not fire retardant, and sparks from the welding rods could ignite the suit; therefore, reduce sparking as much as possible. A welding hood will fit poorly over a full- or half-faced respirator, and your visibility is going to be very limited. Be careful not to ignite any materials, including yourself.

Cleanup of the welding area is important, since the paint on the heavy equipment may contain chromium and lead. Hard steel welding rods contain manganese. If you are using exotic alloy rods, they will contain different kinds of contaminants. You may have to grind down an area to prepare for welding. This area should be wiped down and the surface rinsed before grinding or welding can be accomplished.

13.10. TOOLS AND COMPONENTS USED IN FIELD MAINTENANCE

The type of tools and components to be used in field maintenance must be considered on the basis of ease of decontamination. If they cannot be easily decontaminated, how much will it cost to dispose of the equipment? Sometimes the site will dictate the specific type of equipment that can be used, e.g., nonsparking tools. For example, you may have to decide on rope versus chains. Rope may be cheaper, but chain is easier to decontaminate; woven or spun cable is very difficult to decontaminate. Some cables have a plastic coating, but this coating can deteriorate when in contact with chemicals found at the site. Always handle cables, chains, and rope with gloves, either when moving or when decontamination is necessary.

13.11. POWERED HAND TOOLS

All electrically powered hand tools used in the exclusion zone must be "intrinsically safe" — that is, internally grounded and approved to not generate electrical sparks.

All powered hand tools must also be rated according to National Fire Prevention Association standards.

Any extension electrical cords that are used must also meet safety standards. The cords should be frequently inspected for deterioration of the plastic insulation. Always wear gloves while inspecting the cords, since chemicals can get into the plastic insulation.

Once a tool has entered the Incident Zone or the Exclusion Zone, it should be considered contaminated. Anyone who mishandles or takes a tool home for personal use may spread hazardous contamination to his family, friends, home, or vehicle.

Generally, small powered or air tools are very difficult to decontaminate and may be discarded along with the other hazardous waste.

13.12. CLEAN WORK AREA

Since it is necessary to perform supply or support activities at an incident site or small repairs on equipment, a clean work area is desirable, even in the Incident Zone. This area should be in the safest part of the Incident Zone and have easy access to the resupply area or maintenance shed. A quantity of heavy plastic sheeting should be in this area to be used as a drop cloth, a tool table covering, a covering for the top of a tire tread, a catch basin for leaks from a broken line, or even reinforcement for an emergency berm in heavy chemical or rain runoff.

13.13. CONCLUSION

Maintenance and decontamination of equipment in a hazardous materials situation will generate unusual problems. Personal protective equipment, the hazardous environment, restrictions placed on the types of equipment that can be used in the Incident Zone or the Exclusion Zone, the need to prevent getting hazardous waste upon yourself, not spreading contamination, and preventing taking home any of the contaminants from the incident site are all matters of concern to the incident responder.

You have to learn to perform repairs and decontamination under the worst conceivable limitations of movement and visibility due to the restrictions of the personal protective equipment. While doing all this, you must reduce the hazards to yourself and your workmates.

13.14. CHAPTER REVIEW QUESTIONS

1. Define the following:
   (a) Decontamination
   (b) Simple decontamination
   (c) Decontamination corridor
   (d) Disposal of hazardous waste

2. Describe the problems encountered in decontaminating the following:
   (a) Electrical tools   (b) Personal protective equipment
   (c) Work tools   (d) Emergency response vehicles
   (e) Front end loaders, bull-
       dozers, etc.

3. Discuss the organization of a decontamination zone.

4. List the materials necessary to perform simple decontamination.

5. Runoff from a decontamination zone is considered hazardous waste by the EPA. Describe how you would contain, collect, and dispose of this runoff.

14 RADIATION AND RADIOACTIVITY

14.1. INTRODUCTION

Today, there is considerable fear (some irrational) about radioactivity and almost all forms of radiation. In addition, there is confusion over the differences between the ionizing and nonionizing forms of radiation. This section is intended to make you more knowledgeable about the dangers of, and yet the need for, radiation and radioactivity in the modern world, with emphasis on ionizing radiation. In addition, there might be potential radiation hazards facing you at a hazardous waste site.

14.2. DEFINITIONS

Some of the terms used in this chapter are defined below:

- An *alpha particle* is two protons and two neutrons joined together, that may be emitted from radioactive elements, usually with energy in the range of 4 to 8 million electron volts (MeV) and with a very short range (a few centimeters in air and less than 0.005 mm in aluminum).
- A *beta particle* is an electron (negative charge) or positron (positive charge) that may be emitted from the nucleus of a radioactive element, usually with energy in the range of 0 to 4 MeV and with a range of a few millimeters in tissue.
- *Bremsstrahlung radiation* is electromagnetic radiation produced by the slowing of a charged particle in the vicinity of an atomic nucleus; X-ray in nature.

155

- *Electromagnetic radiation* is composed of waves that travel at the speed of light and possess both a magnetic and an electric character.
- *Electron volt (eV)* is a very small unit of energy equal to about 1.6 × 10⁻¹⁹ joules used in measure of subatomic particle energies.
- *Gamma radiation* is electromagnetic radiation of very short wavelength (very high frequency) and very penetrating energy emitted from the nucleus of radioactive elements, often in conjunction with either alpha or beta emission.
- A *gray (Gy)* is a unit of radiation absorbed dose equal to 1 joule per kilogram of absorbing matter. One gray is 100 rads.
- *Infrared* refers to the region of the electromagnetic spectrum with wavelengths between 0.78 and 300 micrometers (μm) (higher frequency than microwave and lower frequency than visible light).
- *Ionization* is the process of altering the number of electrons around an atom, leaving it with a net electrical charge.
- *Ionizing radiation* is radiation capable of causing the formation of ions generally on the order of greater than 35 electron volts (eV). Examples are alpha, beta, gamma, X-ray, and neutron radiation.
- A *laser* (Light Amplification by Stimulated Emission of Radiation) is a device capable of producing coherent and monochromatic radiation, generally as the result of photon excitation of atomic electrons.
- A *neutron* is a fundamental particle of matter with no electrical charge and approximately the same mass as a proton; as a free particle, it is capable of great penetration (due to lack of a charge) and great biological damage.
- *Nonionizing radiation* is radiation incapable of forming ions due to energy less than that required to remove electrons from an atom. Examples are radio waves, microwaves, visible light, and ultraviolet electromagnetic radiation.
- A *rad* is a unit of absorbed radiation equal to 100 ergs per gram of absorbing material regardless of radiation type. Now replaced by the gray (Gy), which is equal to 1 joule per kilogram. One rad is 0.01 Gy.
- *Radiation* is energy in the form of electromagnetic waves and subatomic particles.
- *Radioactivity* is changes in nuclei of atoms that result in release of energy in the form of particles (alpha, beta, or neutrons) or waves (X-rays or gamma radiation).
- A *rem* is a unit of absorbed radiation equivalent; it relates the type of radiation to potential for damage to living tissue. It has now been replaced by the sievert (Sv). One rem is 0.01 sievert.
- A *sievert (Sv)* is a unit of absorbed dose equivalent; it relates the type

of radiation to potential for damage to living tissue. One sievert is 100 rems.

- *Visible light* is electromagnetic radiation capable of being seen by humans; generally possessing wavelengths between 400 and 800 μm.
- *X-rays* are electromagnetic radiation of very short wavelength (0.006 to 12 μm) and very high energy, capable of penetrating inches of steel or lead and resulting from transitions of inner electrons of heavy atoms.

14.3. TYPES OF RADIATION

Radiation — that is, energy that can move from place to place — is everywhere in the universe. Some forms of radiation can even travel in a vacuum (without air). You are familiar with some forms of radiation like radio, television, and sunlight. There are many others, such as X-rays, gamma radiation, and cosmic radiation. Forms of radiation are present everywhere on the earth and have been from the beginning of time. In fact, many forms of radiation and radioactivity were much stronger in the past than they are today.

In general, radiation may be divided into two broad classes, depending upon the amount of energy and the kind of damage it can do: ionizing radiation and nonionizing radiation. As you will see, these are quite different and must be considered separately.

14.4. ELECTROMAGNETIC RADIATION

The first kind of radiation we are going to consider comprises those forms related to electricity and magnetism. This radiation begins with low frequencies that have relatively little energy and cause less damage to living systems. These low-frequency waves are referred to as nonionizing radiation. The term nonionizing means that this radiation does not have enough energy to knock electrons off atoms (ionization). This does not mean, however, that all forms of nonionizing radiation are harmless.

If we consider all frequencies of those forms of energy related to electricity and magnetism, we find that they have been divided up into groups, generally according to the uses that we have for them, that are collectively called the electromagnetic spectrum (a spectrum is a range of frequencies). The electromagnetic spectrum is shown in Figure 14-1. The lower frequencies possess corresponding low energies (nonionizing). At higher frequencies, the energy increases (ionizing). In general, the higher the frequency (and energy), the greater the hazard to living things. As energy increases, we go from the nonionizing area to the ionizing area at about the region we call "soft" X-rays.

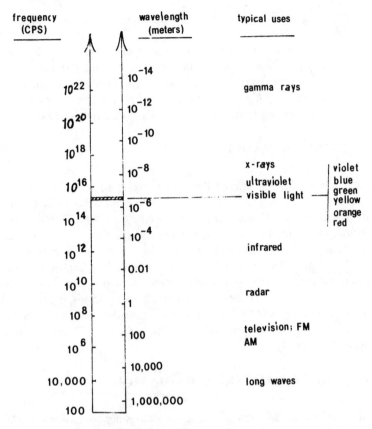

Figure 14-1. The electromagnetic spectrum.

People can control the lower frequencies fairly well, and we use them in many ways. For example, radio and television broadcasting takes place at these lower frequencies (although radio is generally lower in frequency than television). As the frequency increases, we reach the region called infrared (or heat), then the visible light region (which our eyes can see), and then ultraviolet light (which our eyes can't see, hence the term black light). These regions are also used for medical heating, lasers, and ultraviolet sterilization to kill bacteria. Note that small living things (bacteria) can be killed by these frequencies. At this point, we begin to move from the nonionizing to the ionizing regions. As frequency continues to go up, we pass into the frequencies where ionization (removal of electrons from atoms) can take place. The process of removing electrons from atoms can be very harmful to a living creature. The ions (charged particles) that may be formed in living cells can cause cellular change, leading to cancer or damage to genetic material such as DNA, or cellular death.

The ionizing regions are called, in turn, the "soft" X-ray region (having enough energy to go through muscle and skin but not bone); the high-energy X-ray region (enough energy to go through thick steel); and the gamma-radiation region (very high-energy radiation) coming from the nucleus of atoms (radioactivity).

It is these higher-frequency, higher-energy radiations that can seriously harm living creatures. Yet the effects of these forms of radiation can still vary widely. In all cases, it is important to remember that as soon as electromagnetic radiation stops, the effects stop. Any object or person subjected to electromagnetic radiation does not become radioactive and cannot emit radiation themselves. Any damage done by the radiation takes place while the electromagnetic radiation is passing through the person.

What, then, are the harmful effects? The most common effect of electromagnetic radiation is a burn. Most frequencies are capable of heating skin (and other tissue) to the point where burns take place. This can occur down at the radio frequencies as well as up at the ultraviolet frequencies (it is ultraviolet radiation that gives you a sunburn). The extent and depth of the burn depends upon the frequency, with microwave frequencies producing heating effects upon tissue at depth of up to several centimeters (this is how a microwave oven cooks food). At radio and microwave frequencies, this heating can damage the eyes, skin, and reproductive organs. At the higher frequencies (X-rays and above), radiation may pass completely through the person, with his or her cells being damaged or destroyed by the formation of ions (charged atoms) that interfere with normal cellular functioning. In short, the cells are affected because their chemical makeup has been altered, either by heating at lower frequencies or by ionization at higher frequencies.

This brief discussion has considered electromagnetic radiation; we now turn our attention to radioactivity.

14.5. RADIOACTIVITY

In 1896, Henri Becquerel, a French scientist, was studying uranium salts, since he knew that they became luminescent (glowing) when exposed to sunlight. He was also aware that Roentgen had discovered X-rays in 1895, and he thought there might be a connection between X-rays and luminescence. His experiment was to expose uranium salts to decreasing amounts of light and then measure the luminescence by exposing photographic film. To his surprise, he found that the salts could expose the film without being exposed to the light at all! Even when the film was tightly wrapped in heavy paper or even other materials, the film could still be exposed by the uranium salts! Becquerel had discovered natural radioactivity, which had been present since the beginning of time without our knowing about it, since we cannot feel, taste, or otherwise sense it.

Other scientists, chiefly the husband-and-wife team of Pierre and Marie Curie, found other metals in addition to uranium that were radioactive. They discovered polonium (named after Marie's home country of Poland), more than 10 billion times more radioactive than uranium, and radium (named after rays emitted from it), more than twice as radioactive as polonium.

But what was it that exposed the film? It was found that naturally radioactive substances give off radiations, two of a particle nature and one of a wave (electromagnetic) nature. The two particle radiations were called alpha and beta; the wave radiation was called gamma.

14.6. ALPHA RADIATION

Alpha radiation was found to consist of relatively large particles (compared to the other form) weighing about four times as much as a hydrogen atom. The particles were also found to possess a positive electrical charge. Later, scientists realized that the alpha particle was the same as a helium-atom nucleus, consisting of two positive protons and two noncharged neutrons. This alpha particle was produced within the nucleus of heavy, unstable atoms. Because of its size (and energy), an alpha particle cannot travel very far; in fact, it cannot penetrate a sheet of paper or the human skin. Such radiation can therefore be stopped by a very thin sheet of shielding. This does not, however, mean that it is harmless. In reality, alpha radiation can be the most dangerous form of radioactivity if taken internally by ingestion, inhalation, or through a wound, since it expends all of its energy in a very short path. If an alpha-emitting material is breathed into the lungs or swallowed into the digestive tract, damaging ionization can take place (leading to cancer or other health effects).

To protect against alpha radiation, it is necessary to keep the radioactive material outside the body. Since this radiation cannot pass through skin, your skin and clothing will protect you from the radiation; however, if you get such a radioactive material on your clothing or skin, later you could swallow or breathe it. For this reason, around radioactive materials it is vital that respiratory protection be worn and that the skin and clothing be washed down at the end of the work shift. Alpha radiation is only a hazard if taken inside the body.

14.7. BETA RADIATION

Beta particles were found to be relatively small, fast moving, and negatively charged. It was discovered that beta radiation was essentially a stream of electrons, but coming from inside the atom's nucleus unlike the electrons that circle the nucleus like planets around the sun.

Beta radiation can travel much farther than alpha radiation, passing through the skin and even deeper. The first effect of beta exposure is a skin burn, which can become increasingly worse. Later, ionizations can lead to cancer and other cellular disruptions. These beta electrons cause their damage in two ways: first, direct ionization, often leading to the formation of additional (secondary) electrons; and, second, by creation of electromagnetic radiation (called by the German word *bremsstrahlung*, which means *radiation from slowing down*) by the slowing of the beta particle near the atom's nucleus.

Protection from the effects of beta radiation is a little more difficult, since it can be a hazard both from being taken inside the body and from burns and ionization from outside the body. To protect against beta radiation, you must ensure that no radioactive material gets inside the body by washing after each shift and wearing respiratory protection plus wearing thick enough protective clothing to stop the beta particles. However, you should keep in mind that metal shields (aluminum, etc.) can cause an increase in the radiation due to bremsstrahlung, which is similar to gamma radiation.

14.8. GAMMA RADIATION

Gamma radiation was found to be similar to X-rays, which we talked about earlier. Scientists found the only difference to be that X-rays come from the electrons orbiting the nucleus, while gamma rays come from inside the nucleus. Other than their origin, these two radiations can be considered the same. The bremsstrahlung we mentioned under beta radiation is actually a form of X-rays.

Gamma radiation is electromagnetic in nature and can pass through the entire human body or even sheets of lead. Just like X-rays, gamma rays cause intense ionization within living cells, leading to cell death or cancer.

Depending upon the energy, gamma rays can pass through many inches of lead. Protection from gamma radiation depends upon shielding by thick amounts of lead or other material. Gamma radiation is the greatest radioactivity threat from outside the body.

14.9. NEUTRON RADIATION

There is another type of radiation which we have not yet discussed. Neutrons can be emitted from radioactive materials undergoing what is called spontaneous fission (which means the atoms are unstable and break apart into smaller, more stable atoms). While this does take place in nature, the danger is much less than from manmade sources of neutrons, such as atomic weapons or neutron generators. Hopefully, no one will ever have to worry about receiving neutron radiation from the explosion of an atomic bomb. This leaves neutron generators as the primary hazard. Luckily, these devices are found only in a very few industries and are well shielded and isolated from accidental exposure. Very few materials release neutrons.

However, to make our lesson complete, let us talk briefly about neutrons and their health hazards. You may have heard of the so-called neutron bomb, which produces very intense neutron radiation. Such a bomb would be similar to other thermonuclear (hydrogen) bombs except it would be designed to increase neutron generation, since high-energy neutrons can pass through the steel of tank and ship hulls to kill the occupants inside more effectively than even gamma radiation. Neutrons can, however, be stopped by such things as water, some plastics, fuels, and paraffin wax.

Neutrons cause their damage by affecting the nuclei of atoms, causing them to become unstable. Most atoms can be activated by neutrons, and many can become radioactive themselves. Protection from neutrons would generally require a neutron-absorbing shield like water, etc., followed by a lead or other shield, since secondary radiation can be produced by neutrons. Exposure to neutrons should not be expected at a work site unless it is a radiation research facility.

Other types of radiation, such as protons, deuterons, and other heavy ions are only rarely used and only seen in research facilities specifically working on problems in high energy physics. We will not discuss these radiations here.

14.10. PROTECTION FROM RADIATION

As mentioned in each of the cases above, protection from external radiation generally involves shielding — that is, placing something between yourself and the radiation source that absorbs the radiation, bearing in mind that sometimes secondary (subsequent) radiation may be produced by the use of the shield. There are two other principles of protection from external radiation: time and distance. As with most other health hazards, the shorter the time spent near the radiation hazard, the less the danger. Exposure periods should be limited to the shortest possible time. Likewise, the farther you are from the hazard, the better protected you are. You should remain as far from any source of radiation as possible.

These three methods of protection — time, distance, and shielding — will lessen the hazards from working around external radiation, whether ionizing or nonionizing, whether radioactivity or microwaves, whether light waves or X-rays.

Protection from internal radiation is best accomplished by keeping the radioactivity outside the body. Radioactive materials may enter the body through inhalation, ingestion, or wounds. Inhalation protection is afforded by the use of respirators. Ingestion and wound contamination are best prevented by good work habits and cleanliness.

If you know or suspect that you are working around radiation of any kind, you should know the dangers (or lack thereof) and how to protect yourself. Knowledge is your greatest protection.

14.11. EFFECTS OF IONIZING RADIATION

Madame Curie died of leukemia; many early radiation researchers died from one or more types of cancer; radium watch dial painters have exhibited greatly increased rates of bone cancer; uranium miners have exhibited increased lung cancer rates; leukemia increased among the atomic bomb victims in Japan; and children of some of those bomb victims have been born with birth defects, chiefly small body size and mental retardation. Are these "caused" by radiation?

One of the problems associated with the effects of radiation is that radiation does not produce new or strange types of disease (in spite of the 1950s horror movies). All the diseases seen from radiation exposure also occur naturally to many people. Therefore, only statistical analysis can be used to see if exposure to radiation really caused the problem. For example, of the 205,000 military personnel exposed to ionizing radiation during the period of atomic testing, 34,000 will be expected to die of natural cancers. The additional radiation received from the tests might statistically increase this number by 11. Therefore, out of 34,011 dying of cancer, 34,000 would have died regardless of the radiation exposure. It is impossible to determine which of the 34,011 the 11 were.

However, it is well known that very high levels of radiation are harmful. Above some 6 sieverts (600 rems) in a short exposure, about one half of those people exposed will die. Cancers seem to be increased at levels above 1 sievert (100 rem) acutely delivered. Genetic effects in animals may be seen at even lower values, but are even more difficult to determine. What is not known is whether there is a danger at very low levels (say, 1 millisievert [100 millirem] or even 0.1 millisievert [10 millirem]). Today, it is felt that to ensure the safest possible environment, any amount of radiation should be viewed as potentially harmful. The principle of "as low as reasonably achievable," or ALARA, has been the standard for protection of people and the environment from radiation, in order to keep exposures as small as reasonably possible, since we really don't know what the effects of very small amounts might be.

14.12. BACKGROUND RADIATION

Natural radioactivity is all around us.

Cosmic radiation from space can be a potential hazard to space flight crews, as well as to those who take transcontinental air flight routinely. Because of high elevation, cosmic radiation levels in Denver, Albuquerque, and many other locations may be twice as high as for locations at sea level. The average radiation due to cosmic radiation at sea level is about 0.4 millisieverts (40 millirem) each year.

The millisievert and millirem are units of radiation dose that are 1/1000 of a sievert or rem, respectively. One medical chest X-ray will produce between 1 and 2 millisieverts (100 and 200 millirems).

Another source of radiation is the ground. Rocks and soils everywhere contain radioactive materials such as potassium-40 (one of the isotopes of natural potassium), uranium, and radium. One product of the decay of uranium is the gas radon, which is found in the air and in groundwater and may supply as much as one half the total radiation exposure to the population of the U.S. Terrestrial sources, including radon, produce about 2.4 millisieverts (240 millirems) each year, with radon alone responsible for about 2 millisieverts (200 millirems).

The human body is also slightly radioactive, since it contains potassium-40, carbon-14, and uranium or radium taken up from natural food and water. Such sources produce about 0.4 millisievert (40 millirems) each year.

The total natural radiation exposure is about 3 millisieverts (300 millirems) per year. Added to this is an average of some 0.5 millisievert (50 millirem) for medical X-rays and treatment, 0.04 millisievert (4 millirem) from fallout due to past nuclear weapons testing (but this figure is slowly falling), and 0.0005 millisievert (0.05 millirem) from nuclear mining, milling, and nuclear power plants. (Incidentally, the amount of radiation from a coal-fired electric plant is about the same as from a nuclear power plant, due to minute amounts of uranium and radium in the coal being burned.) Clearly, the vast majority of radiation results from natural causes. The total amount of radiation received nonoccupationally is about 3.6 millisieverts (360 millirems) per year.

Radiation workers, including nuclear industries and medical personnel, are currently allowed to receive 0.05 sievert (5 rems) per year of ionizing radiation. While this value is subject to change, no adverse effects have been noted in the 30 years that this standard has been in effect.

14.13. CONCLUSION

Radiations of many kinds are present in the environment and in occupational settings. There are forms of nonionizing radiation such as lasers and intense light that can cause burns and blindness; nonionizing radiation such as microwaves that can damage cells and organs by heating; and ionizing radiations such as alpha, beta, and gamma and X-rays from radioactivity in the environment as well as from industrial and medical sources that can kill cells, cause genetic defects, and increase potential for cancer.

In all cases, you can protect yourself by knowing the differences between these forms of radiation, as well as of each hazard. For protection against external radiation, the principle of time, distance, and shielding is best. For protection against internal radiation, the best principle is prevention of entry into the body.

Remember that there is no hazard from a radar antenna unless the power is turned on; there is no hazard from a laser unless it is turned on; there is no hazard from another person who has been exposed to radiation, even radioactivity, unless the material is still on his or her clothes or body; and there is very little hazard if you know what the dangers are and are protected against them.

While there are hazards from energized lasers, microwaves, and antennas and from natural and manmade radioactivity, there are even greater hazards from ignorance.

Specific information must be given to you about the work site from your employer. Listen, and think, about the dangers and protective measures present. Ask if you have questions.

14.14. CHAPTER REVIEW QUESTIONS

1. Which of the following is the greatest external radiation danger?
   (a) Alpha particles   (b) Beta particles
   (c) Gamma rays   (d) None of these are hazards

2. Which of the following is the greatest internal radiation danger?
   (a) Alpha particles   (b) Beta particles
   (c) Gamma rays   (d) None of these are hazards

3. Which of the following is the heaviest and slowest?
   (a) Alpha particles   (b) Beta particles
   (c) Gamma rays   (d) Neutrons

4. Which of the following is not electromagnetic radiation?
   (a) Microwaves   (b) X-rays
   (c) Neutrons   (d) Gamma rays

5. Which of the following is the best shielding material for gamma rays?
   (a) Plastic   (b) Lead
   (c) Aluminum   (d) Dirt

6. Which of the following is the best shielding material for neutrons?
   (a) Plastic   (b) Lead
   (c) Aluminum   (d) Dirt

7. Bremsstrahlung is which of the following?
   (a) A German drinking mug
   (b) Light rays from distant stars
   (c) Radiation like X-rays caused by charged particles slowing down near atoms
   (d) None of the above

8. What is the allowable occupational annual radiation limit?
   (a) 1 rem   (b) 5 rem
   (c) 10 rem   (d) 500 rem

9. ALARA has what meaning?
   (a) As Long As Radiation Arrives
   (b) Above Limits Allowed by Radiation Authorities
   (c) Always Less than Allowed Regulatory Approval
   (d) As Low As Reasonably Achievable

10. Which of the following contributes the most to annual radiation dose for most people?
    (a) Background radiation
    (b) Nuclear power plants
    (c) Fallout from bomb testing
    (d) Shipping radioactive materials around the country

15   INCIDENT/SITE MONITORING

15.1. INTRODUCTION

In order to safeguard the health and safety of incident responders, of the public residing in the immediate area surrounding the site, and of the environment, monitoring of the incident site for dangerous and unknown hazards has to be performed. At a hazardous materials incident, many unknown gases, vapors, fumes, or other hazards are being emitted into the atmosphere. Identification and determination of the type and quantity of the contaminants is necessary in order to choose the proper entry and protective equipment, and the level of protection that will provide the maximum level for this particular site or incident has to be determined before you begin rescue or cleanup.

But how do you determine the types and quantities of the unknown hazards? Some hazards can be identified by the placards, bill of lading, or containers, but, in many incidents, chemical analysis must be performed to identify the particular chemical hazard. Air-monitoring instruments provide the integral information necessary to fulfill this requirement.

What types of instruments are necessary to accomplish this determination of the hazard? Can one instrument perform this task? In this chapter, we will answer these and other questions about monitoring techniques and instrumentation, and will describe how these instruments provide data that will help response/site personnel.

15.2.   RATIONALE FOR USING MONITORING EQUIPMENT

Ongoing monitoring of atmospheric chemical hazards will indicate the quantity and types of hazards present and the potential or real effects on the environment. This information will determine the incident conditions or any changes in the conditions that would impair the safety of site workers or responders.

In Chapter 3, we discussed the short- and long-term effects of chemical exposures. Air monitoring will determine if there is an immediate risk or a long-term risk to public health, including the health of response workers.

Monitoring equipment will provide information enabling a determination to be made regarding appropriate personal protective equipment and respirator equipment to be used during that specific response or site action. Also, monitoring can help to determine the actions necessary to mitigate the hazards effectively and safely.

15.3. UTILIZATION OF AIR-MONITORING INSTRUMENTS

All test instruments intended for field use should be carefully evaluated for their ability to withstand the rigors of such use.

We can expect to subject the instruments to various types of problems, such as potential contamination, extremely rough handling, rain, mud, heat and cold, and a variety of handling hazards. If their inherent design precludes the adaptability of some test instruments to field use, then we should consider modifications or seek another type of instrument.

If an instrument cannot withstand rough field handling, that instrument would either be deemed unsuitable for usage or would require the implementation of strict handling procedures that might or might not be adaptable to the emergency response or site cleanup procedures.

In this section we will review the various types of field instruments available, their function and reliability, and your actual or potential needs from particular instruments.

In addition to discussing direct-reading instruments, we will also discuss remote-reading instruments and simple wet-chemistry tests that can be used to perform perfunctory or "fingerprint" testing, or what some people call *field hazard categorization*. These tests can place unknown materials into specific hazard classes such as flammable, corrosive, oxidizer, etc. The tests are normally qualitative and not quantitative, although there are some limited quantitative tests that can be performed in the field.

When discussing the various instruments, some "desirable" criteria should be considered for all field instruments. The specifications described below are very general in nature. It is the intent of this manual to provide guidelines to meet the needs of durability, reliability, and ease of usage.

15.4. GENERAL CONSTRUCTION SPECIFICATIONS

Rugged construction is a necessity for instrumentation used during a response or at a site. Test instruments should be in cases that are capable of withstanding a drop of several feet without disrupting the accuracy of the instrument. Critical portions of the test instrument (test probes, knobs, meters, lights, etc.) should be shock protected.

Many manufacturers will encase the meter or knobs to ensure that when they do drop they will be protected. If the instrument is not protected, you might consider using styrofoam to protect the sensitive portions of the instrument. If the instrument is dropped, the styrofoam will cushion some of the impact and may preserve the integrity of the instrument. Any time a test instrument is subjected to a severe shock, the manufacturer's calibration and adjustment procedures should be performed before using the instrument again. Many calibration and zero adjustment procedures can be performed in the field. However it is suggested when you return to quarters or to the site office that the instrument be doublechecked against known standards.

Outer cases or shells are useful and should be considered for the following reasons: ease of carrying; additional shock-absorbent capacity; reduced expense in the event of either chemical or other external contamination that is difficult or impossible to clean (since outer cases or shells are relatively inexpensive compared to the total cost of the instrument); and a lesser chance for damage to the instrument while in storage or being transported.

Ideally, the instrument should be powered by batteries that are rechargeable or that can be obtained from a commercial retail outlet.

If at all possible, instruments requiring power supplies designed just for that particular instrument should be avoided. In the event a battery pack depletes before the need for the instrument is completed, you must wait a considerable amount of time to recharge the power pack unless you have a spare power pack charged and ready to go.

Some instruments use size D batteries. The batteries can be nickel-cadmium rechargeable batteries, but, if the batteries become discharged, they can be replaced with any standard size D flashlight batteries.

Alkaline batteries should last longer than the other batteries, but even if nonalkaline batteries were to be used and only lasted a few hours, you could continue to change batteries, in the event of an emergency, until the task was completed.

Some power packs are designed for a particular instrument even though they consist of size C or D batteries that have been prewired. If that power pack becomes discharged, you can only replace it with another power pack designed for that instrument. Those of you who have used nickel-cadmium rechargeable batteries are aware of the annoying characteristic of battery memory. If you put the battery on charge and "top off" the battery, this memory will only allow you to use the battery for the length of time for which you topped it off or charged it. Many organizations will completely discharge their nickel-cadmium power packs at specific intervals and then completely recharge them. Rechargeable power packs should be routinely cycled.

Without testing equipment, we are "hazard blind." We are unable to measure the actual hazards to which we may be subjected. We experience a gap in our medical monitoring, and it becomes very difficult to make an intelligent decision about the need for particular gear or for evacuation. It is a most uncomfortable experience to be looking at an instrument worth thousands of dollars that cannot be used because of a few dollars worth of batteries. Spares should always be maintained on hand.

Ease of use is another desirable characteristic for any instrument. Avoid instruments that require prolonged daily calibration procedures. Ideally, a battery check and zero-adjust and/or other specific meter dial adjustments in a "clear" area are all that should be needed prior to instrument use.

Audible and visual alarm indicators are desirable. Also, manual reset of alarm conditions can frequently be useful in the event of an intermittent or otherwise infrequent alarm condition.

Avoid instruments that can be easily contaminated. Instruments that use pumps for drawing in samples are probably more easily contaminated than instruments that measure in a static mode.

Moisture traps may be adapted to instruments that use pumps for drawing in their samples. The moisture trap is similar to the fuel filter on a vehicle engine; the trap will capture liquid that is accidentally drawn into the sampling unit. There may be times when you have to sample in the rain, so the filter trap may prevent some very expensive repair bills.

Another requirement for field instruments is their ability to sample and analyze very low contaminant levels and, ideally, to discern among contaminants exhibiting similar characteristics.

When dealing with instruments, it is necessary to consider the following:

Sensitivity is the lowest concentration of a contaminant that an instrument can accurately and repeatedly analyze.

The *operating range* determines the lower and upper use limits of the instrument. It encompasses the sensitivity as its lower limit and the overload point as its upper limit.

Selectivity establishes what contaminants will elicit a response on the instrument. An instrument may have good selectivity, but it may have widely varying sensitivities for a given family of chemicals, for example, aromatics, aliphatics, and amines.

Amplification, often used synonymously (and incorrectly) with sensitivity, deals with the ability of an electronic amplifier to increase small electrical signals from the detector. This capacity may be fixed or variable. However, changing the amplification of the detector does not change its sensitivity.

For optimum field usefulness, an instrument should possess high sensitivity, wide range, high selectivity, and the ability to vary the amplification of the detector signals.

Ease of decontamination of the instrument is a very important factor that you must consider when purchasing it. It would be impractical to purchase an instrument and then have to dispose of it because you cannot clean it after use.

The foregoing are some general physical characteristics desirable in field instruments. Unfortunately, these desirable characteristics are not always available, but many manufacturers have been very cooperative in modifying their stock instruments to meet individual specifications. Frequently, you can provide a manufacturer with a certain specification, and the manufacturer will not only provide what you desire, but may provide other improvements as well.

Most test instruments were originally designed for use in the industrial workplace, not for hazardous materials responses. In recent years, however, field use of many of these instruments has become more common. Most manufacturers of these instruments are still learning how their products are used in the emergency services community. Therefore, manufacturers usually receive your input with an open mind so their products can be improved upon and accepted more readily in the field. You have an opportunity to improve what is currently available and should exercise that opportunity by providing sound input to manufacturers not only on test equipment but also on other safety, containment, and recovery equipment.

15.5. GUIDELINES FOR INSTRUMENT USAGE

During the response situation, there are several common uses of test instruments that assist in determining that there is a hazard and what specific data can be obtained upon which intelligent decisions can be made. In this section we will outline the common uses of the instruments and then a few pointers or procedural guidelines to help you to use the instrument correctly.

Field instruments should provide the following information:

- determination of the existence of a hazard(s) and the parameter(s) of the hazard(s)
- effects of the cleanup or mitigation and/or containment efforts
- extent of dispersion of the hazard by weather, geology, hazard breakdown, etc.
- determination of when the hazard no longer exists

Such basic information can be obtained accurately and safely using a combination of field test instruments and/or some simple chemistry tests.

The primary decision is the selection of which test instrument to use under what conditions. During the response phase, never trust anything to be as it appears; you must confirm the observations with instrumentation or chemistry. You should make your initial approach and assessment of any response as though the hazard were completely unknown. Some sound guidelines for monitoring procedures are as follows:

1. Approach from an upwind, upgradient position.
2. Don protective equipment. This may be a fully encapsulating suit with positive pressure self-contained breathing apparatus.
3. Use the buddy system. Work in pairs. Have another team fully suited up, ready to enter in case you or your partner have difficulty.
4. Enclose your instrument if possible. You could possibly protect the instrument by putting it in a plastic bag with the probe in clean air. This is important because of the difficulty in decontaminating instruments.

There are, of course, exceptions when you know something about the nature of the hazard; for example, a tanker clearly placarded "1203" (gasoline), with flames shooting several feet in the air, or the smell of gasoline so strong you may actually feel ill, does not require that you first monitor for radiation. However, only when you are absolutely certain (and there are few absolutes in hazardous materials) beyond any doubt about what the hazard is should you approach the incident without employing instrumentation.

Since most commonly the hazard is unknown, we must develop a systematic approach to testing. We should measure in the following order:

1. Radioactivity
2. Combustibility
3. Oxygen availability (deficiency)
4. pH (if liquid)
5. Hydrogen sulfide (applicability to be described below)
6. Carbon monoxide (in a fire or a postfire incident)
7. Other chemicals known or suspected of being present

15.6. INHERENT SAFETY

Instruments used to evaluate hazardous materials spills or waste sites must be in themselves safe to use in the hostile environment. For instance, electrical instruments must be constructed in such a manner as to eliminate the possibility of igniting a flammable or combustible mixture.

Instruments should be operable in a hazardous atmosphere. An atmosphere is hazardous if it meets the following criteria:

1. It is a mixture of any flammable material in air whose composition is within the material's flammable range (lower to upper flammable concentrations).
2. A critical volume of the mixture is sufficiently heated by an outside ignition source.
3. The resulting exothermic reaction propagates the flame beyond where it started.

Hazardous atmospheres can be produced by any of the three general types of materials:

- Flammable gases/vapors
- Combustible dusts
- Ignitable fibers

The flammable substance may define the hazard associated with a given product, while the occurrence of release (or how often the material generates a hazardous atmosphere) will dictate the risk involved. There are two types of releases associated with hazardous atmospheres:

1. *Continuous.* These releases exist continuously in an open unconfined area during normal operating conditions
2. *Confined.* These releases exist in closed containers, systems, rooms, or piping, where only ruptures, leaks, or other failures result in a hazardous atmosphere outside the closed system

Being concerned with the aforementioned hazards, we need to ensure that our instruments (a potential ignition source) do not ignite a flammable atmosphere. There are two types of devices we can use to prevent an ignition:

1. *Explosion-proof.* The instrument is encased in a rigidly built container. Explosion-proof instruments allow any explosion to be contained within the specially designed and built enclosure.
2. *Intrinsically safe.* An intrinsically safe device, as defined by the National Electric Code, is incapable of releasing sufficient electrical or thermal energy under normal or even abnormal conditions to cause an explosion.

In order for an instrument to be certified as being intrinsically safe, it must be approved by the Underwriters Laboratories (UL) or Fireman's Mutual (FM) standards laboratory. The UL and FM and the American National Standards Institute (ANSI) working with the National Fire Protection Association (NFPA) have developed test protocols for certifying explosion-proof or intrinsically safe devices.

An electrical device certified under one testing laboratory carries a permanently affixed plate showing the logo of the laboratory granting certification and the Class(es), Division(s), and Group(s) against which it was tested.

Certification means a device has been tested and approved to be used safely around a given class of flammable materials. In order to have an electrical device certified by one of the testing laboratories, a manufacturer must submit a prototype for testing. If the unit passes, it is certified as submitted. Any changes to the unit must also meet the certification-testing agency's approval. The instrument has to be used, maintained, and serviced according to the manufacturer's instruction. Do not use an instrument in an atmosphere for which it has not been certified, since you might cause an explosion or severely damage the instrument.

The test procedures that are used are protocols established by the NFPA and ANSI. Therefore, certification by either organization is acceptable.

15.7. RADIOLOGICAL MONITORING INSTRUMENTS

We first check for radiation at an incident if there is any chance that it might be present. The most common instrument used to monitor for ionizing radiation is the Geiger counter. However, this instrument can only detect beta and gamma radiation and has the characteristic of reading zero when in extremely large radiation fields.

We would test for radiation if the incident involved radioisotopes from a hospital, nuclear fuel for a reactor, or military munitions containing radioactive materials.

Alpha-emitting radioactive materials such as uranium, radium, and plutonium require special instruments not commonly available for responders or site workers.

Even such items as wrist watches or some testing equipment contain small amounts of radioactive materials and could be considered a radioactive hazard if a large enough shipment is involved.

All three forms of ionizing radiation discussed in Chapter 14 require monitoring equipment, since they cannot be felt by humans. If there is a hazardous condition of radiation encountered while you are performing onsite monitoring, then you must proceed with great care. Most emergency response personnel are not equipped to handle radioactive emergencies beyond identification of the hazard, establishment of control zones, and evacuation.

If there is a leak of radioactive materials, an initial approach with a Geiger counter capable of measuring beta and gamma radiation would be the first step. If a determination indicates that there is a radiation leak, the source of the radiation should be identified, the area isolated, evacuation properly handled, and appropriate notifications made. During this type of emergency, the security and isolation of the area are primary.

You may need to request a specialized team to assist you, such as a radiation response team.

15.8. OXYGEN-AVAILABILITY INSTRUMENTS

Oxygen, the most important gas for life, must be checked for when you are responding to a hazardous spill or site. Without sufficient oxygen, life cannot be supported. Terrain variations and unventilated areas may contain insufficient concentrations of oxygen. In addition, oxygen measurements are necessary when combustible gas indicator (CGI) measurements are made, since the level of oxygen in the ambient air will affect the accuracy of the CGI readouts.

The most useful range for the oxygen indicator is the 0% to 25% range. This range will help you determine if an SCBA is necessary.

At sea level, the normal ambient concentration of oxygen in air is 20.93%. If we measure oxygen lower than 19.5%, we should be very concerned. Either our instrument is not functioning correctly or we are in an oxygen-deficient atmosphere. Something could be is displacing the oxygen, such as a combustible, inert, or toxic gas, mist, or vapor. An oxygen availability instrument will not identify the foreign matter displacing the oxygen, but will indicate the presence of a problem.

Many toxic gases can be in an atmosphere in sufficient quantities to cause problems, yet not in sufficient quantities to affect an oxygen-availability meter. The Occupational Safety and Health Administration (OSHA) has established the minimum oxygen availability as 19.5%. Any level under 19.5% requires the use of supplied-air breathing apparatus.

If the ambient air contains 20.5% oxygen and the normal is 20.9%, then what has happened to the missing 0.4%? Has it been replaced by toxic gases? If so, what type and how toxic? Can the toxicity be reduced by using an air-purifying respirator, or do we have to use supplied air? The oxygen-availability meter merely provides you with a piece of the picture. You cannot make an intelligent hazard assessment from the indications of the instrument about the existence of a foreign substance.

Oxygen meters work on the principle of a galvanic cell with a pair of electrodes immersed in an electrolytic solution. This device is referred to as the *oxygen cell*. Current flows through the cell only when molecular oxygen is present. When oxygen enters the cell, it dissociates into ions in the electrolytic solution that collect on the cathode. The movement of these electrons is measured as a current flow on the meter. The meter should be adjusted to 20.9% in a noncontaminated atmosphere. There is a distinction here between calibration and adjustment. When calibrating the instrument, an inert gas such as nitrogen is used and the instrument is zeroed. Then the instrument is adjusted to 20.9%.

Oxygen-availability instruments have some inherent problems. Several gases may "poison" the sensor; carbon monoxide or carbon dioxide from a nearby vehicle could cause the operator to make erroneous adjustment. Also, altitude has to be taken into account when adjusting the instrument, because as you go up in altitude, the oxygen pressure, but not concentration, decreases. If, for example, you were working at sea level and the instrument had been adjusted at that level and you have to respond at a 4000-foot altitude, you could not expect the instrument to read accurately. You would have to recalibrate to a 4000-foot altitude to get an accurate reading.

Another problem is the self-consuming characteristic of the oxygen cell. The electronic activity is continuously causing the cell to degenerate. Oxygen cells can be "regenerated" at the manufacturer, but it would be prudent to keep a spare oxygen cell handy if you are going to be using the instrument frequently.

There are two basic types of oxygen meters on the market. Some of the older units require that you draw air into the instrument by means of a manual pump. This type requires the operator to continuously operate the pump while monitoring the environment. Most of the newer units have battery-operated pumps that draw the air into the unit. Some units are now designed to read the oxygen in the static mode. There are units that will read oxygen only, but some of the newer models will monitor for combustible gases as well as other gases such as hydrogen sulfide or carbon monoxide.

15.9. COMBUSTIBILITY-MONITORING EQUIPMENT

A CGI is a general survey instrument designed to measure the percentage of the lower explosive limit (LEL) of a flammable vapor. The LEL is measured as a percentage of flammable vapor-in-air by volume. For example, the flammability limits of benzene are 1.7% to 7.1%. Concentrations less than 1.7% of benzene in air are too lean to support combustion, while those concentrations in excess of 7.1% are too rich to burn. The lower explosive limit of benzene is 1.7%, the minimum concentration of benzene that would support combustion. The upper explosive limit (UEL) of benzene in air is 7.1%, the maximum concentration of vapor in air for benzene to support combustion. Any concentrations between the LEL and UEL will allow for combustion.

The CGI will measure the LEL percentage of the vapor, provided that the instrument has been properly calibrated and adjusted.

The meter must be used only for measuring the vapor that it was intended to test. For example, Chemical X is a flammable liquid with a flash point of 20 degrees Fahrenheit and has a flammability range of 5% (LEL) to 10% (UEL). Assume that the CGI has been calibrated for Chemical X, the batteries are fully charged, and the instrument is properly adjusted. If the CGI reads 20% on its meter, the instrument is indicating that the amount of Chemical X's vapor in the air is equal to 20% of its LEL of 5%, so that there is 1% vapor of Chemical X in the air. Once the instrument reaches 100%, we know that the amount of Chemical X's vapor in the air is at least 5%, although it could be any number above 5%. It is important to remember that the ambient temperature must be at or above the flash point of the vapor we are attempting to measure or the material may not emit vapors. When older instruments reached full-scale (100%) deflection and the measured vapor went beyond the UEL, the meter would return to zero. This indicated that the vapor was too rich to burn. Newer instruments have been designed so that the samples containing vapor above the UEL will cause the meter to remain at full-scale deflection.

Since it is impossible to calibrate the CGI for each gas, a responder or site worker will have to select one gas for calibration and realize that any other gas measured will have a certain amount of inaccuracy.

Hexane is frequently the calibration gas of choice because its flammability range is close to that of gasoline. Regulatory input has suggested that 20% LEL readings should be a cause for alarm. Other authors agree that it would be realistic to use such a number to be reasonable and prudent. However, when dealing with such a wide variety of factors in inadvertent or deliberate releases of hazardous substances whose properties may or may not be known, we feel that any movement of the meter on a CGI (excepting mechanical movement caused by jarring the instrument) is sufficient cause to assume a flammable problem.

When flammability is indicated, it is necessary to take appropriate precautionary steps. For instance, when working in an area that is cold, but where you expect the air to warm up as the day goes on, a sample of the suspect materials should be collected and very gently heated while measuring with a CGI. This will give an indication of whether the material will emit flammable vapors as the temperature rises.

Operation of a CGI is usually based on the introduction of the questionable gas into a combustion chamber. A heated filament in the chamber initiates combustion of the gas, which in turn produces heat. The amount of heat is proportional to the amount and the type of fuel combusted. Electronically, the combustion chamber contains a wheatstone bridge. Increased heat on the filament of the wheatstone bridge causes increased resistance and unbalances the bridge. The unbalanced bridge causes increased change in the current flow, which in turn causes an instrument movement reflecting a direct indication of percent LEL. In a solid-state instrument, depending upon the design, the vapor is absorbed and/or dissolved into a solid-state bead. Molecular interaction causes increased conductivity, which is again shown as a direct readout of percent LEL.

Different compounds with identical vapor-in-air concentrations may yield different readings on the CGI because of different LELs and/or flash points. CGI instruments can be purchased for the purpose of measuring just the percent of the LEL or, as with oxygen-availability instruments, can include tests for oxygen availability, carbon monoxide, or other capabilities. The CGI has a "combustibility cell," just as the oxygen meter has an "oxygen cell," but the combustibility cell is normally discarded when it fails.

15.10. CORROSIVITY MONITORING

By definition, a corrosive is any material that can destroy human skin tissue, or any liquid that has a severe corrosive effect on steel. Corrosives include acids and bases. In order to monitor the corrosivity of a chemical, we usually determine the pH (percentage of hydrogen ions). Neutral materials have a pH of seven (water is usually used as the standard, although there are natural waters that may not meet the standard). Liquids with a pH of *less* than seven are considered *acidic*. Liquids having a pH of *more* than seven are considered *basic*.

Using litmus paper or pH test paper is a simple way to determine the approximate pH of a liquid. The color change will indicate whether it is an acid or a base. Portable pH meters are available that run on batteries. These are convenient and accurate, but still need to be calibrated before use, whereas test paper is just taken out of a bottle and compared to a color chart.

Personnel handling corrosives must be very careful to prevent splashing or spilling of the material upon themselves. If corrosive materials come in contact with your body, remove any affected clothing and immediately flush the material away with large amounts of water. Even short exposures to some corrosives may have devastating effects on skin tissue.

15.11. SITE MONITORING

Site monitoring must include the following points during the initial evaluation of the scene. Objects such as containers, barrels, cylinders, placards, bills of lading, and type of tank or railroad car will give you some indication as to what the contents may be.

The condition of the objects at the site— intact, ruptured, in an old waste dump, rusted through and falling apart, etc— is very important. We must first determine the physical condition of the hazard. Is it a liquid, gas, or a solid? What is the color or clearness (turbidity)? What is the chemical doing—is it foaming, boiling, smoking, corroding, or vaporizing? What other factors can you see or smell to help you in determining its physical state?

You have to observe the lay of the land to note any natural wind barriers, since these will have a direct effect on where you have to set your monitoring instruments. This information will also aid you if you have to evacuate an area in a safe direction. A determination has to be made regarding the potential routes of dispersion of the hazards, whether by air, groundwater, surface water, or land surface, or by entering the biological chain through animals in the food cycle. Will this dispersion enter a farmer's alfalfa field and contaminate his crop, which in turn will be fed to beef cattle?

Unusual features should be noted when working an incident or hazardous waste site, for example, dead fish in a stream, dead animals near the Incident Zone, dust or spray in the air, and pools of liquids. If you are working a hazardous waste site, look for deteriorating containers, fissures or cracks in a solid surface exposing deeper areas of waste, foams or oils on liquid surfaces, or even gas generation from the site. All these signs can indicate a greater contamination situation. You cannot determine the specific contaminant until testing is performed, but these clues can be of importance to you. Be alert to these clues, and you might head off a disaster if the contaminant is dispersing.

When preparing to monitor an incident, also ask if containers of any reactive, flammable, highly corrosive, etiological, or radioactive materials need to be specifically identified.

Each of these conditions will require specific monitoring techniques.

15.12. AIR SURVEILLANCE

The most obvious place to determine contamination is in the atmosphere around you. By taking airborne samples, we can apply guidelines of safety to the numbers we acquire. The most commonly used guidelines are the Threshold Limit Values (TLV®) of the American Conference of Governmental Industrial Hygienists (ACGIH), which are the "amount of exposure a worker can be exposed to over an eight-hour workday and a forty-hour workweek." These are the best guidelines we have available since there are no values for responders concerned with a massive or acute chemical exposure at an incident scene. There are three categories of TLVs that can aid us at the incident site: the Time-Weighted Average (TWA), the Short-Term Exposure Limit (STEL), and the Ceiling Limit (C). In an earlier chapter, the meaning of these terms was introduced. These factors are beneficial in that they allow us to determine the level and specific type of protection that is indicated. The value that is legally enforceable by OSHA is the Permissible Exposure Limit (PEL). The permissible exposure limits have been generally based upon TLVs. New PELs are now appearing, but the first ones were based upon 1968 TLVs, so they have inherent problems because technology has changed since that time.

The PEL for a substance is the eight-hour time-weighted average or the ceiling concentration below which a worker may be exposed. Not all chemicals have PELs, ceiling limits, or even time-weighted averages.

What happens when there are many chemicals dumped together, causing a synergistic reaction? Some chemicals will mix without adverse reactions; others are highly incompatible and may set off violent reactions. Not every one of the 575,000 common industrial chemicals has been tested, and it is impossible to test all possible combinations. So when there is an incident, combinations may occur that will not follow the guidelines of the TLVs or PELs.

Another term that you may hear is the *Recommended Exposure Limit* (REL). This is a limit of exposure set by the National Institute of Occupational Safety and Health (NIOSH), although it is only a recommendation and is not enforceable. This information is backed by the medical investigatory resources of NIOSH and is considered quite valid.

15.13. BENEFITS OF MONITORING

Monitoring will not only help in the identification and determination of the various types and quantities of chemicals or hazards at the site, but also will help in determining the potential for eye, skin, or inhalation problems. You will remember from the chapter on anatomy and physiology that the skin, eyes, and lungs are very susceptible to chemical exposure. Monitoring will assist in identifying those chemicals that have been found to cause dermatitis, lung disease, or eye irritation. More serious are those chemicals that can cause cancer, mutations, or birth defects.

15.14. MONITORING CONSIDERATIONS

As we mentioned, other considerations have to be taken into account when monitoring. There may be flammable, combustible, explosive, or radioactive materials at the incident or hazardous waste site. During a response, it is critical that a Material Safety Data Sheet (MSDS) be acquired for that particular chemical. Under SARA Title III of Superfund, all states are responsible for having data on chemicals that are used in or travel through their area. These data may be with the local police, fire department, county emergency services director, or emergency medical response personnel.

The MSDS is a valuable tool in that it can give you extensive information regarding flash point, explosive limits or flammability range, toxicity, first aid procedures, and disposal and environmental considerations. You might want to review Chapter 7 to refresh your memory about MSDSs.

15.15. DIRECT-READING INSTRUMENTS

At an incident site, it is imperative that you have instruments that will give you instantaneous readings, as opposed to those that have to be analyzed by a laboratory, such as air sampling.

Instantaneous instruments had their beginning in industry, which needed instruments that would give an immediate indication when an emergency leak occurred.   If there were an accident, these instruments directly gave the airborne concentration to warn the personnel. In an emergency, this information is critical in determining proper response.

The biggest drawback in the use of direct-reading instruments is their built-in problems. Direct-reading instruments can only detect and/or measure specific chemicals, not a wide range of chemicals.   Also, they are generally designed to measure minute amounts and/or to detect very small amounts of airborne contaminants (typically below 1 ppm).   In detecting one type of chemical, the instrument may give false readings due to interference from other substances.   This poses a problem when you are trying to determine what chemical may be in the vehicle, drum, barrel, or tank car.

Two of the instruments most commonly utilized in a response situation are the *Flame Ionization Detector (FID)* and the *Photoionization Detector (PID)*. These instruments are found in mobile laboratories, generally maintained by federal government or commercial response teams.

15.16. AIR-MONITORING INSTRUMENTS

Air-monitoring instruments usually consist of a battery-driven pump that draws air through a cassette containing a specialized filter. The sample is then taken for a specific time period which depends upon the substance being examined. After the sample is taken, generally over an eight-hour work shift, the cassette containing the filter is set to the laboratory for analysis. At the laboratory the filter is removed and the required analyses are performed on it. This type of sampling method does not give instantaneous readings, which is its major disadvantage when you need immediate results.

The lag time from the time the sample is obtained and sent off to a laboratory until the results come back may be hours, if not days. In very large incidents, a mobile laboratory may be dispatched to the scene to perform immediate chemical analysis; unfortunately, this is a rare luxury and not commonly seen.

Sampling will also depend upon the physical state of the contaminants. For example, some chemicals may exist as a gas, vapor, mist, or particulate mixed or bound to other chemicals. During an emergency it is important and necessary to pick the right method to detect and measure the contaminants.

15.17. MOBILE LABORATORY

Generally, a very large response operation or a hazardous waste site will have a mobile laboratory on site. The advantage to having a laboratory at the incident site or hazardous waste site is being able to obtain immediate analysis. The results can be lifesaving; if samples must be sent off site, then by the time the results are obtained, personnel may have received a serious exposure to the chemical or contaminant. Having the laboratory on site can only improve protection of personnel; with immediate analysis, as conditions change, so do the levels of protection.

It is very important to have an ongoing site or incident monitoring program to determine all changes. Preliminary studies will be done at a hazardous waste site, but this luxury does not exist at an incident site. At a hazardous waste site, you will be able to determine what to monitor for and what type of personal protective equipment to wear.

At both incident and hazardous waste sites, it is necessary to perform perimeter monitoring. At a hazardous waste site, fixed monitors may be placed around the site. These monitors are usually out of the area where personal protection is required and will indicate what is happening to the airborne contaminants. Are they reaching the perimeter of the site? If so, are they passing out of the site? If they are, then remedial action can be taken to correct the situation. During the emergency incident, you may not always be able to place monitors around the site; however, this may be necessary in some situations where you want to see if the contaminants are escaping from the incident site. You may have to wait until an environmental response team arrives with extra equipment to assist you in doing the perimeter monitoring.

15.18. PERIODIC MONITORING

Periodic monitoring should be done when there is a change in work-site conditions. It generally is not performed during an emergency situation. The information obtained from periodic monitoring is important to the hazardous waste site operator in that it will identify new contaminants and changes in the work site that warrant testing to see what new conditions are being encountered. Periodic testing is *not* an ongoing testing program, but is warranted in certain situations.

15.19. PERSONAL MONITORING

There are times that the individual needs to use or wear a personal monitor to determine his or her exposure. Personal monitors are just that: personal or worn by the individual. The advantage of personal monitor to a fixed monitor is that the personal monitors will identify and quantify the actual amount of chemical the wearer has received. The drawback to using personal monitors is that they generally are not instantaneous and have to be sent to a laboratory for analysis.

15.20. TYPES OF PERSONAL MONITORS

Passive dosimeters. Passive dosimeters do not need a power supply. They are usually a badge-type device that is hung on your collar. The badge contains a collection or detection media. Some dosimeters will change color to indicate exposure. These are good for quick readings, although they only give a qualitative reading, not a quantitative reading that tells you how much contaminant is actually present. Measuring and testing of the collection media are performed in a laboratory. Passive dosimeters are the more accurate of the two types of personal monitors.

Colorimetric-type devices. Colorimetric-type devices are direct-reading devices. The contaminant is exposed to a color-changing substance. The system basically consists of two parts: a pump and a colorimetric indicator tube. The tube is a clear glass that contains various types of chemicals that will change color depending upon the type of contaminant to which they are exposed. A certain volume of air or a specified number of strokes of air is drawn through the tube, and the tube is then compared to a color chart or is read directly from a scale on the side of the tube. Both passive dosimeters and colorimetric devices have inherent problems such as accuracy and specificity of the contaminant.

Accuracy is based on how well you read the tube scale. The line of the color change or stain may not be very clear. Shelf life will also affect the accuracy, as will other contaminants (which can lead to a false reading). The tubes have an error factor of up to 20%, although this varies from manufacturer to manufacturer and from chemical to chemical.

Colorimetric tape samplers are another passive monitor in that they are chemically impregnated to react to specific chemicals. The degree of color changes indicates the presence of the chemical being tested for. Some tapes will give a rough quantitative result. For instance, pH tapes can be used to give very accurate determinations of the pH in the contaminant.

15.21. GAS AND VAPOR SAMPLING

Gases and vapors are the two forms of hazard that you will most commonly encounter in the air at a hazardous waste site or emergency incident. Samples can easily be obtained by collecting the air in a closed container, such as a flask, bottle, displacement container, syringe, plastic bag, or similar container. Another type of container is the evacuation-type device, which is a sealed vessel containing a vacuum. There is but one entry point to this type of container, so when the entry point is opened, it draws the sample into the container. Another type is the displacement collector, which has two openings for entry of the gas or vapor.

The evacuation-type container is good for short or "grab" samples. These devices are normally used to detect gross gas concentrations, such as methane, propane, or carbon monoxide.

Another type of device that is used to collect gas and vapor samples is a pump or other air-collection device. Examples are a squeeze bulb, piston pump, diaphragm pump, rotary vane pump, exhauster, or aspirator, just to mention a few. All these devices pull the air into the container.

15.22. GAS DETECTORS

Gas-detecting devices differ from the previously mentioned devices in that they are usually electrical or chemical responding devices that are able to determine if you are dealing with methane, propane, or any other gas that could mix with oxygen and form an explosive mixture. They will also determine what type of gas is present, since some of the gases displace oxygen and are thus simple asphyxiants and can cause death.

Oxygen detectors will be able to determine the amount of oxygen in the air. When working in an oxygen-deficient atmosphere, you are working in an Immediately Dangerous to Life and Health (IDLH) condition, so you will need to monitor the oxygen content very closely. The oxygen analyzer will give you a reading in percent of the oxygen in the air. Normal air contains approximately 20.9% oxygen; anything below 19.5% puts you in danger. OSHA requires wearing an SCBA or using an airhose if the oxygen concentration is below 19.5%.

15.23. WEATHER CONSIDERATIONS

In order to maintain the protection of the health and safety of the workers at an emergency or hazardous waste site, weather conditions must be observed. If the wind shifts and begins to blow the contaminant over the countryside, or if heavy rains begin and cause the contaminant to drain across the surrounding land, what will happen? A problem exists not only with the surrounding environment but also in the immediate area of the worker. Temperature, wind speed, rainfall, and moisture can all affect you in this type of situation. We shall discuss each of these.

15.24. TEMPERATURE

Higher temperature can increase the vapor pressure of most chemicals, increasing the rate of permeation through most protective clothing. Temperature changes can also affect worker's health in that there may be an increased chance of hypothermia (cold) or hyperthermia (heat). Both conditions are directly related to temperature.

15.25. WIND SPEED

Increased wind speed during colder weather increases the hypothermia risk from wind-chill problems. Wind obviously can blow contaminants over the surrounding area. Increased wind can stir up dust during any season; this dust can contain contaminated soil or particulate matter. Dust suppression is a concern during high-wind-speed days. All attempts should be made to contain or keep down airborne contamination.

15.26. RAINFALL

Obviously, increased rainfall can create a runoff problem from the site. Water will have to be contained by berming or diking. However, rainfall can sometimes help by putting a lid on dust, open containers, lagoons, exposed soil, etc., thereby preventing vapors from being emitted into the air. Rainfall can also dilute potentially hazardous substances.

But rain also causes problems. Airborne sampling cannot be performed during the rain. If the sampling began prior to the rain, the pumps would have to be collected quickly to protect them.

15.27. MOISTURE

Moisture can interfere with sampling by saturating the filters; however, moisture can keep down airborne particles, especially fumes and smaller particulate matter.

Moisture can mix with the vapors from the acid or alkaline solutions, forming weak acid or alkaline mixtures. These mixtures can cause damage to the respiratory tract, or if strong enough, to metallic parts with which they come in contact.

15.28. BAROMETRIC PRESSURE

Barometric pressure can affect chemical vapor pressures in a manner similar to temperature since, as the pressure goes down, evaporation can increase.

15.29. WEATHER WATCHING

Designated personnel will have the responsibility to monitor weather conditions. There may be times when weather conditions preclude site operations— for example, increased or decreased temperatures, barometric pressure changes affecting vapor pressures and thereby increasing the possibility of explosions, etc. All of these will have to be monitored in order to safeguard personnel.

Weather conditions can also affect the sampling program. It is necessary to note the barometric pressure when calibrating many testing instruments. Adjustments have to be made for barometric pressure changes.

15.30. NATURAL OR BIOLOGICAL HAZARDS

In certain parts of the country, natural hazards may be encountered. Most natural hazards will be in the plant family, such as poison ivy, sumac, or oak. These are found when you are responding to an incident or site in a heavily wooded or grassy area. The primary dangers are from skin contact or inhalation in smoke if there is a fire.

15.31. CONCLUSION

There are various types of equipment and techniques to monitor the emission of chemicals from a hazardous materials incident or waste site. Work schedules, personal protective equipment, and overall site operations will be dependent upon the results obtained by the monitoring instruments. As a responder or site worker, you should have a knowledge of the instruments, especially personal monitoring equipment, that are being used.

15.32. CHAPTER REVIEW QUESTIONS

1. List three primary responsibilities for incident/site monitoring.

2. Identification and determination of the type and quantity of hazards will allow the responder/worker to do what.

3. List two functions of air monitoring.
   (a)   (b)

4. List four requirements for determining the type of field-monitoring instrument to use:
   (a)   (b)
   (c)   (d)

5. When using a field instrument, the instrument should give you what information?
   (a)   (b)
   (c)   (d)

6. Describe four monitoring procedure guidelines.
   (a)   (b)
   (c)   (d)

7. When determining an unknown hazard, what is the first thing that should be identified?

8. What are two factors that you would need to determine if an atmosphere is hazardous?
   (a)   (b)

9. Describe the benefits of monitoring an incident site.

10. List two types of oxygen-monitoring instruments.
    (a)   (b)

16 RESPIRATORY PROTECTION

16.1. INTRODUCTION

Other than the skin, the most important body system that can receive direct insult or injury from chemicals is the respiratory system. In Chapter 11, when we discussed the need to protect the skin with chemical-resistant suits, emphasis was placed on whole-body protection. In this chapter, we will discuss the various types of respiratory protective equipment. We will look at the various types of respirators; how to choose a respirator based upon the level of respiratory protection required by 29 CFR 1910.134; the nature of qualitative and quantitative fit testing; and how to develop a respiratory protection program.

16.2. TYPES OF RESPIRATORS

Various types of respirators are available on the commercial market. Many have specific functions and may not be the correct respirator to use in the emergency response situation or at a hazardous waste site cleanup.

During the emergency response situation and the hazardous waste site cleanup situation the individual may sometimes encounter an atmosphere that is Immediately Dangerous to Life and Health (IDLH). This IDLH condition will determine the level and type of respiratory protection that is needed.

Choosing the proper respiratory protection is as critical as choosing the correct level of chemical-resistant suit. If your respirator is of the wrong type, you may be running the risk of injury or possibly death. This is why it is important to be able to differentiate the various types of respirators and their individual purposes. Not only choosing the proper respirator but also having it fit correctly is also a critical factor. Without a proper fit, you can expose yourself to unnecessary and possibly fatal exposure.

189

There are three basic types of respirators: air-purifying, atmosphere-supplying, and a combination of an air-supplying and air-purifying respirator. Each of these types have specific limitations as to their usage.

16.3.   AIR-PURIFYING RESPIRATORS

As the name implies, the respirator purifies the air by drawing air in through the respirator cartridge, which will remove the impurities that are in the air. The most important part of this system is the cartridge. If you do not have the proper cartridge for the specific contaminant, you may very well be breathing in the contaminant. You have to know the specific type of contaminant, and you have to match the cartridge to the contaminant or chemical group to which the contaminant belongs, for instance, organic vapors.

There are two types of air-purifying respirators: particulate and gas/vapor. The second is a combination type that will remove particles as well as gases or vapors.

The air-purifying respirator depends upon having an adequate concentration of ambient oxygen in the air to breathe. If the air contains less than the acceptable OSHA concentration of oxygen of 19.5%, then an air purifying respirator is *not* to be used.

During an emergency situation, it is very difficult to determine the oxygen concentration in the air unless you have a testing device that will give you that information, such as an oxygen meter, maybe of the direct-reading or digital type. Never assume that the atmosphere is safe to enter without first checking it. If in doubt, do not wear an air-purifying respirator.

Gas masks are considered air-purifying respirators. The gas mask can be either the commercial or the military type. Both are worn only in an atmosphere that has an acceptable level of oxygen in the air.

Cartridge life is a major concern when using air-purifying respirators. The old gas-mask canisters could not be stored for long periods of time before being used. The newer cartridges have a considerable shelf life. But you have to be careful when you utilize a cartridge respirator, because you have no way of knowing how long it will take for breakthrough to occur. A good safety policy is to change the cartridge frequently, based upon the known concentration of the contaminant in the air. Presently, only one type of cartridge can indicate that breakthrough has occurred, but that cartridge is not yet approved by the National Institute for Occupational Safety and Health (NIOSH).

16.4.   POWERED AIR-PURIFYING RESPIRATORS

Powered air-purifying respirators are units that have a motor-powered fan forcing air up to the user. The ambient air is taken in through an intake, which in some cases may be a part of the filter, and forced up into the mask through a filtration device. The mask can be of the half-face or full-face configuration. Systems may have a very tight-fitting mask, or they may have a loose-fitting mask to allow the air to easily flow out around the edges. Another type of powered air-purifying respirator utilizes a helmet. The helmet contains the filter. The user wears a power supply that attaches to his waist belt. Air is drawn up into the helmet, through the filter, and then flows down over the forehead of the wearer. The helmet has a full-face shield that allows the air to flow out. Some helmets have a loose-fitting collar attached to the face shield. This prevents the entry of contaminants up into the breathing zone of the wearer.

Another type of powered air-purifying respirators utilizes a hood instead of a helmet. The only difference is the safety factor provided by the helmet.

16.5. ATMOSPHERE-SUPPLYING RESPIRATORS

This type of respirator supplies its own atmosphere to the wearer, either in the form of air from pressurized cylinders or in the form of air from a compressor that is away from the area of contamination and that supplies air to the wearer through a hose.

Respirators that are attached to a compressor through a hose limit the wearer in mobility. The wearer can only go as far as the hose allows. OSHA also limits the length of the hose to 300 feet to prevent the wearers from overextending themselves. Another limitation of this type of system is that the quality of the air that is being supplied from the compressor must meet grade D breathing-air standards of the American National Standards Institute (ANSI). The compressor also has to supply a pressure of 125 pounds per square inch (psi) with a maximum output of 4 to 15 cubic feet per minute (cfm). Not all compressors can meet these requirements. Compressors that provide air also have to be periodically checked for carbon monoxide, carbon dioxide, oil mist, and the overall quality of the air.

Supplied air-respirators are never to be used in an IDLH situation or when someone is going into an unknown atmosphere. There are units that have a combination system with an escape self-contained breathing apparatus (SCBA) provision. The SCBA unit in this case is usually good for only a 5- to 30-minute duration, or time enough to quickly leave the area.

Supplied-air respirators have two modes of operation, very similar to the operational modes of the SCBA, which will be discussed in the next section. These are the negative-pressure type with a demand feature, where the wearer has to initiate a breath to receive air from the system; or the positive-pressure, continuous-flow type, where the air is always flowing to the wearer. Another less common variation is the positive-pressure demand type.

16.6. SELF-CONTAINED BREATHING APPARATUS

Respirators that use pressurized cylinders to deliver air to the wearer are usually called self-contained breathing apparatus (SCBA). The SCBA provides the wearer with his own portable supply of respirable air. The unit is independent of the surrounding air. There is no need to worry about breakthrough with an SCBA, as with a cartridge respirator; the only real concern is having enough air in your cylinder to allow you to perform whatever function that you are required or are attempting to do. The SCBA offers comparatively free movement over an almost unlimited area. Most SCBAs, excluding the new ultralight units, are generally unsuitable for strenuous work because of their bulk and weight.

We will be discussing SCBAs in detail, since they are the most highly used units, both by emergency response teams and by those who work hazardous waste sites.

16.7. OPERATIONAL MODES OF SCBA

SCBAs may have two modes of operation that distinguish the type of unit. These are positive- or negative-pressure units. The type of demand that is put on the unit can still be differentiated by demand or continuous-flow units. The unit that is now required under NFPA 1500 regulations for fire-fighting personnel is the positive-pressure, continuous-flow type. This protects the fire fighter from taking in a breath or triggering his SCBA, as would be required with a pressure-demand unit. If he takes in a breath he runs the risk of drawing in contaminants around his face mask. The positive-pressure, continuous flow unit precludes this possibility. Air is always flowing out from the face mask.

16.8. OPERATIONAL LIMITATIONS OF SCBA

SCBAs, by OSHA regulations, can weigh no more than 33 pounds. With new technology, manufacturers have been able to reduce the weight to approximately 20 pounds with lighter-weight cylinders. This reduction in weight has reduced the burden or stress of carrying the unit. With all the equipment that a fire fighter has to carry, it is very important to reduce this weight factor.

The only time that a hazardous waste site worker or incident responder would be concerned with extra weight is in a level A or B situation, where SCBAs are required. The emergency response team will be wearing SCBAs because, during an initial response, the concentration of the toxic materials the team may be entering is unknown. During this initial response, the primary concerns are the protection of life and property. You are protecting your safety by wearing an SCBA; you are able to execute entry, rescue victims, and make a faster initial response. Without an SCBA, you could not make this entry without endangering your own life and possibly the lives of others.

The SCBA has what is referred to as a *rated* duration. For example, this means that a 40-cubic-foot cylinder at 1980 psi is rated for 30 minutes. A 45-cubic-foot cylinder at 2216 psi is rated for 40 minutes. Rating indicates the amount of air that is available when the air in the cylinder is reduced to atmospheric pressure for breathing.

The rated duration is one of the most important limiting factors of the SCBA. The following factors are other limitations of the SCBA:

- physical activity of the wearer
- physical condition of the wearer
- breathing rate of the wearer
- level of training or experience the wearer has had in wearing an SCBA
- pressure in the cylinder
- physical age and condition of the SCBA

When dealing with these limitations, we must consider the actual *operational duration*, which is always less than the rated duration. A false sense of security could arise if you were to think that you could get 30 minutes from a cylinder when, in a very strenuous situation, you may only be able to get 15 minutes from the cylinder.

The operational duration is approximately half or less of the rated duration. You can determine your operational duration by doing very strenuous work while wearing the unit during a training drill. This will give you knowledge that may save your life.

16.9.   WEARING AN SCBA DURING A RESPONSE OR ON A SITE

When making an entry into a hazardous waste site or emergency site, it is essential to make a good entry. When making an entry, the following points are critical:

1. Remain calm, do not panic. If you begin to panic, you can become incapable of functioning mentally and physically. Not only will this state hinder your reasoning, it will increase your breathing rate and use up your air much faster.

2. Use the buddy system or work with a partner. This way you can check each other out by staying together or keeping open communications.
3. It may be necessary to stay low. The air is cooler close to the ground. This is important if there is a fire involved. Visibility is also better at this level if there is smoke involved. If you cannot see where you are going, you should feel ahead for holes or obstructions. It may even be more prudent, if smoke is involved and you cannot see the vehicle or incident, not to try to enter until conditions are better.
4. Radio communications are an important safety factor when using SCBAs. Built-in radios are now available for fully encapsulated units. These radios will allow the wearer to communicate with his partner or his command post. The only limitation is range, but if you are in a fully encapsulated suit, it is doubtful that you will ever exceed the range of the radio unit.

16.10. CONSERVATION OF AIR

When you wear an SCBA, the primary concern is how to conserve air. The air in your tank could be the air needed to make one more rescue or to get you out of a hazardous situation. In order to conserve air, you must learn how to breathe slowly and regularly. Do not skip breaths or take extra shallow breaths. You have to learn how to do this by practice during training sessions while performing strenuous exercise that will make you want to breathe faster. Remember, slow down and think about what you are attempting to do.

Stress, fear, and excitement will all cause you to use up more of your air. You must remain calm and try to relax and think logically about what you are doing or attempting to do. Work can be accomplished at a good steady pace; this will not overexert you, nor will it expend all the air in your tank. Another important consideration is the physical stamina and condition of the SCBA wearer, especially if the person is overweight, out-of-shape, or has a medical condition that will impair his or her use of the unit.

Try to conserve air, but never take any chances to conserve air that may endanger you.

16.11. YOUR AIR RUNS OUT!

Many people have died in situations where their air has run out. This usually has occurred in buildings full of smoke. You have to learn to anticipate when your air is going to run out after the low level alarm has sounded. You have to remain calm and relaxed in order to think logically: do not panic. If you are in a smoke-filled building or area, do one of the following:

- Get close to the floor or ground below the smoke level. Disconnect the breathing tube and hold, then move to safety.
- Disconnect the breathing tube and place it under your clothing. Move to safety on hands and knees, or crawl in the prone position, making sure your body does not crimp the tube. This may not be possible during a hazardous materials or waste site operation, since some of the chemicals may lie in air just above the ground.
- Disconnect the breathing tube and cup the connector in the palm of your hand or place a wet rag over the opening. Stay low and move quickly to safety.

16.12. REGULATOR MALFUNCTION

Being a mechanical device, the regulator can malfunction at almost any time. When you have regulator failure, you must react quickly. Follow these rules:

1. Advise your partner.
2. Open the bypass and regulate flow.
3. Close main line valve.
4. Begin moving to safety.

When you have to bypass the main valve, even with a full cylinder, you have approximately 10 minutes of air in the tank.

16.13. EGRESS

When coming out of a hazardous waste site or hazardous incident area, or even an area of radioactive contamination, always remove your facemask last following decontamination. You have to have your clothing decontaminated to the point where removing your facemask will not endanger you. If you remove your mask too soon, serious injury or death could occur.

If your air supply has run out, remove the breathing tube from the regulator and hold the hose end away from your clothing, and continue to breathe while wearing the facemask. Continue doing this until you have been monitored and told that it is safe for you to remove the mask.

16.14. TRAINING

The importance of training in the proper use and care of the SCBA cannot be overemphasized. *You must train*! You have to learn to wear the unit in all types of situations — climbing ladders, crawling, and removing victims trapped in vehicles. All of these may be encountered during a fire fighter's or a first responder's career. Wear your SCBA during strenuous training, work up a good sweat, and see what time limitations you have with your breathing unit. This type of training will help you in developing the stamina necessary to wear the SCBA!

16.15. MEDICAL EXAMINATIONS

Not everyone can wear an SCBA, for example, people who are afraid of being enclosed — suffer from claustrophobia. Many people have pulmonary problems, whether caused from too much smoking or actual pulmonary disease such as asthma, etc. Some people have cardiac or heart problems. Everyone who is going to be using an SCBA on a response team or who is going to be working a hazardous waste site should have baseline medical examinations to determine if wearing the SCBA will cause them harm.

16.16. RESPIRATORY PROTECTION PROGRAM

Anyone who is going to be working in the various respiratory protective devices will have to be part of a respiratory protection program, as required under 29 CFR 1910.134. The respiratory protection program is for your safety and the safety of your workmates who wear respirators or SCBA.

Two major components of the respiratory protection program are qualitative and quantitative fit testing. During the qualitative fit testing, the mask is placed on the face and the wearer is told to cover the intake ports and to inhale. If the mask fits correctly, a tight seal will form between the mask and the skin until the wearer exhales.

It may be necessary when doing the qualitative fit testing to have available a number of masks from various manufacturers and in various sizes. Not everyone's facial characteristics are the same. Some people may have a large jaw, or nose, or have had a broken nose, etc. You have to adjust the mask to the person. People who have beards or mustaches cannot wear half-faced respirators, and those with beards cannot wear an SCBA. Facial hair will not allow a tight seal around the mask; such a seal is critical when working in an environment where the inhalation of chemicals can be lethal.

The quantitative fit test is conducted in a booth and uses banana oil sprayed around the mask. If the mask is incorrectly positioned or does not fit properly, the wearer will smell the odor of bananas. Some people may not be very honest, if they like the smell of bananas, so it may be advisable to use irritant smoke, which will immediately let you know if the mask is not fitting properly.

In order for a unit or workplace to have an acceptable respiratory protection program, the following minimal acceptable standards must be met:

1. Have a written standard operating procedure for selection and usage of respirators.
2. Select a respirator based on the hazards that will be encountered.
3. Have a program for all personnel who will be wearing respiratory equipment to be trained in the proper use of that particular respirator and the limitations and hazards of wearing a respirator.
4. When possible, assign a unit to everyone wearing a respirator for his or her personal use.
5. Have a regular program of cleaning and disinfecting the respirators. Those issued to one worker should be cleaned after daily usage or more often if necessary. Those used by more than one worker should be thoroughly cleaned and disinfected after each use by using a mild solution of bleach to reduce the possibility of transmitting disease.
6. Store respirators in a clean, convenient, and sanitary location, preferably a cabinet.
7. Initiate a preventive maintenance program to look for worn or damaged parts during cleaning, replacing them as needed. SCBAs should be thoroughly inspected once a month. This information should be recorded, and these records kept available for inspection.
8. Maintain for inspection surveillance of work area conditions and degree of responder or worker exposure or stress.
9. Establish periodic inspections of the program to determine the level of effectiveness and what the program is lacking in certain areas.
10. Require physical examinations for persons using respiratory protective equipment. This medical evaluation will identify dangerous situations that could possibly arise from someone wearing a respirator who has a preexisting medical condition.
11. Allow only approved respiratory protective equipment to be worn. This means that the unit has to have either Mine Safety and Health Administration (MSHA) or NIOSH certification. If the unit does not have a certification, then it may not be used. The technical certification (TC) number on the label will indicate whether or not the unit has been certified.

16.17. CONCLUSION

Respiratory protection is one of the two most important systems that can be used in an emergency or hazardous waste situation. Without the proper respiratory protection, you are opening the body to possible harmful or even deadly insult. You have to know what type of chemical environment you are getting into in order to choose the proper respiratory protective device. When in doubt, use a self-contained breathing apparatus (SCBA). Never take unnecessary chances with the SCBA, learn your equipment, and above all, learn how you interface with the equipment!

16.18. CHAPTER REVIEW QUESTIONS

1. List three types of respiratory protection.
   (a)
   (b)
   (c)

2. OSHA requires that you wear an SCBA when the ambient oxygen concentration drops below what level?

3. Is certification required on respiratory protective equipment? Explain your answer.

4. Describe the difference between an air-purifying and an atmosphere-supplying respirator.

5. What type of SCBA must be used during a hazardous materials response/ incident.

6. List four limiting factors of SCBAs.
   (a)   (b)
   (c)   (d)

7. Describe the operational duration of SCBAs.

8. Can parts/pieces from one respirator be interchanged with parts/pieces from a respirator from another manufacturer? Explain your answer.

9. OSHA (29 CFR 1910.134) requires what type of fit testing be performed if a person is to wear a respirator for more than 30 days during one year?

10. Describe the components of a respiratory protection program as required by OSHA.

11. What agency certifies respiratory protective equipment?

17   CONCLUSION

As stated at the beginning of this book, our purpose has been to present many broad topics of interest to responders and hazardous waste site workers. We have never expected that anyone would become an expert in this complex field simply from use of this volume, but we do hope that what you have learned and what you might find here will make you a safer and better worker.

All of us involved with the handling of hazardous materials must ever be mindful that we are protecting the public health and welfare and the environment. The future quality of this planet depends on our doing a good job.

Throughout this book, you may have found more questions than answers. This is not because we failed to give you the answers — it is because many of the answers are either not known or are very site or incident specific. Every hazardous waste site and every response is different. There are often no pat answers. By presenting to you some of the questions you should be asking of your supervisor and of yourself, we hope that you will be more careful.

If the chapters on anatomy or physiology seemed too difficult, this was intentional. We hope you may become interested in seeking more detailed discussions than are possible in this very general book. We also hope that these chapters will serve as a future reference in those cases when you might want to refresh your memory about some related topic of discussion.

Many of the facts presented here will probably be different within a few months or years. This field is new and constantly developing. Supplement this book with your own notes from other training classes, newspapers, and magazines.

Help your fellow workers to be aware of the hazards of these materials. But, above all, don't let fear cause blind ignorance. Hazardous materials will continue to be with us. We must learn to live with them safely. Become a teacher of others with whom you come in contact.

And, finally, thanks for letting us present this material to you.

Safe working!

REFERENCES

1. Brunacini, A.V., *Fire Command* (Quincy, MA: National Fire Protection Association).
2. Brunacini, A.V., and J. David Beageron, *Workbook for Fire Command* (Quincy, MA: National Fire Protection Association).
3. DeMarr, E., *First Responder Materials* (Phoenix, AZ: Arizona Department of Emergency Services).
4. DeMarr, E., Personal communications (Phoenix, AZ: Arizona Department of Emergency Services).
5. *Emergency Response Guidebook, 1987*, U.S. Department of Transportation, U.S. Government Printing Office (1987).
6. *Fire Service Emergency Management Handbook* (Sterling, VA: International Association of Fire Chiefs Foundation, January 1985).
7. *Fundamentals of Industrial Hygiene*, 2nd Ed. (Chicago, IL: National Safety Council, 1985).
8. *Handbook of Hazardous Waste Regulations, Environmental Incident Response* (Madison, CT: Bureau of Law and Business, Inc.,1985).
9. *Hazardous Materials for the First Responder* (Stillwater, OK: International Fire Service Training Association, Fire Protection Publications, Oklahoma State University).
10. *Hazardous Materials Incident Analysis* (Emmettsburg, MD: National Fire Academy, 1984).
11. *Hazardous Materials — Pesticide Challenge* (Emmettsburg, MD: National Fire Academy, 1986).
12. *Incident Command System* (Stillwater, OK: Fire Protection Publications, Oklahoma State University, 1983).
13. *NFPA Code 704-M* (Quincy, MA: National Fire Protection Association).

14. "NIOSH Bulletin, Hazardous Waste Sites and Hazardous Substance Emergencies", U.S. Department of Health and Human Services, National Institutes of Occupational Safety and Health, U.S. Government Printing Office (December 1982).

15. Noll, G. et al., *Hazardous Materials, Managing the Incident* (Stillwater, OK: Fire Protection Publications, Oklahoma State University, 1988).

16. "Occupational Safety and Health Guidance Manual for Hazardous Waste Site Activities", U.S. Department of Health and Human Services, U.S. Government Printing Office (October 1985).

17. "Personal Protective Equipment of Hazardous Materials Incidents: A Selection Guide", U.S. Department of Health and Human Services, U.S. Government Printing Office (October 1984).

18. "Pocket Guide to Chemical Hazards", U.S. Department of Health and Human Services, National Institutes of Occupational Safety and Health, DHS 85-114 (February 1987).

19. *Site Emergency Response Planning* (Washington DC: Chemical Manufacturers Association, 1986).

20. "The Industrial Environment — Its Evaluation and Control", U.S. Department of Health and Human Services, U.S. Government Printing Office (1973).

21. U.S. Department of Labor, Code of Federal Regulations, Titles 20, 29, 40, and 49, U.S. Government Printing Office (latest edition).

22. U.S. Government, Public Law 98-499.

23. U.S. Government, Public Law 96-570.

24. U.S. Government, Public Law 94-580.

APPENDICES

APPENDIX A
INFORMATION RESOURCES

American Chemical Society, *Chemcyclopedia: A Manual of Chemicals, Volume 8*, Washington, DC, 1984.

American Chemical Society, *Chemical Abstracts*, Washington, DC.

Association of American Railroads, *Hazardous Material Emergency Response Plan for Railroads*, National Transportation Board, Washington, DC, 1985.

Association of American Railroad, Bureau of Explosives, *Emergency Handling of Hazardous Materials in Surface Transportation*, Washington, DC, 1987.

Bruncini, Alan V., *Fire Command*, National Fire Protection Association, One Batterymarch Park, Quincy, MA, 1986.

Bruncini, Alan V. and David Beageron, *Workbook for Fire Command*, Batterymarch Park, Quincy, MA, 1987.

Bureau of Law and Business, *Handbook of Hazardous Waste Regulations, Environmental Incident Response*, 64 Wall Street, Madison, CT, 1985.

Chemical Manufacturers Association, *Community Awareness and Emergency Response Program Handbook*, Washington, DC, 1985.

Chemical Manufacturers Association, *Title III Community Awareness Workbook*, Washington, DC, 1987.

Chemical Manufacturers Association, *Community Guide to Title III*, Washington, DC, 1988.

Chemical Manufacturers Association, *Site Emergency Planning*, Washington, DC, 1986.

Clayton, George D. and Florence E. Clayton, eds., *Patty's Industrial Hygiene and Toxicology, Volumes 2A, 2B, and 2C:Toxicology*, Wiley Publishing, New York, NY, 1981.

Code of Federal Regulations, US Government Printing Office.

DeMarr, Ethel, Coordinator, *First Responder Program*, State of Arizona, Office of Emergency Services, 5636 East McDowell, Phoenix, AZ 85008-3495.

Epstein, Samuel, *The Politics of Cancer*, Scribners, New York, NY, 1985.

Hance, B.J., et al., *Improving Dialogue with Communities: A Risk Communication Manual for Government*, New Jersey Department of Environmental Protection, Trenton, NJ, 08625, 1988.

Hawley, Gessner, G., *Condensed Chemical Dictionary*, Van Nostrand, New York, NY, 10th ed., 1981.

International Agency for Research on Cancer, *Monographs*.

International Association of Fire Chiefs Foundation, *Fire Service Emergency Management Handbook*, 101 East Holly Avenue, Unit 10B, Sterling, VA, 22170, 1985.

International Fire Service Association, *Hazardous Materials For the First Responder*, Oklahoma State University, Stillwater, OK, 1988.

International Toxic Waste Action Network Newsletter (ITWAN), 1436 "U" Street, NW, Washington, DC, 20009; (202)462-1177.

Klassen, Curtis D., et al, eds., *Casarett and Doull's Toxicology: The Basic Science of Poisons*, 3rd ed., Macmillan, New York, NY, 1986.

Lewis, Sanford, *Title III Right-to-Know and Chemical Emergency Planning under the Superfund Amendments of 1986; Programs, Issues and Opportunities for the States*, National Center for Policy Alternatives, 2000 Florida Street, Suite 400, Washington, DC, 20009; (202)387-6030, 1988.

Massachusetts Department of Environmental Quality, *Layperson's Guide to Reading MSDSs*, One Winter Street, Boston, MA 02108; (617)292-5993, 1988.

Meister, Richard T., ed., *Farm Chemical Handbook*, 64th ed., Meister, Willoughby, OH, 1988.

Mellan, Ibert, *Industrial Solvents Handbook*, Noyes, Park Ridge, NJ, 1977.

National Fire Academy, *Hazardous Materials Incident Analysis*, Emmettsburg, MD, 1984.

National Fire Academy, *Hazardous Materials: The Pesticide Challenge*, Emmettsburg, MD, 1986.

National Fire Protection Association, *Fire Protection Guide on Hazardous Materials*, 8th ed., National Fire Protection Association, One Batterymarch Park, Quincy, MA, 1984.

National Fire Protection Association, *National Fire Protection Association Code 704-M*, National Fire Protection Association, One Batterymarch Park, Quincy, MA, 1987.

National Response Team, *Hazardous Materials Emergency Planning Guide*, NRT-1, Government Printing Office, Washington, DC, 1987.

National Response Team, *Technical Guide for Hazards Analysis: Emergency Planning for Extremely Hazardous Substances*, Government Printing Office, Washington, DC, 1987.

National Response Team, *Criteria for Review of Hazardous Materials Emergency Planning Guide*, NRT 1-A, Government Printing Office, Washington, DC, 1988.

National Response Team, *National Oil and Hazardous Substances Pollution Contingency Plan*, Government Printing Office, Washington, DC, 1988.

National Safety Council, *Fundamentals of Industrial Hygiene*, 2nd ed., NSC, Chicago, IL, 1985.

National Transportation Safety Board, *Railroad Yard Safety — Hazardous Materials, Emergency Preparedness: Special Investigation Report*, NTSB, Washington, DC, 1985.

National Wildlife Federation, *Reducing the Risk of Chemical Disaster: A Citizen's Guide to the Federal Emergency Planning and Community Right-to-Know Act*, 1400 16th. Street, NW, Washington, DC, 20036; (202)297-6800, 1988.

Noll, Gregory, et al., *Hazardous Materials, Managing the Incident*, Fire Protection Publications, Oklahoma State University, Stillwater, OK, 1988.

Noyes Publications, *Extremely Hazardous Substances: Superfund Chemical Profiles*, Noyes Publications, Park Ridge, NJ, 1988.

Noyes Publications, *Pesticide Fact Handbook*, Noyes Publications, Park Ridge, NJ, 1988.

Oklahoma State University, *Incident Command System*, Fire Protection Publications, Stillwater, OK, 1983.

OMB Watch, *Using Community Right-to-Know: A Guide to the New Federal Law*, 2001 "O" Street, NW, Washington, DC 20036; (202)659-1711, 1988.

Poje, Gerald and Joel Szkrybalo, *Reducing the Risk of Chemical Disaster: A Citizen's Guide to the Federal Emergency Planning and Community Right-to-Know Act*, National Wildlife Federation, 1400 Sixteenth St., NW, Washington, DC, 20036, 1989.

Regional Response Team, *Region IX — Mainland Oil and Hazardous Substance Pollution Contingency Plan*, Environmental Protection Agency, San Francisco, CA, 1988.

Sax, Irving, N., *Dangerous Properties of Industrial Materials*, 6th ed., Van Nostrand, New York, NY, 1984.

Sittig, Marshall, *Handbook of Toxic and Hazardous Chemicals and Carcinogens*, 2nd ed., Noyes Publishing, Park Ridge, NJ, 1985.

Tobey, Scott and Margaret Semnario, *The Right-to-Know — Making It Work: A Resource Manual*, AFL-CIO, 815 Sixteenth St., NW, Washington, DC, 20006, 1988.

United Nations, Consolidated List of Products whose Consumption and/or Sale have been Banned, Withdrawn, Severely Restricted or Not Approved by Governments, 2nd ed., New York, NY, 1987.

U.S. Agency for Toxic Substances and Disease Registry, *Registry of Toxic Effects of Chemical Substances*, Abstracts from literature on reported toxic effects of approximately 90,000 substances.

U.S. Department of Defense, *An Analysis of Strategic Rail Corridor (STRAC-NET) for National Defense*, Military Traffic Management Command Report RND 76-1, U.S. Army, Washington, DC, 1976.

U.S. Department of Health and Human Services, *The Industrial Environment: Its Evaluation and Control*, U.S. Government Printing Office, Washington, DC, 1973.

U.S. Department of Health and Human Services, *Personal Protective Equipment for Hazardous Materials Incidents: A Selective Guide*, U.S. Government Printing Office, Washington, DC, 1984.

U.S. Department of Transportation, *Lessons Learned: A Report on the Lessons Learned from State and Local Experiences in Accident Prevention and Response Planning for Hazardous Materials Transportation*, U.S. Government Printing Office, Washington, DC, 1985.

U.S. Department of Transportation, *Risk Assessment/Vulnerability Users Manual For Small Communities and Rural Areas*, rvsd. ed., DOT/OST/P-34/043, U.S. Government Printing Office, Washington, DC, 1986.

U.S. Department of Transportation, *Community Teamwork; Working Together to Promote Hazardous Materials Transportation Safety, A Guide for Local Officials*, U.S. Government Printing Office, Washington, DC, 1986.

U.S. Department of Transportation, *1987 Emergency Response Guidebook for Initial Response to Hazardous Materials Incidents*, DOT P 5800.4, U.S. Government Printing Office, Washington, DC, 1987.

U.S. Department of Transportation, *Tables of Hazardous Material, Their Description, Proper Shipping Name, Class, Label, Packing, and Other Requirements*, U.S. Government Printing Office, Washington, DC, 1988.

U.S. Environmental Protection Agency, *Fate of Priority Pollutants in Publicly Owned Water Treatment Plants, Final Report*, 2 vols. EPA 4404-82/303, U.S. Government Printing Office, Washington, DC, 1982.

U.S. Environmental Protection Agency, *Chemicals in Your Community, A Guide to the Community Service and Right-to-Know Act*, U.S. Government Printing Office, Washington, DC, 1988.

U.S. Environmental Protection Agency, *RCRA Information on Hazardous Wastes for Publicly Owned Treatment Works*, U.S. Government Printing Office, Washington, DC, 1985.

U.S. Environmental Protection Agency, *User's Guide Overview for Controlling Accidental Releases of Air Toxins*, U.S. Government Printing Office, Washington, DC, July, 1987.
Volume 1, Prevention and Protection Technologies for Controlling Accidental Releases of Air Toxins: August, 1987.
Volume 9, Control of Accidental Releases of Chlorine: August, 1987.
Volume 11, Control of Accidental Releases of Ammonia: August, 1987.
U.S. Environmental Protection Agency, *Characteristics of Hazardous Waste*, 40 CFR Part 261, Subpart C, rvsd. ed., Federal Register, 1 July 1987, U.S. Government Printing Office, Washington, DC, 1987.
U.S. Environmental Protection Agency, *Chemical Profiles for Extremely Hazardous Substances, 366 Profiles*. Also available on disk and in reprint from Noyes Publications (1988) as *Extremely Hazardous Substances: Superfund Chemical Profiles*, 1988.
U.S. Environmental Protection Agency, *Chemicals on Reporting Rule Database (CORR)* (List of chemicals with corresponding OTS regulations pursuant to TSCA, RCRA, CERCLA, AND SARA), 1988.
U.S. Federal Emergency Management Agency, *Federal Assistance Handbook: Emergency Management, Direction and Control Programs*, CPG 1-3, U.S. Government Printing Office, Washington, DC, 1988
Objectives for Local Emergency Management, CPG 1-5.
Guide for the Development of State and Local EOPs, CPG 1-8.
Guide for the Review of State and Local Emergency Operations Plans, CPG 1-8A.
Hazard Identification, Capability Assessment, and Multi-Year Development Plan for Local Governments, CPG 1-35.
Guidance for Developing State and Local Radiological Emergency Response. Plans and Preparedness for Transportation Accidents, FEMA-REP-5.
U.S. National Institute of Occupational Safety and Health, *Occupational Health Guidelines for Chemical Hazards*, NIOSH Pub. 81-123, 1981.
Supplement I - OHG: NIOSH Publ. 88-118.
Supplement II - OHG: NIOSH Publ. 89-104.
U.S. National Institute of Occupational Safety and Health, *Hazardous Waste Sites and Hazardous Substances Emergencies*, U.S. Government Printing Office, Washington, DC, 1982.
U.S. National Institute of Occupational Safety and Health, *Pocket Guide to Chemical Hazards*, DHHS (NIOSH) 85-114, U.S. Government Printing Office, Washington, DC, 1987.
U.S. National Institute of Occupational Safety and Health et al., *Occupational Safety and Health Guidance Manual for Hazardous Waste Site Activities* rsvd. ed., 1985.

212 FUNDAMENTALS OF HAZARDOUS MATERIALS INCIDENTS

U.S. National Institute of Occupational Safety and Health, *Current Intelligence Bulletins,* Occasional publications on current status of knowledge of and recommendations for hazardous materials in the workplace.

U.S. Nuclear Regulatory Agency and Federal Emergency Management Agency, *Criteria for Preparation and Evaluation of Radiological Emergency Response Plans and Preparedness in Support of Nuclear Power Plants,* U.S. Government Printing Office, Washington, DC, (NUREG 0654/ FEMA-REP-1).

U.S. Occupational Safety and Health Administration (OSHA), *Identification, Classification, and Regulation of Potential Occupational Carcinogens,* 29 CFR 1990.

Weast, Robert G., ed., *CRC Handbook of Chemistry and Physics,* 68th ed.,CRC Press, Boca Raton, FL, 1987.

Weiss, G., *Hazardous Chemicals Data Book,* 2nd ed., Noyes Publications, Park Ridge, NJ, 1988.

Windholz, Martha, *The Merck Manual: An Encyclopedia of Chemicals, Drugs, and Biologicals,* 12th ed., Merck, Rahway, NJ, 1986.

Working Group on Community Right-to-Know, *Hazard Assessment and Plume Mapping Documents for LEPCs,* 218 "D" Street, SE, Washington, DC, 20003; (202) 544-2600.

Electronic Databases:

CAMEO II: Developed by NOAA specifically for EPCRA. Computerized maps model the rate and path of toxic gases under various conditions. Characteristics for more than 2600 chemicals included on the database.

CHRIS/HACS (Chemical Hazards Response Information System/Hazard Assessment Computer System): Developed by the U.S. Coast Guard. HACS is a computerized model of the four CHRIS manuals that contain chemical-specific information.

HIT: (Hazard Information Transmission): Provides a digital transmission of the CHEMTREC emergency chemical report to first responders at the scene of a hazardous materials incident.

HSDB (Hazardous Substances Databank): A part of the National Library of Medicine's (NLM) TOXNET. Provides peer-reviewed information on thousands of chemicals, as well as limited medical treatment protocols. Contact the NLM at (301) 496-1131.

NHMIE: (National Hazardous Materials Information Exchange) Electronic bulletin board that provides information on hazardous materials training courses, planning techniques, events and conferences, emergency response experiences, and lessons learned. Call toll-free (800)752-6367.

Planners with personal computer capabilities can access NHMIE by dialing FTS 972-3275 or (312) 972-3275.

NIOSHTIC: Electronic on-line database of occupational health information developed by the National Institute of Occupational Safety and Health.

OHMTADS (Oil and Hazardous Materials Technical Assistance Data System): A computerized data base developed by EPA containing chemical, biological, and toxicological information about hazardous substances. OSC uses OHMTADS to identify unknown chemicals and to learn how to best handle known chemicals.

APPENDIX B
OSHA OFFICES

U.S. Department of Labor
Occupational Safety and Health Administration
Regional Offices

Region I
(CT*, MA, ME, NH, RI, VT*)
16-18 North Street
1 Dock Square Building
Boston, MA 02109
(617) 223-6710

Region II
(NJ, NY*, Puerto Rico*, Virgin Islands*)
1 Astor Plaza, Room 3445
1515 Broadway
New York, NY 10036
(212) 944-3432

Region III
(DC, DE, MD*, PA, VA*, WV)
Gateway Building, Suite 2100
3535 Market Street
Philadelphia, PA 19104
(215) 596-1201

Region IV
(AL, FL, GA, KY*, MS, NC*,
SC*)
1375 Peachtree Street
Suite 586
Atlanta, GA 30367
(404) 347-3573

Region V
(IL, IN*, MI*, MN*, OH, WI)
230 South Dearborn Street
32nd Floor, Room 3244
Chicago, IL 60604
(312) 353-2220

Region VI
(AR, LA, NM*, OK, TX)
525 Griffin Square
Dallas, TX 75202
(214) 767-4731

Region VII
(IA*, KS, MO, NE)
911 Walnut Street
Kansas City, MO 64106
(816) 374-5861

Region IX
(American Samoa, AZ*, CA*,
Guam, HI*, NV*, Pacific
Trust Territories)
P.O. Box 36017
450 Golden Gate Avenue
San Francisco, CA 94102
(415) 556-7260

Region VIII
(CO, MT, ND, SD, UT*, WY*)
Federal Building, Room 1554
1961 Stout Street
Denver, CO 80294
(303) 844-3061

Region X
(AK*, ID, OR*, WA*)
Federal Office Building
Room 6003
909 First Avenue
Seattle, WA 98174
(206) 442-5930

* These states and territories operate their own OSHA-approved job safety
and health programs (except Connecticut and New York, whose plans cover
public employees only).

APPENDIX C
RCRA "D" HAZARDOUS WASTES (WASTES LISTED ACCORDING TO CHARACTERISTICS)

TABLE I—MAXIMUM CONCENTRATION OF CONTAMINANTS FOR CHARACTERISTIC OF EP TOXICITY

EPA hazardous waste number	Contaminant	Maximum concentration (milligrams per liter)
D004	Arsenic	5.0
D005	Barium	100.0
D006	Cadmium	1.0
D007	Chromium	5.0
D008	Lead	5.0
D009	Mercury	0.2
D010	Selenium	1.0
D011	Silver	5.0
D012	Endrin (1,2,3,4,10,10-hexachloro-1,7-epoxy-1,4,4a,5,6,7,8,8a-octahydro-1,4-endo, endo-5,8-dimethano-naphthalene.	0.02
D013	Lindane (1,2,3,4,5,6-hexa- chlorocyclohexane, gamma isomer.	0.4
D014	Methoxychlor (1,1,1-Trichloro-2,2-bis [p-methoxyphenyl]ethane).	10.0
D015	Toxaphene ($C_{10}H_{10}Cl_8$, Technical chlorinated camphene, 67–69 percent chlorine).	0.5
D016	2,4-D, (2,4-Dichlorophenoxyacetic acid).	10.0
D017	2,4,5-TP Silvex (2,4,5-Trichlorophenoxypropionic acid).	1.0

APPENDIX D
RCRA "F" HAZARDOUS WASTES (WASTES LISTED ACCORDING TO SOURCE)

Industry and EPA hazardous waste No.	Hazardous waste	Hazard code
Generic: F001	The following spent halogenated solvents used in degreasing: Tetrachloroethylene, trichloroethylene, methylene chloride, 1,1,1-trichloroethane, carbon tetrachloride, and chlorinated fluorocarbons; all spent solvent mixtures/blends used in degreasing containing, before use, a total of ten percent or more (by volume) of one or more of the above halogenated solvents or those solvents listed in F002, F004, and F005; and still bottoms from the recovery of these spent solvents and spent solvent mixtures.	(T)
F002	The following spent halogenated solvents: Tetrachloroethylene, methylene chloride, trichloroethylene, 1,1,1-trichloroethane, chlorobenzene, 1,1,2-trichloro-1,2,2-trifluoroethane, ortho-dichlorobenzene, trichlorofluoromethane, and 1,1,2-trichloroethane; all spent solvent mixtures/blends containing, before use, a total of ten percent or more (by volume) of one or more of the above halogenated solvents or those listed in F001, F004, or F005; and still bottoms from the recovery of these spent solvents and spent solvent mixtures.	(T)
F003	The following spent non-halogenated solvents: Xylene, acetone, ethyl acetate, ethyl benzene, ethyl ether, methyl isobutyl ketone, n-butyl alcohol, cyclohexanone, and methanol; all spent solvent mixtures/blends containing, before use, only the above spent non-halogenated solvents; and all spent solvent mixtures/blends containing, before use, one or more of the above non-halogenated solvents, and, a total of ten percent or more (by volume) of one or more of those solvents listed in F001, F002, F004, and F005; and still bottoms from the recovery of these spent solvents and spent solvent mixtures.	(I)*
F004	The following spent non-halogenated solvents: Cresols and cresylic acid, and nitrobenzene; all spent solvent mixtures/blends containing, before use, a total of ten percent or more (by volume) of one or more of the above non-halogenated solvents or those solvents listed in F001, F002, and F005; and still bottoms from the recovery of these spent solvents and spent solvent mixtures.	(T)
F005	The following spent non-halogenated solvents: Toluene, methyl ethyl ketone, carbon disulfide, isobutanol, pyridine, benzene, 2-ethoxyethanol, and 2-nitropropane; all spent solvent mixtures/blends containing, before use, a total of ten percent or more (by volume) of one or more of the above non-halogenated solvents or those solvents listed in F001, F002, or F004; and still bottoms from the recovery of these spent solvents and spent solvent mixtures.	(I,T)
F006	Wastewater treatment sludges from electroplating operations except from the following processes: (1) Sulfuric acid anodizing of aluminum; (2) tin plating on carbon steel; (3) zinc plating (segregated basis) on carbon steel; (4) aluminum or zinc-aluminum plating on carbon steel; (5) cleaning/stripping associated with tin, zinc and aluminum plating on carbon steel; and (6) chemical etching and milling of aluminum.	(T)

Industry and EPA hazardous waste No.	Hazardous waste	Hazard code
F019	Wastewater treatment sludges from the chemical conversion coating of aluminum	(T)
F007	Spent cyanide plating bath solutions from electroplating operations	(R, T)
F008	Plating bath residues from the bottom of plating baths from electroplating operations where cyanides are used in the process.	(R, T)
F009	Spent stripping and cleaning bath solutions from electroplating operations where cyanides are used in the process.	(R, T)
F010	Quenching bath residues from oil baths from metal heat treating operations where cyanides are used in the process.	(R, T)
F011	Spent cyanide solutions from salt bath pot cleaning from metal heat treating operations.	(R, T)
F012	Quenching waste water treatment sludges from metal heat treating operations where cyanides are used in the process.	(T)
F024	Wastes, including but not limited to, distillation residues, heavy ends, tars, and reactor clean-out wastes from the production of chlorinated aliphatic hydrocarbons, having carbon content from one to five, utilizing free radical catalyzed processes. [This listing does not include light ends, spent filters and filter aids, spent dessicants, wastewater, wastewater treatment sludges, spent catalysts, and wastes listed in § 261.32.].	(T)
F020	Wastes (except wastewater and spent carbon from hydrogen chloride purification) from the production or manufacturing use (as a reactant, chemical intermediate, or component in a formulating process) of tri- or tetrachlorophenol, or of intermediates used to produce their pesticide derivatives. (This listing does not include wastes from the production of Hexachlorophene from highly purified 2,4,5-trichlorophenol.).	(H)
F021	Wastes (except wastewater and spent carbon from hydrogen chloride purification) from the production or manufacturing use (as a reactant, chemical intermediate, or component in a formulating process) of pentachlorophenol, or of intermediates used to produce its derivatives.	(H)
F022	Wastes (except wastewater and spent carbon from hydrogen chloride purification) from the manufacturing use (as a reactant, chemical intermediate, or component in a formulating process) of tetra-, penta-, or hexachlorobenzenes under alkaline conditions.	(H)
F023	Wastes (except wastewater and spent carbon from hydrogen chloride purification) from the production of materials on equipment previously used for the production or manufacturing use (as a reactant, chemical intermediate, or component in a formulating process) of tri- and tetrachlorophenols. (This listing does not include wastes from equipment used only for the production or use of Hexachlorophene from highly purified 2,4,5-trichlorophenol.).	(H)
F026	Wastes (except wastewater and spent carbon from hydrogen chloride purification) from the production of materials on equipment previously used for the manufacturing use (as a reactant, chemical intermediate, or component in a formulating process) of tetra-, penta-, or hexachlorobenzene under alkaline conditions.	(H)
F027	Discarded unused formulations containing tri-, tetra-, or pentachlorophenol or discarded unused formulations containing compounds derived from these chlorophenols. (This listing does not include formulations containing Hexachlorophene sythesized from prepurified 2,4,5-trichlorophenol as the sole component.).	(H)
F028	Residues resulting from the incineration or thermal treatment of soil contaminated with EPA Hazardous Waste Nos. F020, F021, F022, F023, F026, and F027.	(T)

*(I,T) should be used to specify mixtures containing ignitable and toxic constituents.

APPENDIX E
RCRA "P" HAZARDOUS WASTES
(EXTREMELY TOXIC WASTES)

Haz-ardous waste No.	Chemical abstracts No.	Substance
P023	107-20-0	Acetaldehyde, chloro-
P002	591-08-2	Acetamide, N-(aminothioxomethyl)-
P057	640-19-7	Acetamide, 2-fluoro-
P058	62-74-8	Acetic acid, fluoro-, sodium salt
P002	591-08-2	1-Acetyl-2-thiourea
P003	107-02-8	Acrolein
P070	116-06-3	Aldicarb
P004	309-00-2	Aldrin
P005	107-18-6	Allyl alcohol
P006	20859-73-8	Aluminum phosphide (R,T)
P007	2763-96-4	5-(Aminomethyl)-3-isoxazolol
P008	504-24-5	4-Aminopyridine
P009	131-74-8	Ammonium picrate (R)
P119	7803-55-6	Ammonium vanadate
P099	506-61-6	Argentate(1-), bis(cyano-C)-, potassium
P010	7778-39-4	Arsenic acid H_3AsO_4
P012	1327-53-3	Arsenic oxide As_2O_3
P011	1303-28-2	Arsenic oxide As_2O_5
P011	1303-28-2	Arsenic pentoxide
P012	1327-53-3	Arsenic trioxide
⌐038	692-42-2	Arsine, diethyl-
P036	696-28-6	Arsonous dichloride, phenyl-
P054	151-56-4	Aziridine
P067	75-55-8	Aziridine, 2-methyl-
P013	542-62-1	Barium cyanide
P024	106-47-8	Benzenamine, 4-chloro-
P077	100-01-6	Benzenamine, 4-nitro-
P028	100-44-7	Benzene, (chloromethyl)-
P042	51-43-4	1,2-Benzenediol, 4-[1-hydroxy-2-(methylamino)ethyl]-, (R)-
P046	122-09-8	Benzeneethanamine, alpha,alpha-dimethyl-
P014	108-98-5	Benzenethiol
P001	[1] 81-81-2	2H-1-Benzopyran-2-one, 4-hydroxy-3-(3-oxo-1-phenylbutyl)-, & salts, when present at concentrations greater than 0.3%
P028	100-44-7	Benzyl chloride
P015	7440-41-7	Beryllium
P017	598-31-2	Bromoacetone
P018	357-57-3	Brucine

221

Haz-ardous waste No.	Chemical abstracts No.	Substance
P045	39196-18-4	2-Butanone, 3,3-dimethyl-1-(methylthio)-, O-[methylamino)carbonyl] oxime
P021	592-01-8	Calcium cyanide
P021	592-01-8	Calcium cyanide Ca(CN)₂
P022	75-15-0	Carbon disulfide
P095	75-44-5	Carbonic dichloride
P023	107-20-0	Chloroacetaldehyde
P024	106-47-8	p-Chloroaniline
P026	5344-82-1	1-(o-Chlorophenyl)thiourea
P027	542-76-7	3-Chloropropionitrile
P029	544-92-3	Copper cyanide
P029	544-92-3	Copper cyanide Cu(CN)
P030	....	Cyanides (soluble cyanide salts), not otherwise specified
P031	460-19-5	Cyanogen
P033	506-77-4	Cyanogen chloride
P033	506-77-4	Cyanogen chloride (CN)Cl
P034	131-89-5	2-Cyclohexyl-4,6-dinitrophenol
P016	542-88-1	Dichloromethyl ether
P036	696-28-6	Dichlorophenylarsine
P037	60-57-1	Dieldrin
P038	692-42-2	Diethylarsine
P041	311-45-5	Diethyl-p-nitrophenyl phosphate
P040	297-97-2	O,O-Diethyl O-pyrazinyl phosphorothioate
P043	55-91-4	Diisopropylfluorophosphate (DFP)
P004	309-00-2	1,4,5,8-Dimethanonaphthalene, 1,2,3,4,10,10-hexa- chloro-1,4,4a,5,8,8a,-hexahydro-, (1alpha,4alpha,4abeta,5alpha,8alpha,8abeta)-
P060	465-73-6	1,4,5,8-Dimethanonaphthalene, 1,2,3,4,10,10-hexa- chloro-1,4,4a,5,8,8a-hexahydro-, (1alpha,4alpha,4abeta,5beta,8beta,8abeta)-
P037	60-57-1	2,7:3,6-Dimethanonaphth[2,3-b]oxirene, 3,4,5,6,9,9-hexachloro-1a,2,2a,3,6,6a,7,7a-octahydro-, (1aalpha,2beta,2aalpha,3beta,6beta,6aalpha,7beta, 7aalpha)-
P051	¹ 72-20-8	2,7:3,6-Dimethanonaphth [2,3-b]oxirene, 3,4,5,6,9,9-hexachloro-1a,2,2a,3alpha,6alpha,6abeta,7beta, 7aalpha)-, & metabolites
P044	60-51-5	Dimethoate
P046	122-09-8	alpha,alpha-Dimethylphenethylamine
P047	¹ 534-52-1	4,6-Dinitro-o-cresol, & salts
P048	51-28-5	2,4-Dinitrophenol
P020	88-85-7	Dinoseb
P085	152-16-9	Diphosphoramide, octamethyl-
P111	107-49-3	Diphosphoric acid, tetraethyl ester
P039	298-04-4	Disulfoton
P049	541-53-7	Dithiobiuret
P050	115-29-7	Endosulfan
P088	145-73-3	Endothall
P051	72-20-8	Endrin
P051	72-20-8	Endrin, & metabolites
P042	51-43-4	Epinephrine
P031	460-19-5	Ethanedinitrile
P066	16752-77-5	Ethanimidothioic acid, N-[[(methylamino)carbonyl]oxy]-, methyl ester
P101	107-12-0	Ethyl cyanide
P054	151-56-4	Ethyleneimine
P097	52-85-7	Famphur
P056	7782-41-4	Fluorine
P057	640-19-7	Fluoroacetamide
P058	62-74-8	Fluoroacetic acid, sodium salt
P065	628-86-4	Fulminic acid, mercury(2+) salt (R,T)
P059	76-44-8	Heptachlor
P062	757-58-4	Hexaethyl tetraphosphate
P116	79-19-6	Hydrazinecarbothioamide
P068	60-34-4	Hydrazine, methyl-
P063	74-90-8	Hydrocyanic acid
P063	74-90-8	Hydrogen cyanide
P096	7803-51-2	Hydrogen phosphide
P060	465-73-6	Isodrin

Hazardous waste No.	Chemical abstracts No.	Substance
P007	2763-96-4	3(2H)-Isoxazolone, 5-(aminomethyl)-
P092	62-38-4	Mercury, (acetato-O)phenyl-
P065	628 86-4	Mercury fulminate (R,T)
P082	62-75-9	Methanamine, N-methyl-N-nitroso-
P064	624-83-9	Methane, isocyanato-
P016	542-88-1	Methane, oxybis[chloro-
P112	509-14-8	Methane, tetranitro- (R)
P118	75-70-7	Methanethiol, trichloro-
P050	115-29-7	6,9-Methano-2,4,3-benzodioxathiepin, 6,7,8,9,10,10-hexachloro-1,5,5a,6,9,9a-hexahydro-, 3-oxide
P059	76-44-8	4,7-Methano-1H-indene, 1,4,5,6,7,8,8-heptachloro-3a,4,7,7a-tetrahydro-
P066	16752-77-5	Methomyl
P068	60-34-4	Methyl hydrazine
P064	624-83-9	Methyl isocyanate
P069	75-86-5	2-Methyllactonitrile
P071	298-00-0	Methyl parathion
P072	86-88-4	alpha-Naphthylthiourea
P073	13463-39-3	Nickel carbonyl
P073	13463-39-3	Nickel carbonyl Ni(CO)₄, (T-4)-
P074	557-19-7	Nickel cyanide
P074	557-19-7	Nickel cynaide Ni(CN)₂
P075	¹ 54-11-5	Nicotine, & salts
P076	10102-43-9	Nitric oxide
P077	100-01-6	p-Nitroaniline
P078	10102-44-0	Nitrogen dioxide
P076	10102-43-9	Nitrogen oxide NO
P078	10102-44-0	Nitrogen oxide NO₂
P081	55-63-0	Nitroglycerine (R)
P082	62-75-9	N-Nitrosodimethylamine
P084	4549-40-0	N-Nitrosomethylvinylamine
P085	152-16-9	Octamethylpyrophosphoramide
P087	20816-12-0	Osmium oxide OsO₄, (T-4)-
P087	20816-12-0	Osmium tetroxide
P088	145-73-3	7-Oxabicyclo[2.2.1]heptane-2,3-dicarboxylic acid
P089	56-38-2	Parathion
P034	131-89-5	Phenol, 2-cyclohexyl-4,6-dinitro-
P048	51-28-5	Phenol, 2,4-dinitro-
P047	¹ 534-52-1	Phenol, 2-methyl-4,6-dinitro-, & salts
P020	88-85-7	Phenol, 2-(1-methylpropyl)-4,6-dinitro-
P009	131-74-8	Phenol, 2,4,6-trinitro-, ammonium salt (R)
P092	62-38-4	Phenylmercury acetate
P093	103-85-5	Phenylthiourea
P094	298-02-2	Phorate
P095	75-44-5	Phosgene
P096	7803-51-2	Phosphine
P041	311-45-5	Phosphoric acid, diethyl 4-nitrophenyl ester
P039	298-04-4	Phosphorodithioic acid, O,O-diethyl S-[2-(ethylthio)ethyl] ester
P094	298-02-2	Phosphorodithioic acid, O,O-diethyl S-[(ethylthio)methyl] ester
P044	60-51-5	Phosphorodithioic acid, O,O-dimethyl S-[2-(methylamino)-2-oxoethyl] ester
P043	55-91-4	Phosphorofluoridic acid, bis(1-methylethyl) ester
P089	56-38-2	Phosphorothioic acid, O,O-diethyl O-(4-nitrophenyl) ester
P040	297-97-2	Phosphorothioic acid, O,O-diethyl O-pyrazinyl ester
P097	52-85-7	Phosphorothioic acid, O-[4-[(dimethylamino)sulfonyl]phenyl] O,O-dimethyl ester
P071	298-00-0	Phosphorothioic acid, O,O,-dimethyl O-(4-nitrophenyl) ester
P110	78-00-2	Plumbane, tetraethyl-
P098	151-50-8	Potassium cyanide
P098	151-50-8	Potassium cyanide K(CN)
P099	506-61-6	Potassium silver cyanide
P070	116-06-3	Propanal, 2-methyl-2-(methylthio)-, O-[(methylamino)carbonyl]oxime
P101	107-12-0	Propanenitrile

Hazardous waste No.	Chemical abstracts No.	Substance
P027	542–76–7	Propanenitrile, 3-chloro-
P069	75–86–5	Propanenitrile, 2-hydroxy-2-methyl-
P081	55–63–0	1,2,3-Propanetriol, trinitrate (R)
P017	598–31–2	2-Propanone, 1-bromo-
P102	107–19–7	Propargyl alcohol
P003	107–02–8	2-Propénal
P005	107–18–6	2-Propen-1-ol
P067	75–55–8	1,2-Propylenimine
P102	107–19–7	2-Propyn-1-ol
P008	504–24–5	4-Pyridinamine
P075	¹ 54–11–5	Pyridine, 3-(1-methyl-2-pyrrolidinyl)-, (S)-, & salts
P114	12039–52–0	Selenious acid, dithallium(1+) salt
P103	630–10–4	Selenourea
P104	506–64–9	Silver cyanide
P104	506–64–9	Silver cyanide Ag(CN)
P105	26628–22–8	Sodium azide
P106	143–33–9	Sodium cyanide
P106	143–33–9	Sodium cyanide Na(CN)
P107	1314–96–1	Strontium sulfide SrS
P108	¹ 57–24–9	Strychnidin-10-one, & salts
P018	357–57–3	Strychnidin-10-one, 2,3-dimethoxy-
P108	¹ 57–24–9	Strychnine, & salts
P115	7446–18–6	Sulfuric acid, dithallium(1+) salt
P109	3689–24–5	Tetraethyldithiopyrophosphate
P110	78–00–2	Tetraethyl lead
P111	107–49–3	Tetraethyl pyrophosphate
P112	509–14–8	Tetranitromethane (R)
P062	757–58–4	Tetraphosphoric acid, hexaethyl ester
P113	1314–32–5	Thallic oxide
P113	1314–32–5	Thallium oxide Tl_2O_3
P114	12039–52–0	Thallium(I) selenite
P115	7446–18–6	Thallium(I) sulfate
P109	3689–24–5	Thiodiphosphoric acid, tetraethyl ester
P045	39196–18–4	Thiofanox
P049	541–53–7	Thioimidodicarbonic diamide $[(H_2N)C(S)]_2NH$
P014	108–98–5	Thiophenol
P116	79–19–6	Thiosemicarbazide
P026	5344–82–1	Thiourea, (2-chlorophenyl)-
P072	86–88–4	Thiourea, 1-naphthalenyl-
P093	103–85–5	Thiourea, phenyl-
P123	8001–35–2	Toxaphene
P118	75–70–7	Trichloromethanethiol
P119	7803–55–6	Vanadic acid, ammonium salt
P120	1314–62–1	Vanadium oxide V_2O_5
P120	1314–62–1	Vanadium pentoxide
P084	4549–40–0	Vinylamine, N-methyl-N-nitroso-
P001	¹ 81–81–2	Warfarin, & salts, when present at concentrations greater than 0.3%
P121	557–21–1	Zinc cyanide
P121	557–21–1	Zinc cyanide $Zn(CN)_2$
P122	1314–84–7	Zinc phosphide Zn_3P_2, when present at concentrations greater than 10% (R,T)

¹ CAS Number given for parent compound only.

APPENDIX F
RCRA "U" HAZARDOUS WASTES
(TOXIC WASTES)

Hazardous waste No.	Chemical abstracts No.	Substance
U001	75–07–0	Acetaldehyde (I)
U034	75–87–6	Acetaldehyde, trichloro-
U187	62–44–2	Acetamide, N-(4-ethoxyphenyl)-
U005	53–96–3	Acetamide, N-9H-fluoren-2-yl-
U240	¹ 94–75–7	Acetic acid, (2,4-dichlorophenoxy)-, salts & esters
U112	141–78–6	Acetic acid ethyl ester (I)
U144	301–04–2	Acetic acid, lead(2+) salt
U214	563–68–8	Acetic acid, thallium(1+) salt
see F027	93–76–،	Acetic acid, (2,4,5-trichlorophenoxy)-
U002	67–64–1	Acetone (I)
U003	75–05–8	Acetonitrile (I,T)
U004	98–86–2	Acetophenone
U005	53–96–3	2-Acetylaminofluorene
U006	75–36–5	Acetyl chloride (C,R,T)
U007	79–06–1	Acrylamide
U008	79–10–7	Acrylic acid (I)
U009	107–13–1	Acrylonitrile
U011	61–82–5	Amitrole
U012	62–53–3	Aniline (I,T)
U136	75–60–5	Arsinic acid, dimethyl-
U014	492–80–8	Auramine
U015	115–02–6	Azaserine
U010	50–07–7	Azirino[2',3':3,4]pyrrolo[1,2-a]indole-4,7-dione, 6-amino-8-[[(aminocarbonyl)oxy]methyl]-1,1a,2,8,8a,8b-hexahydro-8a-methoxy-5-methyl-, [1aS-(1aalpha, 8beta,8aalpha,8balpha)]-
U157	56–49–5	Benz[j]aceanthrylene, 1,2-dihydro-3-methyl-
U016	225–51–4	Benz[c]acridine
U017	98–87–3	Benzal chloride
U192	23950–58–5	Benzamide, 3,5-dichloro-N-(1,1-dimethyl-2-propynyl)-
U018	56–55–3	Benz[a]anthracene
U094	57–97–6	Benz[a]anthracene, 7,12-dimethyl-
U012	62–53–3	Benzenamine (I,T)
U014	492–80–8	Benzenamine, 4,4'-carbonimidoylbis[N,N-dimethyl-
U049	3165–93–3	Benzenamine, 4-chloro-2-methyl-, hydrochloride
U093	60–11–7	Benzenamine, N,N-dimethyl-4-(phenylazo)-
U328	95–53–4	Benzenamine, 2-methyl-
U353	106–49–0	Benzenamine, 4-methyl-

Haz-ardous waste No.	Chemical abstracts No.	Substance
U158	101–14–4	Benzenamine, 4,4'-methylenebis[2-chloro-
U222	636–21–5	Benzenamine, 2-methyl-, hydrochloride
U181	99–55–8	Benzenamine, 2-methyl-5-nitro-
U019	71–43–2	Benzene (I,T)
U038	510–15–6	Benzeneacetic acid, 4-chloro-alpha-(4-chlorophenyl)-alpha-hydroxy-, ethyl ester
U030	101–55–3	Benzene, 1-bromo-4-phenoxy-
U035	305–03–3	Benzenebutanoic acid, 4-[bis(2-chloroethyl)amino]-
U037	108–90–7	Benzene, chloro-
U221	25376–45–8	Benzenediamine, ar-methyl-
U028	117–81–7	1,2-Benzenedicarboxylic acid, bis(2-ethylhexyl) ester
U069	84–74–2	1,2-Benzenedicarboxylic acid, dibutyl ester
U088	84–66–2	1,2-Benzenedicarboxylic acid, diethyl ester
U102	131–11–3	1,2-Benzenedicarboxylic acid, dimethyl ester
U107	117–84–0	1,2-Benzenedicarboxylic acid, dioctyl ester
U070	95–50–1	Benzene, 1,2-dichloro-
U071	541–73–1	Benzene, 1,3-dichloro-
U072	106–46–7	Benzene, 1,4-dichloro-
U060*	72–54–8	Benzene, 1,1'-(2,2-dichloroethylidene)bis[4-chloro-
U017	98–87–3	Benzene, (dichloromethyl)-
U223	26471–62–5	Benzene, 1,3-diisocyanatomethyl- (R,T)
U239	1330–20–7	Benzene, dimethyl- (I,T)
U201	108–46–3	1,3-Benzenediol
U127	118–74–1	Benzene, hexachloro-
U056	110–82–7	Benzene, hexahydro- (I)
U220	108–88–3	Benzene, methyl-
U105	121–14–2	Benzene, 1-methyl-2,4-dinitro-
U106	606–20–2	Benzene, 2-methyl-1,3-dinitro-
U055	98–82–8	Benzene, (1-methylethyl)- (I)
U169	98–95–3	Benzene, nitro-
U183	608–93–5	Benzene, pentachloro-
U185	82–68–8	Benzene, pentachloronitro-
U020	98–09–9	Benzenesulfonic acid chloride (C,R)
U020	98–09–9	Benzenesulfonyl chloride (C,R)
U207	95–94–3	Benzene, 1,2,4,5-tetrachloro-
U061	50–29–3	Benzene, 1,1'-(2,2,2-trichloroethylidene)bis[4-chloro-
U247	72–43–5	Benzene, 1,1'-(2,2,2-trichloroethylidene)bis[4- methoxy-
U023	98–07–7	Benzene, (trichloromethyl)-
U234	99–35–4	Benzene, 1,3,5-trinitro-
U021	92–87–5	Benzidine
U202	¹81–07–2	1,2-Benzisothiazol-3(2H)-one, 1,1-dioxide, & salts
U203	94–59–7	1,3-Benzodioxole, 5-(2-propenyl)-
U141	120–58–1	1,3-Benzodioxole, 5-(1-propenyl)-
U090	94–58–6	1,3-Benzodioxole, 5-propyl-
U064	189–55–9	Benzo[rst]pentaphene
U248	¹81–81–2	2H-1-Benzopyran-2-one, 4-hydroxy-3-(3-oxo-1-phenyl-butyl)-, & salts, when present at concentrations of 0.3% or less
U022	50–32–8	Benzo[a]pyrene
U197	106–51–4	p-Benzoquinone
U023	98–07–7	Benzotrichloride (C,R,T)
U085	1464–53–5	2,2'-Bioxirane
U021	92–87–5	[1,1'-Biphenyl]-4,4'-diamine
U073	91–94–1	[1,1'-Biphenyl]-4,4'-diamine, 3,3'-dichloro-
U091	119–90–4	[1,1'-Biphenyl]-4,4'-diamine, 3,3'-dimethoxy-
U095	119–93–7	[1,1'-Biphenyl]-4,4'-diamine, 3,3'-dimethyl-
U225	75–25–2	Bromoform
U030	101–55–3	4-Bromophenyl phenyl ether
U128	87–68–3	1,3-Butadiene, 1,1,2,3,4,4-hexachloro-
U172	924–16–3	1-Butanamine, N-butyl-N-nitroso-
U031	71–36–3	1-Butanol (I)
U159	78–93–3	2-Butanone (I,T)
U160	1338–23–4	2-Butanone, peroxide (R,T)
U053	4170–30–3	2-Butenal
U074	764–41–0	2-Butene, 1,4-dichloro- (I,T)

Haz-ardous waste No.	Chemical abstracts No.	Substance
U143	303-34-4	2-Butenoic acid, 2-methyl-, 7-[[2,3-dihydroxy-2-(1-methoxyethyl)-3-methyl-1-oxobutoxy]methyl]-2,3,5,7a-tetrahydro-1H-pyrrolizin-1-yl ester, [1S-[1alpha(Z),7(2S*,3R*),7aalpha]]-
U031	71-36-3	n-Butyl alcohol (I)
U136	75-60-5	Cacodylic acid
U032	13765-19-0	Calcium chromate
U238	51-79-6	Carbamic acid, ethyl ester
U178	615-53-2	Carbamic acid, methylnitroso-, ethyl ester
U097	79-44-7	Carbamic chloride, dimethyl-
U114	¹ 111-54-6	Carbamodithioic acid, 1,2-ethanediylbis-, salts & esters
U062	2303-16-4	Carbamothioic acid, bis(1-methylethyl)-, S-(2,3-dichloro-2-propenyl) ester
U215	6533-73-9	Carbonic acid, dithallium(1+) salt
U033	353-50-4	Carbonic difluoride
U156	79-22-1	Carbonochloridic acid, methyl ester (I,T)
U033	353-50-4	Carbon oxyfluoride (R,T)
U211	56-23-5	Carbon tetrachloride
U034	75-87-6	Chloral
U035	305-03-3	Chlorambucil
U036	57-74-9	Chlordane, alpha & gamma isomers
U026	494-03-1	Chlornaphazin
U037	108-90-7	Chlorobenzene
U038	510-15-6	Chlorobenzilate
U039	59-50-7	p-Chloro-m-cresol
U042	110-75-8	2-Chloroethyl vinyl ether
U044	67-66-3	Chloroform
U046	107-30-2	Chloromethyl methyl ether
U047	91-58-7	beta-Chloronaphthalene
U048	95-57-8	o-Chlorophenol
U049	3165-93-3	4-Chloro-o-toluidine, hydrochloride
U032	13765-19-0	Chromic acid H₂CrO₄, calcium salt
U050	218-01-9	Chrysene
U051	....	Creosote
U052	1319-77-3	Cresol (Cresylic acid)
U053	4170-30-3	Crotonaldehyde
U055	98-82-8	Cumene (I)
U246	506-68-3	Cyanogen bromide (CN)Br
U197	106-51-4	2,5-Cyclohexadiene-1,4-dione
U056	110-82-7	Cyclohexane (I)
U129	58-89-9	Cyclohexane, 1,2,3,4,5,6-hexachloro-, (1alpha,2alpha,3beta,4alpha,5alpha,6beta)-
U057	108-94-1	Cyclohexanone (I)
U130	77-47-4	1,3-Cyclopentadiene, 1,2,3,4,5,5-hexachloro-
U058	50-18-0	Cyclophosphamide
U240	¹ 94-75-7	2,4-D, salts & esters
U059	20830-81-3	Daunomycin
U060	72-54-8	DDD
U061	50-29-3	DDT
U062	2303-16-4	Diallate
U063	53-70-3	Dibenz[a,h]anthracene
U064	189-55-9	Dibenzo[a,i]pyrene
U066	96-12-8	1,2-Dibromo-3-chloropropane
U069	84-74-2	Dibutyl phthalate
U070	95-50-1	o-Dichlorobenzene
U071	541-73-1	m-Dichlorobenzene
U072	106-46-7	p-Dichlorobenzene
U073	91-94-1	3,3'-Dichlorobenzidine
U074	764-41-0	1,4-Dichloro-2-butene (I,T)
U075	75-71-8	Dichlorodifluoromethane
U078	75-35-4	1,1-Dichloroethylene
U079	156-60-5	1,2-Dichloroethylene
U025	111-44-4	Dichloroethyl ether
U027	108-60-1	Dichloroisopropyl ether
U024	111-91-1	Dichloromethoxy ethane

Haz-ardous waste No.	Chemical abstracts No.	Substance
U081	120-83-2	2,4-Dichlorophenol
U082	87-65-0	2,6-Dichlorophenol
U084	542-75-6	1,3-Dichloropropene
U085	1464-53-5	1,2:3,4-Diepoxybutane (I,T)
U108	123-91-1	1,4-Diethyleneoxide
U028	117-81-7	Diethylhexyl phthalate
U086	1615-80-1	N,N'-Diethylhydrazine
U087	3288-58-2	O,O-Diethyl S-methyl dithiophosphate
U088	84-66-2	Diethyl phthalate
U089	56-53-1	Diethylstilbesterol
U090	94-58-6	Dihydrosafrole
U091	119-90-4	3,3'-Dimethoxybenzidine
U092	124-40-3	Dimethylamine (I)
U093	60-11-7	p-Dimethylaminoazobenzene
U094	57-97-6	7,12-Dimethylbenz[a]anthracene
U095	119-93-7	3,3'-Dimethylbenzidine
U096	80-15-9	alpha,alpha-Dimethylbenzylhydroperoxide (R)
U097	79-44-7	Dimethylcarbamoyl chloride
U098	57-14-7	1,1-Dimethylhydrazine
U099	540-73-8	1,2-Dimethylhydrazine
U101	105-67-9	2,4-Dimethylphenol
U102	131-11-3	Dimethyl phthalate
U103	77-78-1	Dimethyl sulfate
U105	121-14-2	2,4-Dinitrotoluene
U106	606-20-2	2,6-Dinitrotoluene
U107	117-84-0	Di-n-octyl phthalate
U108	123-91-1	1,4-Dioxane
U109	122-66-7	1,2-Diphenylhydrazine
U110	142-84-7	Dipropylamine (I)
U111	621-64-7	Di-n-propylnitrosamine
U041	106-89-8	Epichlorohydrin
U001	75-07-0	Ethanal (I)
U174	55-18-5	Ethanamine, N-ethyl-N-nitroso-
U155	91-80-5	1,2-Ethanediamine, N,N-dimethyl-N'-2-pyridinyl-N'-(2-thienylmethyl)-
U067	106-93-4	Ethane, 1,2-dibromo-
U076	75-34-3	Ethane, 1,1-dichloro-
U077	107-06-2	Ethane, 1,2-dichloro-
U131	67-72-1	Ethane, hexachloro-
U024	111-91-1	Ethane, 1,1'-[methylenebis(oxy)]bis[2-chloro-
U117	60-29-7	Ethane, 1,1'-oxybis-(I)
U025	111-44-4	Ethane, 1,1'-oxybis[2-chloro-
U184	76-01-7	Ethane, pentachloro-
U208	630-20-6	Ethane, 1,1,1,2-tetrachloro-
U209	79-34-5	Ethane, 1,1,2,2-tetrachloro-
U218	62-55-5	Ethanethioamide
U226	71-55-6	Ethane, 1,1,1-trichloro-
U227	79-00-5	Ethane, 1,1,2-trichloro-
U359	110-80-5	Ethanol, 2-ethoxy-
U173	1116-54-7	Ethanol, 2,2'-(nitrosoimino)bis-
U004	98-86-2	Ethanone, 1-phenyl-
U043	75-01-4	Ethene, chloro-
U042	110-75-8	Ethene, (2-chloroethoxy)-
U078	75-35-4	Ethene, 1,1-dichloro-
U079	156-60-5	Ethene, 1,2-dichloro-, (E)-
U210	127-18-4	Ethene, tetrachloro-
U228	79-01-6	Ethene, trichloro-
U112	141-78-6	Ethyl acetate (I)
U113	140-88-5	Ethyl acrylate (I)
U238	51-79-6	Ethyl carbamate (urethane)
U117	60-29-7	Ethyl ether (I)
U114	¹ 111-54-6	Ethylenebisdithiocarbamic acid, salts & esters
U067	106-93-4	Ethylene dibromide
U077	107-06-2	Ethylene dichloride
U359	110-80-5	Ethylene glycol monoethyl ether
U115	75-21-8	Ethylene oxide (I,T)

Hazardous waste No.	Chemical abstracts No.	Substance
U116	96-45-7	Ethylenethiourea
U076	75-34-3	Ethylidene dichloride
U118	97-63-2	Ethyl methacrylate
U119	62-50-0	Ethyl methanesulfonate
U120	206-44-0	Fluoranthene
U122	50-00-0	Formaldehyde
U123	64-18-6	Formic acid (C,T)
U124	110-00-9	Furan (I)
U125	98-01-1	2-Furancarboxaldehyde (I)
U147	108-31-6	2,5-Furandione
U213	109-99-9	Furan, tetrahydro-(I)
U125	98-01-1	Furfural (I)
U124	110-00-9	Furfuran (I)
U206	18883-66-4	Glucopyranose, 2-deoxy-2-(3-methyl-3-nitrosoureido)-, D-
U206	18883-66-4	D-Glucose, 2-deoxy-2-[[(methylnitrosoamino)-carbonyl]amino]-
U126	765-34-4	Glycidylaldehyde
U163	70-25-7	Guanidine, N-methyl-N'-nitro-N-nitroso-
U127	118-74-1	Hexachlorobenzene
U128	87-68-3	Hexachlorobutadiene
U130	77-47-4	Hexachlorocyclopentadiene
U131	67-72-1	Hexachloroethane
U132	70-30-4	Hexachlorophene
U243	1888-71-7	Hexachloropropene
U133	302-01-2	Hydrazine (R,T)
U086	1615-80-1	Hydrazine, 1,2-diethyl-
U098	57-14-7	Hydrazine, 1,1-dimethyl-
U099	540-73-8	Hydrazine, 1,2-dimethyl-
U109	122-66-7	Hydrazine, 1,2-diphenyl-
U134	7664-39-3	Hydrofluoric acid (C,T)
U134	7664-39-3	Hydrogen fluoride (C,T)
U135	7783-06-4	Hydrogen sulfide
U135	7783-06-4	Hydrogen sulfide H₂S
U096	80-15-9	Hydroperoxide, 1-methyl-1-phenylethyl- (R)
U116	96-45-7	2-Imidazolidinethione
U137	193-39-5	Indeno[1,2,3-cd]pyrene
U190	85-44-9	1,3-Isobenzofurandione
U140	78-83-1	Isobutyl alcohol (I,T)
U141	120-58-1	Isosafrole
U142	143-50-0	Kepone
U143	303-34-4	Lasiocarpine
U144	301-04-2	Lead acetate
U146	1335-32-6	Lead, bis(acetato-O)tetrahydroxytri-
U145	7446-27-7	Lead phosphate
U146	1335-32-6	Lead subacetate
U129	58-89-9	Lindane
U163	70-25-7	MNNG
U147	108-31-6	Maleic anhydride
U148	123-33-1	Maleic hydrazide
U149	109-77-3	Malononitrile
U150	148-82-3	Melphalan
U151	7439-97-6	Mercury
U152	126-98-7	Methacrylonitrile (I, T)
U092	124-40-3	Methanamine, N-methyl- (I)
U029	74-83-9	Methane, bromo-
U045	74-87-3	Methane, chloro- (I, T)
U046	107-30-2	Methane, chloromethoxy-
U068	74-95-3	Methane, dibromo-
U080	75-09-2	Methane, dichloro-
U075	75-71-8	Methane, dichlorodifluoro-
U138	74-88-4	Methane, iodo-
U119	62-50-0	Methanesulfonic acid, ethyl ester
U211	56-23-5	Methane, tetrachloro-
U153	74-93-1	Methanethiol (I, T)
U225	75-25-2	Methane, tribromo-

Haz-ardous waste No.	Chemical abstracts No.	Substance
U044	67-66-3	Methane, trichloro-
U121	75-69-4	Methane, trichlorofluoro-
U036	57-74-9	4,7-Methano-1H-indene, 1,2,4,5,6,7,8,8-octachloro-2,3,3a,4,7,7a-hexahydro-
U154	67-56-1	Methanol (I)
U155	91-80-5	Methapyrilene
U142	143-50-0	1,3,4-Metheno-2H-cyclobuta[cd]pentalen-2-one, 1,1a,3,3a,4,5,5,5a,5b,6-decachlorooctahydro-
U247	72-43-5	Methoxychlor
U154	67-56-1	Methyl alcohol (I)
U029	74-83-9	Methyl bromide
U186	504-60-9	1-Methylbutadiene (I)
U045	74-87-3	Methyl chloride (I,T)
U156	79-22-1	Methyl chlorocarbonate (I,T)
U226	71-55-6	Methyl chloroform
U157	56-49-5	3-Methylcholanthrene
U158	101-14-4	4,4'-Methylenebis(2-chloroaniline)
U068	74-95-3	Methylene bromide
U080	75-09-2	Methylene chloride
U159	78-93-3	Methyl ethyl ketone (MEK) (I,T)
U160	1338-23-4	Methyl ethyl ketone peroxide (R,T)
U138	74-88-4	Methyl iodide
U161	108-10-1	Methyl isobutyl ketone (I)
U162	80-62-6	Methyl methacrylate (I,T)
U161	108-10-1	4-Methyl-2-pentanone (I)
U164	56-04-2	Methylthiouracil
U010	50-07-7	Mitomycin C
U059	20830-81-3	5,12-Naphthacenedione, 8-acetyl-10-[(3-amino-2,3,6-trideoxy)-alpha-L-lyxo-hexopyranosyl)oxy]-7,8,9,10-tetrahydro-6,8,11-trihydroxy-1-methoxy-, (8S-cis)-
U167	134-32-7	1-Naphthalenamine
U168	91-59-8	2-Naphthalenamine
U026	494-03-1	Naphthalenamine, N,N'-bis(2-chloroethyl)-
U165	91-20-3	Naphthalene
U047	91-58-7	Naphthalene, 2-chloro-
U166	130-15-4	1,4-Naphthalenedione
U236	72-57-1	2,7-Naphthalenedisulfonic acid, 3,3'-[(3,3'-dimethyl[1,1'-biphenyl]-4,4'-diyl)bis(azo)bis[5-amino-4-hydroxy]-, tetrasodium salt
U166	130-15-4	1,4-Naphthoquinone
U167	134-32-7	alpha-Naphthylamine
U168	91-59-8	beta-Naphthylamine
U217	10102-45-1	Nitric acid, thallium(1+) salt
U169	98-95-3	Nitrobenzene (I,T)
U170	100-02-7	p-Nitrophenol
U171	79-46-9	2-Nitropropane (I,T)
U172	924-16-3	N-Nitrosodi-n-butylamine
U173	1116-54-7	N-Nitrosodiethanolamine
U174	55-18-5	N-Nitrosodiethylamine
U176	759-73-9	N-Nitroso-N-ethylurea
U177	684-93-5	N-Nitroso-N-methylurea
U178	615-53-2	N-Nitroso-N-methylurethane
U179	100-75-4	N-Nitrosopiperidine
U180	930-55-2	N-Nitrosopyrrolidine
U181	99-55-8	5-Nitro-o-toluidine
U193	1120-71-4	1,2-Oxathiolane, 2,2-dioxide
U058	50-18-0	2H-1,3,2-Oxazaphosphorin-2-amine, N,N-bis(2-chloroethyl)tetrahydro-, 2-oxide
U115	75-21-8	Oxirane (I,T)
U126	765-34-4	Oxiranecarboxyaldehyde
U041	106-89-8	Oxirane, (chloromethyl)-
U182	123-63-7	Paraldehyde
U183	608-93-5	Pentachlorobenzene
U184	76-01-7	Pentachloroethane
U185	82-68-8	Pentachloronitrobenzene (PCNB)
See F027	87-86-5	Pentachlorophenol
U161	108-10-1	Pentanol, 4-methyl-
U186	504-60-9	1,3-Pentadiene (I)

Haz-ardous waste No.	Chemical abstracts No.	Substance
U187	62-44-2	Phenacetin
U188	108-95-2	Phenol
U048	95-57-8	Phenol, 2-chloro-
U039	59-50-7	Phenol, 4-chloro-3-methyl-
U081	120-83-2	Phenol, 2,4-dichloro-
U082	87-65-0	Phenol, 2,6-dichloro-
U089	56-53-1	Phenol, 4,4'-(1,2-diethyl-1,2-ethenediyl)bis-, (E)-
U101	105-67-9	Phenol, 2,4-dimethyl-
U052	1319-77-3	Phenol, methyl-
U132	70-30-4	Phenol, 2,2'-methylenebis[3,4,6-trichloro-
U170	100-02-7	Phenol, 4-nitro-
See F027	87-86-5	Phenol, pentachloro-
See F027	58-90-2	Phenol, 2,3,4,6-tetrachloro-
See F027	95-95-4	Phenol, 2,4,5-trichloro-
See F027	88-06-2	Phenol, 2,4,6-trichloro-
U150	148-82-3	L-Phenylalanine, 4-[bis(2-chloroethyl)amino]-
U145	7446-27-7	Phosphoric acid, lead(2+) salt (2:3)
U087	3288-58-2	Phosphorodithioic acid, O,O-diethyl S-methyl ester
U189	1314-80-3	Phosphorus sulfide (R)
U190	85-44-9	Phthalic anhydride
U191	109-06-8	2-Picoline
U179	100-75-4	Piperidine, 1-nitroso-
U192	23950-58-5	Pronamide
U194	107-10-8	1-Propanamine (I,T)
U111	621-64-7	1-Propanamine, N-nitroso-N-propyl-
U110	142-84-7	1-Propanamine, N-propyl- (I)
U066	96-12-8	Propane, 1,2-dibromo-3-chloro-
U083	78-87-5	Propane, 1,2-dichloro-
U149	109-77-3	Propanedinitrile
U171	79-46-9	Propane, 2-nitro- (I,T)
U027	108-60-1	Propane, 2,2'-oxybis[2-chloro-
U193	1120-71-4	1,3-Propane sultone
See F027	93-72-1	Propanoic acid, 2-(2,4,5-trichlorophenoxy)-
U235	126-72-7	1-Propanol, 2,3-dibromo-, phosphate (3:1)
U140	78-83-1	1-Propanol, 2-methyl- (I,T)
U002	67-64-1	2-Propanone (I)
U007	79-06-1	2-Propenamide
U084	542-75-6	1-Propene, 1,3-dichloro-
U243	1888-71-7	1-Propene, 1,1,2,3,3,3-hexachloro-
U009	107-13-1	2-Propenenitrile
U152	126-98-7	2-Propenenitrile, 2-methyl- (I,T)
U008	79-10-7	2-Propenoic acid (I)
U113	140-88-5	2-Propenoic acid, ethyl ester (I)
U118	97-63-2	2-Propenoic acid, 2-methyl-, ethyl ester
U162	80-62-6	2-Propenoic acid, 2-methyl-, methyl ester (I,T)
U194	107-10-8	n-Propylamine (I,T)
U083	78-87-5	Propylene dichloride
U148	123-33-1	3,6-Pyridazinedione, 1,2-dihydro-
U196	110-86-1	Pyridine
U191	109-06-8	Pyridine, 2-methyl-
U237	66-75-1	2,4-(1H,3H)-Pyrimidinedione, 5-[bis(2-chloroethyl)amino]-
U164	56-04-2	4(1H)-Pyrimidinone, 2,3-dihydro-6-methyl-2-thioxo-
U180	930-55-2	Pyrrolidine, 1-nitroso-
U200	50-55-5	Reserpine
U201	108-46-3	Resorcinol
U202	¹ 81-07-2	Saccharin, & salts
U203	94-59-7	Safrole
U204	7783-00-8	Selenious acid
U204	7783-00-8	Selenium dioxide

Haz-ardous waste No.	Chemical abstracts No.	Substance
U205	7488–56–4	Selenium sulfide
U205	7488–56–4	Selenium sulfide SeS_2 (R,T)
U015	115–02–6	L-Serine, diazoacetate (ester)
See F027	93–72–1	Silvex (2,4,5-TP)
U206	18883–66–4	Streptozotocin
U103	77–78–1	Sulfuric acid, dimethyl ester
U189	1314–80–3	Sulfur phosphide (R)
See F027	93–76–5	2,4,5-T
U207	95–94–3	1,2,4,5-Tetrachlorobenzene
U208	630–20–6	1,1,1,2-Tetrachloroethane
U209	79–34–5	1,1,2,2-Tetrachloroethane
U210	127–18–4	Tetrachloroethylene
See F027	58–90–2	2,3,4,6-Tetrachlorophenol
U213	109–99–9	Tetrahydrofuran (I)
U214	563–68–8	Thallium(I) acetate
U215	6533–73–9	Thallium(I) carbonate
U216	7791–12–0	Thallium(I) chloride
U216	7791–12–0	Thallium chloride TIcl
U217	10102–45–1	Thallium(I) nitrate
U218	62–55–5	Thioacetamide
U153	74–93–1	Thiomethanol (I,T)
U244	137–26–8	Thioperoxydicarbonic diamide [$(H_2N)C(S)$]$_2S_2$, tetramethyl-
U219	62–56–6	Thiourea
U244	137–26–8	Thiram
U220	108–88–3	Toluene
U221	25376–45–8	Toluenediamine
U223	26471–62–5	Toluene diisocyanate (R,T)
U328	95–53–4	o-Toluidine
U353	106–49–0	p-Toluidine
U222	636–21–5	o-Toluidine hydrochloride
U011	61–82–5	1H-1,2,4-Triazol-3-amine
U227	79–00–5	1,1,2-Trichloroethane
U228	79–01–6	Trichloroethylene
U121	75–69–4	Trichloromonofluoromethane
See F027	95–95–4	2,4,5-Trichlorophenol
See F027	88–06–2	2,4,6-Trichlorophenol
U234	99–35–4	1,3,5-Trinitrobenzene (R,T)
U182	123–63–7	1,3,5-Trioxane, 2,4,6-trimethyl-
U235	126–72–7	Tris(2,3-dibromopropyl) phosphate
U236	72–57–1	Trypan blue
U237	66–75–1	Uracil mustard
U176	759–73–9	Urea, N-ethyl-N-nitroso-
U177	684–93–5	Urea, N-methyl-N-nitroso-
U043	75–01–4	Vinyl chloride
U248	[1] 81–81–2	Warfarin, & salts, when present at concentrations of 0.3% or less
U239	1330–20–7	Xylene (I)
U200	50–55–5	Yohimban-16-carboxylic acid, 11,17-dimethoxy-18-[(3,4,5-trimethoxybenzoyl)oxy]-, methyl ester, (3beta,16beta,17alpha,18beta,20alpha)-
U249	1314–84–7	Zinc phosphide Zn_3P_2, when present at concentrations of 10% or less

[1] CAS Number given for parent compound only.

APPENDIX G
NUCLEAR REGULATORY
COMMISSION OFFICES

Region I
631 Park Avenue
King of Prussia, PA 19406
(215) 337-5000

Region II
101 Marietta Street, N.W.
Suite 2900
Atlanta, GA 30323
(404) 221-4503

Region III
799 Roosevelt Road
Glen Ellyn, IL 60137
(312) 790-5500

Region IV
611 Ryan Plaza Drive
Suite 1000
Arlington, TX 76011
(817) 860-8100

Region V
1450 Marie Lane
Suite 210
Walnut Creek, CA 94596
(415) 943-3700

APPENDIX H
THE LIST OF EXTREMELY HAZARDOUS SUBSTANCES AND COMMUNITY RIGHT-TO-KNOW REGULATIONS

APPENDIX A.—THE LIST OF EXTREMELY HAZARDOUS SUBSTANCES AND THEIR THRESHOLD PLANNING QUANTITIES

[Alphabetical Order]

CAS No.	Chemical name	Notes	Reportable quantity* (pounds)	Threshold planning quantity (pounds)
75–86–5	Acetone Cyanohydrin		10	1,000
1752–30–3	Acetone Thiosemicarbazide	e	1	1,000/10,000
107–02–8	Acrolein		1	500
79–06–1	Acrylamide	d, l	5,000	1,000/10,000
107–13–1	Acrylonitrile	d, l	100	10,000
814–68–6	Acrylyl Chloride	e, h	1	100
111–69–3	Adiponitrile	e, l	1	1,000
116–06–3	Aldicarb	c	1	100/10,000
309–00–2	Aldrin	d	1	500/10,000
107–18–6	Allyl Alcohol		100	1,000
107–11–9	Allylamine	e	1	500
20859–73–8	Aluminum Phosphide	b	100	500
54–62–6	Aminopterin	e	1	500/10,000
78–53–5	Amiton	e	1	500
3734–97–2	Amiton Oxalate	e	1	100/10,000
7664–41–7	Ammonia	l	100	500
16919–58–7	Ammonium Chloroplatinate	a, e	1	10,000
300–62–9	Amphetamine	e	1	1,000
62–53–3	Aniline	d, l	5,000	1,000
88–05–1	Aniline, 2,4,6-Trimethyl-	e	1	500
7783–70–2	Antimony Pentafluoride	e	1	500
1397–94–0	Antimycin A	c, e	1	1,000/10,000
86–88–4	ANTU		100	500/10,000
1303–28–2	Arsenic Pentoxide	d	5,000	100/10,000
1327–53–3	Arsenous Oxide	d, h	5,000	100/10,000
7784–34–1	Arsenous Trichloride	d	5,000	500
7784–42–1	Arsine	e	1	100
2642–71–9	Azinphos-Ethyl	e	1	100/10,000
86–50–0	Azinphos-Methyl		1	10/10,000
1405–87–4	Bacitracin	a, e	1	10,000
98–87–3	Benzal Chloride	d	5,000	500
98–16–8	Benzenamine, 3-(Trifluoromethyl)-	e	1	500
100–14–1	Benzene, 1-(Chloromethyl)-4-Nitro-	e	1	500/10,000
98–05–5	Benzenearsonic Acid	e	1	10/10,000
98–09–9	Benzenesulfonyl Chloride	a	100	10,000
3615–21–2	Benzimidazole, 4,5-Dichloro-2-(Trifluoromethyl)-	e, g	1	500/10,000
98–07–7	Benzotrichloride	d	1	100
100–44–7	Benzyl Chloride	d	100	500
140–29–4	Benzyl Cyanide	e, h	1	500

235

13398   Federal Register / Vol. 52, No. 77 / Wednesday, April 22, 1987 / Rules and Regulations

APPENDIX A.—THE LIST OF EXTREMELY HAZARDOUS SUBSTANCES AND THEIR THRESHOLD PLANNING QUANTITIES—Continued

[Alphabetical Order]

CAS No.	Chemical name	Notes	Reportable quantity* (pounds)	Threshold planning quantity (pounds)
15271-41-7	Bicyclo[2.2.1]Heptane-2-Carbonitrile, 5-Chloro-6-((((Methylamino)Carbonyl)Oxy)Imino)-, (1s-(1-alpha, 2-beta, 4-alpha, 5-alpha, 6E))-.	e	1	500/10,000
534-07-6	Bis(Chloromethyl) Ketone	e	1	10/10,000
4044-65-9	Bitoscanate	e	1	500/10,000
10294-34-5	Boron Trichloride	e	1	500
7637-07-2	Boron Trifluoride	e	1	500
353-42-4	Boron Trifluoride Compound With Methyl Ether (1:1)	e	1	1,000
28772-56-7	Bromadiolone	e	1	100/10,000
7726-95-6	Bromine	e, l	1	500
106-99-0	Butadiene	a, e	1	10,000
109-19-3	Butyl Isovalerate	a, e	1	10,000
111-34-2	Butyl Vinyl Ether	a, e	1	10,000
1306-19-0	Cadmium Oxide	e	1	100/10,000
2223-93-0	Cadmium Stearate	c, e	1	1,000/10,000
7778-44-1	Calcium Arsenate	d	1,000	500/10,000
8001-35-2	Camphechlor	d	1	500/10,000
56-25-7	Cantharidin	e	1	100/10,000
51-83-2	Carbachol Chloride	e	1	500/10,000
26419-73-8	Carbamic Acid, Methyl-, O-(((2,4-Dimethyl-1, 3-Dithiolan-2-yl)Methylene)Amino)-	e	1	100/10,000
1563-66-2	Carbofuran	e	10	10/10,000
75-15-0	Carbon Disulfide	l	100	10,000
786-19-6	Carbophenothion	e	1	500
2244-16-8	Carvone	a, e	1	10,000
57-74-9	Chlordane	d	1	1,000
470-90-6	Chlorfenvinfos	e	1	500
7782-50-5	Chlorine		10	100
24934-91-6	Chlormephos	e	1	500
999-81-5	Chlormequat Chloride	e, h	1	100/10,000
107-20-0	Chloroacetaldehyde	a	1,000	10,000
79-11-8	Chloroacetic Acid	e	1	100/10,000
107-07-3	Chloroethanol	e	1	500
627-11-2	Chloroethyl Chloroformate	e	1	1,000
67-66-3	Chloroform	d, l	5,000	10,000
542-88-1	Chloromethyl Ether	d, h	1	100
107-30-2	Chloromethyl Methyl Ether	c, d	1	100
3691-35-8	Chlorophacinone	e	1	100/10,000
1982-47-4	Chloroxuron	e	1	500/10,000
21923-23-9	Chlorthiophos	e, h	1	500
10025-73-7	Chromic Chloride	e	1	1/10,000
7440-48-4	Cobalt	e	1	10,000
62207-76-5	Cobalt, ((2,2'-(1,2-Ethanediylbis (Nitrilomethylidyne))Bis(6-Fluorophenolato))(2-)-N,N',O,O')-,.	e	1	100/10,000
10210-68-1	Cobalt Carbonyl	e, h	1	10/10,000
64-86-8	Colchicine	e, h	1	10/10,000
117-52-2	Coumafuryl	a, e	1	10,000
56-72-4	Coumaphos	e	10	100/10,000
5836-29-3	Coumatetralyl	e	1	500/10,000
95-48-7	Cresol, o-	d	1,000	1,000/10,000
535-89-7	Crimidine	e	1	100/10,000
4170-30-3	Crotonaldehyde		100	1,000
123-73-9	Crotonaldehyde, (E)-		100	1,000
506-68-3	Cyanogen Bromide		1,000	500/10,000
506-78-5	Cyanogen Iodide		1,000	1,000/10,000
2636-26-2	Cyanophos	e	1	1,000
675-14-9	Cyanuric Fluoride	e	1	100
66-81-9	Cycloheximide	e	1	100/10,000
108-91-8	Cyclohexylamine	e, l	1	10,000
287-92-3	Cyclopentane	a, e	1	10,000
633-03-4	C. I. Basic Green 1	a, e	1	10,000
17702-41-9	Decaborane(14)	e	1	500/10,000
8065-48-3	Demeton	e	1	500
919-86-8	Demeton-S-Methyl	e	1	500
10311-84-9	Dialifor	e	1	100/10,000
19287-45-7	Diborane	e	1	100
84-74-2	Dibutyl Phthalate	a	10	10,000
8023-53-8	Dichlorobenzalkonium Chloride	a, e	1	10,000
111-44-4	Dichloroethyl Ether	d	1	10,000

Federal Register / Vol. 52, No. 77 / Wednesday, April 22, 1987 / Rules and Regulations   13399

APPENDIX A.—THE LIST OF EXTREMELY HAZARDOUS SUBSTANCES AND THEIR THRESHOLD PLANNING QUANTITIES—Continued

[Alphabetical Order]

CAS No.	Chemical name	Notes	Reportable quantity* (pounds)	Threshold planning quantity (pounds)
149-74-6	Dichloromethylphenylsilane	e	1	1,000
62-73-7	Dichlorvos		10	1,000
141-66-2	Dicrotophos	e	1	100
1464-53-5	Diepoxybutane	d	1	500
814-49-3	Diethyl Chlorophospate	e, h	1	500
1642-54-2	Diethylcarbamazine Citrate	e	1	100/10,000
93-05-0	Diethyl-p-Phenylenediamine	a, e	1	10,000
71-63-6	Digitoxin	c, e	1	100/10,000
2238-07-5	Diglycidyl Ether	e	1	1,000
20830-75-5	Digoxin	e, h	1	10/10,000
115-26-4	Dimefox	e	1	500
60-51-5	Dimethoate		10	500/10,000
2524-03-0	Dimethyl Phosphorochlondothioate	e	1	500
131-11-3	Dimethyl Phthalate	a	5,000	10,000
77-78-1	Dimethyl Sulfate	d	1	500
75-18-3	Dimethyl Sulfide	e	1	100
75-78-5	Dimethyldichlorosilane	e, h	1	500
57-14-7	Dimethylhydrazine	d	1	1,000
99-98-9	Dimethyl-p-Phenylenediamine	e	1	10/10,000
644-64-4	Dimetilan	e	1	500/10,000
534-52-1	Dinitrocresol		10	10/10,000
88-85-7	Dinoseb		1,000	100/10,000
1420-07-1	Dinoterb	e	1	500/10,000
117-84-0	Dioctyl Phthalate	a	5,000	10,000
78-34-2	Dioxathion	e	1	500
646-06-0	Dioxolane	a, e	1	10,000
82-66-6	Diphacinone	e	1	10/10,000
152-16-9	Diphosphoramide, Octamethyl-		100	100
298-04-4	Disulfoton		1	500
514-73-8	Dithiazanine Iodide	e	1	500/10,000
541-53-7	Dithiobiuret		100	100/10,000
316-42-7	Emetine, Dihydrochloride	e, h	1	1/10,000
115-29-7	Endosulfan		1	10/10,000
2778-04-3	Endothion	e	1	500/10,000
72-20-8	Endrin		1	500/10,000
106-89-8	Epichlorohydrin	d, l	1,000	1,000
2104-64-5	EPN	e	1	100/10,000
50-14-6	Ergocalciferol	c, e	1	1,000/10,000
379-79-3	Ergotamine Tartrate	e	1	500/10,000
1622-32-8	Ethanesulfonyl Chloride, 2-Chloro-	e	1	500
10140-87-1	Ethanol, 1,2-Dichloro-, Acetate	e	1	1,000
563-12-2	Ethion		10	1,000
13194-48-4	Ethoprophos	e	1	1,000
538-07-8	Ethylbis(2-Chloroethyl)Amine	e, h	1	500
371-62-0	Ethylene Fluorohydrin	c, e, h	1	10
75-21-8	Ethylene Oxide	d, l	1	1,000
107-15-3	Ethylenediamine		5,000	10,000
151-56-4	Ethyleneimine	d	1	500
2235-25-8	Ethylmercuric Phosphate	a, e	1	10,000
542-90-5	Ethylthiocyanate	e	1	10,000
22224-92-6	Fenamiphos	e	1	10/10,000
122-14-5	Fenitrothion	e	1	500
115-90-2	Fensulfothion	e, h	1	500
4301-50-2	Fluenetil	e	1	100/10,000
7782-41-4	Fluorine		10	500
640-19-7	Fluoroacetamide	j	100	100/10,000
144-49-0	Fluoroacetic Acid	e	1	10/10,000
359-06-8	Fluoroacetyl Chloride	c, e	1	10
51-21-8	Fluorouracil	e	1	500/10,000
944-22-9	Fonofos	e	1	500
50-00-0	Formaldehyde	d, l	1,000	500
107-16-4	Formaldehyde Cyanohydrin	e, h	1	1,000
23422-53-9	Formetanate Hydrochloride	e,h	1	500/10,000
2540-82-1	Formothion	e	1	100
17702-57-7	Formparanate	e	1	100/10,000
21548-32-3	Fosthietan	e	1	500

13400 Federal Register / Vol. 52, No. 77 / Wednesday, April 22, 1987 / Rules and Regulations

APPENDIX A.—THE LIST OF EXTREMELY HAZARDOUS SUBSTANCES AND THEIR THRESHOLD PLANNING QUANTITIES—Continued

[Alphabetical Order]

CAS No.	Chemical name	Notes	Reportable quantity* (pounds)	Threshold planning quantity (pounds)
3878-19-1	Fubendazole	e	1	100/10,000
110-00-9	Furan		100	500
13450-90-3	Gallium Trichloride	e	1	500/10,000
77-47-4	Hexachlorocyclopentadiene	d, h	1	100
1335-87-1	Hexachloronaphthalene	a, e	1	10,000
4835-11-4	Hexamethylenediamine, N,N'-Dibutyl-	e	1	500
302-01-2	Hydrazine	d	1	1,000
74-90-8	Hydrocyanic Acid		10	100
7647-01-0	Hydrogen Chloride (Gas Only)	e, l	1	500
7664-39-3	Hydrogen Fluoride		100	100
7722-84-1	Hydrogen Peroxide (Conc >52%)	e, l	1	1,000
7783-07-5	Hydrogen Selenide	e	1	10
7783-06-4	Hydrogen Sulfide	l	100	500
123-31-9	Hydroquinone		1	500/10,000
53-86-1	Indomethacin	a, e	1	10,000
10025-97-5	Iridium Tetrachloride	a, e	1	10,000
13463-40-6	Iron, Pentacarbonyl-	e	1	100
297-78-9	Isobenzan	e	1	100/10,000
78-82-0	Isobutyronitrile	e, h	1	1,000
102-36-3	Isocyanic Acid, 3,4-Dichlorophenyl Ester	e	1	500/10,000
465-73-6	Isodrin		1	100/10,000
55-91-4	Isofluorphate	c	100	100
4098-71-9	Isophorone Diisocyanate	b, e	1	100
108-23-6	Isopropyl Chloroformate	e	1	1,000
625-55-8	Isopropyl Formate	e	1	500
119-38-0	Isopropylmethylpyrazolyl Dimethylcarbamate	e	1	500
78-97-7	Lactonitrile	e	1	1,000
21609-90-5	Leptophos	e	1	500/10,000
541-25-3	Lewisite	c, e, h	1	10
58-89-9	Lindane	d	1	1,000/10,000
7580-67-8	Lithium Hydride	b, e	1	100
109-77-3	Malononitrile		1,000	500/10,000
12108-13-3	Manganese, Tricarbonyl Methylcyclopentadienyl	e, h	1	100
51-75-2	Mechlorethamine	c, e	1	10
950-10-7	Mephosfolan	e	1	500
1600-27-7	Mercuric Acetate	e	1	500/10,000
7487-94-7	Mercuric Chloride	e	1	500/10,000
21908-53-2	Mercuric Oxide	e	1	500/10,000
108-67-8	Mesitylene	a, e	1	10,000
10476-95-6	Methacrolein Diacetate	e	1	1,000
760-93-0	Methacrylic Anhydride	e	1	500
126-98-7	Methacrylonitrile	h	1	500
920-46-7	Methacryloyl Chloride	e	1	100
30674-80-7	Methacryloyloxyethyl Isocyanate	e, h	1	100
10265-92-6	Methamidophos	e	1	100/10,000
558-25-8	Methanesulfonyl Fluoride	e	1	1,000
950-37-8	Methidathion	e	1	500/10,000
2032-65-7	Methiocarb		10	500/10,000
16752-77-5	Methomyl	h	100	500/10,000
151-38-2	Methoxyethylmercuric Acetate	e	1	500/10,000
80-63-7	Methyl 2-Chloroacrylate	e	1	500
74-83-9	Methyl Bromide	l	1,000	1,000
79-22-1	Methyl Chloroformate	d, h	1,000	500
624-92-0	Methyl Disulfide	e	1	100
60-34-4	Methyl Hydrazine		10	500
624-83-9	Methyl Isocyanate	f	1	500
556-61-6	Methyl Isothiocyanate	b, e	1	500
74-93-1	Methyl Mercaptan		100	500
3735-23-7	Methyl Phenkapton	e	1	500
676-97-1	Methyl Phosphonic Dichloride	b, e	1	100
556-64-9	Methyl Thiocyanate	e	1	10,000
78-94-4	Methyl Vinyl Ketone	e	1	10
502-39-6	Methylmercuric Dicyanamide	e	1	500/10,000
75-79-6	Methyltrichlorosilane	e, h	1	500
1129-41-5	Metolcarb	e	1	100/10,000
7786-34-7	Mevinphos		10	500

Federal Register / Vol. 52, No. 77 / Wednesday, April 22, 1987 / Rules and Regulations   **13401**

APPENDIX A.—THE LIST OF EXTREMELY HAZARDOUS SUBSTANCES AND THEIR THRESHOLD PLANNING QUANTITIES—Continued

[Alphabetical Order]

CAS No.	Chemical name	Notes	Reportable quantity* (pounds)	Threshold planning quantity (pounds)
315–18–4	Mexacarbate		1,000	500/10,000
50–07–7	Mitomycin C	d	1	500/10,000
6923–22–4	Monocrotophos	e	1	10/10,000
2763–96–4	Muscimol	a, h	1,000	10,000
505–60–2	Mustard Gas	e, h	1	500
7440–02–0	Nickel	a, d	1	10,000
13463–39–3	Nickel Carbonyl	d	1	1
54–11–5	Nicotine	c	100	100
65–30–5	Nicotine Sulfate	e	1	100/10,000
7697–37–2	Nitric Acid		1,000	1,000
10102–43–9	Nitric Oxide	c	10	100
98–95–3	Nitrobenzene	l	1,000	10,000
1122–60–7	Nitrocyclohexane	e	1	500
10102–44–0	Nitrogen Dioxide		10	100
62–75–9	Nitrosodimethylamine	d, h	1	1,000
991–42–4	Norbormide	e	1	100/10,000
0	Organorhodium Complex (PMN–82–147)	e	1	10/10,000
65–86–1	Orotic Acid	a, e	1	10,000
20816–12–0	Osmium Tetroxide	a	1,000	10,000
630–60–4	Ouabain	c, e	1	100/10,000
23135–22–0	Oxamyl	e	1	100/10,000
78–71–7	Oxetane, 3,3-Bis(Chloromethyl)-	l	e	500
2497–07–6	Oxydisulfoton	e, h	1	500
10028–15–6	Ozone		1	100
1910–42–5	Paraquat	e	1	10/10,000
2074–50–2	Paraquat Methosulfate	e	1	10/10,000
56–38–2	Parathion	c, d	1	100
298–00–0	Parathion-Methyl	c	100	100/10,000
12002–03–8	Paris Green	d	100	500/10,000
19624–22–7	Pentaborane	e	1	500
76–01–7	Pentachloroethane	a, d	1	10,000
87–86–5	Pentachlorophenol	a, d	10	10,000
2570–26–5	Pentadecylamine	e	1	100/10,000
79–21–0	Peracetic Acid	e	1	500
594–42–3	Perchloromethylmercaptan		100	500
108–95–2	Phenol		1,000	500/10,000
97–18–7	Phenol, 2,2'-Thiobis(4,6-Dichloro-	e	1	100/10,000
4418–66–0	Phenol, 2,2'-Thiobis(4-Chloro-6-Methyl-Phenol, 2,2'-Thiobis (4-Chloro-6-Methyl)-	e	1	100/10,000
64–00–0	Phenol, 3-(1-Methylethyl)-, Methylcarbamate	e	1	500/10,000
58–36–6	Phenoxarsine, 10,10'-Oxydi-	e	1	500/10,000
696–28–6	Phenyl Dichloroarsine	d, h	1	500
59–88–1	Phenylhydrazine Hydrochloride	e	1	1,000/10,000
62–38–4	Phenylmercury Acetate		100	500/10,000
2097–19–0	Phenylsilatrane	e, h	1	100/10,000
103–85–5	Phenylthiourea		100	100/10,000
298–02–2	Phorate		10	10
4104–14–7	Phosacetim	e	1	100/10,000
947–02–4	Phosfolan	e	1	100/10,000
75–44–5	Phosgene	l	10	10
732–11–6	Phosmet	e	1	10/10,000
13171–21–6	Phosphamidon	e	1	100
7803–51–2	Phosphine		100	500
2703–13–1	Phosphonothioic Acid, Methyl-, O-Ethyl O-(4-(Methylthio)Phenyl) Ester	e	1	500
50782–69–9	Phosphonothioic Acid, Methyl-, S-(2-(Bis(1-Methylethyl)Amino)Ethyl O-Ethyl Ester	e	1	100
2665–30–7	Phosphonothioic Acid, Methyl-, O-(4-Nitrophenyl) O-Phenyl Ester	e	1	500
3254–63–5	Phosphoric Acid, Dimethyl 4-(Methylthio) Phenyl Ester	e	1	500
2587–90–8	Phosphorothioic Acid, O,O-Dimethyl-S-(2-Methylthio) Ethyl Ester	c, e, g	1	500
7723–14–0	Phosphorus	b, h	1	100
10025–87–3	Phosphorus Oxychloride	d	1,000	500
10026–13–8	Phosphorus Pentachloride	b, e	1	500
1314–56–3	Phosphorus Pentoxide	b, e	1	10
7719–12–2	Phosphorus Trichloride		1,000	1,000
84–80–0	Phylloquinone	a, e	1	10,000
57–47–6	Physostigmine	e	1	100/10,000
57–64–7	Physostigmine, Salicylate (1:1)	e	1	100/10,000
124–87–8	Picrotoxin	e	1	500/10,000

13402 Federal Register / Vol. 52, No. 77 / Wednesday, April 22, 1987 / Rules and Regulations

APPENDIX A.—THE LIST OF EXTREMELY HAZARDOUS SUBSTANCES AND THEIR THRESHOLD PLANNING QUANTITIES—Continued

[Alphabetical Order]

CAS No.	Chemical name	Notes	Reportable quantity* (pounds)	Threshold planning quantity (pounds)
110–89–4	Piperidine	e	1	1,000
5281–13–0	Piprotal	e	1	100/10,000
23505–41–1	Pinmifos-Ethyl	e	1	1,000
10025–65–7	Platinous Chloride	a, e	1	10,000
13454–96–1	Platinum Tetrachloride	a, e	1	10,000
10124–50–2	Potassium Arsenite	d	1,000	500/10,000
151–50–8	Potassium Cyanide	b	10	100
506–61–6	Potassium Silver Cyanide	b	1	500
2631–37–0	Promecarb	e, h	1	500/10,000
106–96–7	Propargyl Bromide	e	1	10
57–57–8	Propiolactone, Beta-	e	1	500
107–12–0	Propionitrile		10	500
542–76–7	Propionitrile, 3-Chloro-		1,000	1,000
70–69–9	Propiophenone, 4-Amino-	e, g	1	100/10,000
109–61–5	Propyl Chloroformate	e	1	500
1331–17–5	Propylene Glycol, Allyl Ether	a, e	1	10,000
75–56–9	Propylene Oxide	l	100	10,000
75–55–8	Propyleneimine	d	1	10,000
2275–18–5	Prothoate	e	1	100/10,000
95–63–6	Pseudocumene	a, e	1	10,000
129–00–0	Pyrene	c	5,000	1,000/10,000
140–76–1	Pyridine, 2-Methyl-5-Vinyl-	e	1	500
504–24–5	Pyridine, 4-Amino-	h	1,000	500/10,000
1124–33–0	Pyridine, 4-Nitro-, 1-Oxide	e	1	500/10,000
53558–25–1	Pyriminil	e, h	1	100/10,000
10049–07–7	Rhodium Trichloride	a, e	1	10,000
14167–18–1	Salcomine	e	1	500/10,000
107–44–8	Sarin	e, h	1	10
7783–00–8	Selenious Acid		10	1,000/10,000
7791–23–3	Selenium Oxychloride	e	1	500
563–41–7	Semicarbazide Hydrochloride	e	1	1,000/10,000
3037–72–7	Silane, (4-Aminobutyl)Diethoxymethyl-	e	1	1,000
128–56–3	Sodium Anthraquinone-1-Sulfonate	a, e	1	10,000
7631–89–2	Sodium Arsenate	d	1,000	1,000/10,000
7784–46–5	Sodium Arsenite	d	1,000	500/10,000
26628–22–8	Sodium Azide (Na(N3))	b	1,000	500
124–65–2	Sodium Cacodylate	e	1	100/10,000
143–33–9	Sodium Cyanide (Na(CN))	b	10	100
62–74–8	Sodium Fluoroacetate		10	10/10,000
131–52–2	Sodium Pentachlorophenate	e	1	100/10,000
13410–01–0	Sodium Selenate	e	1	100/10,000
10102–18–8	Sodium Selenite	h	100	100/10,000
10102–20–2	Sodium Tellurite	e	1	500/10,000
900–95–8	Stannane, Acetoxytriphenyl-	e, g	1	500/10,000
57–24–9	Strychnine	e	10	100/10,000
60–41–3	Strychnine, Sulfate	e	1	100/10,000
3689–24–5	Sulfotep		100	500
3569–57–1	Sulfoxide, 3-Chloropropyl Octyl	e	1	500
7446–09–5	Sulfur Dioxide	e, l	1	500
7783–60–0	Sulfur Tetrafluoride	e	1	100
7446–11–9	Sulfur Trioxide	b, e	1	100
7664–93–9	Sulfur Acid		1,000	1,000
77–81–6	Tabun	c, e, h	1	10
13494–80–9	Tellurium	e	1	500/10,000
7783–80–4	Tellurium Hexafluoride	e, k	1	100
107–49–3	TEPP		10	100
13071–79–9	Terbufos	e, h	1	100
78–00–2	Tetraethyllead	c, d	10	100
597–64–8	Tetraethyltin	c, e	1	100
75–74–1	Tetramethyllead	c, e, l	1	100
509–14–8	Tetranitromethane		10	500
1314–32–5	Thallic Oxide	a	100	10,000
10031–59–1	Thallium Sulfate	h	100	100/10,000
6533–73–9	Thallous Carbonate	c, h	100	100/10,000
7791–12–0	Thallous Chloride	c, h	100	100/10,000

APPENDIX A.—THE LIST OF EXTREMELY HAZARDOUS SUBSTANCES AND THEIR THRESHOLD PLANNING QUANTITIES—Continued

[Alphabetical Order]

CAS No.	Chemical name	Notes	Reportable quantity* (pounds)	Threshold planning quantity (pounds)
2757-18-8	Thallous Malonate	c, e, h	1	100/10,000
7446-18-6	Thallous Sulfate		100	100/10,000
2231-57-4	Thiocarbazide	e	1	1,000/10,000
21564-17-0	Thiocyanic Acid, 2-(Benzothiazolylthio)Methyl Ester	a, e	1	10,000
39196-18-4	Thiofanox		100	100/10,000
640-15-3	Thiometon	a, e	1	10,000
297-97-2	Thionazin		100	500
108-98-5	Thiophenol		100	500
79-19-6	Thiosemicarbazide		100	100/10,000
5344-82-1	Thiourea, (2-Chlorophenyl)-		100	100/10,000
614-78-8	Thiourea, (2-Methylphenyl)-	e	1	500/10,000
7550-45-0	Titanium Tetrachloride	e	1	100
584-84-9	Toluene 2,4-Diisocyanate		100	500
91-08-7	Toluene 2,6-Diisocyanate		100	100
110-57-6	Trans-1,4-Dichlorobutene	e	1	500
1031-47-6	Triamiphos	e	1	500/10,000
24017-47-8	Triazofos	e	1	500
76-02-8	Trichloroacetyl Chloride	e	1	500
115-21-9	Trichloroethylsilane	e, h	1	500
327-98-0	Trichloronate	e, k	1	500
98-13-5	Trichlorophenylsilane	e, h	1	500
52-68-6	Trichlorophon	a	100	10,000
1558-25-4	Trichloro(Chloromethyl)Silane	e	1	100
27137-85-5	Trichloro(Dichlorophenyl)Silane	e	1	500
998-30-1	Triethoxysilane	e	1	500
75-77-4	Trimethylchlorosilane	e	1	1,000
824-11-3	Trimethylolpropane Phosphite	e, h	1	100/10,000
1066-45-1	Trimethyltin Chloride	e	1	500/10,000
639-58-7	Triphenyltin Chloride	e	1	500/10,000
555-77-1	Tris(2-Chloroethyl)Amine	e, h	1	100
2001-95-8	Valinomycin	c, e	1	1,000/10,000
1314-62-1	Vanadium Pentoxide		1,000	100/10,000
108-05-4	Vinyl Acetate Monomer	d, l	5,000	1,000
3048-64-4	Vinylnorbornene	a, e	1	10,000
81-81-2	Warfarin		100	500/10,000
129-06-6	Warfarin Sodium	e, h	1	100/10,000
28347-13-9	Xylylene Dichloride	e	1	100/10,000
58270-08-9	Zinc, Dichloro(4,4-Dimethyl-5((((Methylamino) Carbonyl)Oxy)Imino)Pentanenitrile)-,(T-4)-	e	1	100/10,000
1314-84-7	Zinc Phosphide	b	100	500

*Only the statutory or final RQ is shown. For more information, see 40 CFR Table 302.4

Notes:

a This chemical does not meet acute toxicity criteria. Its TPQ is set at 10,000 pounds.

b This material is a reactive solid. The TPQ does not default to 10,000 pounds for non-powder, non-molten, non-solution form.

c The calculated TPQ changed after technical review as described in the technical support document.

d Indicates that the RQ is subject to change when the assessment of potential carcinogenicity and/or other toxicity is completed.

e Statutory reportable quantity for purposes of notification under SARA sect 304(a)(2).

f The statutory 1 pound reportable quantity for methyl isocyanate may be adjusted in a future rulemaking action.

g New chemicals added that were not part of the original list of 402 substances.

h Revised TPQ based on new or re-evaluated toxicity data.

j TPQ is revised to its calculated value and does not change due to technical review as in proposed rule.

k The TPQ was revised after proposal due to calculation error.

l Chemicals on the original list that do not meet toxicity criteria but because of their high production volume and recognized toxicity are considered chemicals of concern ("Other chemicals").

APPENDIX B.—THE LIST OF EXTREMELY HAZARDOUS SUBSTANCES AND THEIR THRESHOLD PLANNING QUANTITIES

[CAS Number Order]

CAS No.	Chemical name	Notes	Reportable quantity* (pounds)	Threshold planning quantity (pounds)
0	Organorhodium Complex (PMN-82-147)	e	1	10/10,000
50-00-0	Formaldehyde	d, l	1,000	500
50-07-7	Mitomycin C	d	1	500/10,000
50-14-6	Ergocalciferol	c, e	1	1,000/10,000

13404   Federal Register / Vol. 52. No. 77 / Wednesday. April 22. 1987 / Rules and Regulations

APPENDIX B.—THE LIST OF EXTREMELY HAZARDOUS SUBSTANCES AND THEIR THRESHOLD PLANNING QUANTITIES—Continued

[CAS Number Order]

CAS No.	Chemical name	Notes	Reportable quantity* (pounds)	Threshold planning quantity (pounds)
51–21–8	Fluorouracil	e	1	500/10,000
51–75–2	Mechlorethamine	c, e	1	10
51–83–2	Carbachol Chloride	e	1	500/10,000
52–68–6	Trichlorophon	a, e	100	10,000
53–86–1	Indomethacin	a, e	1	10,000
54–11–5	Nicotine	c	100	100
54–62–6	Aminopterin	e	1	500/10,000
55–91–4	Isofluorphate	c	100	100
56–25–7	Cantharidin	e	1	100/10,000
56–38–2	Parathion	c, d	1	100
56–72–4	Coumaphos		10	100/10,000
57–14–7	Dimethylhydrazine	d	1	1,000
57–24–9	Strychnine	c	10	100/10,000
57–47–6	Physostigmine	e	1	100/10,000
57–57–8	Propiolactone, Beta-	e	1	500
57–64–7	Physostigmine, Salicylate (1:1)	e	1	100/10,000
57–74–9	Chlordane	d	1	1,000
58–36–6	Phenoxarsine, 10,10'-Oxydi-	e	1	500/10,000
58–89–9	Lindane	d	1	1,000/10,000
59–88–1	Phenylhydrazine Hydrochloride	e	10	500
60–34–4	Methyl Hydrazine		10	500
60–41–3	Strychnine, Sulfate	e	10	100/10,000
60–51–5	Dimethoate		10	500/10,000
62–38–4	Phenylmercury Acetate		100	500/10,000
62–53–3	Aniline	d, I	5,000	1,000
62–73–7	Dichlervos		10	1,000
62–74–8	Sodium Fluoroacetate		10	10/10,000
62–75–9	Nitrosodimethylamine	d, h	1	1,000
64–00–6	Phenol, 3-(1-Methylethyl)-, Methylcarbamate	e, h	1	500/10,000
64–86–8	Colchicine	e, h	1	10/10,000
65–30–5	Nicotine Sulfate	e	1	100/10,000
65–86–1	Orotic Acid	a, e	1	10,000
66–81–9	Cyclohexamide	e	1	100/10,000
67–66–3	Chloroform	d, I	5,000	10,000
70–69–9	Propiophenone, 4-Amino-	e, g	1	100/10,000
71–63–6	Digitoxin	c, e	1	100/10,000
72–20–8	Endrin	c, e	1	500/10,000
74–83–9	Methyl bromide	I	1,000	1,000
74–90–8	Hydrocyanic Acid		10	100
74–93–1	Methyl Mercaptan		100	500
75–15–0	Carbon Disulfide	I	100	10,000
75–18–3	Dimethyl Sulfide	e	1	100
75–21–8	Ethylene Oxide	d, I	1	1,000
75–44–5	Phosgene	I	10	10
75–55–8	Propylenemine	d	1	10,000
75–56–9	Propylene Oxide	I	100	10,000
75–74–1	Tetramethyllead	c, e, I	1	100
75–77–4	Trimethylchlorosilane	e	1	1,000
75–78–5	Dimethyldichlorosilane	e, h	1	500
75–79–6	Methyltrichlorosilane	e, h	1	500
75–86–5	Acetone Cyanohydrin		10	1,000
76–01–7	Pentachloroethane	a, d	1	10,000
76–02–8	Trichloroacetyl Chloride	e	1	500
77–47–4	Hexachlorocyclopentadiene	d, h	1	100
77–78–1	Dimethyl Sulfate	d	1	500
77–81–6	Tabun	c, e, h	1	10
78–00–2	Tetraethyllead	c, d	10	100
78–34–2	Dioxathion	e	1	500
78–53–5	Amiton	e	1	500
78–71–7	Oxetane, 3,3-Bis(Chloromethyl)-	e, h	1	1,000
78–82–0	Isobutyronitrile	e	1	10
78–94–4	Methyl Vinyl Ketone	e	1	1,000
78–97–7	Lactonitrile	d, I	5,000	1,000/10,000
79–06–1	Acrylamide	e	1	100/10,000
79–11–8	Chloroacetic Acid	e	100	100/10,000
79–19–6	Thiosemicarbazide			

Federal Register / Vol. 52, No. 77 / Wednesday, April 22, 1987 / Rules and Regulations   13405

APPENDIX B.—THE LIST OF EXTREMELY HAZARDOUS SUBSTANCES AND THEIR THRESHOLD PLANNING QUANTITIES—Continued

[CAS Number Order]

CAS No.	Chemical name	Notes	Reportable quantity* (pounds)	Threshold planning quantity (pounds)
79-21-0	Peracetic Acid	e	1	500
79-22-1	Methyl Chloroformate	d, h	1,000	500
80-63-7	Methyl 2-Chloroacrylate	e	1	500
81-81-2	Warfarin		100	500/10,000
82-66-6	Diphacinone	e	1	10/10,000
84-74-2	Dibutyl Phthalate	a	10	10,000
84-80-0	Phylloquinone	a, e	1	10,000
86-50-0	Azinphos-Methyl		1	10/10,000
86-88-4	ANTU		100	500/10,000
87-86-5	Pentachlorophenol	a, d	10	10,000
88-05-1	Aniline, 2,4,6-Trimethyl-	e	1	500
88-85-7	Dinoseb		1,000	100/10,000
91-08-7	Toluene 2,6-Diisocyanate		100	100
93-05-0	Diethyl-p-Phenylenediamine	a, e	1	10,000
95-48-7	Cresol, o-	d	1,000	1,000/10,000
95-63-6	Pseudocumene	a, e	1	10,000
97-18-7	Phenol, 2,2'-Thiobis(4,6-Dichloro-(4,6-dichloro)-	e	1	100/10,000
98-05-5	Benzenearsonic Acid	e	1	10/10,000
98-07-7	Benzotrichloride	e	1	100
98-09-9	Benzenesulfonyl Chloride	d	1	100
98-13-5	Trichlorophenylsilane	a	100	10,000
98-16-8	Benzenamine, 3-(Trifluoromethyl)-	e, h	1	500
98-87-3	Benzal Chloride	e	1	500
98-95-3	Nitrobenzene	d	5,000	500
99-98-9	Dimethyl-p-Phenylenediamine	l	1,000	10,000
100-14-1	Benzene, 1-(Chloromethyl)-4-Nitro-	e	1	10/10,000
100-44-7	Benzyl Chloride	e	1	500/10,000
102-36-3	Isocyanic Acid, 3,4-Dichlorophenyl Ester	d	100	500
103-85-5	Phenylthiourea	e	1	500/10,000
106-89-8	Epichlorohydrin		100	100/10,000
106-96-7	Propargyl Bromide	d, l	1,000	1,000
106-99-0	Butadiene	e	1	10
107-02-8	Acrolein	a, e	1	10,000
107-07-3	Chloroethanol	e	1	500
107-11-9	Allylamine	e	1	500
107-12-0	Propionitrile		10	500
107-13-1	Acrylonitrile	d, l	100	10,000
107-15-3	Ethylenediamine		5,000	10,000
107-16-4	Formaldehyde Cyanohydrin	e, h	1	1,000
107-18-6	Allyl Alcohol		100	1,000
107-20-0	Chloroacetaldehyde	a	1,000	10,000
107-30-2	Chloromethyl Methyl Ether	c, d	1	100
107-44-8	Sarin	e, h	1	10
107-49-3	TEPP		10	100
108-05-4	Vinyl Acetate Monomer	d, l	5,000	1,000
108-23-6	Isopropyl Chloroformate	e	1	1,000
108-67-8	Mesitylene	a, e	1	10,000
108-91-8	Cyclohexylamine	e, l	1	10,000
108-95-2	Phenol		1,000	500/10,000
108-98-5	Thiophenol		100	500
109-19-3	Butyl Isovalerate	a, e	1	10,000
109-61-5	Propyl Chloroformate	e	1	500
109-77-3	Malononitrile		1,000	500/10,000
110-00-9	Furan		100	500
110-57-6	Trans-1,4-Dichlorobutene	e	1	500
110-89-4	Piperidine	e	1	1,000
111-34-2	Butyl Vinyl Ether	a, e	1	10,000
111-44-4	Dichloroethyl Ether	d	1	10,000
111-69-3	Adiponitrile	e, l	1	1,000
115-21-9	Trichloroethylsilane	e, h	1	500
115-26-4	Dimefox	e	1	500
115-29-7	Endosulfan		1	10/10,000
115-90-2	Fensulfothion	e, h	1	500
116-06-3	Aldicarb	c	1	100/10,000
117-52-2	Coumafuryl	a, e	1	10,000
117-84-0	Dioctyl Phthalate	a	5,000	10,000
119-38-0	Isopropylmethylpyrazolyl Dimethylcarbamate	e	1	500

13406   Federal Register / Vol. 52, No. 77 / Wednesday, April 22, 1987 / Rules and Regulations

APPENDIX B.—THE LIST OF EXTREMELY HAZARDOUS SUBSTANCES AND THEIR THRESHOLD PLANNING QUANTITIES—Continued

[CAS Number Order]

CAS No.	Chemical name	Notes	Reportable quantity* (pounds)	Threshold planning quantity (pounds)
122–14–5	Fenitrothion	e	1	500
123–31–9	Hydroquinone	i	1	500/10,000
123–73–9	Crotonaldehyde, (E)-		100	1,000
124–65–2	Sodium Cacodylate	e	1	100/10,000
124–87–8	Picrotoxin	e	1	500/10,000
126–98–7	Methacrylonitrile	h	1	500
128–56–3	Sodium Anthraquinone-1-Sulfonate	a, e	1	10,000
129–00–0	Pyrene	c	5,000	1,000/10,000
129–06–6	Warfarin Sodium	e, h	1	100/10,000
131–11–3	Dimethyl Phthalate	a	5,000	10,000
131–52–2	Sodium Pentachlorophenate	e	1	100/10,000
140–29–4	Benzyl Cyanide	e, h	1	500
140–76–1	Pyridine, 2-Methyl-5-Vinyl-	e	1	500
141–66–2	Dicrotophos	e	1	100
143–33–9	Sodium Cyanide (Na(CN))	b	10	100
144–49–0	Fluoroacetic Acid	e	1	10/10,000
149–74–6	Dichloromethylphenylsilane	e	1	1,000
151–38–2	Methoxyethylmercuric Acetate	e	1	500/10,000
151–50–8	Potassium Cyanide	b	10	100
151–56–4	Ethyleneimine	d	1	500
152–16–9	Diphosphoramide, Octamethyl-		100	100
287–92–3	Cyclopentane	a, e	1	10,000
297–78–9	Isobenzan	e	1	100/10,000
297–97–2	Thionazin		100	500
298–00–0	Parathion-Methyl	c	100	100/10,000
298–02–2	Phorate		10	10
298–04–4	Disulfoton		1	500
300–62–9	Amphetamine	e	1	1,000
302–01–2	Hydrazine	d	1	1,000
309–00–2	Aldrin		1	500/10,000
315–18–4	Mexacarbate		1,000	500/10,000
316–42–7	Emetine, Dihydrochloride	e, h	1	1/10,000
327–98–0	Trichloronate	e, k	1	500
353–42–4	Boron Trifluoride Compound With Methyl Ether (1:1)	c, e	1	1,000
359–06–8	Fluoroacetyl Chloride	c, e, h	1	10
371–62–0	Ethylene Fluorohydrin		1	10
379–79–3	Ergotamine Tartrate	e	1	500/10,000
465–73–6	Isodrin		1	100/10,000
470–90–6	Chlorfenvinfos	e	1	500
502–39–6	Methylmercuric Dicyanamide	e	1	500/10,000
504–24–5	Pyridine, 4-Amino-	h	1,000	500/10,000
505–60–2	Mustard Gas	a, h	1	500
506–61–6	Potassium Silver Cyanide	b	1	500
506–68–3	Cyanogen Bromide		1,000	500/10,000
506–78–5	Cyanogen Iodide	e	1	1,000/10,000
509–14–8	Tetranitromethane		10	500
514–73–8	Dithiazanine Iodide	e	1	500/10,000
534–07–6	Bis(Chloromethyl) Ketone	e	1	10/10,000
534–52–1	Dinitrocresol		10	10/10,000
535–89–7	Crimidine	e	1	100/10,000
538–07–8	Ethylbis(2-Chloroethyl)Amine	a, b	1	500
541–25–3	Lewisite	c, e, h	1	10
541–53–7	Dithiobiuret		100	100/10,000
542–76–7	Propionitrile, 3-Chloro-		1,000	1,000
542–88–1	Chloromethyl Ether	d, h	1	100
542–90–5	Ethylthiocyanate	e	1	10,000
555–77–1	Tris(2-Chloroethyl)Amine	a, h	1	100
556–61–6	Methyl Isothiocyanate	b, e	1	500
556–64–9	Methyl Thiocyanate	e	1	10,000
558–25–8	Methanesulfonyl Fluoride	e	1	1,000
563–12–2	Ethion		10	1,000
563–41–7	Semicarbazide Hydrochloride	e	1	1,000/10,000
584–84–9	Toluene 2,4-Diisocyanate		100	500
594–42–3	Perchloromethylmercaptan		100	500
597–64–8	Tetraethyltin	c, e	1	100

Federal Register / Vol. 52, No. 77 / Wednesday, April 22, 1987 / Rules and Regulations 13407

APPENDIX B.—THE LIST OF EXTREMELY HAZARDOUS SUBSTANCES AND THEIR THRESHOLD PLANNING QUANTITIES—Continued

[CAS Number Order]

CAS No.	Chemical name	Notes	Reportable quantity* (pounds)	Threshold planning quantity (pounds)
614-78-8	Thiourea, (2-Methylphenyl)-	e	1	500/10,000
624-83-9	Methyl Isocyanate	f	1	500
624-92-0	Methyl Disulfide	e	1	100
625-55-8	Isopropyl Formate	e	1	500
627-11-2	Chloroethyl Chloroformate	e	1	1,000
630-60-4	Ouabain	c, e	1	100/10,000
633-03-4	C. I. Basic Green 1	a, e	1	10,000
639-58-7	Triphenyltin Chloride	e	1	500/10,000
640-15-3	Thiometon	a, e	1	10,000
640-19-7	Fluoroacetamide	j	100	100/10,000
644-64-4	Dimetilan	e	1	500/10,000
646-06-0	Dioxolane	a, e	1	10,000
675-14-9	Cyanuric Fluoride	e	1	100
676-97-1	Methyl Phosphonic Dichloride	b, e	1	100
696-28-6	Phenyl Dichloroarsine	d, h	1	500
732-11-6	Phosmet	e	1	10/10,000
760-93-0	Methacrylic Anhydride	e	1	500
786-19-6	Carbophenothion	e	1	500
814-49-3	Diethyl Chlorophosphate	e, h	1	500
814-68-6	Acrylyl Chloride	e, h	1	100
824-11-3	Trimethylolpropane Phosphite	e, h	1	100/10,000
900-95-8	Stannane, Acetoxytriphenyl-	e, g	1	500/10,000
919-86-8	Demeton-S-Methyl	e	1	500
920-46-7	Methacryloyl Chloride	e	1	100
944-22-9	Fonofos	e	1	500
947-02-4	Phosfolan	e	1	100/10,000
950-10-7	Mephosfolan	e	1	500
950-37-8	Methidathion	e	1	500/10,000
991-42-4	Norbormide	e	1	100/10,000
998-30-1	Triethoxysilane	e	1	500
999-81-5	Chlormequat Chloride	e, h	1	100/10,000
1031-47-6	Triamiphos	e	1	500/10,000
1066-45-1	Trimethyltin Chloride	e	1	500/10,000
1122-60-7	Nitrocyclohexane	e	1	500
1124-33-0	Pyridine, 4-Nitro-, 1-Oxide	e	1	500/10,000
1129-41-5	Metolcarb	e	1	100/10,000
1303-28-2	Arsenic Pentoxide	d	5,000	100/10,000
1306-19-0	Cadmium Oxide	e	1	100/10,000
1314-32-5	Thallic Oxide	e	100	10,000
1314-56-3	Phosphorus Pentoxide	b, e	1	10
1314-62-1	Vanadium Pentoxide		1,000	100/10,000
1314-84-7	Zinc Phosphide	b	100	500
1327-53-3	Arsenous Oxide	d, h	5,000	100/10,000
1331-17-5	Propylene Glycol, Allyl Ether	a, e	1	10,000
1335-87-1	Hexachloronaphthalene	a, e	1	10,000
1397-94-0	Antimycin A	c, e	1	1,000/10,000
1405-87-4	Bacitracin	a, e	1	10,000
1420-07-1	Dinoterb	e	1	500/10,000
1464-53-5	Diepoxybutane	d	1	500
1558-25-4	Trichloro(Chloromethyl)Silane	e	1	100
1563-66-2	Carbofuran		10	10/10,000
1600-27-7	Mercuric Acetate	e	1	500/10,000
1622-32-8	Ethanesulfonyl Chloride, 2-Chloro-	e	1	500
1642-54-2	Diethylcarbamazine Citrate	e	1	100/10,000
1752-30-3	Acetone Thiosemicarbazide	e	1	1,000/10,000
1910-42-5	Paraquat	e	1	10/10,000
1982-47-4	Chloroxuron	e	1	500/10,000
2001-95-8	Valinomycin	c, e	1	1,000/10,000
2032-65-7	Methiocarb		10	500/10,000
2074-50-2	Paraquat Methosulfate	e	1	10/10,000
2097-19-0	Phenylsilatrane	a, h	1	100/10,000
2104-64-5	EPN	e	1	100/10,000
2223-93-0	Cadmium Stearate	c, e	1	1,000/10,000
2231-57-4	Thiocarbazide	e	1	1,000/10,000
2235-25-8	Ethylmercuric Phosphate	a, e	1	10,000
2238-07-5	Diglycidyl Ether	e	1	1,000
2244-16-8	Carvone	a, e	1	10,000

13408   Federal Register / Vol. 52, No. 77 / Wednesday, April 22, 1987 / Rules and Regulations

APPENDIX B.—THE LIST OF EXTREMELY HAZARDOUS SUBSTANCES AND THEIR THRESHOLD PLANNING QUANTITIES—Continued

[CAS Number Order]

CAS No.	Chemical name	Notes	Reportable quantity* (pounds)	Threshold planning quantity (pounds)
2275-18-5	Prothoate	e	1	100/10,000
2497-07-6	Oxydisulfoton	e, h	1	500
2524-03-0	Dimethyl Phosphorochloridothioate	e	1	500
2540-82-1	Formothion	e	1	100
2570-26-5	Pentadecylamine	e	1	100/10,000
2587-90-8	Phosphorothioic Acid, O,O-Dimethyl-S-(2-Methylthio) Ethyl Ester	c, e, g	1	500
2631-37-0	Promecarb	e, h	1	500/10,000
2636-26-2	Cyanophos	e	1	1,000
2642-71-9	Azinphos-Ethyl	e	1	100/10,000
2665-30-7	Phosphonothioic Acid, Methyl-,O-(4-Nitrophenyl) O-Phenyl Ester	e	1	500
2703-13-1	Phosphonothioic Acid, Methyl-,O-Ethyl O-(4-(Methylthio)Phenyl) Ester	e	1	500
2757-18-8	Thallous Malonate	c, e, h	1	100/10,000
2763-96-4	Muscimol	a, h	1,000	10,000
2778-04-3	Endothion	e	1	500/10,000
3037-72-7	Silane, (4-Aminobutyl)Diethoxymethyl-	e	1	1,000
3048-64-4	Vinylnorbornene	a, e	1	10,000
3254-63-5	Phosphoric Acid, Dimethyl 4-(Methylthio) Phenyl Ester	e	1	500
3569-57-1	Sulfoxide, 3-Chloropropyl Octyl	e	1	500
3615-21-2	Benzimidazole, 4,5-Dichloro-2-(Trifluoromethyl)-	e, g	1	500/10,000
3689-24-5	Sulfotep		100	500
3691-35-8	Chlorophacinone	e	1	100/10,000
3734-97-2	Amiton Oxalate	e	1	100/10,000
3735-23-7	Methyl Phenkapton	e	1	500
3878-19-1	Fuberidazole	e	1	100/10,000
4044-65-9	Bitoscanate	e	1	500/10,000
4098-71-9	Phosacetim	b, e	1	100
4104-14-7	Phosacetim	e	1	100/10,000
4170-30-3	Crotonaldehyde		100	1,000
4301-50-2	Fluenetil	e	1	100/10,000
4418-66-0	Phenol, 2,2'-Thiobis(4-Chloro-6-Methyl)-	e	1	100/10,000
4835-11-4	Hexamethylenediamine, N,N'-Dibutyl-	e	1	500
5281-13-0	Piprotal	e	1	100/10,000
5344-82-1	Thiourea, (2-Chlorophenyl)-		100	100/10,000
5836-29-3	Coumatetralyl	e	1	500/10,000
6533-73-9	Thallous Carbonate	c, h	100	100/10,000
6923-22-4	Monocrotophos	e	1	10/10,000
7440-02-0	Nickel	a, d	1	10,000
7440-48-4	Cobalt	a, e	1	10,000
7446-09-5	Sulfur Dioxide	e, l	1	500
7446-11-9	Sulfur Trioxide	b, e	1	100
7446-18-6	Thallous Sulfate		100	100/10,000
7487-94-7	Mercuric Chloride	e	1	500/10,000
7550-45-0	Titanium Tetrachloride	e	1	100
7580-67-8	Lithium Hydride	b, e	1	100
7631-89-2	Sodium Arsenate	d	1,000	1,000/10,000
7637-07-2	Boron Trifluoride	e	1	500
7647-01-0	Hydrogen Chloride (Gas Only)	e, l	1	500
7664-39-3	Hydrogen Fluoride		100	100
7664-41-7	Ammonia	l	100	500
7664-93-9	Sulfuric Acid		1,000	1,000
7697-37-2	Nitric Acid		1,000	1,000
7719-12-2	Phosphorus Trichloride		1,000	1,000
7722-84-1	Hydrogen Peroxide (Conc >52%)	e, l	1	1,000
7723-14-0	Phosphorus	b, h	1	100
7726-95-6	Bromine	e, l	1	500
7778-44-1	Calcium Arsenate	d	1,000	500/10,000
7782-41-4	Fluorine	k	10	500
7782-50-5	Chlorine		10	100
7783-00-8	Selenious Acid		10	1,000/10,000
7783-06-4	Hydrogen Sulfide	l	100	500
7783-07-5	Hydrogen Selenide	e	1	10
7783-60-0	Sulfur Tetrafluoride	e	1	100
7783-70-2	Antimony Pentafluoride	e	1	500
7783-80-4	Tellurium Hexafluoride	e, k	1	100
7784-34-1	Arsenous Trichloride	d	5,000	500

APPENDIX H   247

Federal Register / Vol. 52, No. 77 / Wednesday, April 22, 1987 / Rules and Regulations   13409

APPENDIX B.—THE LIST OF EXTREMELY HAZARDOUS SUBSTANCES AND THEIR THRESHOLD PLANNING QUANTITIES—Continued

[CAS Number Order]

CAS No.	Chemical name	Notes	Reportable quantity* (pounds)	Threshold planning quantity (pounds)
7784-42-1	Arsine	e	1	100
7784-46-5	Sodium Arsenite	d	1,000	500/10,000
7786-34-7	Mevinphos		10	500
7791-12-0	Thallous Chloride	c, h	100	100/10,000
7791-23-3	Selenium Oxychloride	e	1	500
7803-51-2	Phosphine		100	500
8001-35-2	Camphechlor	d	1	500/10,000
8023-53-8	Dichlorobenzalkonium Chloride	a, e	1	10,000
8065-48-3	Demeton	e	1	500
10025-65-7	Plantinous Chloride	a, e	1	10,000
10025-73-7	Chromic Chloride	e	1	1/10,000
10025-87-3	Phosphorus Oxychloride	d	1,000	500
10025-97-5	Iridium Tetrachloride	a, e	1	10,000
10026-13-8	Phosphorus Pentachloride	b, e	1	500
10028-15-6	Ozone	e	1	100
10031-59-1	Thallium Sulfate	h	100	100/10,000
10049-07-7	Rhodium Trichloride	a, e	1	10,000
10102-18-8	Sodium Selenite	h	100	100/10,000
10102-20-2	Sodium Tellurite		1	500/10,000
10102-43-9	Nitric Oxide	c	10	100
10102-44-0	Nitrogen Dioxide		10	100
10124-50-2	Potassium Arsenite	d	1,000	500/10,000
10140-87-1	Ethanol, 1,2-Dichloro-, Acetate	e	1	1,000
10210-68-1	Cobalt Carbonyl	e, h	1	10/10,000
10265-92-6	Methamidophos	e	1	100/10,000
10294-34-5	Boron Trichloride	e	1	500
10311-84-9	Dialifor	e	1	100/10,000
10476-95-6	Methacrolein Diacetate	e	1	1,000
12002-03-8	Paris Green	d	100	500/10,000
12108-13-3	Manganese, Tricarbonyl Methylcyclopentadienyl	e, h	1	100
13071-79-9	Terbufos	e, h	1	100
13171-21-6	Phosphamidon	e	1	100
13194-48-4	Ethoprophos	e	1	1,000
13410-01-0	Sodium Selenate	e	1	100/10,000
13450-90-3	Gallium Trichloride	e	1	500/10,000
13454-96-1	Platinum Tetrachloride	a, e	1	10,000
13463-39-3	Nickel Carbonyl	d	1	1
13463-40-6	Iron, Pentacarbonyl	e	1	100
13494-80-9	Tellurium	e	1	500/10,000
14167-18-1	Salcomine	e	1	500/10,000
15271-41-7	Bicyclo [2.2.1] Heptane-2-Carbonitrile, 5-Chloro-6-(((Methylamino)Carbonyl)Oxy)Imino)-, (1s-(1-alpha, 2-beta, 4-alpha, 5-alpha, 6E))-.	e	1	500/10,000
16752-77-5	Methomyl	h	100	500/10,000
16919-58-7	Ammonium Chloroplatinate	a, e	1	10,000
17702-41-9	Decaborane(14)	e	1	500/10,000
17702-57-7	Formparanate	e	1	100/10,000
19287-45-7	Diborane	e	1	100
19624-22-7	Pentaborane	e	1	500
20816-12-0	Osmium Tetroxide	a	1,000	10,000
20830-75-5	Digoxin	e, h	1	10/10,000
20859-73-8	Aluminum Phosphide	b	100	500
21548-32-3	Fosthietan	e	1	500
21564-17-0	Thiocyanic Acid, 2-(Benzothiazolylthio)Methyl Ester	a, e	1	10,000
21609-90-5	Leptophos	e	1	500/10,000
21908-53-2	Mercuric Oxide		1	500/10,000
21923-23-9	Chlorthiophos	e, h	1	500
22224-92-6	Fenamiphos	e	1	10/10,000
23135-22-0	Oxamyl	e	1	100/10,000
23422-53-9	Formetanate Hydrochloride	e,h	1	500/10,000
23505-41-1	Pirimifos-Ethyl	e	1	1,000
24017-47-8	Triazofos	e	1	500
24934-91-6	Chlormephos	e	1	500
26419-73-8	Carbamic Acid, Methyl-, O-(((2,4-Dimethyl-1, 3-Dithiolan-2-yl)Methylene)Amino)-	e	1	100/10,000
26628-22-8	Sodium Azide (Na(N₃))	b	1,000	500
27137-85-5	Trichloro(Dichlorophenyl)Silane	e	1	500
28347-13-9	Xylylene Dichloride	e	1	100/10,000
28772-56-7	Bromadiolone	e	1	100/10,000

13410   Federal Register / Vol. 52, No. 77 / Wednesday, April 22, 1987 / Rules and Regulations

APPENDIX B.—THE LIST OF EXTREMELY HAZARDOUS SUBSTANCES AND THEIR THRESHOLD PLANNING QUANTITIES—Continued

[CAS Number Order]

CAS No.	Chemical name	Notes	Reportable quantity* (pounds)	Threshold planning quantity (pounds)
30674-80-7	Methacryloyloxyethyl Isocyanate....................	e, h	1	100
39196-18-4	Thiofanox		100	100/10,000
50782-69-9	Phosphonothioic Acid, Methyl-, S-(2-(Bis(1-Methylethyl)Amino)Ethyl) O-Ethyl Ester	e	1	100
53558-25-1	Pyriminil	e, h	1	100/10,000
58270-08-9	Zinc, Dichloro(4,4-Dimethyl-5((((Methylamino)Carbonyl)Oxy)Imino)Pentanenitrile)-, (T-4)-..	e	1	100/10,000
62207-76-5	Cobalt, ((2,2'-(1,2-Ethanediylbis(Nitrilomethylidyne))Bis(6-Fluorophenolato))(2-)-N,N',O,O')-.	e	1	100/10,000

* Only the statutory or final RQ is shown. For more information, see 40 CFR Table 302.4.

Notes:
a This chemical does not meet acute toxicity criteria. Its TPQ is set at 10,000 pounds.
b This material is a reactive solid. The TPQ does not default to 10,000 pounds for non-powder, non-molten, non-solution form.
c The calculated TPQ changed after technical review as described in the technical support document.
d Indicates that the RQ is subject to change when the assessment of potential carcinogenicity and/or other toxicity is completed.
e Statutory reportable quantity for purposes of notification under SARA sect 304(a)(2).
f The statutory 1 pound reportable quantity for methyl isocyanate may be adjusted in a future rulemaking action.
g New chemicals added that were not part of the original list of 402 substances.
h Revised TPQ based on new or re-evaluated toxicity data.
j TPQ is revised to its calculated value and does not change due to technical review as in proposed rule.
k The TPQ was revised after proposal due to calculation error.
l Chemicals on the original list that do not meet the toxicity criteria but because of their high production volume and recognized toxicity are considered chemicals of concern ("Other chemicals").

[FR Doc. 87-9089 Filed 4-20-87; 11:24 am]
BILLING CODE 6560-50-M

March 23, 1988

Revision to SARA TITLE III CONSOLIDATED CHEMICAL LIST

IN ACCORDANCE WITH 52 FR 48072,48073 (DECEMBER 17, 1987) AND 53 FR 5574 (FEBRUARY 25, 1988), THE FOLLOWING CHEMICALS ARE REMOVED FROM THE SECTION 302: EXTREMELY HAZARDOUS SUBSTANCES.

CAS Number	Chemical Name
16919-58-7	AMMONIUM CHLOROPLATINATE
1404-87-4	BACITRACIN
98-09-9	BENZENESULFONYL CHLORIDE
106-99-0	BUTADIENE
109-19-3	BUTYL ISOVALERATE
111-34-2	BUTYL VINYL ETHER
633-03-4	C.I. BASIC GREEN 1
2244-16-8	CARVONE
107-20-0	CHLOROACETALDEHYDE
7440-48-4	COBALT
117-52-2	COUMAPHOS
287-92-3	CYCLOPENTANE
84-72-2	DIBUTYL PHTHALATE
8023-53-8	DICHLOROBENZALKONIUM CHLORIDE
131-11-3	DIMETHYL PHTHALATE
117-84-0	DIOCTYL PHTHALATE
2235-25-8	ETHYLMERCURIC PHOSPHATE
1335-87-1	HEXACHLORONAPHTHALENE
53-86-1	INDOMETHACIN
10025-97-5	IRIDIUM TETRACHLORIDE
108-67-8	MESITYLENE
7440-02-0	NICKEL
65-86-1	OROTIX ACID
20816-12-0	OSMIUM TETROXIDE
76-01-7	PENTACHLOROETHANE
87-86-5	PENTACHLOROPHENOL
1331-17-5	PROPYLENE OXIDE, ALLYL ETHER
95-63-6	PSEUDOCUMENE OR 1,2,4 TRIMETHYLBENZENE
10049-07-7	RHODIUM TRICHLORIDE
128-56-3	SODIUM ANTHRAQUINONE-1-SULFONATE
21564-17-0	THIOCYANIC ACID 2(BENZOTHIAZOLTHIO)METHYL ETHER
640-15-3	THIOMETON
52-68-6	TRICHLOROPHON
95-63-6	TRIMETHYLBENZENE
3048-64-4	VINYLNORBORNENE

IF ANY OF THE CHEMICALS WERE SUBJECT TO OTHER PARTS OF SARA TITLE III, THEY WILL CONTINUE TO BE COVERED BY THE OTHER SECTIONS.

AUTHORITY: Secs. 311, 312, 324, 325, 328, 329 of Pub. L. 99-499, 100 Stat. 1613, 42 U.S.C. 11011, 11012, 11024, 11025, 11028, 11029.

SOURCE: 52 FR 38364, Oct. 15, 1987, unless otherwise noted.

§ 370.1

Subpart A—General Provisions

§ 370.1 Purpose.

These regulations establish reporting requirements which provide the public with important information on the hazardous chemicals in their communities for the purpose of enhancing community awareness of chemical hazards and facilitating development of State and local emergency response plans.

§ 370.2 Definitions.

"*Commission*" means the State emergency response commission, or the Governor if there is no commission, for the State in which the facility is located.

"*Committee*" means the local emergency planning committee for the emergency planning district in which the facility is located.

"*Environment*" includes water, air, and land and the interrelationship that exists among and between water, air, and land and all living things.

"*Extremely hazardous substance*" means a substance listed in the Appendices to 40 CFR Part 355, Emergency Planning and Notification.

"*Facility*" means all buildings, equipment, structures, and other stationary items that are located on a single site or on contiguous or adjacent sites and that are owned or operated by the same person (or by any person which controls, is controlled by, or under common control with, such person). For purposes of emergency release notification, the term includes motor vehicles, rolling stock, and aircraft.

"*Hazard category*" means any of the following:

(1) "Immediate (acute) health hazard," including "highly toxic," "toxic," "irritant," "sensitizer," "corrosive," (as defined under § 1910.1200 of Title 29 of the Code of Federal Regulations) and other hazardous chemicals that cause an adverse effect to a target organ and which effect usually occurs rapidly as a result of short term exposure and is of short duration;

(2) "Delayed (chronic) health hazard," including "carcinogens" (as defined under § 1910.1200 of Title 29 of the Code of Federal Regulations)

and other hazardous chemicals that cause an adverse effect to a target organ and which effect generally occurs as a result of long term exposure and is of long duration;

(3) "Fire hazard," including "flammable," combustible liquid," "pyrophoric," and "oxidizer" (as defined under § 1910.1200 of Title 29 of the Code of Federal Regulations);

(4) "Sudden release of pressure," including "explosive" and "compressed gas" (as defined under § 1910.1200 of Title 29 of the Code of Federal Regulations); and

(5) "Reactive," including "unstable reactive," "organic peroxide," and "water reactive" (as defined under § 1910.1200 of Title 29 of the Code of Federal Regulations).

"*Hazardous chemical*" means any hazardous chemical as defined under § 1910.1200(c) of Title 29 of the Code of Federal Regulations, except that such term does not include the following substances:

(1) Any food, food additive, color additive, drug, or cosmetic regulated by the Food and Drug Administration.

(2) Any substance present as a solid in any manufactured item to the extent exposure to the substance does not occur under normal conditions of use.

(3) Any substance to the extent it is used for personal, family, or household purposes, or is present in the same form and concentration as a product packaged for distribution and use by the general public.

(4) Any substance to the extent it is used in a research laboratory or a hospital or other medical facility under the direct supervision of a technically qualified individual.

(5) Any substance to the extent it is used in routine agricultural operations or is a fertilizer held for sale by a retailer to the ultimate customer.

"*Inventory form*" means the Tier I and Tier II emergency and hazardous chemical inventory forms set forth in Subpart D of this Part

"*Material Safety Data Sheet*" or "MSDS" means the sheet required to be developed under § 1910.1200(g) of Title 29 of the Code of Federal Regulations.

Environmental Protection Agency   **§ 370.20**

"Person" means any individual, trust, firm, joint stock company, corporation (including a government corporation), partnership, association, State, municipality, commission, political subdivision of State, or interstate body.

"Present in the same form and concentration as a product packaged for distribution and use by the general public" means a substance packaged in a similar manner and present in the same concentration as the substance when packaged for use by the general public, whether or not it is intended for distribution to the general public or used for the same purpose as when it is packaged for use by the general public.

"State" means any State of the United States, the District of Columbia, the Commonwealth of Puerto Rico, Guam, American Samoa, the United States Virgin Islands, the Northern Mariana Islands, and any other territory or possession over which the United States has jurisdiction.

"TPQ" means the threshold planning quantity for an extremely hazardous substance as defined in 40 CFR Part 355.

§ 370.5 Penalties.

(a) *MSDA reporting.* Any person other than a governmental entity who violates any requirement of § 370.21 shall be liable for civil and administrative penalties of not more than $10,000 for each violation.

(b) *Inventory reporting.* Any person other than a governmental entity who violates any requirement of § 370.25 shall be liable for civil and administrative penalties of not more than $25,000 for each violation.

(c) *Continuing violations.* Each day a violation described in paragraph (a) or (b) of this section continues shall constitute a separate violation.

Subpart B—Reporting Requirements

§ 370.20 Applicability.

(a) *General.* The requirements of this subpart apply to any facility that is required to prepare or have available a material safety data sheet (or MSDS) for a hazardous chemical

under the Occupational Safety and Health Act of 1970 and regulations promulgated under that Act.

(b) *Minimum threshold levels.* Except as provided in paragraph (b)(3) of this section, the minimum threshold level for reporting under this subpart shall be according to the following schedule.

(1) The owner or operator of a facility subject to this Subpart shall submit an MSDS:

(i) On or before October 17, 1987 (or 3 months after the facility first becomes subject to this subpart), for all hazardous chemicals present at the facility in amounts equal to or greater than 10,000 pounds, or that are extremely hazardous substances present at the facility in an amount greater than or equal to 500 pounds (or 55 gallons) or the TPQ, whichever is less, and

(ii) On or before October 17, 1989 (or 2 years and 3 months after the facility first becomes subject to this Subpart), for all hazardous chemicals present at the facility between 10,000 and zero pounds for which an MSDS has not yet been submitted.

(2) The owner or operator of a facility subject to this Subpart shall submit the Tier I form:

(i) On or before March 1, 1988 (or March 1 of the first year after the facility first becomes subject to this Subpart), covering all hazardous chemicals present at the facility during the preceding calendar year in amounts equal to or greater than 10,000 pounds, or that are extremely hazardous substances present at the facility in an amount greater than or equal to 500 pounds (or 55 gallons) or the TPQ, whichever is less, and

(ii) On or before March 1, 1989 (or March 1 of the second year after the facility first becomes subject to this Subpart), covering all hazardous chemicals present at the facility during the preceding calendar year in amounts equal to or greater than 10,000 pounds, or that are extremely hazardous substances present at the facility in an amount greater than or equal to 500 pounds (or 55 gallons) or the TPQ, whichever is less, and

(iii) On or before March 1990 (or March 1 of the third year after the facility first becomes subject to this Subpart), and annually thereafter, covering all hazardous chemicals present at the facility during the preceding calendar year in amounts equal to or greater than zero pounds or that are extremely hazardous substances present at the facility in an amount equal to or greater than 500 pounds (or 55 gallons) or the TPQ, whichever is less.

(3) The minimum threshold for reporting in response to requests for submission of an MSDS or a Tier II form pursuant to §§ 370.21(d) and 370.25(c) of this part shall be zero.

§ 370.21 MSDS reporting.

(a) *Basic requirement.* The owner or operator of a facility subject to this Subpart shall submit an MSDS for each hazardous chemical present at the facility according to the minimum threshold schedule provided in paragraph (b) of § 370.20 to the committee, the commission, and the fire department with jurisdiction over the facility.

(b) *Alternative reporting.* In lieu of the submission of an MSDS for each hazardous chemical under paragraph (a) of this section, the owner or operator may submit the following:

(1) A list of the hazardous chemicals for which the MSDS is required, grouped by hazard category as defined under § 370.2 of this Part;

(2) The chemical or common name of each hazardous chemical as provided on the MSDS; and

(3) Except for reporting of mixtures under § 370.28(a)(2), any hazardous component of each hazardous chemical as provided on the MSDS.

(c) *Supplemental reporting.* (1) The owner or operator of a facility that has submitted an MSDS under this section shall provide a revised MSDS to the committee, the commission, and the fire department with jurisdiction over the facility within three months after discovery of significant new information concerning the hazardous chemical for which the MSDS was submitted.

(2) After October 17, 1987, the owner or operator of a facility subject to this section shall submit an MSDS for a hazardous chemical pursuant to para-

graph (a) of this section or a list pursuant to paragraph (b) of this section within three months after the owner or operator is first required to prepare or have available the MSDS or after a hazardous chemical requiring an MSDS becomes present in an amount exceeding the threshold established in § 370.20(b).

(d) *Submission of MSDS upon request.* The owner or operator of a facility that has not submitted the MSDS for a hazardous chemical present at the facility shall submit the MSDS for any such hazardous chemical to the committee upon its request. The MSDS shall be submitted within 30 days of the receipt of such request.

§ 370.25 Inventory reporting.

(a) *Basic requirement.* The owner or operator of a facility subject to this Subpart shall submit an inventory form to the commission, the committee, and the fire department with jurisdiction over the facility. The inventory form containing Tier I information on hazardous chemicals present at the facility during the preceding calendar year above the threshold levels established in § 370.20(b) shall be submitted on or before March 1 of each year, beginning in 1988.

(b) *Alternative reporting.* With respect to any specific hazardous chemical at the facility, the owner or operator may submit a Tier II form in lieu of the Tier I information.

(c) *Submission of Tier II information.* The owner or operator of a facility subject to this Section shall submit the Tier II form to the commission, committee, or the fire department having jurisdiction over the facility upon request of such persons. The Tier II form shall be submitted within 30 days of the receipt of each request.

(d) *Fire department inspection.* The owner or operator of a facility that has submitted an inventory form under this section shall allow on-site inspection by the fire department having jurisdiction over the facility upon request of the department, and shall provide to the department specific location information on hazardous chemicals at the facility.

Environmental Protection Agency § 370.40

§ 370.28 Mixtures.

(a) *Basic reporting.* The owner or operator of a facility may meet the reporting requirements of §§ 370.21 (MSDS reporting) and 370.25 (inventory form reporting) of this Subpart for a hazardous chemical that is a mixture of hazardous chemicals by:

(1) Providing the required information on each component in the mixture which is a hazardous chemical, or

(2) Providing the required information on the mixture itself, so long as the reporting of mixtures by a facility under § 370.21 is in the same manner as under § 370.25, where practicable.

(b) *Calculation of the quantity.* (1) if the reporting is on each component of the mixture which is a hazardous chemical, then the concentration of the hazardous chemical, in weight percent (greater than 1% or 0.1% if carcinogenic) shall be multiplied by the mass (in pounds) of the mixture to determine the quantity of the hazardous chemical in the mixture.

(2) If the reporting is on the mixture itself, the total quantity of the mixture shall be reported.

Subpart C—Public Access and Availability of Information

§ 370.30 Requests for information.

(a) *Request for MSDS information.* (1) Any person may obtain an MSDS with respect to a specific facility by submitting a written request to the committee.

(2) If the committee does not have in its possession the MSDS requested in paragraph (a)(1) of this section, it shall request a submission of the MSDS from the owner or operator of the facility that is the subject of the request.

(b) *Requests for Tier II information.* (1) Any person may request Tier II information with respect to a specific facility by submitting a written request to the commission or committee in accordance with the requirements of this section.

(2) If the committee or commission does not have in its possession the Tier II information requested in paragraph (b)(1) of this section, it shall request a submission of the Tier II form from the owner or operator of the facility that is the subject of the request, provided that the request is from a State or local official acting in his or her official capacity or the request is limited to hazardous chemicals stored at the facility in an amount in excess of 10,000 pounds.

(3) If the request under paragraph (b)(1) of this section does not meet the requirements of paragraph (b)(2) of this section, the committee or commission may request submission of the Tier II form from the owner or operator of the facility that is the subject of the request if the request under paragraph (b)(1) of this section includes a general statement of need.

§ 370.31 Provision of information.

All information obtained from an owner or operator in response to a request under this subpart and any requested Tier II form or MSDS otherwise in possession of the commission or the committee shall be made available to the person submitting the request under this Subpart; provided upon request of the owner or operator, the commission or committee shall withhold from disclosure the location of any specific chemical identified in the Tier II form.

Subpart D—Inventory Forms

§ 370.40 Tier I emergency and hazardous chemical inventory form.

(a) The form set out in paragraph (b) of this section shall be completed and submitted as required in § 370.25(a). In lieu of the form set out in paragraph (b) of this section, the facility owner or operator may submit a State or local form that contains identical content.

(b) Tier I Emergency and Hazardous Chemical Inventory Form.

§ 370.40

40 CFR Ch. I (7-1-88 Edition)

Page _____ of _____ pages
Form Approved OMB No. 2050-007?

| Tier One | EMERGENCY AND HAZARDOUS CHEMICAL INVENTORY *Aggregate Information by Hazard Type* | FOR OFFICIAL USE ONLY | ID # |
| | | | Date Received |

Important: Read instructions before completing form

Reporting Period   From January 1 to December 31, 19____

Facility Identification

Name _____

Street Address _____

City _____ State _____ Zip _____

SIC Code [ ][ ][ ][ ]   Dun & Brad Number [ ][ ]-[ ][ ][ ]-[ ][ ][ ]

Owner/Operator

Name _____

Mail Address _____

Phone ( ) _____

Emergency Contacts

Name _____

Title _____

Phone ( ) _____

24 Hour Phone ( ) _____

Name _____

Title _____

Phone ( ) _____

24 Hour Phone ( ) _____

| Hazard Type | Max Amount* | Average Daily Amount* | Number of Days On-Site | General Location | Check if site plan is attached |

Physical Hazards

Fire [ ][ ]   [ ][ ]   [ ][ ] _____

Sudden Release of Pressure [ ][ ]   [ ][ ]   [ ][ ] _____

Reactivity [ ][ ]   [ ][ ]   [ ][ ] _____

Health Hazards

Immediate (acute) [ ][ ]   [ ][ ]   [ ][ ] _____

Delayed (Chronic) [ ][ ]   [ ][ ]   [ ][ ] _____

Certification *(Read and sign after completing all sections)*

I certify under penalty of law that I have personally examined and am familiar with the information submitted in this and all attached documents, and that based on my inquiry of those individuals responsible for obtaining the information, I believe that the submitted information is true, accurate and complete.

_____
Name and official title of owner/operator OR owner/operator's authorized representative

Signature _____ Date signed _____

| * Reporting Ranges | Range Value | Weight Range in Pounds | |
		From...	To...
	00	0	99
	01	100	999
	02	1000	9,999
	03	10,000	99,999
	04	100,000	999,999
	05	1,000,000	9,999,999
	06	10,000,000	49,999,999
	07	50,000,000	99,999,999
	08	100,000,000	499,999,999
	09	500,000,000	999,999,999
	10	1 billion	higher than 1 billion

Environmental Protection Agency   § 370.40

TIER ONE INSTRUCTIONS

GENERAL INFORMATION

Submission of this form is required by Title III of the Superfund Amendments and Reauthorization Act of 1986, Section 312, Public Law 99-499.

The purpose of this form is to provide State and local officials and the public with information on the general types and locations of hazardous chemicals present at your facility during the past year.

YOU MUST PROVIDE ALL INFORMATION REQUESTED ON THIS FORM.

You may substitute the Tier Two form for this Tier One form. (The Tier Two form provides detailed information and must be submitted in response to a specific request from State or local officials.)

WHO MUST SUBMIT THIS FORM

Section 312 of Title III requires that the owner or operator of a facility submit this form if, under regulations implementing the Occupational Safety and Health Act of 1970, the owner or operator is required to prepare or have available Material Safety Data Sheets (MSDS) for hazardous chemicals present at the facility. MSDS requirements are specified in the Occupational Safety and Health Administration (OSHA) Hazard Communication Standard, found in Title 29 of the Code of Federal Regulations at §1910.1200.

WHAT CHEMICALS ARE INCLUDED

You must report the information required on this form for every hazardous chemical for which you are required to prepare or have available an MSDS under the Hazard Communication Standard. However, OSHA regulations and Title III exempt some chemicals from reporting.

Section 1910.1200(b) of the OSHA regulations currently provides the following exemptions:

(i) Any hazardous waste as such term is defined by the Solid Waste Disposal Act, as amended (42 U.S.C. 6901 et seq.) when subject to regulations issued under that Act;

(ii) Tobacco or tobacco products;

(iii) Wood or wood products;

(iv) "Articles"- defined under §1910.1200 (b) as a manufactured item;

• Which is formed to a specific shape or design during manufacture;

• Which has end use function(s) dependent in whole or in part upon the shape or design during end use; and

• Which does not release, or otherwise result in exposure to a hazardous chemical under normal conditions of use.

(v) Food, drugs, cosmetics or alcoholic beverages in a retail establishment which are packaged for sale to consumers;

(vi) Foods, drugs, or cosmetics intended for personal consumption by employees while in the workplace;

(vii) Any consumer product or hazardous substance, as those terms are defined in the Consumer Product Safety Act (15 U.S.C. 1251 et seq.) respectively, where the employer can demonstrate it is used in the workplace in the same manner as normal consumer use, and which use results in a duration and frequency of exposure which is not greater than exposures experienced by consumers; and

(viii) Any drug, as that term is defined in the Federal Food, Drug, and Cosmetic Act (21 U.S.C. 301 et seq.), when it is in solid, final form for direct administration to the patient (i.e., tablets or pills).

In addition, Section 311(e) of Title III excludes the following substances:

(i) Any food, food additive, color additive, drug, or cosmetic regulated by the Food and Drug Administration;

(ii) Any substance present as a solid in any manufactured item to the extent exposure to the substance does not occur under normal conditions of use;

(iii) Any substance to the extent it is used for personal, family, or household purposes, or is present in the same form and concentration as a product packaged for distribution and use by the general public;

(iv) Any substance to the extent it is used in a research laboratory or a hospital or other medical facility under the direct supervision of a technically qualified individual;

(v) Any substance to the extent it is used in routine agricultural operations or is a fertilizer held for sale by a retailer to the ultimate customer.

Also, minimum reporting thresholds have been established under Title III, Section 312. You need to report only those hazardous chemicals that were present at your facility at any time during the preceding calendar year at or above the levels listed below:

• January to December 1987
   (or first year of reporting) ...10,000 lbs.

• January to December 1988
   (or second year of reporting) ...10,000 lbs.

• January to December 1989
   (or third year of reporting) ...zero lbs.*

   * EPA will publish the final threshold, effective in the third year, after additional analysis.

• For extremely hazardous substances...500 lbs. or the threshold planning quantity, whichever is less, from the first year of reporting and thereafter.

WHEN TO SUBMIT THIS FORM

Beginning March 1, 1988, owners or operators must submit the Tier One form (or substitute the Tier Two form) on or before March 1 of every year.

§ 370.40   40 CFR Ch. I (7-1-88 Edition)

INSTRUCTIONS

Please read these instructions carefully. Print or type all responses.

WHERE TO SUBMIT THIS FORM

Send one completed inventory form to each of the following organizations:

1. Your State emergency planning commission

2. Your local emergency planning committee

3. The fire department with jurisdiction over your facility.

PENALTIES

Any owner or operator of a facility who fails to submit or supplies false Tier One information shall be liable to the United States for a civil penalty of up to $25,000 for each such violation. Each day a violation continues shall constitute a separate violation. In addition, any citizen may commence a civil action on his or her own behalf against any owner or operator who fails to submit Tier One information.

You may use the Tier Two form as a worksheet for completing Tier One. Filling in the Tier Two chemical information section should help you assemble your Tier One responses.

If your responses require more than one page, fill in the page number at the top of the form.

REPORTING PERIOD

Enter the appropriate calendar year, beginning January 1 and ending December 31.

FACILITY IDENTIFICATION

Enter the complete name of your facility (and company identifier where appropriate).

Enter the full street address or state road. If a street address is not available, enter other appropriate identifiers that describe the physical location of your facility (e.g., longitude and latitude). Include city, state, and zip code.

Enter the primary Standard Industrial Classification (SIC) code and the Dun & Bradstreet number for your facility. The financial officer of your facility should be able to provide the Dun & Bradstreet number. If your firm does not have this information, contact the state or regional office of Dun & Bradstreet to obtain your facility number or have one assigned.

OWNER/OPERATOR

Enter the owner's or operator's full name, mailing address, and phone number.

EMERGENCY CONTACT

Enter the name, title, and work phone number of at least one local person or office that can act as a referral if emergency responders need assistance in responding to a chemical accident at the facility.

Provide an emergency phone number where such emergency information will be available 24 hours a day, every day.

PHYSICAL AND HEALTH HAZARDS

Descriptions, Amounts, and Locations

This section requires aggregate information on chemicals by hazard categories as defined in 40 CFR 370.3. The two health hazard categories and three physical hazard categories are a consolidation of the 23 hazard categories defined in the OSHA Hazard Communication Standard, 29 CFR 1910.1200. For each hazard type, indicate the total amounts and general locations of all applicable chemicals present at your facility during the past year.

- What units should I use?

  Calculate all amounts as *weight in pounds*. To convert gas or liquid volume to weight in pounds, multiply by an appropriate density factor.

- What about mixtures?

  If a chemical is part of a mixture, *you have the option* of reporting either the weight of the entire mixture or only the portion of the mixture that is a particular hazardous chemical. For example, if a hazardous solution weighs 100 lbs. but is composed of only 5% of a particular hazardous chemical, you can indicate either 100 lbs. of the mixture or 5 lbs. of the chemical).

  Select the option consistent with your Section 311 reporting of the chemical on the MSDS or list of MSDS chemicals.

- Where do I count a chemical that is a fire reactivity physical hazard and an immediate (acute) health hazard?

  Add the chemical's weight to your totals for all three hazard categories and include its location in all three hazard categories. Many chemicals fall into more than one hazard category, which results in double-counting.

MAXIMUM AMOUNT

The amounts of chemicals you have on hand may vary throughout the year. The peak weights -- greatest single-day weights during the year -- are added together in this column to determine the maximum weight for each hazard type. Since the peaks for different chemicals often occur on different days, this maximum amount will seem artificially high.

To complete this and the following sections, you may choose to use the Tier Two form as a worksheet.

To determine the Maximum Amount:

1. List all of your hazardous chemicals individually.

2. For each chemical...

   a. Indicate all physical and health hazards that the chemical presents. Include all chemicals, even if they are present for only a short period of time during the year.

Environmental Protection Agency   **§ 370.40**

b. Estimate the maximum weight in pounds that was present at your facility on any single day of the reporting period.

3. For each hazard type -- beginning with Fire and repeating for all physical and health hazard types...

   a. Add the maximum weights of all chemicals you indicated as the particular hazard type.

   b. Look at the Reporting Ranges at the bottom of the Tier One form. Find the appropriate range value code.

   c. Enter this range value as the Maximum Amount.

EXAMPLE:

You are using the Tier Two form as a worksheet and have listed raw weights in pounds for each of your hazardous chemicals. You have marked an X in the Immediate (acute) hazard column for phenol and sulfuric acid. The maximum amount raw weight you listed were 10,000 lbs. and 50 lbs. respectively. You add these together to reach a total of 10,050 lbs. Then you look at the Reporting Range at the bottom of your Tier One form and find that the value of 03 corresponds to 10,050 lbs. Enter 03 as your Maximum Amount for Immediate (acute) hazards materials.

You also marked an X in the Fire hazard box for phenol. When you calculate your Maximum Amount totals for fire hazards, add the 10,000 lb. weight again.

AVERAGE DAILY AMOUNT

This column should represent the average daily amount of chemicals of each hazard type that were present at your facility at any point during the year.

To determine this amount:

1. List all of your hazardous chemicals individually (same as for Maximum Amount).

2. For each chemical...

   a. Indicate all physical and health hazards that the chemical presents (same as for Maximum Amount).

   b. Estimate the average weight in pounds that was present at your facility throughout the year. To do this, total all daily weights and divide by the number of days the chemical was present on the site.

3. For each hazard type -- beginning with Fire and repeating for all physical and health hazards...

   a. Add the average weights of all chemicals you indicated for the particular hazard type.

   b. Look at the Reporting Ranges at the bottom of the Tier One form. Find the appropriate range value code.

   c. Enter this range value as the Average Daily Amount.

EXAMPLE:

You are using the Tier Two form, and have marked an X in the Immediate (acute) hazard column for nicotine and phenol. Nicotine is present at your facility 100 days during the year, and the sum of the daily weights is 100,000 lbs. By dividing 100,000 lbs. by 100 days on-site, you calculate an Average Daily Amount of 1,000 lbs. for nicotine. Phenol is present at your facility 50 days during the year, and the sum of the daily weights is 10,000 lbs. By dividing 10,000 lbs. by 50 days on-site, you calculate an Average Daily Amount of 200 lbs. for phenol. You then add the two average daily amounts together to reach a total of 1,200 lbs. Then you look at the Reporting Range on your Tier One form and find that the value 02 corresponds to 1,200 lbs. Enter 02 as your Average Daily Amount for Immediate (acute) Hazard.

You also marked an X in the Fire hazard column for phenol. When you calculate your Average Daily Amount for fire hazards, use the 200 lb. weight again.

NUMBER OF DAYS ON-SITE

Enter the greatest number of days that a single chemical within that hazard category was present on-site.

EXAMPLE:

At your facility, nicotine is present for 100 days and phosgene is present for 150 days. Enter 150 in the space provided.

GENERAL LOCATION

Enter the general location within your facility where each hazard may be found. General locations should include the names or identifications of buildings, tank fields, lots, sheds, or other such areas.

For each hazard type, list the locations of all applicable chemicals. As an alternative you may also attach a site plan and list the site coordinates related to the appropriate locations. If you do so, check the Site Plan box.

EXAMPLE:

On your worksheet you have marked an X in the Fire hazard column for acetone and butane. You noted that these are kept in steel drums in Room C of the Main Building, and in pressurized cylinders in Storage Shed 13, respectively. You could enter Main Building and Storage Shed 13 as the General Locations of your fire hazards. However, you choose to attach a site plan and list coordinates. Check the Site Plan box at the top of the column and enter site coordinates for the Main Building and Storage Shed 13 under General Locations.

If you need more space to list locations, attach an additional Tier One form and continue your list on the proper line. Number all pages.

CERTIFICATION

This must be completed by the owner or operator or the officially designated representative of the owner or operator. Enter your full name and official title. Sign your name and enter the current date.

§ 370.41

§ 370.41   Tier II emergency and hazardous chemical inventory form.

(a) The form set out in paragraph (b) of this section must be completed and submitted as required in § 370.25(c). In lieu of the form set out in paragraph (b) of this section, the facility owner or operator may submit a State or local form that contains identical content.

(b) Tier II Emergency and Hazardous Chemical Inventory Form.

Environmental Protection Agency   § 370.41

Form Approved OMB No. 2060-0072

Tier Two

EMERGENCY AND HAZARDOUS CHEMICAL INVENTORY

Specific Information by Chemical

Facility Identification

Name

Street Address

City   State   Zip

SIC Code   Dun & Brad Number

FOR OFFICIAL USE ONLY   ID #   Date Received

Owner/Operator Name

Name

Mail Address

Phone ( )

Emergency Contact

Name   Title

Phone ( )   24 Hr Phone ( )

Name   Title

Phone ( )   24 Hr Phone ( )

Reporting Period From January 1 to December 31, 19___

Important: Read all instructions before completing form

Chemical Description	Physical and Health Hazards (check all that apply)	Inventory			Storage Codes and Locations (Non-Confidential)
		Max. Daily Amount (code)	Avg. Daily Amount (code)	No. of Days On-site (days)	Storage Code Storage Locations

CAS [ ] Trade Secret

Chem. Name

Check all that apply: Pure Mix Solid Liquid Gas

Fire / Sudden Release of Pressure / Reactivity / Immediate (acute) / Delayed (chronic)

CAS [ ] Trade Secret

Chem. Name

Check all that apply: Pure Mix Solid Liquid Gas

Fire / Sudden Release of Pressure / Reactivity / Immediate (acute) / Delayed (chronic)

CAS [ ] Trade Secret

Chem. Name

Check all that apply: Pure Mix Solid Liquid Gas

Fire / Sudden Release of Pressure / Reactivity / Immediate (acute) / Delayed (chronic)

Certification (Read and sign after completing all sections)

I certify under penalty of law that I have personally examined and am familiar with the information submitted in this and all attached documents, and that based on my inquiry of those individuals responsible for obtaining the information, I believe that the submitted information is true, accurate, and complete.

Name and official title of owner/operator OR owner/operator's authorized representative   Signature   Date signed

Optional Attachments (Check one)

[ ] I have attached a site plan

[ ] I have attached a list of site coordinate abbreviations

§ 370.41

Tier Two

EMERGENCY AND HAZARDOUS CHEMICAL INVENTORY

Specific Information by Chemical

Facility Identification

Name
Street Address
City   State   Zip
SIC Code   Dun & Brad Number

FOR OFFICIAL USE ONLY
ID #
Date Received

Owner/Operator Name

Name
Mail Address

Emergency Contact

Name   Title
Phone ( )   24 Hr. Phone ( )
Name   Title
Phone ( )   24 Hr. Phone ( )

Page ____ of ____ pages
Form Approved OMB No. 2050-0072

Important: Read all instructions before completing form

Confidential Location Information Sheet

Reporting Period From January 1 to December 31, 19____

Storage Codes and Locations
(Confidential)

Storage Codes   *Storage Locations*

CAS #   Chem. Name

CAS #   Chem. Name

CAS #   Chem. Name

Certification *(Read and sign after completing all sections)*

I certify under penalty of law that I have personally examined and am familiar with the information submitted in this and all attached documents, and that based on my inquiry of those individuals responsible for obtaining the information, I believe that the submitted information is true, accurate, and complete.

Name and official title of owner/operator OR owner/operator's authorized representative   Signature   Date signed

Optional Attachments *(Check one)*

☐ I have attached a site plan
☐ I have attached a list of site coordinate abbreviations

Environmental Protection Agency   **§ 370.41**

TIER TWO INSTRUCTIONS

GENERAL INFORMATION

Submission of this Tier Two form (when requested) is required by Title III of the Superfund Amendments and Reauthorization Act of 1986, Section 312, Public Law 99-499. The purpose of this Tier Two form is to provide State and local officials and the public with specific information on hazardous chemicals present at your facility during the past year.

YOU MUST PROVIDE ALL INFORMATION REQUESTED ON THIS FORM TO FULFILL TIER TWO REPORTING REQUIREMENTS.

This form may also be used as a worksheet for completing the Tier One form or may be submitted in place of the Tier One form.

WHO MUST SUBMIT THIS FORM

Section 312 of Title III requires that the owner or operator of a facility submit this Tier Two form if so requested by a State emergency planning commission, a local emergency planning committee, or a fire department with jurisdiction over the facility.

This request may apply to the owner or operator of any facility that is required, under regulations implementing the Occupational Safety and Health Act of 1970, to prepare or have available a Material Safety Data Sheet (MSDS) for a hazardous chemical present at the facility. MSDS requirements are specified in the Occupational Safety and Health Administration (OSHA) Hazard Communications Standard, found in Title 29 of the Code of Federal Regulations at § 1910.1200.

WHAT CHEMICALS ARE INCLUDED

You must report the information required on this form for each hazardous chemical for which Tier Two information is requested. However, OSHA regulations and Title III exempt some chemicals from reporting.

Section 1910.1200(b) of the OSHA regulations currently provides the following exemptions:

(i) Any hazardous waste as such term is defined by the Solid Waste Disposal Act as amended (42 U.S.C. 6901 et seq.) when subject to regulations issued under that Act;

(ii) Tobacco or tobacco products;

(iii) Wood or wood products;

(iv) "Articles" - defined under § 1910.1200(b) as a manufactured item:

• Which is formed to a specific shape or design during manufacture;

• Which has end use function(s) dependent in whole or in part upon the shape or design during end use; and

• Which does not release, or otherwise result in exposure to a hazardous chemical under normal conditions of use.

(v) Food, drugs, cosmetics or alcoholic beverages in a retail establishment which are packaged for sale to consumers;

(vi) Foods, drugs, or cosmetics intended for personal consumption by employees while in the workplace.

(vii) Any consumer product or hazardous substance, as those terms are defined in the Consumer Product Safety Act (15 U.S.C. 1251 et seq.) respectively, where the employer can demonstrate it is used in the workplace in the same manner as normal consumer use, and which use results in a duration and frequency of exposure which is not greater than exposures experienced by consumers

(viii) Any drug, as that term is defined in the Federal Food, Drug, and Cosmetic Act (21 U.S.C. 301 et seq.), when it is in solid, final form for direct administration to the patient (i.e., tablets or pills).

In addition, Section 311(e) of Title III excludes the following substances:

(i) Any food, food additive, color additive, drug, or cosmetic regulated by the Food and Drug Administration;

(ii) Any substance present as a solid in any manufactured item to the extent exposure to the substance does not occur under normal conditions of use;

(iii) Any substance to the extent it is used for personal, family, or household purposes, or is present in the same form and concentration as a product packaged for distribution and use by the general public;

(iv) Any substance to the extent it is used in a research laboratory or a hospital or other medical facility under the direct supervision of a technically qualified individual;

(v) Any substance to the extent it is used in routine agricultural operations or is a fertilizer held for sale by a retailer to the ultimate customer.

Also, minimum reporting thresholds have been established for Tier One under Title III, Section 312. You need to report only those hazardous chemicals that were present at your facility at any time during the preceding calendar year at or above the levels listed below:

• January to December 1987 (or first year of reporting) ...10,000 lbs.

• January to December 1988 (or second year of reporting) ...10,000 lbs.

• January to December 1989 (or third year of reporting) ...zero lbs. *

    * EPA will publish the final threshold, effective in the third year, after additional analysis.

• For extremely hazardous substances...500 lbs. or the threshold planning quantity, whichever is less, from the first year of reporting and thereafter.

A requesting official may limit the responses required under Tier Two by specifying particular chemicals or groups of chemicals. Such requests apply to hazardous chemicals regardless of established thresholds.

INSTRUCTIONS

Please read these instructions carefully. Print or type all responses.

WHEN TO SUBMIT THIS FORM

Owners or operators must submit the Tier Two form to the requesting agency within 30 days of receipt of a written request from an authorized official.

WHERE TO SUBMIT THIS FORM

Send the completed Tier Two form to the requesting agency.

PENALTIES

Any owner or operator who violates any Tier Two reporting requirements shall be liable to the United States for a civil penalty of up to $25,000 for each such violation. Each day a violation continues shall constitute a separate violation.

You may use the Tier Two form as a worksheet for completing the Tier One form. Filling in the Tier Two Chemical Information section should help you assemble your Tier One responses.

If your responses require more than one page, fill in the page number at the top of the form.

REPORTING PERIOD

Enter the appropriate calendar year, beginning January 1 and ending December 31.

FACILITY IDENTIFICATION

Enter the full name of your facility (and company identifier where appropriate).

Enter the full street address or state road. If a street address is not available, enter other appropriate identifiers that describe the physical location of your facility (e.g., longitude and latitude). Include city, state, and zip code.

Enter the primary Standard Industrial Classification (SIC) code and the Dun & Bradstreet number for your facility. The financial officer of your facility should be able to provide the Dun & Bradstreet number. If your firm does not have this information, contact the state or regional office of Dun & Bradstreet to obtain your facility number or have one assigned.

OWNER/OPERATOR

Enter the owner's or operator's full name, mailing address, and phone number.

EMERGENCY CONTACT

Enter the name, title, and work phone number of at least one local person or office who can act as a referral if emergency responders need assistance in responding to a chemical accident at the facility.

Provide an emergency phone number where such emergency chemical information will be available 24 hours a day, every day.

CHEMICAL INFORMATION: Description, Hazards, Amounts, and Locations

The main section of the Tier Two form requires specific information on amounts and locations of hazardous chemicals, as defined in the OSHA Hazard Communication Standard.

- What units should I use?

  Calculate all amounts as *weight in pounds*. To convert gas or liquid volume to weight in pounds, multiply by an appropriate density factor.

- What about mixtures?

  If a chemical is part of a mixture, you have the option of reporting either the weight of the entire mixture or only the portion of the mixture that is a particular hazardous chemical (e.g., if a hazardous solution weighs 100 lbs. but is composed of only 5% of a particular hazardous chemical, you can indicate either 100 lbs. of the mixture or 5 lbs. of the chemical).

  Select the option consistent with your Section 311 reporting of the chemical on the MSDS or list of MSDS chemicals.

CHEMICAL DESCRIPTION

1. Enter the Chemical Abstract Service number (CAS#).

   For mixtures, enter the CAS number of the mixture as a whole if it has been assigned a number distinct from its components. For a mixture that has no CAS number, leave this item blank or report the CAS numbers of as many constituent chemicals as possible.

   If you are withholding the name of a chemical in accordance with criteria specified in Title III, Section 322, enter the generic chemical class (e.g., list toluene diisocyanate as organic isocyanate) and check the box marked Trade Secret. Trade secret information should be submitted to EPA and must include a substantiation. Please refer to Section 322 of Title III for detailed information on how to comply with trade secret requests.

2. Enter the chemical name or common name of each hazardous chemical.

3. Circle ALL applicable descriptors: pure or mixture, and solid, liquid, or gas.

EXAMPLE:

You have pure chlorine gas on hand, as well as two mixtures that contain liquid chlorine. You write "chlorine" and enter the CAS#. Then you circle "pure" *and* "mix" -- as well as "liq" *and* "gas".

Environmental Protection Agency

§ 370.41

PHYSICAL AND HEALTH HAZARDS

For each chemical you have listed, check all the physical and health hazard boxes that apply. These hazard categories are defined in 40 CFR 370.3. The two health hazard categories and three physical hazard categories are a consolidation of the 23 hazard categories defined in the OSHA Hazard Communication Standard, 29 CFR 910.1200.

MAXIMUM AMOUNT

1. For each hazardous chemical, estimate the greatest amount present at your facility on any single day during the reporting period.
2. Find the appropriate range value code in Table I.
3. Enter this range value as the Maximum Amount.

Table I REPORTING RANGES

Range Value	Weight Range in Pounds From...	To...
00	0	99
01	100	999
02	1,000	9,999
03	10,000	99,999
04	100,000	999,999
05	1,000,000	9,999,999
06	10,000,000	49,999,999
07	50,000,000	99,999,999
08	100,000,000	499,999,999
09	500,000,000	999,999,999
10	1 billion	higher than 1 billion

If you are using this form as a worksheet for completing Tier One, enter the actual weight in pounds in the shaded space below the response blocks. Do this for both Maximum Amount and Average Daily Amount.

EXAMPLE:

You received one large shipment of a solvent mixture last year. The shipment filled your 5,000-gallon storage tank. You know that the solvent contains 10% benzene, which is a hazardous chemical.

You figure that 10% of 5,000 gallons is 500 gallons. You also know that the density of benzene is 7.29 pounds per gallon, so you multiply 500 by 7.29 to get a weight of 3,645 pounds.

Then you look at Table I and find that the range value 02 corresponds to 3,645. You enter 02 as the Maximum Amount.

(If you are using the form as a worksheet for completing a Tier One form, you should write 3,645 in the shaded area.)

AVERAGE DAILY AMOUNT

1. For each hazardous chemical, estimate the average weight in pounds that was present at your facility during the year.

   To do this, total all daily weights and divide by the number of days the chemical was present on the site.
2. Find the appropriate range value in Table I.
3. Enter this range value as the Average Daily Amount.

EXAMPLE:

The 5,000-gallon shipment of solvent you received last year was gradually used up and completely gone in 315 days. The sum of the daily volume levels in the tank is 929,250 gallons. By dividing 929,250 gallons by 315 days on-site, you calculate an average daily amount of 2,950 gallons.

You already know that the solvent contains 10% benzene, which is a hazardous chemical. Since 10% of 2,950 is 295, you figure that you had an average of 295 gallons of benzene. You also know that the density of benzene is 7.29 pounds per gallon, so you multiply 295 by 7.29 to get a weight of 2,150 pounds.

Then you look at Table I and find that the range value 02 corresponds to 2,150. You enter 02 as the Average Daily Amount.

(If you are using the form as a worksheet for completing a Tier One form, you should write 2,150 in the shaded area.)

NUMBER OF DAYS ON-SITE

Enter the number of days that the hazardous chemical was found on-site.

EXAMPLE:

The solvent composed of 10% benzene was present for 315 days at your facility. Enter 315 in the space provided.

STORAGE CODES AND STORAGE LOCATIONS

List all non-confidential chemical locations in this column, along with storage types/conditions associated with each location.

Storage Codes: Indicate the types and conditions of storage present.

  a. Look at Table II. For each location, find the appropriate storage type(s). Enter the corresponding code(s) in front of the parentheses.
  b. Look at Table III. For each storage type, find the temperature and pressure conditions. Enter the applicable pressure code in the first space within the parentheses. Enter the applicable temperature code in the last space within the parentheses.

§ 370.41

Table II – STORAGE TYPES

CODES	Types of Storage
A	Above ground tank
B	Below ground tank
C	Tank inside building
D	Steel drum
E	Plastic or non-metallic drum
F	Can
G	Carboy
H	Silo
I	Fiber drum
J	Bag
K	Box
L	Cylinder
M	Glass bottles or jugs
N	Plastic bottles or jugs
O	Tote bin
P	Tank wagon
Q	Rail car
R	Other

Table III – TEMPERATURE AND PRESSURE
CONDITIONS

CODES	Storage Conditions
	(PRESSURE)
1	Ambient pressure
2	Greater than ambient pressure
3	Less than ambient pressure
	(TEMPERATURE)
4	Ambient temperature
5	Greater than ambient temperature
6	Less than ambient temperature but not cryogenic
7	Cryogenic conditions

EXAMPLE:

The benzene in the main building is kept in a tank inside the building, at ambient pressure and less than ambient temperature.

Table II shows you that the code for a tank inside a building is C. Table III shows you that code for ambient pressure is 1, and the code for less than ambient temperature is 6.

You enter: C(1,6)

Storage Locations:

Provide a brief description of the precise location of the chemical, so that emergency responders can locate the area easily. You may find it advantageous to provide the optional site plan or site coordinates as explained below.

For each chemical, indicate at a minimum the building or lot. Additionally, where practical, the room or area may be indicated. You may respond in narrative form with appropriate site coordinates or abbreviations.

If the chemical is present in more than one building, lot, or area location, continue your responses down the page as needed. If the chemical exists everywhere at the plant site simultaneously, you may report that the chemical is ubiquitous at the site.

Optional attachments: If you choose to attach one of the following, check the appropriate Attachments box at the bottom of the Tier Two form.

a. A site plan with site coordinates indicated for buildings, lots, areas, etc. throughout your facility.

b. A list of site coordinate abbreviations that correspond to buildings, lots, areas, etc. throughout your facility.

EXAMPLE:

You have benzene in the main room of the main building, and in tank 2 in tank field 10. You attach a site plan with coordinates as follows: main building = G-2, tank field 10 = B-6. Fill in the Storage Location as follows:

B-6 [Tank 2   ] G-2 [Main Room]

Under Title III, Section 324, you may elect to withhold location information on a specific chemical from disclosure to the public. If you choose to do so:

● Enter the word "confidential" in the Non-Confidential Location section of the Tier Two form.

● On a separate Tier Two Confidential Location Information Sheet, enter the name and CAS# of each chemical for which you are keeping the location confidential.

● Enter the appropriate location and storage information, as described above for non-confidential locations.

● Attach the Tier Two Confidential Location Information Sheet to the Tier Two form. This separates confidential locations from other information that will be disclosed to the public.

CERTIFICATION.

This must be completed by the owner or operator or the officially designated representative of the owner or operator. Enter your full name and official title. Sign your name and enter the current date.

Environmental Protection Agency   **§ 372.3**

PART 372—TOXIC CHEMICAL RELEASE REPORTING: COMMUNITY RIGHT-TO-KNOW

AUTHORITY: 42 U.S.C. 11013, 11028.
SOURCE: 53 FR 4525, Feb. 16, 1988, unless otherwise noted.

Subpart A—General Provisions

§ 372.1 Scope and purpose.

This part sets forth requirements for the submission of information relating to the release of toxic chemicals under section 313 of Title III of the Superfund Amendments and Reauthorization Act of 1986. The information collected under this part is intended to inform the general public and the communities surrounding covered facilities about releases of toxic chemicals, to assist research, to aid in the development of regulations, guidelines, and standards, and for other purposes. This part also sets forth requirements for suppliers to notify persons to whom they distribute mixtures or trade name products containing toxic chemicals that they contain such chemicals.

§ 372.3 Definitions.

Terms defined in sections 313(b)(1)(c) and 329 of Title III and not explicitly defined herein are used with the meaning given in Title III. For the purpose of this part:

"*Acts*" means Title III.

"*Article*" means a manufactured item: (1) Which is formed to a specific shape or design during manufacture; (2) which has end use functions dependent in whole or in part upon its shape or design during end use; and (3) which does not release a toxic chemical under normal conditions of processing or use of that item at the facility or establishments.

"*Customs territory of the United States*" means the 50 States, the District of Columbia, and Puerto Rico.

"*EPA*" means the United States Environmental Protection Agency.

"*Establishment*" means an economic unit, generally at a single physical location, where business is conducted or where services or industrial operations are performed.

"*Facility*" means all buildings, equipment, structures, and other stationary items which are located on a single site or on contiguous or adjacent sites and which are owned or operated by the same person (or by any person which controls, is controlled by, or under common control with such person). A facility may contain more than one establishment.

"*Full-time employee*" means 2,000 hours per year of full-time equivalent employment. A facility would calculate the number of full-time employees by totaling the hours worked during the calendar year by all employees, including contract employees, and dividing that total by 2,000 hours.

"*Import*" means to cause a chemical to be imported into the customs territory of the United States. For purposes of this definition, "to cause" means to intend that the chemical be imported and to control the identity of the imported chemical and the amount to be imported.

"*Manufacture*" means to produce, prepare, import, or compound a toxic chemical. Manufacture also applies to a toxic chemical that is produced coincidentally during the manufacture,

§ 372.5

processing, use, or disposal of another chemical or mixture of chemicals, including a toxic chemical that is separated from that other chemical or mixture of chemicals as a byproduct, and a toxic chemical that remains in that other chemical or mixture of chemicals as an impurity.

"*Mixture*" means any combination of two or more chemicals, if the combination is not, in whole or in part, the result of a chemical reaction. However, if the combination was produced by a chemical reaction but could have been produced without a chemical reaction, it is also treated as a mixture. A mixture also includes any combination which consists of a chemical and associated impurities.

"*Otherwise use*" or "*use*" means any use of a toxic chemical that is not covered by the terms "manufacture" or "process" and includes use of a toxic chemical contained in a mixture or trade name product. Relabeling or redistributing a container of a toxic chemical where no repackaging of the toxic chemical occurs does not constitute use or processing of the toxic chemical.

"*Process*" means the preparation of a toxic chemical, after its manufacture, for distribution in commerce:

(1) In the same form or physical state as, or in a different form or physical state from, that in which it was received by the person so preparing such substance, or

(2) As part of an article containing the toxic chemical. Process also applies to the processing of a toxic chemical contained in a mixture or trade name product.

"*Release*" means any spilling, leaking, pumping, pouring, emitting, emptying, discharging, injecting, escaping, leaching, dumping, or disposing into the environment (including the abandonment or discarding of barrels, containers, and other closed receptacles) of any toxic chemical.

"*Senior management official*" means an official with management responsibility for the person or persons completing the report, or the manager of environmental programs for the facility or establishments, or for the corporation owning or operating the facility or establishments responsible for certi-

fying similar reports under other environmental regulatory requirements.

"*Title III*" means Title III of the Superfund Amendments and Reauthorization Act of 1986, also titled the Emergency Planning and Community Right-To-Know Act of 1986.

"*Toxic chemical*" means a chemical or chemical category listed in § 372.65.

"*Trade name product*" means a chemical or mixture of chemicals that is distributed to other persons and that incorporates a toxic chemical component that is not identified by the applicable chemical name or Chemical Abstracts Service Registry number listed in § 372.65.

§ 372.5   Persons subject to this part.

Owners and operators of facilities described in §§ 372.22 and 372.45 are subject to the requirements of this part. If the owner and operator of a facility are different persons, only one need report under § 372.17 or provide a notice under § 372.45 for each toxic chemical in a mixture or trade name product distributed from the facility. However, if no report is submitted or notice provided, EPA will hold both the owner and the operator liable under section 325(c) of Title III, except as provided in §§ 372.38(e) and 372.45(g).

§ 372.10   Recordkeeping.

(a) Each person subject to the reporting requirements of this part must retain the following records for a period of 3 years from the date of the submission of a report under § 372.30:

(1) A copy of each report submitted by the person under § 372.30.

(2) All supporting materials and documentation used by the person to make the compliance determination that the facility or establishments is a covered facility under § 372.22 or § 372.45.

(3) Documentation supporting the report submitted under § 372.30 including:

(i) Documentation supporting any determination that a claimed allowable exemption under § 372.38 applies.

(ii) Data supporting the determination of whether a threshold under

268 FUNDAMENTALS OF HAZARDOUS MATERIALS INCIDENTS

§ 372.25 applies for each toxic chemical.

(iii) Documentation supporting the calculations of the quantity of each toxic chemical released to the environment or transferred to an off-site location.

(iv) Documentation supporting the use indications and quantity on site reporting for each toxic chemical, including dates of manufacturing, processing, or use.

(v) Documentation supporting the basis of estimate used in developing any release or off-site transfer estimates for each toxic chemical.

(vi) Receipts or manifests associated with the transfer of each toxic chemical in waste to off-site locations.

(vii) Documentation supporting reported waste treatment methods, estimates of treatment efficiencies, ranges of influent concentration to such treatment, the sequential nature of treatment steps, if applicable, and the actual operating data, if applicable, to support the waste treatment efficiency estimate for each toxic chemical.

(b) Each person subject to the notification requirements of this part must retain the following records for a period of 3 years from the date of the submission of a notification under § 372.45.

(1) All supporting materials and documentation used by the person to determine whether a notice is required under § 372.45.

(2) All supporting materials and documentation used in developing each required notice under § 372.45 and a copy of each notice.

(c) Records retained under this section must be maintained at the facility to which the report applies or from which a notification was provided. Such records must be readily available for purposes of inspection by EPA.

§ 372.18 Compliance and enforcement.

Violators of the requirements of this part shall be liable for a civil penalty in an amount not to exceed $25,000 each day for each violation as provided in section 325(c) of Title III.

Subpart B—Reporting Requirements

§ 372.22 Covered facilities for toxic chemical release reporting.

A facility that meets all of the following criteria for a calendar year is a covered facility for that calendar year and must report under § 372.30.

(a) The facility has 10 or more full-time employees.

(b) The facility is in Standard Industrial Classification Codes 20 through 39 (as in effect on January 1, 1987) by virtue of the fact that it meets one of the following criteria:

(1) The facility is an establishment with a primary SIC code of 20 through 39.

(2) The facility is a multi-establishment complex where all establishments have a primary SIC code of 20 through 39.

(3) The facility is a multi-establishment complex in which one of the following is true:

(i) The sum of the value of products shipped and/or produced from those establishments that have a primary SIC code of 20 through 39 is greater than 50 percent of the total value of all products shipped and/or produced from all establishments at the facility.

(ii) One establishment having a primary SIC code of 20 through 39 contributes more in terms of value of products shipped and/or produced than any other establishment within the facility.

(c) The facility manufactured (including imported), processed, or otherwise used a toxic chemical in excess of an applicable threshold quantity of that chemical set forth in § 372.25.

§ 372.25 Thresholds for reporting.

The threshold amounts for purposes of reporting under § 372.30 for toxic chemicals are as follows:

(a) With respect to a toxic chemical manufactured (including imported) or processed at a facility during the following calendar years:

1987—75,000 pounds of the chemical manufactured or processed for the year.
1988—50,000 pounds of the chemical manufactured or processed for the year.

1989 and thereafter—25,000 pounds of the chemical manufactured or processed for the year.

(b) With respect to a chemical otherwise used at a facility, 10,000 pounds of the chemical used for the applicable calendar year.

(c) With respect to activities involving a toxic chemical at a facility, when more than one threshold applies to the activities, the owner or operator of the facility must report if it exceeds any applicable threshold and must report on all activities at the facility involving the chemical, except as provided in § 372.38.

(d) When a facility manufactures, processes, or otherwise uses more than one member of a chemical category listed in § 372.65(c), the owner or operator of the facility must report if it exceeds any applicable threshold for the total volume of all the members of the category involved in the applicable activity. Any such report must cover all activities at the facility involving members of the category.

(e) A facility may process or otherwise use a toxic chemical in a recycle/reuse operation. To determine whether the facility has processed or used more than an applicable threshold of the chemical, the owner or operator of the facility shall count the amount of the chemical added to the recycle/reuse operation during the calendar year. In particular, if the facility starts up such an operation during a calendar year, or in the event that the contents of the whole recycle/reuse operation are replaced in a calendar year, the owner or operator of the facility shall also count the amount of the chemical placed into the system at these times.

(f) A toxic chemical may be listed in § 372.65 with the notation that only persons who manufacture the chemical, or manufacture it by a certain method, are required to report. In that case, only owners or operators of facilities that manufacture that chemical as described in § 372.65 in excess of the threshold applicable to such manufacture in § 372.25 are required to report. In completing the reporting form, the owner or operator is only required to account for the quantity of the chemical so manufactured and re-

leases associated with such manufacturing, but not releases associated with subsequent processing or use of the chemical at that facility. Owners and operators of facilities that solely process or use such a chemical are not required to report for that chemical.

(g) A toxic chemical may be listed in § 372.65 with the notation that it is in a specific form (e.g., fume or dust, solution, or friable) or of a specific color (e.g., yellow or white). In that case, only owners or operators of facilities that manufacture, process, or use that chemical in the form or of the color. specified in § 372.65 in excess of the threshold applicable to such activity in § 372.25 are required to report. In completing the reporting form, the owner or operator is only required to account for the quantity of the chemical manufactured, processed, or used in the form or color specified in § 372.65 and for releases associated with the chemical in that form or color. Owners or operators of facilities that solely manufacture, process, or use such a chemical in a form or color other than those specified by § 372.65 are not required to report for that chemical.

(h) Metal compound categories are listed in § 372.65(c). For purposes of determining whether any of the thresholds specified in § 372.25 are met for metal compound category, the owner or operator of a facility must make the threshold determination based on the total amount of all members of the metal compound category manufactured, processed, or used at the facility. In completing the release portion of the reporting form for releases of the metal compounds, the owner or operator is only required to account for the weight of the parent metal released. Any contribution to the mass of the release attributable to other portions of each compound in the category is excluded.

§ 372.30 Reporting requirements and schedule for reporting.

(a) For each toxic chemical known by the owner or operator to be manufactured (including imported), processed, or otherwise used in excess of an applicable threshold quantity in

§ 372.25 at its covered facility described in § 372.22 for a calendar year, the owner or operator must submit to EPA and to the State in which the facility is located a completed EPA Form R (EPA Form 9350-1) in accordance with the instructions in Subpart E.

(b)(1) The owner or operator of a covered facility is required to report as described in paragraph (a) of this section on a toxic chemical that the owner or operator knows is present as a component of a mixture or trade name product which the owner or operator receives from another person, if that chemical is imported, processed, or otherwise used by the owner or operator in excess of an applicable threshold quantity in § 372.25 at the facility as part of that mixture or trade name product.

(2) The owner or operator knows that a toxic chemical is present as a component of a mixture or trade name product (i) if the owner or operator knows or has been told the chemical identity or Chemical Abstracts Service Registry Number of the chemical and the identity or Number corresponds to an identity or Number in § 372.65, or (ii) if the owner or operator has been told by the supplier of the mixture or trade name product that the mixture or trade name product contains a toxic chemical subject to section 313 of the Act or this part.

(3) To determine whether a toxic chemical which is a component of a mixture or trade name product has been imported, processed, or otherwise used in excess of an applicable threshold in § 372.25 at the facility, the owner or operator shall consider only the portion of the mixture or trade name product that consists of the toxic chemical and that is imported, processed, or otherwise used at the facility, together with any other amounts of the same toxic chemical that the owner or operator manufactures, imports, processes, or otherwise uses at the facility as follows:

(i) If the owner or operator knows the specific chemical identity of the toxic chemical and the specific concentration at which it is present in the mixture or trade name product, the owner or operator shall determine the weight of the chemical imported, proc-

essed, or otherwise used as part of the mixture or trade name product at the facility and shall combine that with the weight of the toxic chemical manufactured (including imported) processed, or otherwise used at the facility other than as part of the mixture or trade name product. After combining these amounts, if the owner or operator determines that the toxic chemical was manufactured, processed, or otherwise used in excess of an applicable threshold in § 372.25, the owner or operator shall report the specific chemical identity and all releases of the toxic chemical on EPA Form R in accordance with the instructions in Subpart E.

(ii) If the owner or operator knows the specific chemical identity of the toxic chemical and does not know the specific concentration at which the chemical is present in the mixture or trade name product, but has been told the upper bound concentration of the chemical in the mixture or trade name product, the owner or operator shall assume that the toxic chemical is present in the mixture or trade name product at the upper bound concentration, shall determine whether the chemical has been manufactured, processed, or otherwise used at the facility in excess of an applicable threshold as provided in paragraph (b)(3)(i) of this section, and shall report as provided in paragraph (b)(3)(i) of this section.

(iii) If the owner or operator knows the specific chemical identity of the toxic chemical, does not know the specific concentration at which the chemical is present in the mixture or trade name product, has not been told the upper bound concentration of the chemical in the mixture or trade name product, and has not otherwise developed information on the composition of the chemical in the mixture or trade name product, then the owner or operator is not required to factor that chemical in that mixture or trade name product into threshold and release calculations for that chemical.

(iv) If the owner or operator has been told that a mixture or trade name product contains a toxic chemical, does not know the specific chemical identity of the chemical and knows the specific concentration at which it is present in the mixture or trade name product, the owner or operator shall determine the weight of the chemical imported, processed, or otherwise used as part of the mixture or trade name product at the facility. Since the owner or operator does not know the specific identity of the toxic chemical, the owner or operator shall make the threshold determination only for the weight of the toxic chemical in the mixture or trade name product. If the owner or operator determines that the toxic chemical was imported, processed, or otherwise used as part of the mixture or trade name product in excess of an applicable threshold in § 372.25, the owner or operator shall report the generic chemical name of the toxic chemical, or a trade name if the generic chemical name is not known, and all releases of the toxic chemical on EPA Form R in accordance with the instructions in Subpart E.

(v) If the owner or operator has been told that a mixture or trade name product contains a toxic chemical, does not know the specific chemical identity of the chemical, and does not know the specific concentration at which the chemical is present in the mixture or trade name product, but has been told the upper bound concentration of the chemical in the mixture or trade name product, the owner or operator shall assume that the toxic chemical is present in the mixture or trade name product at the upper bound concentration, shall determine whether the chemical has been imported, processed, or otherwise used at the facility in excess of an applicable threshold as provided in paragraph (b)(3)(iv) of this section, and shall report as provided in paragraph (b)(3)(iv) of this section.

(vi) If the owner or operator has been told that a mixture or trade name product contains a toxic chemical, does not know the specific chemical identity of the chemical, does not know the specific concentration at which the chemical is present in the mixture or trade name product, including information they have themselves developed, and has not been told the upper bound concentration of

the chemical in the mixture or trade name product, the owner or operator is not required to report with respect to that toxic chemical.

(c) A covered facility may consist of more than one establishment. The owner or operator of such a facility at which a toxic chemical was manufactured (including imported), processed, or otherwise used in excess of an applicable threshold may submit a separate Form R for each establishment or for each group of establishments within the facility to report the activities involving the toxic chemical at each establishment or group of establishments, provided that activities involving that toxic chemical at all the establishments within the covered facility are reported. If each establishment or group of establishments files separate reports then for all other chemicals subject to reporting at that facility they must also submit separate reports. However, an establishment or group of establishments does not have to submit a report for a chemical that is not manufactured (including imported), processed, otherwise used, or released at that establishment or group of establishments.

(d) Each report under this section for activities involving a toxic chemical that occurred during a calendar year at a covered facility must be submitted on or before July 1 of the next year. The first such report for calendar year 1987 activities must be submitted on or before July 1, 1988.

(e) For reports applicable to activities for calendar years 1987, 1988, and 1989 only, the owner or operator of a covered facility may report releases of a specific toxic chemical to an environmental medium, or transfers of wastes containing a specific toxic chemical to an off-site location, of less than 1,000 pounds using the ranges provided in the form and instructions in Subpart E. For reports applicable to activities in calendar year 1990 and beyond, these ranges may not be used.

[53 FR 4525, Feb. 16, 1988; 53 FR 12748, Apr. 18, 1988]

§ 372.38   Exemptions.

(a) *De minimis concentrations of a toxic chemical in a mixture.* If a toxic chemical is present in a mixture of chemicals at a covered facility and the toxic chemical is in a concentration in the mixture which is below 1 percent of the mixture, or 0.1 percent of the mixture in the case of a toxic chemical which is a carcinogen as defined in 29 CFR 1910.1200(d)(4), a person is not required to consider the quantity of the toxic chemical present in such mixture when determining whether an applicable threshold has been met under § 372.25 or determining the amount of release to be reported under § 372.30. This exemption applies whether the person received the mixture from another person or the person produced the mixture, either by mixing the chemicals involved or by causing a chemical reaction which resulted in the creation of the toxic chemical in the mixture. However, this exemption applies only to the quantity of the toxic chemical present in the mixture. If the toxic chemical is also manufactured (including imported), processed, or otherwise used at the covered facility other than as part of the mixture or in a mixture at higher concentrations, in excess of an applicable threshold quantity set forth in § 372.25, the person is required to report under § 372.30.

(b) *Articles.* If a toxic chemical is present in an article at a covered facility, a person is not required to consider the quantity of the toxic chemical present in such article when determining whether an applicable threshold has been met under § 372.25 or determining the amount of release to be reported under § 372.30. This exemption applies whether the person received the article from another person or the person produced the article. However, this exemption applies only to the quantity of the toxic chemical present in the article. If the toxic chemical is manufactured (including imported), processed, or otherwise used at the covered facility other than as part of the article, in excess of an applicable threshold quantity set forth in § 372.25, the person is required to report under § 372.30. Persons potentially subject to this exemption should carefully review the definitions of "article" and "release" in § 372.3. If a release of a toxic chemical occurs as a

result of the processing or use of an item at the facility, that item does not meet the definition of "article."

(c) *Uses.* If a toxic chemical is used at a covered facility for a purpose described in this paragraph (c), a person is not required to consider the quantity of the toxic chemical used for such purpose when determining whether an applicable threshold has been met under § 372.25 or determining the amount of releases to be reported under § 372.30. However, this exemption only applies to the quantity of the toxic chemical used for the purpose described in this paragraph (c). If the toxic chemical is also manufactured (including imported), processed, or otherwise used at the covered facility other than as described in this paragraph (c), in excess of an applicable threshold quantity set forth in § 372.25, the person is required to report under § 372.30.

(1) Use as a structural component of the facility.

(2) Use of products for routine janitorial or facility grounds maintenance. Examples include use of janitorial cleaning supplies, fertilizers, and pesticides similar in type or concentration to consumer products.

(3) Personal use by employees or other persons at the facility of foods, drugs, cosmetics, or other personal items containing toxic chemicals, including supplies of such products within the facility such as in a facility operated cafeteria, store, or infirmary.

(4) Use of products containing toxic chemicals for the purpose of maintaining motor vehicles operated by the facility.

(5) Use of toxic chemicals present in process water and non-contact cooling water as drawn from the environment or from municipal sources, or toxic chemicals present in air used either as compressed air or as part of combustion.

(d) *Activities in laboratories.* If a toxic chemical is manufactured, processed, or used in a laboratory at a covered facility under the supervision of a technically qualified individual as defined in § 720.3(ee) of this title, a person is not required to consider the quantity so manufactured, processed, or used when determining whether an

§ 372.45

applicable threshold has been met under § 372.25 or determining the amount of release to be reported under § 372.30. This exemption does not apply in the following cases:

(1) Specialty chemical production.

(2) Manufacture, processing, or use of toxic chemicals in pilot plant scale operations.

(3) Activities conducted outside the laboratory.

(e) *Certain owners of leased property.* The owner of a covered facility is not subject to reporting under § 372.30 if such owner's only interest in the facility is ownership of the real estate upon which the facility is operated. This exemption applies to owners of facilities such as industrial parks, all or part of which are leased to persons who operate establishments within SIC code 20 through 39 where the owner has no other business interest in the operation of the covered facility.

(f) *Reporting by certain operators of establishments on leased property such as industrial parks.* If two or more persons, who do not have any common corporate or business interest (including common ownership or control), operate separate establishments within a single facility, each such person shall treat the establishments it operates as a facility for purposes of this part. The determinations in § 372.22 and § 372.25 shall be made for those establishments. If any such operator determines that its establishment is a covered facility under § 372.22 and that a toxic chemical has been manufactured (including imported), processed, or otherwise used at the establishment in excess of an applicable threshold in § 372.25 for a calendar year, the operator shall submit a report in accordance with § 372.30 for the establishment. For purposes of this paragraph (f), a common corporate or business interest includes ownership, partnership, joint ventures, ownership of a controlling interest in one person by the other, or ownership of a controlling interest in both persons by a third person.

Subpart C—Supplier Notification Requirement

§ 372.45 Notification about toxic chemicals.

(a) Except as provided in paragraphs (c), (d), and (e) of this section and § 372.65, a person who owns or operates a facility or establishment which:

(1) Is in Standard Industrial Classification codes 20 through 39 as set forth in paragraph (b) of § 372.22,

(2) Manufactures (including imports) or processes a toxic chemical, and

(3) Sells or otherwise distributes a mixture or trade name product containing the toxic chemical, to (i) a facility described in § 372.22, or (ii) to a person who in turn may sell or otherwise distributes such mixture or trade name product to a facility described in § 372.22(b), must notify each person to whom the mixture or trade name product is sold or otherwise distributed from the facility or establishment in accordance with paragraph (b) of this section.

(b) The notification required in paragraph (a) of this section shall be in writing and shall include:

(1) A statement that the mixture or trade name product contains a toxic chemical or chemicals subject to the reporting requirements of section 313 of Title III of the Superfund Amendments and Reauthorization Act of 1986 and 40 CFR Part 372.

(2) The name of each toxic chemical, and the associated Chemical Abstracts Service registry number of each chemical if applicable, as set forth in § 372.65.

(3) The percent by weight of each toxic chemical in the mixture or trade name product.

(c) Notification under this section shall be provided as follows:

(1) For a mixture or trade name product containing a toxic chemical listed in § 373.65 with an effective date of January 1, 1987, the person shall provide the written notice described in paragraph (b) of this section to each recipient of the mixture or trade name product with at least the first shipment of each mixture or trade name

product to each recipient in each calendar year beginning January 1, 1989.

(2) For a mixture or trade name product containing a toxic chemical listed in § 372.65 with an effective date of January 1, 1989 or later, the person shall provide the written notice described in paragraph (b) of this section to each recipient of the mixture or trade name product with at least the first shipment of the mixture or trade name product to each recipient in each calendar year beginning with the applicable effective date.

(3) If a person changes a mixture or trade name product for which notification was previously provided under paragraph (b) of this section by adding a toxic chemical, removing a toxic chemical, or changing the percent by weight of a toxic chemical in the mixture or trade name product, the person shall provide each recipient of the changed mixture or trade name product a revised notification reflecting the change with the first shipment of the changed mixture or trade name product to the recipient.

(4) If a person discovers (i) that a mixture or trade name product previously sold or otherwise distributed to another person during the calendar year of the discovery contains one or more toxic chemicals and (ii), that any notification providied to such other persons in that calendar year for the mixture or trade name product either did not properly identify any of the toxic chemicals or did not accurately present the percent by weight of any of the toxic chemicals in the mixture or trade name product, the person shall provide a new notification to the recipient within 30 days of the discovery which contains the information described in paragraph (b) of this section and identifies the prior shipments of the mixture or product in that calendar year to which the new notification applies.

(5) If a Material Safety Data Sheet (MSDS) is required to be prepared and distributed for the mixture or trade name product in accordance with 29 CFR 1910.1200, the notification must be attached to or otherwise incorporated into such MSDS. When the notification is attached to the MSDS, the notice must contain clear instructions that the notifications must not be detached from the MSDS and that any copying and redistribution of the MSDS shall include copying and redistribution of the notice attached to copies of the MSDS subsequently redistributed.

(d) Notifications are not required in the following instances:

(1) If a mixture or trade name product contains no toxic chemical in excess of the applicable de minimis concentration as specified in § 372.38(a).

(2) If a mixture or trade name product is one of the following:

(i) An "article" as defined in § 372.3

(ii) Foods, drugs, cosmetics, alcoholic beverages, tobacco, or tobacco products packaged for distribution to the general public.

(iii) Any consumer product as the term is defined in the Consumer Product Safety Act (15 U.S.C. 1251 *et seq.*) packaged for distribution to the general public.

(e) If the person considers the specific identity of a toxic chemical in a mixture or trade name product to be a trade secret under provisions of 29 CFR 1910.1200, the notice shall contain a generic chemical name that is descriptive of that toxic chemical.

(f) If the person considers the specific percent by weight composition of a toxic chemical in the mixture or trade name product to be a trade secret under applicable State law or under the Restatement of Torts section 757, comment b, the notice must contain a statement that the chemical is present at a concentration that does not exceed a specified upper bound concentration value. For example, a mixture contains 12 percent of a toxic chemical. However, the supplier considers the specific concentration of the toxic chemical in the product to be a trade secret. The notice would indicate that the toxic chemical is present in the mixture in a concentration of no more than 15 percent by weight. The upper bound value chosen must be no larger than necessary to adequately protect the trade secret.

(g) A person is not subject to the requirements of this section to the extent the person does not know that the facility or establishment(s) is sell-

§ 372.65

ing or otherwise distributing a toxic chemical to another person in a mixture or trade name product. However, for purposes of this section, a person has such knowledge if the person receives a notice under this section from a supplier of a mixture or trade name product and the person in turn sells or otherwise distributes that mixture or trade name product to another person.

(h) If two or more persons, who do not have any common corporate or business interest (including common ownership or control), as described in § 372.38(f), operate separate establishments within a single facility, each such persons shall treat the establishment(s) it operates as a facility for purposes of this section. The determination under paragraph (a) of this section shall be made for those establishments.

[53 FR 4525, Feb. 16, 1988; 53 FR 12748, Apr. 18, 1988]

Subpart D—Specific Toxic Chemical Listings

§ 372.65 Chemicals and chemical categories to which this part applies.

The requirements of this part apply to the following chemicals and chemical categories. This section contains three listings. Paragraph (a) of this section is an alphabetical order listing of those chemicals that have an associated Chemical Abstracts Service (CAS) Registry number. Paragraph (b) of this section contains a CAS number order list of the same chemicals listed in paragraph (a) of this section. Paragraph (c) of this section contains the chemical categories for which reporting is required. These chemical categories are listed in alphabetical order and do not have CAS numbers. Each listing identifies the effective date for reporting under § 372.30.

(a) *Alphabetical listing.*

Chemical name	CAS No.	Effective date
Acetaldehyde	75-07-0	01/01/87
Acetamide	60-35-5	01/01/87
Acetone	67-64-1	01/01/87
Acetonitrile	75-05-8	01/01/87
2-Acetylaminofluorene	53-96-3	01/01/87
Acrolein	107-02-8	01/01/87
Acrylamide	79-06-1	01/01/87
Acrylic acid	79-10-7	01/01/87
Acrylonitrile	107-13-1	01/01/87
Aldrin[1,4:5,8-Dimethanonaphthalene,1,2,3,4,10,10-hexachloro-1,4,4a,5,8,8a-hexahydro-(1 alpha.,4 alpha.,4a beta.,5 alpha.,8 alpha., 8a beta.)-]	309-00-2	01/01/87
Allyl chloride	107-05-1	01/01/87
Aluminum (fume or dust)	7429-90-5	01/01/87
Aluminum oxide	1344-28-1	01/01/87
2-Aminoanthraquinone	117-79-3	01/01/87
4-Aminoazobenzene	60-09-3	01/01/87
4-Aminobiphenyl	92-67-1	01/01/87
1-Amino-2-methylanthraquinone	82-28-0	01/01/87
Ammonia	7664-41-7	01/01/87
Ammonium nitrate (solution)	6484-52-2	01/01/87
Ammonium sulfate (solution)	7783-20-2	01/01/87
Aniline	62-53-3	01/01/87
o-Anisidine	90-04-0	01/01/87
p-Anisidine	104-94-9	01/01/87
o-Anisidine hydrochloride	134-29-2	01/01/87
Anthracene	120-12-7	01/01/87
Antimony	7440-36-0	01/01/87
Arsenic	7440-38-2	01/01/87
Asbestos (friable)	1332-21-4	01/01/87
Barium	7440-39-3	01/01/87
Benzal chloride	98-87-3	01/01/87
Benzamide	55-21-0	01/01/87
Benzene	71-43-2	01/01/87
Benzidine	92-87-5	01/01/87
Benzoic trichloride (Benzotrichloride)	98-07-7	01/01/87
Benzoyl chloride	98-88-4	01/01/87
Benzoyl peroxide	94-36-0	01/01/87
Benzyl chloride	100-44-7	01/01/87
Beryllium	7440-41-7	01/01/87

Environmental Protection Agency §372.65

Chemical name	CAS No	Effective date
Biphenyl	92-52-4	01/01/87
Bis(2-chloroethyl) ether	111-44-4	01/01/87
Bis(chloromethyl) ether	542-88-1	01/01/87
Bis(2-chloro-1-methylethyl) ether	108-60-1	01/01/87
Bis(2-ethylhexyl) adipate	103-23-1	01/01/87
Bromoform (Tribromomethane)	75-25-2	01/01/87
Bromomethane (Methyl bromide)	74-83-9	01/01/87
1,3-Butadiene	106-99-0	01/01/87
Butyl acrylate	141-32-2	01/01/87
n-Butyl alcohol	71-36-3	01/01/87
sec-Butyl alcohol	78-92-2	01/01/87
tert-Butyl alcohol	75-65-0	01/01/87
Butyl benzyl phthalate	85-68-7	01/01/87
1,2-Butylene oxide	106-88-7	01/01/87
Butyraldehyde	123-72-8	01/01/87
C.I. Acid Blue 9, diammonium salt	2650-18-2	01/01/87
C.I. Acid Blue 9, disodium salt	3844-45-9	01/01/87
C.I. Acid Green 3	4680-78-8	01/01/87
C.I. Basic Green 4	569-64-2	01/01/87
C.I. Basic Red 1	989-38-8	01/01/87
C.I. Direct Black 38	1937-37-7	01/01/87
C.I. Direct Blue 6	2602-46-2	01/01/87
C.I. Direct Brown 95	16071-86-6	01/01/87
C.I. Disperse Yellow 3	2832-40-8	01/01/87
C.I. Food Red 5	3761-53-3	01/01/87
C.I. Food Red 15	81-88-9	01/01/87
C.I. Solvent Orange 7	3118-97-6	01/01/87
C.I. Solvent Yellow 3	97-56-3	01/01/87
C.I. Solvent Yellow 14	842-07-9	01/01/87
C.I. Solvent Yellow 34 (Auramine)	492-80-8	01/01/87
C.I. Vat Yellow 4	128-66-5	01/01/87
Cadmium	7440-43-9	01/01/87
Calcium cyanamide	156-62-7	01/01/87
Captan[1H-Isoindole-1,3(2H)-dione,3a,4,7,7a-tetrahydro-2-[(trichloromethyl)thio]-]	133-06-2	01/01/87
Carbaryl [1-Naphthalenol, methylcarbamate]	63-25-2	01/01/87
Carbon disulfide	75-15-0	01/01/87
Carbon tetrachloride	56-23-5	01/01/87
Carbonyl sulfide	463-58-1	01/01/87
Catechol	120-80-9	01/01/87
Chloramben [Benzoic acid,3-amino-2,5-dichloro-]	133-90-4	01/01/87
Chlordane [4,7-Methanoindan,1,2,4,5,6,7,8,8-octachloro-2,3,3a,4,7,7a-hexahydro-]	57-74-9	01/01/87
Chlorine	7782-50-5	01/01/87
Chlorine dioxide	10049-04-4	01/01/87
Chloroacetic acid	79-11-8	01/01/87
2-Chloroacetophenone	532-27-4	01/01/87
Chlorobenzene	108-90-7	01/01/87
Chlorobenzilate [Benzeneacetic acid, 4-chloro-.alpha.-(4-chlorophenyl)-alpha-hydroxy-, ethyl ester]	510-15-6	01/01/87
Chloroethane (Ethyl chloride)	75-00-3	01/01/87
Chloroform	67-66-3	01/01/87
Chloromethane (Methyl chloride)	74-87-3	01/01/87
Chloromethyl methyl ether	107-30-2	01/01/87
Chloroprene	126-99-8	01/01/87
Chlorothalonil [1,3-Benzenedicarbonitrile,2,4,5,6-tetrachloro-]	1897-45-6	01/01/87
Chromium	7440-47-3	01/01/87
Cobalt	7440-48-4	01/01/87
Copper	7440-50-8	01/01/87
p-Cresidine	120-71-8	01/01/87
Cresol (mixed isomers)	1319-77-3	01/01/87
m-Cresol	108-39-4	01/01/87
o-Cresol	95-48-7	01/01/87
p-Cresol	106-44-5	01/01/87
Cumene	98-82-8	01/01/87
Cumene hydroperoxide	80-15-9	01/01/87
Cupferron[Benzeneamine, N-hydroxy-N-nitroso, ammonium salt]	135-20-6	01/01/87
Cyclohexane	110-82-7	01/01/87
2,4-D [Acetic acid, (2,4-dichlorophenoxy)-]	94-75-7	01/01/87
Decabromodiphenyl oxide	1163-19-5	01/01/87
Diallate [Carbamothioic acid, bis(1-methylethyl)-, S-(2,3-dichloro-2-propenyl) ester]	2303-16-4	01/01/87
2,4-Diaminoanisole	615-05-4	01/01/87
2,4-Diaminoanisole sulfate	39156-41-7	01/01/87
4,4'-Diaminodiphenyl ether	101-80-4	01/01/87
Diaminotoluene (mixed isomers)	25376-45-8	01/01/87

§ 372.65   40 CFR Ch. I (7-1-88 Edition)

Chemical name	CAS No.	Effective date
2,4-Diaminotoluene	95-80-7	01/01/87
Diazomethane	334-88-3	01/01/87
Dibenzofuran	132-64-9	01/01/87
1,2-Dibromo-3-chloropropane (DBCP)	96-12-8	01/01/87
1,2-Dibromoethane (Ethylene dibromide)	106-93-4	01/01/87
Dibutyl phthalate	84-74-2	01/01/87
Dichlorobenzene (mixed isomers)	25321-22-6	01/01/87
1,2-Dichlorobenzene	95-50-1	01/01/87
1,3-Dichlorobenzene	541-73-1	01/01/87
1,4-Dichlorobenzene	106-46-7	01/01/87
3,3'-Dichlorobenzidine	91-94-1	01/01/87
Dichlorobromomethane	75-27-4	01/01/87
1,2-Dichloroethane (Ethylene dichloride)	107-06-2	01/01/87
1,2-Dichlorethylene	540-59-0	01/01/87
Dichloromethane (Methylene chloride)	75-09-2	01/01/87
2,4-Dichlorophenol	120-83-2	01/01/87
1,2-Dichloropropane	78-87-5	01/01/87
1,3-Dichloropropylene	542-75-6	01/01/87
Dichlorvos [Phosphoric acid, 2,2-dichloroethenyl dimethyl ester]	62-73-7	01/01/87
Dicofol [Benzenemethanol,4-chloro-.alpha.-(4-chlorophenyl)-.alpha.-(trichloromethyl)-]	115-32-2	01/01/87
Diepoxybutane	1464-53-5	01/01/87
Diethanolamine	111-42-2	01/01/87
Di-(2-ethylhexyl) phthalate (DEHP)	117-81-7	01/01/87
Diethyl phthalate	84-66-2	01/01/87
Diethyl sulfate	64-67-5	01/01/87
3,3'-Dimethoxybenzidine	119-90-4	01/01/87
4-Dimethylaminoazobenzene	60-11-7	01/01/87
3,3'-Dimethylbenzidine (o-Tolidine)	119-93-7	01/01/87
Dimethylcarbamyl chloride	79-44-7	01/01/87
1,1-Dimethyl hydrazine	57-14-7	01/01/87
2,4-Dimethylphenol	105-67-9	01/01/87
Dimethyl phthalate	131-11-3	01/01/87
Dimethyl sulfate	77-78-1	01/01/87
4,6-Dinitro-o-cresol	534-52-1	01/01/87
2,4-Dinitrophenol	51-28-5	01/01/87
2,4-Dinitrotoluene	121-14-2	01/01/87
2,6-Dinitrotoluene	606-20-2	01/01/87
n-Dioctyl phthalate	117-84-0	01/01/87
1,4-Dioxane	123-91-1	01/01/87
1,2-Diphenylhydrazine (Hydrazobenzene)	122-66-7	01/01/87
Epichlorohydrin	106-89-8	01/01/87
2-Ethoxyethanol	110-80-5	01/01/87
Ethyl acrylate	140-88-5	01/01/87
Ethylbenzene	100-41-4	01/01/87
Ethyl chloroformate	541-41-3	01/01/87
Ethylene	74-85-1	01/01/87
Ethylene glycol	107-21-1	01/01/87
Ethyleneimine(Aziridine)	151-56-4	01/01/87
Ethylene oxide	75-21-8	01/01/87
Ethylene thiourea	96-45-7	01/01/87
Fluometuron [Urea, N,N-dimethyl-N'-[3-(trifluoromethyl)phenyl]-]	2164-17-2	01/01/87
Formaldehyde	50-00-0	01/01/87
Freon 113 [Ethane, 1,1,2-trichloro-1,2,2-trifluoro-]	76-13-1	01/01/87
Heptachlor[1,4,5,6,7,8,8-Heptachloro-3a,4,7,7a-tetrahydro-4,7-methano-1H-indene]	76-44-8	01/01/87
Hexachlorobenzene	118-74-1	01/01/87
Hexachloro-1,3-butadiene	87-68-3	01/01/87
Hexachlorocyclopentadiene	77-47-4	01/01/87
Hexachloroethane	67-72-1	01/01/87
Hexachloronaphthalene	1335-87-1	01/01/87
Hexamethylphosphoramide	680-31-9	01/01/87
Hydrazine	302-01-2	01/01/87
Hydrazine sulfate	10034-93-2	01/01/87
Hydrochloric acid	7647-01-0	01/01/87
Hydrogen cyanide	74-90-8	01/01/87
Hydrogen fluoride	7664-39-3	01/01/87
Hydroquinone	123-31-9	01/01/87
Isobutyraldehyde	78-84-2	01/01/87
Isopropyl alcohol (Only persons who manufacture by the strong acid process are subject, no supplier notifiction)	67-63-0	01/01/87
4,4'-Isopropylidenediphenol	80-05-7	01/01/87
Lead	7439-92-1	01/01/87
Lindane [Cyclohexane, 1,2,3,4,5,6-hexachloro-(1 alpha,2 alpha,3 beta,4 alpha,5 alpha,6 beta-)-]	58-89-9	01/01/87

Environmental Protection Agency   **§ 372.65**

Chemical name	CAS No.	Effective date
Maleic anhydride	108-31-6	01/01/87
Maneb [Carbamodithioic acid, 1,2-ethanediylbis-, manganese complex]	12427-38-2	01/01/87
Manganese	7439-96-5	01/01/87
Melamine	108-78-1	01/01/87
Mercury	7439-97-6	01/01/87
Methanol	67-56-1	01/01/87
Methoxychlor [Benzene, 1,1'-(2,2,2-trichloroethylidene)bis[4-methoxy-]	72-43-5	01/01/87
2-Methoxyethanol	109-86-4	01/01/87
Methyl acrylate	96-33-3	01/01/87
Methyl tert-butyl ether	1634-04-4	01/01/87
4,4'-Methylenebis(2-chloroaniline) (MBOCA)	101-14-4	01/01/87
4,4'-Methylenebis(N,N-dimethyl) benzenamine	101-61-1	01/01/87
Methylenebis(phenylisocyanate) (MBI)	101-68-8	01/01/87
Methylene bromide	74-95-3	01/01/87
4,4'-Methylenedianiline	101-77-9	01/01/87
Methyl ethyl ketone	78-93-3	01/01/87
Methyl hydrazine	60-34-4	01/01/87
Methyl iodide	74-88-4	01/01/87
Methyl isobutyl ketone	108-10-1	01/01/87
Methyl isocyanate	624-83-9	01/01/87
Methyl methacrylate	80-62-6	01/01/87
Michler's ketone	90-94-8	01/01/87
Molybdenum trioxide	1313-27-5	01/01/87
Mustard gas [Ethane, 1,1'-thiobis[2-chloro-]	505-60-2	01/01/87
Naphthalene	91-20-3	01/01/87
alpha-Naphthylamine	134-32-7	01/01/87
beta-Naphthylamine	91-59-8	01/01/87
Nickel	7440-02-0	01/01/87
Nitric acid	7697-37-2	01/01/87
Nitrilotriacetic acid	139-13-9	01/01/87
5-Nitro-o-anisidine	99-59-2	01/01/87
Nitrobenzene	98-95-3	01/01/87
4-Nitrobiphenyl	92-93-3	01/01/87
Nitrofen [Benzene, 2,4-dichloro-1-(4-nitrophenoxy)-]	1836-75-5	01/01/87
Nitrogen mustard [2-Chloro-N-(2-chloroethyl)-N-methylethanamine]	51-75-2	01/01/87
Nitroglycerin	55-63-0	01/01/87
2-Nitrophenol	88-75-5	01/01/87
4-Nitrophenol	100-02-7	01/01/87
2-Nitropropane	79-46-9	01/01/87
p-Nitrosodiphenylamine	156-10-5	01/01/87
N,N-Dimethylaniline	121-69-7	01/01/87
N-Nitrosodi-n-butylamine	924-16-3	01/01/87
N-Nitrosodiethylamine	55-18-5	01/01/87
N-Nitrosodimethylamine	62-75-9	01/01/87
N-Nitrosodiphenylamine	86-30-6	01/01/87
N-Nitrosodi-n-propylamine	621-64-7	01/01/87
N-Nitrosomethylvinylamine	4549-40-0	01/01/87
N-Nitrosomorpholine	59-89-2	01/01/87
N-Nitroso-N-ethylurea	759-73-9	01/01/87
N-Nitroso-N-methylurea	684-93-5	01/01/87
N-Nitrosonornicotine	16543-55-8	01/01/87
N-Nitrosopiperidine	100-75-4	01/01/87
Octachloronaphthalene	2234-13-1	01/01/87
Osmium tetroxide	20816-12-0	01/01/87
Parathion [Phosphorothioic acid, O,O-diethyl-O-(4-nitrophenyl) ester]	56-38-2	01/01/87
Pentachlorophenol (PCP)	87-86-5	01/01/87
Peracetic acid	79-21-0	01/01/87
Phenol	108-95-2	01/01/87
p-Phenylenediamine	106-50-3	01/01/87
2-Phenylphenol	90-43-7	01/01/87
Phosgene	75-44-5	01/01/87
Phosphoric acid	7664-38-2	01/01/87
Phosphorus (yellow or white)	7723-14-0	01/01/87
Phthalic anhydride	85-44-9	01/01/87
Picric acid	88-89-1	01/01/87
Polychlorinated biphenyls (PCBs)	1336-36-3	01/01/87
Propane sultone	1120-71-4	01/01/87
beta-Propiolactone	57-57-8	01/01/87
Propionaldehyde	123-38-6	01/01/87
Propoxur [Phenol, 2-(1-methylethoxy)-, methylcarbamate]	114-26-1	01/01/87
Propylene (Propene)	115-07-1	01/01/87
Propyleneimine	75-55-8	01/01/87
Propylene oxide	75-56-9	01/01/87

§ 372.65   40 CFR Ch. I (7-1-88 Edition)

Chemical name	CAS No.	Effective date
Pyridine	110-86-1	01/01/8*
Quinoline	91-22-5	01/01/8*
Quinone	106-51-4	01/01/8*
Quintozene [Pentachloronitrobenzene]	82-68-8	01/01/8*
Saccharin (only persons who manufacture are subject, no supplier notification) [1,2-Benzisothiazol-3(2H)-one,1,1-dioxide]	81-07-2	01/01/8*
Safrole	94-59-7	01/01/8*
Selenium	7782-49-2	01/01/8*
Silver	7440-22-4	01/01/8*
Sodium hydroxide (solution)	1310-73-2	01/01/8*
Sodium sulfate (solution)	7757-82-6	01/01/8*
Styrene	100-42-5	01/01/8*
Styrene oxide	96-09-3	01/01/8*
Sulfuric acid	7664-93-9	01/01/8*
Terephthalic acid	100-21-0	01/01/8*
1,1,2,2-Tetrachloroethane	79-34-5	01/01/8*
Tetrachloroethylene (Perchloroethylene)	127-18-4	01/01/8*
Tetrachlorvinphos [Phosphoric acid, 2-chloro-1-(2,4,5-trichlorophenyl)ethenyl dimethyl ester]	961-11-5	01/01/8*
Thallium	7440-28-0	01/01/87
Thioacetamide	62-55-5	01/01/8*
4,4'-Thiodianiline	139-65-1	01/01/8*
Thiourea	62-56-6	01/01/87
Thorium dioxide	1314-20-1	01/01/87
Titanium tetrachloride	7550-45-0	01/01/87
Toluene	108-88-3	01/01/87
Toluene-2,4-diisocyanate	584-84-9	01/01/87
Toluene-2,6-diisocyanate	91-08-7	01/01/87
o-Toluidine	95-53-4	01/01/87
o-Toluidine hydrochloride	636-21-5	01/01/87
Toxaphene	8001-35-2	01/01/87
Triaziquone [2,5-Cyclohexadiene-1,4-dione,2,3,5-tris(1-aziridinyl)-]	68-76-8	01/01/87
Trichlorfon [Phosphoric acid, (2,2,2-trichloro-1-hydroxyethyl)-, dimethyl ester]	52-68-6	01/01/87
1,2,4-Trichlorobenzene	120-82-1	01/01/87
1,1,1-Trichloroethane (Methyl chloroform)	71-55-6	01/01/87
1,1,2-Trichloroethane	79-00-5	01/01/87
Trichloroethylene	79-01-6	01/01/87
2,4,5-Trichlorophenol	95-95-4	01/01/87
2,4,6-Trichlorophenol	88-06-2	01/01/87
Trifluralin [Benzeneamine, 2,6-dinitro-N,N-dipropyl-4-(trifluoromethyl)-1]	1582-09-8	01/01/87
1,2,4-Trimethylbenzene	95-63-6	01/01/87
Tris(2,3-dibromopropyl) phosphate	126-72-7	01/01/87
Urethane (Ethyl carbamate)	51-79-6	01/01/87
Vanadium (fume or dust)	7440-62-2	01/01/87
Vinyl acetate	108-05-4	01/01/87
Vinyl bromide	593-60-2	01/01/87
Vinyl chloride	75-01-4	01/01/87
Vinylidene chloride	75-35-4	01/01/87
Xylene (mixed isomers)	1330-20-7	01/01/87
m-Xylene	108-38-3	01/01/87
o-Xylene	95-47-6	01/01/87
p-Xylene	106-42-3	01/01/87
2,6-Xylidine	87-62-7	01/01/87
Zinc (fume or dust)	7440-66-6	01/01/87
Zineb [Carbamodithioic acid, 1,2-ethanediylbis-, zinc complex]	12122-67-7	01/01/87

(b) CAS Number listing.

CAS No.	Chemical name	Effective date
50-00-0	Formaldehyde	01/01/87
51-28-5	2,4-Dinitrophenol	01/01/87
51-75-2	Nitrogen mustard [2-Chloro-N-(2-chloroethyl)-N-methylethanamine]	01/01/87
51-79-6	Urethane (Ethyl carbamate)	01/01/87
52-68-6	Trichlorfon [Phosphoric acid, (2,2,2-trichloro-1-hydroxyethyl)-dimethyl ester]	01/01/87
53-96-3	2-Acetylaminofluorene	01/01/87
55-18-5	N-Nitrosodiethylamine	01/01/87
55-21-0	Benzamide	01/01/87
55-63-0	Nitroglycerin	01/01/87

Environmental Protection Agency §372.65

CAS No.	Chemical name	Effective date
56-23-5	Carbon tetrachloride	01/01/87
56-38-2	Parathion [Phosphorothioic acid, O,O-diethyl-O-(4-nitrophenyl)ester]	01/01/87
57-14-7	1,1-Dimethyl hydrazine	01/01/87
57-57-8	beta-Propiolactone	01/01/87
57-74-9	Chlordane [4,7-Methanoindan, 1,2,4,5,6,7,8,8-octachloro-2,3,3a,4,7,7a-hexahydro-]	01/01/87
58-89-9	Lindane [Cyclohexane, 1,2,3,4,5,6-hexachloro-(1.alpha.,2.alpha.,3.beta.,4.alpha.,5.alpha.,6.beta.)-]	01/01/87
59-89-2	N-Nitrosomorpholine	01/01/87
60-09-3	4-Aminoazobenzene	01/01/87
60-11-7	4-Dimethylaminoazobenzene	01/01/87
60-34-4	Methyl hydrazine	01/01/87
60-35-5	Acetamide	01/01/87
62-53-3	Aniline	01/01/87
62-55-5	Thioacetamide	01/01/87
62-56-6	Thiourea	01/01/87
62-73-7	Dichlorvos [Phosphoric acid, 2,2-dichloroethenyl dimethyl ester]	01/01/87
62-75-9	N-Nitrosodimethylamine	01/01/87
63-25-2	Carbaryl [1-Naphthalenol, methylcarbamate]	01/01/87
64-67-5	Diethyl sulfate	01/01/87
67-56-1	Methanol	01/01/87
67-63-0	Isopropyl alcohol (only persons who manufacture by the strong acid process are subject, supplier notification not required.)	01/01/87
67-64-1	Acetone	01/01/87
67-66-3	Chloroform	01/01/87
67-72-1	Hexachloroethane	01/01/87
68-76-8	Triaziquone [2,5-Cyclohexadiene-1,4-dione,2,3,5-tris(1-aziridinyl)-]	01/01/87
71-36-3	n-Butyl alcohol	01/01/87
71-43-2	Benzene	01/01/87
71-55-6	1,1,1-Trichloroethane (Methyl chloroform)	01/01/87
72-43-5	Methoxychlor [Benzene, 1,1'-(2,2,2,-trichloroethylidene)bis [4-methoxy-]	01/01/87
74-83-9	Bromomethane (Methyl bromide)	01/01/87
74-85-1	Ethylene	01/01/87
74-87-3	Chloromethane (Methyl chloride)	01/01/87
74-88-4	Methyl iodide	01/01/87
74-90-8	Hydrogen cyanide	01/01/87
74-95-3	Methylene bromide	01/01/87
75-00-3	Chloroethane (Ethyl chloride)	01/01/87
75-01-4	Vinyl chloride	01/01/87
75-05-8	Acetonitrile	01/01/87
75-07-0	Acetaldehyde	01/01/87
75-09-2	Dichloromethane (Methylene chloride)	01/01/87
75-15-0	Carbon disulfide	01/01/87
75-21-8	Ethylene oxide	01/01/87
75-25-2	Bromoform (Tribromomethane)	01/01/87
75-27-4	Dichlorobromomethane	01/01/87
75-35-4	Vinylidene chloride	01/01/87
75-44-5	Phosgene	01/01/87
75-55-8	Propyleneimine	01/01/87
75-56-9	Propylene oxide	01/01/87
75-65-0	tert-Butyl alcohol	01/01/87
77-13-1	Freon 113 [Ethane, 1,1,2-trichloro-1,2,2-trifluoro-]	01/01/87
76-44-8	Heptachlor [1,4,5,6,7,8,8-Heptachloro-3a,4,7,7a-tetrahydro-4,7-methano-1H-indene]	01/01/87
77-47-4	Hexachlorocyclopentadiene	01/01/87
77-78-1	Dimethyl sulfate	01/01/87
78-84-2	Isobutyraldehyde	01/01/87
78-87-5	1,2-Dichloropropane	01/01/87
78-92-2	sec-Butyl alcohol	01/01/87
78-93-3	Methyl ethyl ketone	01/01/87
79-00-5	1,1,2-Trichloroethane	01/01/87
79-01-6	Trichloroethylene	01/01/87
79-06-1	Acrylamide	01/01/87
79-10-7	Acrylic acid	01/01/87
79-11-8	Chloroacetic acid	01/01/87
79-21-0	Peracetic acid	01/01/87
79-34-5	1,1,2,2-Tetrachloroethane	01/01/87
79-44-7	Dimethylcarbamyl chloride	01/01/87
79-46-9	2-Nitropropane	01/01/87
80-05-7	4,4'-Isopropylidenediphenol	01/01/87
80-15-9	Cumene hydroperoxide	01/01/87
80-62-6	Methyl methacrylate	01/01/87
81-07-2	Saccharin (only persons who manufacture are subject, no supplier notification) [1,2-Benzisothiazol-3(2H)-one,1,1-dioxide]	01/01/87
81-88-9	C.I. Food Red 15	01/01/87

CAS No	Chemical name	Effective date
82-28-0	1-Amino-2-methylanthraquinone	01/01/87
82-68-8	Quintozene [Pentachloronitrobenzene]	C12
84-66-2	Diethyl phthalate	01/01/87
84-74-2	Dibutyl phthalate	01/01/87
85-44-9	Phthalic anhydride	01/01/87
85-68-7	Butyl benzyl phthalate	01/01/87
86-30-6	N-Nitrosodiphenylamine	01/01/87
87-62-7	2,6-Xylidine	01/01/87
87-68-3	Hexachloro-1,3-butadiene	01/01/87
87-86-5	Pentachlorophenol (PCP)	01/01/87
88-06-2	2,4,6-Trichlorophenol	01/01/87
88-75-5	2-Nitrophenol	01/01/87
88-89-1	Picric acid	01/01/87
90-04-0	o-Anisidine	01/01/87
90-43-7	2-Phenylphenol	01/01/87
90-94-8	Michler's ketone	01/01/87
91-08-7	Toluene-2,6-diisocyanate	01/01/87
91-20-3	Naphthalene	01/01/87
91-22-5	Quinoline	01/01/87
91-59-8	beta-Naphthylamine	01/01/87
91-94-1	3,3'-Dichlorobenzidine	01/01/87
92-52-4	Biphenyl	01/01/87
92-67-1	4-Aminobiphenyl	01/01/87
92-87-5	Benzidine	01/01/87
92-93-3	4-Nitrobiphenyl	01/01/87
94-36-0	Benzoyl peroxide	01/01/87
94-59-7	Safrole	01/01/87
94-75-7	2,4-D [Acetic acid, (2,4-dichlorophenoxy)-]	01/01/87
95-47-6	o-Xylene	01/01/87
95-48-7	o-Cresol	01/01/87
95-50-1	1,2-Dichlorobenzene	01/01/87
95-53-4	o-Toluidine	01/01/87
95-63-6	1,2,4-Trimethylbenzene	01/01/87
95-80-7	2,4-Diaminotoluene	01/01/87
95-95-4	2,4,5-Trichlorophenol	01/01/87
96-09-3	Styrene oxide	01/01/87
96-12-8	1,2-Dibromo-3-chloropropane (DBCP)	01/01/87
96-33-3	Methyl acrylate	01/01/87
96-45-7	Ethylene thiourea	01/01/87
97-56-3	C.I. Solvent Yellow 3	01/01/87
98-07-7	Benzoic trichloride (Benzotrichloride)	01/01/87
98-82-8	Cumene	01/01/87
98-87-3	Benzal chloride	01/01/87
98-88-4	Benzoyl chloride	01/01/87
98-95-3	Nitrobenzene	01/01/87
99-59-2	5-Nitro-o-anisidine	01/01/87
100-02-7	4-Nitrophenol	01/01/87
100-21-0	Terephthalic acid	01/01/87
100-41-4	Ethylbenzene	01/01/87
100-42-5	Styrene	01/01/87
100-44-7	Benzyl chloride	01/01/87
100-75-4	N-Nitrosopiperidine	01/01/87
101-14-4	4,4'-Methylenebis(2-chloroaniline) (MBOCA)	01/01/87
101-61-1	4,4'-Methylenebis(N,N-dimethyl)benzenamine	01/01/87
101-68-8	Methylenebis(phenylisocyanate) (MBI)	01/01/87
101-77-9	4,4'-Methylenedianiline	01/01/87
101-80-4	4,4'-Diaminodiphenyl ether	01/01/87
103-23-1	Bis(2-ethylhexyl) adipate	01/01/87
104-94-9	p-Anisidine	01/01/87
105-67-9	2,4-Dimethylphenol	01/01/87
106-42-3	p-Xylene	01/01/87
106-44-5	p-Cresol	01/01/87
106-46-7	1,4-Dichlorobenzene	01/01/87
106-50-3	p-Phenylenediamine	01/01/87
106-51-4	Quinone	01/01/87
106-88-7	1,2-Butylene oxide	01/01/87
106-89-8	Epichlorohydrin	01/01/87
106-93-4	1,2-Dibromoethane (Ethylene dibromide)	01/01/87
106-99-0	1,3-Butadiene	01/01/87
107-02-8	Acrolein	01/01/87
107-05-1	Allyl chloride	01/01/87
107-06-2	1,2-Dichloroethane (Ethylene dichloride)	01/01/87
107-13-1	Acrylonitrile	01/01/87

Environmental Protection Agency § 372.65

CAS No.	Chemical name	Effective date
107-21-1	Ethylene glycol	01/01/87
107-30-2	Chloromethyl methyl ether	01/01/87
108-05-4	Vinyl acetate	01/01/87
108-10-1	Methyl isobutyl ketone	01/01/87
108-31-6	Maleic anhydride	01/01/87
108-38-3	m-Xylene	01/01/87
108-39-4	m-Cresol	01/01/87
108-60-1	Bis(2-chloro-1-methylethyl)ether	01/01/87
108-78-1	Melamine	01/01/87
108-88-3	Toluene	01/01/87
108-90-7	Chlorobenzene	01/01/87
108-95-2	Phenol	01/01/87
109-86-4	2-Methoxyethanol	01/01/87
110-80-5	2-Ethoxyethanol	01/01/87
110-82-7	Cyclohexane	01/01/87
110-86-1	Pyridine	01/01/87
111-42-2	Diethanolamine	01/01/87
111-44-4	Bis(2-chloroethyl) ether	01/01/87
114-26-1	Propoxur [Phenol, 2-(1-methylethoxy)-, methylcarbamate]	01/01/87
115-07-1	Propylene (Propene)	01/01/87
115-32-2	Dicofol [Benzenemethanol, 4-chloro-.alpha.-(4-chlorophenyl)-.alpha.-(trichloromethyl)-]	01/01/87
117-79-3	2-Aminoanthraquinone	01/01/87
117-81-7	Di(2-ethylhexyl) phthalate (DEHP)	01/01/87
117-84-0	n-Dioctyl phthalate	01/01/87
118-74-1	Hexachlorobenzene	01/01/87
119-90-4	3,3'-Dimethoxybenzidine	01/01/87
119-93-7	3,3'-Dimethylbenzidine (o-Tolidine)	01/01/87
120-12-7	Anthracene	01/01/87
120-71-8	p-Cresidine	01/01/87
120-80-9	Catechol	01/01/87
120-82-1	1,2,4-Trichlorobenzene	01/01/87
120-83-2	2,4-Dichlorophenol	01/01/87
121-14-2	2,4-Dinitrotoluene	01/01/87
121-69-7	N,N-Dimethylaniline	01/01/87
122-66-7	1,2-Diphenylhydrazine (Hydrazobenzene)	01/01/87
123-31-9	Hydroquinone	01/01/87
123-38-6	Propionaldehyde	01/01/87
123-72-8	Butyraldehyde	01/01/87
123-91-1	1,4-Dioxane	01/01/87
126-72-7	Tris-2,3-dibromopropyl) phosphate	01/01/87
126-99-8	Chloroprene	01/01/87
127-18-4	Tetrachloroethylene (Perchloroethylene)	01/01/87
128-66-5	C.I. Vat Yellow 4	01/01/87
131-11-3	Dimethyl phthalate	01/01/87
132-64-9	Dibenzofuran	01/01/87
133-06-2	Captan [1H-Isoindole-1,3(2H)-dione,3a,4,7,7a-tetrahydro-2-[(trichloromethyl)thio]-]	01/01/87
133-90-4	Chloramben [Benzoic acid, 3-amino-2,5-dichloro-]	01/01/87
134-29-2	o-Anisidine hydrochloride	01/01/87
134-32-7	alpha-Naphthylamine	01/01/87
135-20-6	Cupferron [Benzeneamine, N-hydroxy-N-nitroso, ammonium salt]	01/01/87
139-13-9	Nitrilotriacetic acid	01/01/87
139-65-1	4,4'-Thiodianiline	01/01/87
140-88-5	Ethyl acrylate	01/01/87
141-32-2	Butyl acrylate	01/01/87
151-56-4	Ethyleneimine (Aziridine)	01/01/87
156-10-5	p-Nitrosodiphenylamine	01/01/87
156-62-7	Calcium cyanamide	01/01/87
302-01-2	Hydrazine	01/01/87
309-00-2	Aldrin[1,4:5,8-Dimethanonaphthalene,1,2,3,4,10,10-hexachloro-1,4,4a,5,8,8a-hexahydro-(1.alpha.,4.alpha.,4a.beta.,5.alpha., 8.alpha.,8a.beta.)-]	01/01/87
334-88-3	Diazomethane	01/01/87
463-58-1	Carbonyl sulfide	01/01/87
492-80-8	C.I. Solvent Yellow 34 (Auramine)	01/01/87
505-60-2	Mustard gas [Ethane, 1,1'-thiobis[2-chloro-]	01/01/87
510-15-6	Chlorobenzilate[Benezeneacetic acid, 4-chloro-.alpha.-(4-chlorophenyl)-.alpha.,-hydroxy-, ethyl ester]	01/01/87
532-27-4	2-Chloroacetophenone	01/01/87
534-52-1	4,6-Dinitro-o-cresol	01/01/87
540-59-0	1,2-Dichloroethylene	01/01/87
541-41-3	Ethyl chloroformate	01/01/87
541-73-1	1,3-Dichlorobenzene	01/01/87
542-75-6	1,3-Dichloropropylene	01/01/87
542-88-1	Bis(chloromethyl) ether	01/01/87

§ 372.65

CAS No	Chemical name	Effective date
569-64-2	C.I. Basic Green 4	01/01/87
606-20-2	2,6-Dinitrotoluene	01/01/87
615-05-4	2,4-Diaminoanisole	01/01/87
621-64-7	N-Nitrosodi-n-propylamine	01/01/87
624-83-9	Methyl isocyanate	01/01/87
636-21-5	o-Toluidine hydrochloride	01/01/87
680-31-9	Hexamethylphosphoramide	01/01/87
684-93-5	N-Nitroso-N-methylurea	01/01/87
759-73-9	N-Nitroso-N-ethylurea	01/01/87
842-07-9	C.I. Solvent Yellow 14	01/01/87
924-16-3	N-Nitrosodi-n-butylamine	01/01/87
961-11-5	Tetrachlorvinphos [Phosphoric acid, 2-chloro-1-(2,4,5-trichlorophenyl)ethenyl dimethyl ester]	01/01/87
989-38-8	C.I. Basic Red 1	01/01/87
1120-71-4	Propane sultone	01/01/87
1163-19-5	Decabromodiphenyl oxide	01/01/87
1310-73-2	Sodium hydroxide (solution)	01/01/87
1313-27-5	Molybdenum trioxide	01/01/87
1314-20-1	Thorium dioxide	01/01/87
1319-77-3	Cresol (mixed isomers)	01/01/87
1330-20-7	Xylene (mixed isomers)	01/01/87
1332-21-4	Asbestos (friable)	01/01/87
1335-87-1	Hexachloronaphthalene	01/01/87
1336-36-3	Polychlorinated biphenyls (PCBs)	01/01/87
1344-28-1	Aluminum oxide	01/01/87
1464-53-5	Diepoxybutane	01/01/87
1582-09-8	Trifluralin [Benzeneamine, 2,6-dinitro-N,N-dipropyl-4-(trifluoromethyl)-]	01/01/87
1634-04-4	Methyl tert-butyl ether	01/01/87
1836-75-5	Nitrofen [Benzene, 2,4-dichloro-1-(4-nitrophenoxy)-]	01/01/87
1897-45-6	Chlorothalonil [1,3-Benzenedicarbonitrile,2,4,5,6-tetrachloro-]	01/01/87
1937-37-7	C.I. Direct Black 38	01/01/87
2164-17-2	Fluometuron [Urea, N,N-dimethyl-N'-[3-(trifluoromethyl)phenyl]-]	01/01/87
2234-13-1	Octachloronaphthalene	01/01/87
2303-16-4	Diallate [Carbamothioic acid, bis(1-methylethyl)-, S-(2,3-dichloro-2-propenyl)ester]	01/01/87
2602-46-2	C.I. Direct Blue 6	01/01/87
2650-18-2	C.I. Acid Blue 9, diammonium salt	01/01/87
2832-40-8	C.I. Disperse Yellow 3	01/01/87
3118-97-6	C.I. Solvent Orange 7	01/01/87
3761-53-3	C.I. Food Red 5	01/01/87
3844-45-9	C.I. Acid Blue 9, disodium salt	01/01/87
4549-40-0	N-Nitrosomethylvinylamine	01/01/87
4680-78-8	C.I. Acid Green 3	01/01/87
6484-52-2	Ammonium nitrate (solution)	01/01/87
7429-90-5	Aluminum (fume or dust)	01/01/87
7439-92-1	Lead	01/01/87
7439-96-5	Manganese	01/01/87
7439-97-6	Mercury	01/01/87
7440-02-0	Nickel	01/01/87
7440-22-4	Silver	01/01/87
7440-28-0	Thallium	01/01/87
7440-36-0	Antimony	01/01/87
7440-38-2	Arsenic	01/01/87
7440-39-3	Barium	01/01/87
7440-41-7	Beryllium	01/01/87
7440-43-9	Cadmium	01/01/87
7440-47-3	Chromium	01/01/87
7440-48-4	Cobalt	01/01/87
7440-50-8	Copper	01/01/87
7440-62-2	Vanadium (fume or dust)	01/01/87
7440-66-6	Zinc (fume or dust)	01/01/87
7550-45-0	Titanium tetrachloride	01/01/87
7647-01-0	Hydrochloric acid	01/01/87
7664-38-2	Phosphoric acid	01/01/87
7664-39-3	Hydrogen fluoride	01/01/87
7664-41-7	Ammonia	01/01/87
7664-93-9	Sulfuric acid	01/01/87
7697-37-2	Nitric acid	01/01/87
7723-14-0	Phosphorus (yellow or white)	01/01/87
7757-82-6	Sodium sulfate (solution)	01/01/87
7782-49-2	Selenium	01/01/87
7782-50-5	Chlorine	01/01/87
7783-20-2	Ammonium sulfate (solution)	01/01/87
8001-35-2	Toxaphene	01/01/87

Environmental Protection Agency   § 372.65

CAS No	Chemical name	Effective date
10034-93-2	Hydrazine sulfate	01/01/87
10049-04-4	Chlorine dioxide	01/01/87
12122-67-7	Zineb [Carbamodithioic acid, 1,2-ethanediylbis-, zinc complex]	01/01/87
12427-38-2	Maneb [Carbamodithioic acid, 1,2-ethanediylbis-, manganese complex]	01/01/87
16071-86-6	C.I. Direct Brown 95	01/01/87
16543-55-8	N-Nitrosonornicotine	01/01/87
20816-12-0	Osmium tetroxide	01/01/87
25321-22-6	Dichlorobenzene (mixed isomers)	01/01/87
25376-45-8	Diaminotoluene (mixed isomers)	01/01/87
39156-41-7	2,4-Diaminoanisole sulfate	01/01/87

(c) *Chemical categories in alphabetical order.*

Category name	Effective date
Antimony Compounds: Includes any unique chemical substance that contains antimony as part of that chemical's infrastructure	01/01/87
Arsenic Compounds: Includes any unique chemical substance that contains arsenic as part of that chemical's infrastructure	01/01/87
Barium Compounds: Includes any unique chemical substance that contains barium as part of that chemical's infrastructure	01/01/87
Beryllium Compounds: Includes any unique chemical substance that contains beryllium as part of that chemical's infrastructure	01/01/87
Cadmium Compounds: Includes any unique chemical substance that contains cadmium as part of that chemical's infrastructure	01/01/87
Chlorophenols	01/01/87

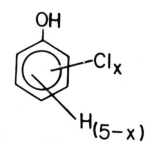

Where x = 1 to 5

Category name	Effective date
Chromium Compounds: Includes any unique chemical substance that contains chromium as part of that chemical's infrastructure	01/01/87
Cobalt Compounds: Includes any unique chemical substance that contains cobalt as part of that chemical's infrastructure	01/01/87
Copper Compounds: Includes any unique chemical substance that contains copper as part of that chemical's infrastructure	01/01/87
Cyanide Compounds: X⁺CN⁻ where X = H⁺ or any other group where a formal dissociation can be made. For example KCN, or Ca(CN)₂	01/01/87
Glycol Ethers: Includes mono- and di- ethers of ethylene glycol, diethylene glycol, and triethylene glycol	01/01/87

§ 372.65   40 CFR Ch. I (7-1-88 Edition)

Category name	Effective date
$R\ (OCH\cdot CH\cdot)n\ ^{nk}$	
Where:	
n - 1, 2, or 3	
R - alkyl or aryl groups	
R - R - H, or groups which, when removed, yield glycol ethers with the structure	
$R\ (OCH)_{2(n)n-nn}$	
Polymers are excluded from this category.	
Lead Compounds: Includes any unique chemical substance that contains lead as part of that chemical's infrastructure	01/01/87
Manganese Compounds: Includes any unique chemical substance that contains manganese as part of that chemical's infrastructure	01/01/87
Mercury Compounds: Includes any unique chemical substance that contains mercury as part of that chemical's infrastructure	01/01/87
Nickel Compounds: Includes any unique chemical substance that contains nickel as part of that chemical's infrastructure	01/01/87
Polybrominated Biphenyls (PBBs)	01/01/87

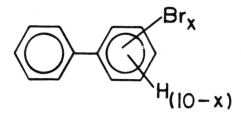

Where x = 1 to 10

Category name	Effective date
Selenium Compounds: Includes any unique chemical substance that contains selenium as part of that chemical's infrastructure	01/01/87
Silver Compounds: Includes any unique chemical substance that contains silver as part of that chemical's infrastructure	01/01/87
Thallium Compounds: Includes any unique chemical substance that contains thallium as part of that chemical's infrastructure	01/01/87
Zinc Compounds: Includes any unique chemical substance that contains zinc as part of that chemical's infrastructure	01/01/87

[53 FR 4525, Feb. 16, 1988; 53 FR 12748, Apr. 18, 1988, as amended at 53 FR 23112, June 20, 1988]

Environmental Protection Agency   § 372.85

Subpart E—Forms and Instructions

§ 372.85   Toxic chemical release reporting form and instructions.

(a) *Reporting form.*

Form Approved OMB No.: __2070-0093__

Approval Expires: __01/91__

(Important: Type or print; read instructions before completing form.)   Page 1 of 5

U.S. Environmental Protection Agency

♦EPA   TOXIC CHEMICAL RELEASE INVENTORY REPORTING FORM

EPA FORM
R

Section 313, Title III of The Superfund Amendments and Reauthorization Act of 1986

(This space for EPA use only.)

PART I.   FACILITY IDENTIFICATION INFORMATION

| 1. | 1.1 Does this report contain trade secret information? ☐ Yes (Answer 1.2) ☐ No (Do not answer 1.2) | 1.2 Is this a sanitized copy? ☐ Yes ☐ No | 1.3 Reporting Year |

2. CERTIFICATION (Read and sign after completing all sections.)

I hereby certify that I have reviewed the attached documents and that, to the best of my knowledge and belief, the submitted information is true and complete and that the amounts and values in this report are accurate based on reasonable estimates using data available to the preparers of this report.

Name and official title of owner/operator or senior management official

Signature   Date signed

3. FACILITY IDENTIFICATION

3.1	Facility or Establishment Name			This report contains information for: (check one)
	Street Address			
	City	County	3.2	a. ☐ An entire covered facility b. ☐ Part of a covered facility.
	State	Zip Code		

3.3	Technical Contact	Telephone Number (include area code) ( ) –
3.4	Public Contact	Telephone Number (include area code) ( ) –

3.5	a. SIC Code b. c.

| 3.6 | Latitude Deg. Min. Sec.   Longitude Deg. Min. Sec. |

Where to send completed forms:

U.S. Environmental Protection Agency
P.O. Box 70266
Washington, DC 20024-0266
Attn: Toxic Chemical Release Inventory

3.7	Dun & Bradstreet Number(s) a. b.
3.8	EPA Identification Number (RCRA I.D. No.) a. b.
3.9	NPDES Permit Number(s) a. b.
3.10	Name of Receiving Stream(s) or Water Body(s) a. b. c.
3.11	Underground Injection Well Code (UIC) Identification No.

4. PARENT COMPANY INFORMATION

4.1	Name of Parent Company
4.2	Parent Company's Dun & Bradstreet No.

EPA Form 9350-1 (1-88)

§ 372.85   40 CFR Ch. I (7-1-88 Edition)

(Important: Type or print; read instructions before completing form.)   Page 2 of 5

| EPA FORM **R**
PART II.   OFF-SITE LOCATIONS TO WHICH TOXIC
CHEMICALS ARE TRANSFERRED IN WASTES | (This space for EPA use only.) |

1. PUBLICLY OWNED TREATMENT WORKS (POTW)

Facility Name

Street Address

| City | County |
| State | Zip |

2. OTHER OFF-SITE LOCATIONS - Number these locations sequentially on this and any additional page of this form you use.

☐ Other off-site location

EPA Identification Number (RCRA I.D. No.)

Facility Name

Street Address

| City | County |
| State | Zip |

Is location under control of reporting facility or parent company?   ☐ Yes   ☐ No

☐ Other off-site location

EPA Identification Number (RCRA I.D. No.)

Facility Name

Street Address

| City | County |
| State | Zip |

Is location under control of reporting facility or parent company?   ☐ Yes   ☐ No

☐ Other off-site location

EPA Identification Number (RCRA I.D. No.)

Facility Name

Street Address

| City | County |
| State | Zip |

Is location under control of reporting facility or parent company?   ☐ Yes   ☐ No

☐ Check if additional pages of Part II are attached.

EPA Form 9350-1(1-88)

Environmental Protection Agency   **§ 372.85**

(Important: Type or print; read instructions before completing form.)   Page 3 of 5

EPA FORM R

PART III. CHEMICAL SPECIFIC INFORMATION

(This space for EPA use only.)

---

1. CHEMICAL IDENTITY

1.1	☐ Trade Secret (Provide a generic name in 1.4 below. Attach substantiation form to this submission.)
1.2	CAS # ☐☐☐☐☐☐ - ☐☐ - ☐ (Use leading zeros if CAS number does not fill space provided.)
1.3	Chemical or Chemical Category Name
1.4	Generic Chemical Name (Complete only if 1.1 is checked.)

2.	**MIXTURE COMPONENT IDENTITY** (Do not complete this section if you have completed Section 1.) Generic Chemical Name Provided by Supplier (Limit the name to a maximum of 70 characters (e.g., numbers, letters, spaces, punctuation)).

---

3. ACTIVITIES AND USES OF THE CHEMICAL AT THE FACILITY (Check all that apply.)

3.1	**Manufacture:**	a. ☐	Produce	b. ☐	Import	c. ☐	For on-site use/processing
		d. ☐	For sale/ distribution	e. ☐	As a byproduct	f. ☐	As an impurity
3.2	**Process:**	a. ☐	As a reactant	b. ☐	As a formulation component	c. ☐	As an article component
		d. ☐	Repackaging only				
3.3	**Otherwise Used:**	a. ☐	As a chemical processing aid	b. ☐	As a manufacturing aid	c. ☐	Ancillary or other use

---

4. MAXIMUM AMOUNT OF THE CHEMICAL ON SITE AT ANY TIME DURING THE CALENDAR YEAR

☐☐ (enter code)

---

5. RELEASES OF THE CHEMICAL TO THE ENVIRONMENT

You may report releases of less than 1,000 lbs. by checking ranges under A.1.

		A. Total Release (lbs/yr)				B. Basis of Estimate (enter code)	
		A.1 Reporting Ranges			A.2 Enter Estimate		
		0	1–499	500–999			
5.1 Fugitive or non-point air emissions	5.1a					5.1b ☐	
5.2 Stack or point air emissions	5.2a					5.2b ☐	
5.3 Discharges to water 5.3.1 ☐	5.3.1a					5.3.1b ☐	C. % From Stormwater 5.3.1c
(Enter letter code from Part I Section 3.10 for streams(s).) 5.3.2 ☐	5.3.2a					5.3.2b ☐	5.3.2c
5.3.3 ☐	5.3.3a					5.3.3b ☐	5.3.3c
5.4 Underground injection	5.4a					5.4b ☐	
5.5 Releases to land 5.5.1 ☐☐☐ (enter code)	5.5.1a					5.5.1b ☐	
5.5.2 ☐☐☐ (enter code)	5.5.2a					5.5.2b ☐	
5.5.3 ☐☐☐ (enter code)	5.5.3a					5.5.3b ☐	

☐ (Check if additional information is provided on Part IV-Supplemental Information.)

EPA Form 9350-1 (1-88)

EPA FORM R, Part III (Continued)   Page 4 of 5

6. TRANSFERS OF THE CHEMICAL IN WASTE TO OFF-SITE LOCATIONS

You may report transfers of less than 1,000 lbs. by checking ranges under A.1.	A. Total Transfers (lbs/yr)				B. Basis of Estimate (enter code)	C. Type of Treatment/ Disposal (enter code)
	A.1 Reporting Ranges			A.2 Enter Estimate		
	0	1–199	500–999			
6.1 Discharge to POTW					6.1b ☐	
6.2 Other off-site location (Enter block number from Part II, Section 2.) ☐					6.2b ☐	6.2c ☐☐☐
6.3 Other off-site location (Enter block number from Part II, Section 2.) ☐					6.3b ☐	6.3c ☐☐☐
6.4 Other off-site location (Enter block number from Part II, Section 2.) ☐					6.4b ☐	6.4c ☐☐☐

☐ (Check if additional information is provided on Part IV-Supplemental Information)

7. WASTE TREATMENT METHODS AND EFFICIENCY

A. General Wastestream (enter code)	B. Treatment Method (enter code)	C. Range of Influent Concentration (enter code)	D. Sequential Treatment? (check if applicable)	E. Treatment Efficiency Estimate	F. Based on Operating Data? Yes No
7.1a ☐	7.1b ☐☐☐	7.1c ☐	7.1d ☐	7.1e %	7.1f ☐ ☐
7.2a ☐	7.2b ☐☐☐	7.2c ☐	7.2d ☐	7.2e %	7.2f ☐ ☐
7.3a ☐	7.3b ☐☐☐	7.3c ☐	7.3d ☐	7.3e %	7.3f ☐ ☐
7.4a ☐	7.4b ☐☐☐	7.4c ☐	7.4d ☐	7.4e %	7.4f ☐ ☐
7.5a ☐	7.5b ☐☐☐	7.5c ☐	7.5d ☐	7.5e %	7.5f ☐ ☐
7.6a ☐	7.6b ☐☐☐	7.6c ☐	7.6d ☐	7.6e %	7.6f ☐ ☐
7.7a ☐	7.7b ☐☐☐	7.7c ☐	7.7d ☐	7.7e %	7.7f ☐ ☐
7.8a ☐	7.8b ☐☐☐	7.8c ☐	7.8d ☐	7.8e %	7.8f ☐ ☐
7.9a ☐	7.9b ☐☐☐	7.9c ☐	7.9d ☐	7.9e %	7.9f ☐ ☐
7.10a ☐	7.10b ☐☐☐	7.10c ☐	7.10d ☐	7.10e %	7.10f ☐ ☐
7.11a ☐	7.11b ☐☐☐	7.11c ☐	7.11d ☐	7.11e %	7.11f ☐ ☐
7.12a ☐	7.12b ☐☐☐	7.12c ☐	7.12d ☐	7.12e %	7.12f ☐ ☐
7.13a ☐	7.13b ☐☐☐	7.13c ☐	7.13d ☐	7.13e %	7.13f ☐ ☐
7.14a ☐	7.14b ☐☐☐	7.14c ☐	7.14d ☐	7.14e %	7.14f ☐ ☐

☐ (Check if additional information is provided on Part IV-Supplemental Information.)

8. OPTIONAL INFORMATION ON WASTE MINIMIZATION

(Indicate actions taken to reduce the amount of the chemical being released from the facility. See the instructions for coded items and an explanation of what information to include.)

A. Type of modification (enter code)	B. Quantity of the chemical in the waste stream prior to treatment/disposal			C. Index	D. Reason for action (enter code)
	Current reporting year (lbs/yr)	Prior year (lbs/yr)	Or percent change		
☐☐	_____	_____	_____ %	☐☐	☐☐

Environmental Protection Agency   **§ 372.85**

(Inportant: Type or print; read instructions before completing form.)   Page 5 of 5

EPA FORM **R**
PART IV. SUPPLEMENTAL INFORMATION

Use this section if you need additional space for answers to questions in Parts I and III.
Number or letter this information sequentially from prior sections (e.g., D.E, F, or 5.54, 5.55).

(This space for EPA use only.)

ADDITIONAL INFORMATION ON FACILITY IDENTIFICATION (Part I – Section 3)

3.5   SIC Code

3.7   Dun & Bradstreet Number(s)

3.8   EPA Identification Number(s) RCRA I.D. No.)

3.9   NPDES Permit Number(s)

3.10 Name of Receiving Stream(s) or Water Body(s)

ADDITIONAL INFORMATION ON RELEASES TO LAND (Part III – Section 5.5)

Releases to Land	A. Total Release (lbs/yr)			B. Basis of Estimate (enter code)
	A.1 Reporting Ranges 0 1–499 500–999		A.2 Enter Estimate	
5.5___ [ ] (enter code)	5.5___ a			5.5___ b [ ]
5.5___ [ ] (enter code)	5.5___ a			5.5___ b [ ]
5.5___ [ ] (enter code)	5.5___ a			5.5___ b [ ]

ADDITIONAL INFORMATION ON OFF-SITE TRANSFER (Part III – Section 6)

	A.Total Transfers (lbs/yr)			B. Basis of Estimate (enter code)	C. Type of Treatment/ Disposal (enter code)
	A.1 Reporting Ranges 0 1–499 500–999		A.2 Enter Estimate		
6.___ Discharge to POTW	6.___ a			6.___ b [ ]	
6.___ Other off-site location (Enter block number from Part II, Section 2.) [ ]	6.___ a			6.___ b [ ]	6.___ c.
6.___ Other off-site location (Enter block number from Part II, Section 2.) [ ]	6.___ a			6.___ b [ ]	6.___ c.

ADDITIONAL INFORMATION ON WASTE TREATMENT (Part III – Section 7)

A. General Wastestream (enter code)	B. Treatment Method (enter code)	C. Range of Influent Concentration (enter code)	D. Sequential Treatment? (check if applicable)	E. Treatment Efficiency Estimate	F. Based on Operating Data? Yes No
7.___ a [ ]	7.___ b	7.___ c [ ]	7.___ d [ ]	7.___ e ___%	7.___ f [ ] [ ]
7.___ a [ ]	7.___ b	7.___ c [ ]	7.___ d [ ]	7.___ e ___%	7.___ f [ ] [ ]
7.___ a [ ]	7.___ b	7.___ c [ ]	7.___ d [ ]	7.___ e ___%	7.___ f [ ] [ ]
7.___ a [ ]	7.___ b	7.___ c [ ]	7.___ d [ ]	7.___ e ___%	7.___ f [ ] [ ]
7.___ a [ ]	7.___ b	7.___ c [ ]	7.___ d [ ]	7.___ e ___%	7.___ f [ ] [ ]

EPA Form 9350-1(1-88)

§ 372.85

b. Instructions.

INSTRUCTIONS FOR COMPLETING EPA FORM R, THE TOXIC CHEMICAL RELEASE REPORTING FORM

GENERAL INFORMATION

A complete report Form R must be submitted for each toxic chemical manufactured, processed, or otherwise used at each covered facility as prescribed in the reporting rule in 40 CFR Part 372. These instructions supplement and elaborate on the requirements in the reporting rule. Together with the reporting rule, they constitute the reporting requirements, you should read both before attempting to complete Form R. All references in these instructions are to sections in the reporting rule unless otherwise indicated.

The Toxic Chemical Release Reporting Form, EPA Form R, consists of four parts:

- Part I, Facility Identification Information;
- Part II, Off-Site Locations to Which Toxic Chemicals are Transferred in Wastes;
- Part III, Chemical Specific Information; and
- Part IV, Supplemental Information.

Form R is designed so that a majority of the information required in Part I and III of the information required in Part II should be the same for each chemical reported by your facility. If the information in Parts I and II are identical for two or more chemicals, you may submit photostatic copies of those parts for those chemicals as long as each Part I has an original signature on the certification statement. Part III must be completed separately for each chemical. Part IV provides additional space, if needed, to complete the information required by the preceding sections of the form.

A complete report for any listed toxic chemical that is not claimed trade secret consists of the following completed parts:

- Part I with an original signature on the certification statement (Section 2);
- Part II;
- Part III (Section 8 is optional); and
- If applicable, Part IV.

A complete report for a toxic chemical claimed trade secret includes all of the above items plus the following:

- A completed trade secret substantiation form;
- A "sanitized" version of the report in which the chemical identity items (Part III, Sections 1.2 and 1.3) have been left blank but in which a generic chemical name has been supplied (Part III, Section 1.4); and
- A "sanitized" version of the trade secret substantiation form.

WHEN THE REPORT MUST BE SUBMITTED

The report for any calendar year must be submitted on or before July 1 of the following year (e.g., the report for calendar year 1987, January through December, must be submitted on or before July 1, 1988).

WHERE TO SEND THE REPORT

Submit reports, including reports containing trade secret claims (i.e., sanitized) to:

U.S. Environmental Protection Agency
P.O. Box 70266
Washington, D.C. 20024-0266
Attn: Toxic Chemical Release Inventory

In addition, you must send a copy of the report to the State (State of the U.S., the District of Columbia, the Commonwealth of Puerto Rico, Guam, American Samoa, the U.S. Virgin Islands, the Northern Mariana Islands, and any other territory or possession over which the U.S. has jurisdiction) in which the facility is located. States will provide addresses where the copies of the reports are to be sent. Copies of the report sent to the State should be the "sanitized," non-trade-secret version of the report, unless the State specifically requests otherwise. For additional information, refer to the discussion of trade secret/confidentiality claims in the instructions for completing Part III, Section 1, of the form.

HOW TO OBTAIN FORMS AND OTHER INFORMATION

Additional copies of EPA Form R and related guidance documents may be obtained from:

Emergency Planning and Community Right-to-Know Hotline
U.S. Environmental Protection Agency
WH-562A
401 M Street, S.W.
Washington, D.C. 20460
(800) 535-0202
(202) 479-2449 (Washington D.C. and Alaska)

INSTRUCTIONS FOR COMPLETING SPECIFIC SECTIONS OF EPA FORM R

The following are specific instructions for completing each section of EPA Form R. The number designations of the parts and sections of these instructions correspond to those in Form R unless otherwise indicated.

PART I. FACILITY IDENTIFICATION INFORMATION

1.1 Does This Report Contain Trade Secret Information?

You must answer this question only after you have completed the rest of the report. The specific identity of the toxic chemical being reported in Part III, Sections 1.2 and 1.3 may be designated as trade secret. If you are making a trade secret claim, answer by marking the "yes" box and proceed to Section 1.2. (See Part III, Section 1, of these instructions for specific instructions on trade secrecy claims.) If the answer is no, proceed to Section 1.3.

1.2 Is This a Sanitized Copy?

You must answer this question only after you have completed the rest of the report. Answer yes if this copy of the report is the public "sanitized" version of a report where the chemical identity is claimed trade secret in Part III, Section 1.4 of the report. Otherwise, answer no.

Environmental Protection Agency § 372.85

1.3 Reporting Year

In Section 1.3, you must enter the year to which the reported information applies, not the year in which you are submitting the report.

2. Certification

The certification statement must be signed by the owner or operator, or a senior official with management responsibility for the person (or persons) completing the form. The owner, operator, or official must certify the accuracy and completeness of the information reported on the form by signing and dating the certification statement. Each report must contain an original signature. Print or type the name and title of the person who signs the statement in the space provided. This certification statement applies to all the information supplied on the form and should be signed only after the form has been completed.

3. Facility Identification

3.1 Facility Name and Location

You must enter the name of your facility (plant site name or appropriate facility designation), street address, city, county, state, and zip code in the space provided. You may not use a post office box number for this location information. The address provided should be the location where the chemicals are manufactured, processed, or otherwise used.

3.2 Full or Partial Facility Indication

You must indicate whether your report is for the covered facility as a whole or for part of a covered facility. Check box a. if the report contains information about a chemical for an entire covered facility. Check box b. if the report contains information about a chemical but for only part of a covered facility.

The SIC code system classifies business "establishments," which are defined as "distinct and separate economic activities [which] are performed at a single physical location." Under section 372.30(c) of the reporting rule, you may choose to submit a separate Form R for each establishment, or for groups of establishments, in your covered facility. This allows you the option of reporting separately on the activities involving a toxic chemical at each establishment, or group of establishments (e.g., part of a covered facility), rather than submitting a single Form R for that chemical for the entire facility. You may do this provided that all releases of the toxic chemical from the entire covered facility are reported. However, if an establishment or group of establishments does not manufacture, process, otherwise use, or release a toxic chemical, then you do not have to submit a report on that chemical from that establishment or group of establishments.

3.3 Technical Contact

You must enter the name and telephone number (including area code) of a technical representative whom EPA or State officials may contact for clarification of the information reported on the form. This person does not have to be the person who prepares the report or signs the certification statement. However, this person must have detailed knowledge of the report to be able to respond to questions.

3.4 Public Contact

You must enter the name and telephone number of a person who can respond to questions from the public about the report. If you choose to designate the same person as both the technical and the public contact, enter "same as 3.3" in this space. If no public contact is designated in Section 3.4, EPA will treat the technical contact as the public contact.

3.5 Standard Industrial Classification (SIC) Code

You must enter the appropriate 4-digit primary Standard Industrial Classification (SIC) code for your facility. If the report covers more than one establishment, enter the primary 4 digit SIC code for each establishment. You are only required to enter SIC codes for establishments within the facility that fall within SIC codes 20 through 39 as identified in section 372.22 of the reporting rule. Use the Supplemental Information sheet (Part IV) if you need to enter more than three SIC codes.

3.6 Latitude and Longitude

Enter the latitudinal and longitudinal coordinates of your facility. You must supply the latitude and longitude for calendar year 1987 reports if the information is readily available to you. Sources of these data include EPA permits (e.g., NPDES permits), county property records, facility blueprints, and site plans. If these geographic coordinates are not readily available to you for calendar year 1987 reports, enter not applicable [N/A]. All facilities are required to provide this information in reports submitted for the calendar year 1988 and subsequent years. Use leading place holding zeros.

3.7 Facility Dun and Bradstreet Number

You must enter the number assigned by Dun and Bradstreet for your facility or each establishment within your facility. This may be available from your facility's financial office. If none of your establishments have been assigned Dun and Bradstreet Numbers, indicate this in Section 3.7 by entering not applicable [N/A] in box a. If only some of your establishments have been assigned Dun and Bradstreet numbers, indicate this in Section 3.7 by entering those numbers. Use leading place holding zeros. For more than two establishments, use the Supplemental Information sheet (Part IV).

3.8 EPA Identification Number

If your facility has been assigned EPA Identification Numbers, you must enter those numbers. The EPA I.D. Number is a 12-digit number assigned to facilities covered by hazardous waste regulations under the Resource Conservation and Recovery Act (RCRA). Facilities not covered by RCRA are not likely to have an assigned EPA I.D. Number. If your facility does not have an EPA I.D. Number, enter not applicable [N/A] in box a. If your facility has more than two numbers, use the Supplemental Information sheet (Part IV). Use leading place holding zeros.

3.9 NPDES Permit Numbers

You must enter the numbers of any permits your facility holds under the National Pollutant Discharge Elimination System (NPDES). This 9-digit permit number is assigned to your facility by EPA or the State under the authority of the Clean Water Act. If your facility has more than two permits, use the Supplemental Information sheet (Part IV). Use leading place holding zeros. If your facility does not have a permit, enter not applicable [N/A] in box a.

3.10 Name of Receiving Stream or Water Body

You must enter the name of each surface water body or receiving stream to which chemicals being reported are directly discharged. Report the name of each receiving stream or water body as it appears on the NPDES permit for the facility. Enter not applicable [N/A] to any unneeded spaces. If your facility discharges the toxic chemical to more than three receiving streams or water bodies, use the Supplemental Information sheet (Part IV).

3.11 Underground Injection Well Code (UIC) Identification Number

If your facility has a permit to inject chemical-containing waste which includes any toxic chemical being reported into Class 1 deep wells, you must enter the Underground Injection Control (UIC) 12-digit identification number assigned by EPA or by the State under the authority of the Safe Drinking Water Act. If your facility does not hold such a permit, enter not applicable [N/A] in this space.

4. Parent Company Information

You must provide information on your parent company. For purposes of this form, parent company is defined as a company which directly owns at least 50 percent of the voting stock of another company.

4.1 Name of Parent Company

You must enter the name of the corporation or other business entity that is your parent company. If you have no parent company, enter not applicable [N/A].

4.2 Parent Company's Dun & Bradstreet Number

If applicable, you must enter the Dun and Bradstreet Number for your parent company. The number may be obtained from the treasurer or financial officer of the company. If your parent company does not have a Dun and Bradstreet number, enter not applicable [N/A]. Use leading place holding zeros.

PART II. OFF-SITE LOCATIONS TO WHICH TOXIC CHEMICALS ARE TRANSFERRED IN WASTES

This section requires a listing of all off-site locations to which you transfer wastes containing the toxic chemical. The information that you enter in this section relates to data to be reported in Part III, Section 6, of the form. List only publicly owned treatment works (POTW) and treatment or disposal facilities. Do not list locations to which products containing the toxic chemical are shipped for sale or distribution in commerce or for further use. Also, do not list locations to which wastes containing the chemical are sold or sent for recovery, recycling, or reuse of the toxic chemical.

1. Publicly Owned Treatment Works (POTW)

You must enter the name and address of the POTW to which your facility discharges wastewater containing any toxic chemical you are reporting. If you do not discharge wastewater containing reported toxic chemicals to a POTW, enter not applicable [N/A]. If you discharge wastewater containing toxic chemicals to more than one POTW, use additional copies of Part II.

2. Other Off-Site Locations

In the block next to the heading "Other off-site location," enter a number. For the first such off-site location enter "1" in the block. Continue numbering the off-site locations in ascending order. This is the block number required by Part III, Section 6. If your facility transfers the toxic chemical to more than three off-site locations, use additional copies of Part II and continue numbering these locations in ascending order.

In the spaces provided, you must enter the name and address of each location (other than POTWs) to which you ship or transfer wastes containing the toxic chemical. Also enter the RCRA I.D. Number (EPA I.D. Number) for each such location, if known to you. Such information may be found on the Uniform Hazardous Waste Manifest which is required by RCRA regulations.

You must also indicate in the space provided whether the location is owned or controlled by your facility or your parent company.

PART III. CHEMICAL SPECIFIC INFORMATION

1.1 Trade Secret Block

If you are claiming chemical identity as a trade secret, you must mark the trade secret claim box in Section 1.1. In addition, you must attach a completed trade secret substantiation form to the report, as set forth in the trade secret rule in 40 CFR Part 350. When the chemical identity is claimed trade secret, you must also provide a generic name in Section 1.4.

Note: If you complete and submit your Toxic Chemical Release Inventory Reporting Form before the trade secret rule is in effect, you are still required to substantiate your claim that the specific chemical identity is a trade secret. Accordingly, you should follow the provisions of the proposed trade secret rule and use the proposed trade secret substantiation form which appeared in the FEDERAL REGISTER of October 15, 1987 (52FR 38312-38377).

1.2 CAS Registry Number

You must enter the Chemical Abstracts Service (CAS) registry number that appears in section 372.65 of the reporting rule for the chemical being reported. Use leading place holding zeros. If you are reporting one of the chemical categories in section 372.65 of the rule (e.g., copper compounds), enter [N/A] in the CAS number space. CAS numbers are cross-referenced with an alphabetical list of trade names and chemical names in section 372.65 of the rule.

1.3 Chemical or Chemical Category Name

You must enter in the space provided the name of the chemical or chemical category as it is listed in section 372.65 of the reporting rule. Only use names listed in section 372.65.

1.4 Generic Chemical Name

You must complete Section 1.4 if you are claiming the specific chemical identity of the toxic chemical as a trade secret and have marked the trade secret block in

Section 1.1. The generic chemical name must be descriptive of the chemical structure. You must limit the generic name to seventy characters (e.g., numbers, letters, spaces, punctuation) or less.

2. Mixture Component Identity

Use this section to report a mixture component that you know exceeds a threshold. Do not complete this section if you have completed Section 1 of Part III.

You may have received a mixture or trade name product from another person which you were told contains a section 313 toxic chemical, but that person did not tell you the specific chemical name or CAS number of the toxic chemical. The person may have given you a generic chemical name and the percentage composition of the toxic chemical in the mixture or trade name product under section 372.45 of the reporting rule. If you determine that you have imported, processed, or otherwise used the toxic chemical in the mixture or trade name product in excess of an applicable threshold, you must enter the generic chemical name given to you by your supplier in Section 2. If your supplier did not give you a generic chemical name, you must enter the name by which the supplier identified the chemical to you. (In some cases, this may be the same as the name of the mixture or trade name product.)

For example, your facility uses 20,000 pounds of a solvent which your supplier has told you contains eighty percent "chlorocyclocarbon," his generic name for a chemical subject to reporting under section 313. You therefore know that you have exceeded the use threshold for this toxic chemical. You would enter the name "chlorocyclocarbon," in the space provided in Section 2.

3. Activities and Uses of the Chemical at the Facility

This section requires an indication of whether the chemical is manufactured (including imported), processed, or otherwise used at the facility for which the form is being filed and the general nature of such activities and uses at the facility during the calendar year. Report activities that take place only at your facility, not activities that take place at other facilities involving your products. You must mark all of the appropriate blocks in this Section that apply to the activities at your facility. Refer to the definitions of "manufacture," "process," and "otherwise used" in section 372.3 of the reporting rule for explanations supplementing those provided below.

3.1 Manufacture

a. Produce.
A chemical included in this category is produced at the facility.

b. Import.
A chemical included in this category is imported to the facility.

c. For on-site use/processing.
A chemical included in this category is manufactured and then further processed or otherwise used at the same facility.

d. For sale/distribution.
A chemical in this category is manufactured specifically for sale or distribution outside the manufacturing facility.

e. As a byproduct.
A chemical in this category is produced coincidentally

during the production, processing, use, or disposal of another chemical substance or mixture, and following its production, is separated from that other chemical substance or mixture.

f. As an impurity.
A chemical in this category is produced coincidentally with another chemical substance, and is processed, used, or distributed with it.

3.2 Process (incorporative-type activities)

a. As reactant.
A natural or synthetic chemical used in chemical reactions for the manufacture of another chemical substance or of a product. Includes, but is not limited to, feedstocks, raw materials, intermediates, and initiators.

b. As a formulation component.
A chemical added to a product or product mixture prior to further distribution of the product that aids the performance of the product in its use. Examples include, but are not limited to, additives, dyes, reaction diluents, initiators, solvents, inhibitors, emulsifiers, surfactants, lubricants, flame retardants, and rheological modifiers.

c. As an article component.
A chemical substance that becomes an integral component of an article distributed for industrial, trade, or consumer use.

d. Repackaging only.
Processing or preparation of a chemical or product mixture for distribution in commerce in a different form, state, or quantity.

3.3 Otherwise Used (non-incorporative-type activities)

a. As a chemical processing aid.
A chemical that is added to a reaction mixture to aid in the manufacture or synthesis of another chemical substance but does not intentionally remain in or become part of the product or product mixture. Examples of such chemicals include, but are not limited to, process solvents, catalysts, inhibitors, initiators, reaction terminators, and solution buffers.

b. As a manufacturing aid.
A chemical whose function is to aid the manufacturing process but does not become part of the resulting product. Examples include, but are not limited to, lubricants, metalworking fluids, coolants, refrigerants, and hydraulic fluids.

c. Ancillary or other use.
A chemical in this category is used at a facility for purposes other than as a chemical processing aid or manufacturing aid as described above. Includes, but is not limited to, cleaners, degreasers, lubricants, and fuels.

4. Maximum Amount of the Chemical On Site at Any Time During the Calendar Year

You must insert the appropriate code (see below) that indicates the maximum quantity of the chemical (in storage tanks, process vessels, on-site shipping containers, etc.) at your facility at any time during the calendar year. If the chemical was present at several locations within your facility, use the maximum <u>total</u> amount present at the entire facility at any one time.

Range Code	Weight Range in Pounds From...	To....
01	0	99
02	100	999
03	1,000	9,999
04	10,000	99,999
05	100,000	999,999
06	1,000,000	9,999,999
07	10,000,000	49,999,999
08	50,000,000	99,999,999
09	100,000,000	499,999,999
10	500,000,000	999,999,999
11	1 billion	more than 1 billion

If the toxic chemical was present at your facility as part of a mixture or trade name product, to determine the maximum quantity of the chemical present at the facility you must calculate only the weight of the toxic chemical, not the weight of the entire mixture or trade name product. See section 372.30(b) of the reporting rule for further information on how to calculate the weight of the chemical in the mixture or trade name product.

5. Releases of the Chemical to the Environment

In Section 5 you must account for the total aggregate releases of the toxic chemical from your facility to the environment for the calendar year. Releases to the environment include emissions to the air, discharges to surface waters, and releases to land and underground injection wells.

All air releases of the chemical from the facility must be covered. In case of doubt about whether an air release is a point or non-point release, it is important that the release be included as one or the other rather than omitted. Do not enter information on individual emission points or releases. Enter only the total release.

5.1 Fugitive or non-point air emissions.

These are releases to the air that are not released through stacks, vents, ducts, pipes, or any other confined air stream. You must include (1) fugitive equipment leaks from valves, pump seals, flanges, compressors, sampling connections, open-ended lines, etc.; (2) evaporative losses from surface impoundments; (3) releases from building ventilation systems; and (4) any other fugitive or non-point air emissions.

5.2 Stack or point air emissions.

These are releases to the air that are through stacks, vents, ducts, pipes, or other confined air streams. You must include storage tank emissions. Air releases from control equipment would generally fall in this category.

5.3 Discharges to water

You must enter the applicable letter code for the receiving stream or water body from Section 3.10 of Part I of the form. Also, you must enter the total annual amount of the chemical released from all discharge points at the facility to each receiving stream or water body. You must include process outfalls such as pipes and open trenches, releases from on-site wastewater treatment systems; and the contribution from stormwater runoff if applicable (see instructions for column C below). Do not include "indirect" discharges to surface waters such as to a POTW or off-site wastewater treatment facility. These must be reported in Section 6.

5.4 Underground injection

You must enter the total annual amount of the chemical that was injected into wells, including Class I and other types, at the facility.

5.5 Releases to land

You must report quantities of the chemical that were landfilled, impounded, or otherwise disposed of at the facility. Do not report land disposal at off-site locations in this section. You must enter the appropriate disposal code from the following list:

Disposal Codes

D02 Landfill
D03 Land Treatment/Application/Farming
D05 Surface Impoundment (to be closed as a Landfill)
D99 Other Disposal

Three lines are provided in this section of the form to accommodate various types of land disposal.

For the purpose of this form, a surface impoundment is considered "final disposal." Quantities of the chemical released to surface impoundments that are used merely as part of a wastewater treatment process generally must not be reported in this section of the form. However, if the impoundment accumulates sludges containing the chemical, you must include an estimate in this section unless the sludges are removed and otherwise disposed of (in which case they should be reported under the appropriate section of the form). For the purposes of this reporting, storage tanks are not considered to be a type of disposal and are not to be reported in this section of the form.

A. Total Release

Only releases of the toxic chemical to the environment for the calendar year are to be reported in this section of the form. The total releases from your facility do not include transfers or shipments of the chemical from your facility for sale or distribution in commerce or of wastes to other facilities for treatment or disposal (see Section 6.1). Both routine releases, such as fugitive air emissions, and accidental or non-routine releases, such as chemical spills, must be included in your estimate of the quantity released.

A.1 Reporting Ranges

For reports submitted for calendar years 1987, 1988, and 1989 only, you may take advantage of range reporting for releases that are less than 1,000 pounds for the year to an environmental medium. You may mark one of the three boxes, 0, 1-499, or 500-999, corresponding to releases of the chemical to any environmental medium (i.e., any line item); however, you do not have to use these range check boxes. You have the option of providing a specific figure in column A.2 as described below.

For releases of 1,000 pounds or more for the year to any medium, you must provide an estimate in pounds per year in column A.2. Any estimate provided in column A.2 is required to be accurate to no more than two significant figures. Beginning with reports for calendar year 1990, you may not use ranges to report; you must report in column A.2.

Environmental Protection Agency   **§ 372.85**

A.2 Enter Estimates

You must provide your estimates of releases in pounds for the year in column A.2. This estimate is required to be accurate to no more than two significant figures.

Calculating Releases

To provide the release information required in both Sections A.1 and A.2 in this section of the form, you must use all readily available data (including relevant monitoring data and emissions measurements) collected at your facility pursuant to other provisions of law or as part of routine plant operations, to the extent you have it for the toxic chemical.

When relevant monitoring data or emission measurements are not readily available, reasonable estimates of the amounts released must be made using published emission factors, material balance calculations, or engineering calculations. You may not use emission factors or calculations to estimate releases if more accurate data are available.

No additional monitoring or measurement of the quantities or concentrations of any toxic chemical released into the environment, or of the frequency of such releases, is required for the purpose of completing this form, beyond that which is required under other provisions of law or regulation or as part of routine plant operations.

You must estimate as accurately as possible the quantity in pounds of the chemical or chemical category that is released annually to each environmental medium. Do not include the quantity of components of a waste stream other than the toxic chemical in this estimate.

If the toxic chemical was present at your facility as part of a mixture or trade name product, you must calculate the releases of the chemical only. Do not include releases of the other components of the mixture or trade name product. If you only know about or are only able to estimate the releases of the mixture or trade name product as a whole, you must assume that the toxic chemical is released in proportion to its concentration in the mixture or trade name product. See section 372.30(b) of the reporting rule for further information on how to calculate the concentration and weight of the chemical in the mixture or trade name product.

If you are reporting a chemical category listed in section 372.65(c) of the reporting rule rather than a specific chemical, you must combine the release data for all chemicals in the listed chemical category (e.g., all glycol ethers or all chlorophenols) and report the aggregate amount for that chemical category. Do not report releases of each individual chemical in that category separately. For example, if your facility releases 3,000 pounds per year of 2-chlorophenol, 4,000 pounds per year of 3-chlorophenol, and 4,000 pounds per year of 4-chlorophenol, you should report that your facility releases 11,000 pounds per year of chlorophenols. (Other than for listed chemical categories in section 372.65(c) of the rule, each form must report for an individual chemical.

Listed chemicals with the qualifier "solution," such as sodium sulfate, in concentrations of 1 percent (or 0.1 percent in the case of a carcinogen) or greater, must be factored into threshold and release calculations, because threshold and release amounts relate to the amount of chemical in solution, not the amount of solution.

For metal compound categories (e.g., chromium

compounds), report releases of only the parent metal For example, a user of various inorganic chromium salts would report the total chromium released in each waste type regardless of the chemical form (e.g., as the original salts, chromium ion, oxide, etc.), and exclude any contribution to mass made by other species in the molecule.

B. Basis of Estimate

For each release estimate you are required to indicate the principal method by which the quantity was derived. Enter the letter code to identify the method which applies to the largest portion of the total estimated quantity.

For example, if 40 percent of stack emissions of the reported substance were derived using monitoring data, 30 percent by mass balance, and 30 percent by emission factors, you would enter the code letter "M" for monitoring. The codes are as follows:

M - Based on monitoring data or measurements for the toxic chemical as released to the environment and/or off-site facility.

C - Based on mass balance calculations, such as calculation of the amount of the toxic chemical in streams entering and leaving process equipment.

E - Based on published emission factors, such as those relating release quantity to throughput or equipment type (e.g., air emissions factors)

O - Based on other approaches such as engineering calculations (e.g., estimating volatilization or solubility using published mathematical formulas) or best engineering judgment. This would include applying an estimated removal efficiency to a wastestream even if the composition of stream before treatment was fully characterized by monitoring data.

If the monitoring data, mass balance, or emission factor used to estimate the release is not specific to the toxic chemical, the estimate should be reported as based on engineering calculations or judgment.

C. Percent From Stormwater

This column only relates to Section 5.3 - Discharges to Water. The quantity of the chemical released to any receiving stream or water body in each box in column A must include the amount contributed by stormwater runoff from the facility which contains the chemical. In addition, the percentage of the total quantity (by weight) of the chemical contributed by stormwater must be entered in column C. If your facility has monitoring data on the chemical and an estimate of flow rate, you must use this data to determine percent stormwater.

If your facility does not have periodic measurements of stormwater releases of the chemical but has submitted chemical specific monitoring data in permit applications, then these data must be used to calculate the percent contribution from stormwater. Flow rate data can be estimated by multiplying the annual amount of rainfall times the land area times the runoff coefficient. The runoff coefficient represents the fraction of rainfall that does not infiltrate into the ground but runs off as stormwater. The runoff coefficient is directly related to the land uses located in the drainage area and ranges from 0.5-0.8 for light industrial areas and 0.6-0.9 for heavy industrial areas. Site specific determinations can be calculated using the following formula:

(1 x fractional paved or roofed area) + (0.2 x fractional grass area) + (0.3 x fractional graveled area) = site runoff coefficient

If you have monitored stormwater but did not detect the chemical, enter zero (0) in this space. If your facility has no stormwater monitoring data for the of the chemical, enter no data [N/D] in this space on the form.

6. Transfers of the Chemical in Waste to Off-Site Locations

You must report in this section the total annual quantity of the chemical sent to any of the off-site disposal, treatment, or storage facilities for which you have provided an address in Part II.

Line 6.1 is for transfers to a POTW. Lines 6.2 through 6.4 are provided for transfers to other off-site locations, including privately owned wastewater treatment facilities.

Enter, from Section 2 of Part II, the block number that corresponds to the off-site location to which you transferred waste containing the chemical. If you need additional space (i.e., you ship waste to more than three off-site locations), check the box at the bottom of Section 6 and use the Supplemental Information sheet (Part IV).

A. Total Transferred

You must follow the instructions for providing estimates as presented in the instructions for column A of Section 5 above. You must enter the amount in pounds of only the toxic chemical that is being transferred; do not enter the total poundage of wastes, including mixtures or trade name products containing the chemical. As with Section 5, you may report in ranges only for calendar years 1987, 1988, and 1989.

B. Basis of Estimate

You must identify the basis for your estimate. Follow the instructions and use the same codes as presented in the instructions for column B of Section 5.

C. Type of Treatment/Disposal

You must enter one of the following codes to identify the type of treatment or disposal method used by the off-site ocation for the chemical being reported. You may have his information in your copy of EPA Form SO, Item S of he Annual/Biennial Hazardous Waste Treatment, Storage, and Disposal Report (RCRA). Applicable codes for this section are as follows:

M10 Storage Only
M20 Reuse as Fuel/Fuel Blending
M40 Solidification/Stabilization
M50 Incineration/Thermal Treatment
M61 Wastewater Treatment (Excluding POTW)
M69 Other Treatment
M71 Underground Injection
M72 Landfill/Disposal Surface Impoundment
M73 Land Treatment
M79 Other Land Disposal
M90 Other Off-Site Management
M91 Transfer to Waste Broker
M99 Unknown

7. Waste Treatment Methods and Efficiency

In Section 7, you must provide the following information

related to the chemical whose releases are being reported: (A) the general wastestream types containing the chemical being reported; (B) the waste treatment methods (if any) used on all wastestreams containing the chemical; (C) the range of concentrations of the chemical in the influent to the treatment method (D) whether sequential treatment is used; (E) the efficiency or effectiveness of each treatment method in removing the chemical; and (F) whether the treatment efficiency figure was based on actual operating data. You must use a separate line in Section 7 for each treatment method used on a wastestream. This section is to be used to report only treatment of wastestreams at your facility, not treatment off-site.

A. General Wastestream

For each waste treatment method report you must indicate the type of wastestream containing the chemical that is treated. Enter the letter code that corresponds to the general wastestream type:

A = Gaseous (including gases, vapors, airborne particulates)
W = Wastewater (aqueous waste)
L = Liquid waste (non-aqueous waste)
S = Solid waste (including sludges and slurries)

If a waste is a mixture of water and organic liquid, you must report it under wastewater unless the organic content exceeds 50 percent. Slurries and sludges containing water must be reported as solid waste if they contain appreciable amounts of dissolved solids, or solids that may settle, such that the viscosity or density of the waste is considerably different from that of process wastewater.

B. Treatment Method

Codes for treatment methods are included in Table I of these instructions. You must enter the code for each treatment method used on a wastestream containing the toxic chemical, regardless of whether this treatment method actually removes the specific chemical. Treatment methods must be reported by type of waste being treated (i.e., gaseous wastes, aqueous wastes, liquid non-aqueous wastes, and solids).

Wastestreams containing the chemical may have a single source or may be aggregates of many sources. For example, process water from several pieces of equipment at your facility may be combined prior to treatment. Report treatment methods that apply to the aggregate wastestream as well as treatment methods that apply to individual wastestreams. If your facility treats various wastewater streams containing the chemical in different ways, the different treatment methods must each be listed separately.

Your facility may have several pieces of equipment performing a similar service. It is not necessary to enter four lines of data to cover four scrubber units, for example, if all four are treating wastes of similar character (e.g., sulfuric acid mist emissions), have similar influent concentrations, and have similar removal efficiencies. If, however, any of these parameters differ from one unit to the next, each scrubber must be listed separately.

C. Range of Influent Concentration

The form requires an indication of the range of concentration of the toxic chemical in the wastestream (i.e., the influent) as it typically enters the treatment equipment. You must enter in the space provided one of

Environmental Protection Agency   **§ 372.85**

the following code numbers corresponding to the concentration of the chemical in the influent:

1 = Greater than 1 percent
2 = 100 parts per million (0.01 percent) to 1 percent (10,000 parts per million)
3 = 1 part per million to 100 parts per million
4 = 1 part per billion to 1 part per million
5 = Less than 1 part per billion

Note that parts per million (ppm) is milligrams/kilogram (mass/mass) for solids and liquids; cubic centimeters/cubic meter (volume/volume) for gases; milligrams/liter for solutions or dispersions of the chemical in water; and milligrams of chemical/kilogram of air for particulates in air. If you have particulate concentrations (at standard temperature and pressure) as grains/cubic foot of air, multiply by 1766.6 to convert to parts per million: if in mg/m^3, multiply by 0.773 to obtain ppm. (Note: Factors are for standard conditions of 0°C (32°F) and 760 mmHg atmospheric pressure).

D. Sequential Treatment?

You may use various treatment steps in a sequence but only be able to estimate the treatment efficiency of the overall treatment process. If this is the case, you must enter codes for all of the treatment steps in the process. You must check the column D "sequential treatment?" box for all of these steps in the same sequence. With respect to information that must be supplied in columns C and E, you must provide the range of influent concentrations (column C) in connection with the first step of the sequential treatment. Then provide the treatment efficiency (column E) in connection with the last step in the treatment. You do not need to complete C or E for any intermediate step in the sequence.

E. Treatment Efficiency Estimate

In the space provided, you must enter the number that indicates the percentage of the toxic chemical that is removed from the wastestream. The treatment efficiency (expressed as percent removal) represents any destruction, biological degradation, chemical conversion, or physical removal of the chemical from the wastestream being treated. This efficiency must represent the mass or weight percentage of chemical destroyed or removed, not just changes in volume or concentration of the chemical or the wastestream. The efficiency indicated for a treatment method must refer only to the percent conversion or removal of the listed toxic chemical from the wastestream, not the percent conversion or removal of other wastestream constituents (alone or together with the listed chemical), and not the general efficiency of the method for any wastestream.

For some treatments, the percent removal will represent removal by several mechanisms, as in secondary wastewater treatment, where a chemical may evaporate, be biodegraded, or be physically removed in the sludge.

Percent removal must be calculated as follows:

$$\frac{(I - E)}{I} \times 100$$

where I = mass of the chemical in the influent wastestream and E = mass of the chemical in the effluent wastestream.

The mass or weight of chemical in the wastestream being treated must be calculated by multiplying the concentration (by weight) of the chemical in the wastestream times the flow rate. When calculating or

estimating percent removal efficiency for various wastestreams, the percent removal must compare the gaseous effluent from treatment to the gaseous influent, the aqueous effluent from treatment to the aqueous influent, and similarly for organic liquid and solid waste. However, some treatment methods may not result in a comparable form of effluent wastestream. Examples are incineration or solidification of wastewater. In these cases, the percent removal of the chemical from the influent wastestream would be reported as 100 percent because the wastestream does not exist in a comparable form after treatment.

Some of the treatments listed in Table 1 do not destroy, chemically convert, or physically remove the chemical from its wastestream. Some examples include fuel blending and evaporation. For these treatment methods, an efficiency of zero must be reported.

For metal compounds, the reportable concentration and treatment efficiency must be calculated based on the weight of the parent metal and not the weight of the metal compounds. Metals are not destroyed but can only be physically removed or chemically converted from one form into another. The treatment efficiency reported must only represent physical removal of the parent metal from the wastestream, not the percent chemical conversion of the metal compound. If a listed treatment method converts but does not remove a metal (e.g., chromium reduction), the method must be reported, but the treatment efficiency must be reported as zero.

All data available at your facility must be utilized to calculate treatment efficiency and influent chemical concentration. You are not required to collect any new data for the purposes of this reporting requirement. If data are lacking, estimates must be made using best engineering judgment or other methods.

F. Based on Operating Data?

This column requires you to indicate "Yes" or "No" to whether the treatment efficiency estimate is based on actual operating data. For example, you would check "Yes" if the estimate is based on monitoring of influent and effluent wastes under typical operating conditions. If the efficiency estimate is based on published data for similar processes or on equipment supplier's literature, or if you otherwise estimated either the influent or effluent waste comparison or the flow rate, you must check "No."

8. Optional Information on Waste Minimization

Information provided in Part III, Section 8, of the form is optional. This section allows you to identify waste minimization efforts relating to the reported toxic chemical that may not have been reflected in your responses to previous sections of the form.

A. Type of modification

Enter one code from the following list that best describes the type of waste minimization activity:

M1 - Recycling/reuse on-site.
M2 - Recycling/reuse off-site.
M3 - Equipment/technology modifications.
M4 - Process procedure modifications.
M5 - Reformulation/redesign of product.
M6 - Substitution of raw materials.
M7 - Improved housekeeping training, inventory control.
M8 - Other waste minimization technique.

B. Quantity of the chemical in the wastestream prior to treatment/disposal

Enter the pounds of the toxic chemical in wastes in the reporting year and the pounds in wastes in the year prior to implementing waste minimization (the "base year"). Alternatively, to protect confidential information, you may wish to enter only the percentage by which the weight of the chemical in the wastes has changed. This figure (percentage) may be calculated using the following formula:

$$\frac{\text{toxic chemical in wastes in reporting year} - \text{toxic chemical in wastes in prior year}}{\text{toxic chemical in wastes in prior year}} \times 100$$

The resulting figure may be either negative or positive.

C. Index

Enter the ratio of reporting-year production to production in the base year. This index should be calculated to most closely reflect activities involving the chemical. Examples of acceptable indices include:

- Chemical produced in 1987/chemical produced in 1986.
- Paint produced in 1987/paint produced in 1986.
- Appliances coated in 1987/appliances coated in 1986.
- Square feet of solar collector fabricated in 1987/square feet of solar collector fabricated in 1986.
- Value of sales in 1987/value of sales in 1986.

For example, a company manufactures 200,000 pounds of a chemical in 1986 and 250,000 pounds of the same chemical in 1987. The index figure to report would be 1.3 (1.25 rounded). The index provides a means for users of the data to distinguish the effects of changes in business activity from the effects specifically of waste minimization efforts. It is not necessary to indicate the units on which the index is based.

D. Reason for action

Finally, enter the codes from the following list that best describe the reason for initiating the waste minimization effort:

R1 - Regulatory requirement for the waste.
R2 - Reduction of treatment/disposal costs.
R3 - Other process cost reduction.
R4 - Self-initiated program.
R5 - Other (e.g., discontinuation of product, occupational safety, etc.).

Environmental Protection Agency   § 372.85

TABLE I
TREATMENT CODES

AIR EMISSIONS TREATMENT

A01 Flare
A02 Condenser
A03 Scrubber
A04 Absorber
A05 Electrostatic Precipitator
A06 Mechanical Separation
A07 Other Air Emission Treatment

BIOLOGICAL TREATMENT

B11 Biological Treatment -- Aerobic
B21 Biological Treatment -- Anaerobic
B31 Biological Treatment -- Facultative
B99 Biological Treatment -- Other

CHEMICAL TREATMENT

C01 Chemical Precipitation -- Lime or Sodium
     Hydroxide
C02 Chemical Precipitation -- Sulfide
C09 Chemical Precipitation -- Other
C11 Neutralization
C21 Chromium Reduction
C31 Complexed Metals Treatment (other than pH
     Adjustment)
C41 Cyanide Oxidation -- Alkaline Chlorination
C42 Cyanide Oxidation -- Electrochemical
C43 Cyanide Oxidation -- Other
C44 General Oxidation (including Disinfection) --
     Chlorination
C45 General Oxidation (including Disinfection) --
     Ozonation
C46 General Oxidation (including Disinfection) --
     Ozonation
C99 Other Chemical Treatment

INCINERATION/THERMAL TREATMENT

F01 Liquid Injection
F11 Rotary Kiln with Liquid Injection Unit
F19 Other Rotary Kiln
F31 Two Stage
F41 Fixed Hearth
F42 Multiple Hearth
F51 Fluidized Bed
F61 Infra-Red
F71 Fume/Vapor
F81 Pyrolytic Destructor
F82 Wet Air Oxidation
F83 Thermal Drying/Dewatering
F99 Other Incineration/Thermal Treatment

SOLIDIFICATION/STABILIZATION

G01 Cement Processes (including Silicates)
G09 Other Pozzolonic Processes (including Silicates)
G11 Asphaltic Processes
G21 Thermoplastic Techniques
G99 Other Solidification Processes

PHYSICAL TREATMENT

P01 Equalization
P09 Other Blending
P11 Settling/Clarification
P12 Filtration
P13 Sludge Dewatering (non-thermal)
P14 Air Flotation
P15 Oil Skimming
P16 Emulsion Breaking -- Thermal
P17 Emulsion Breaking -- Chemical
P18 Emulsion Breaking -- Other
P19 Other Liquid Phase Separation
P21 Adsorption -- Carbon
P22 Adsorption -- Ion Exchange (other than for
     recovery/reuse)
P23 Adsorption -- Resin
P29 Adsorption -- Other
P31 Reverse Osmosis (other than for recovery/reuse)
P41 Stripping -- Air
P42 Stripping -- Steam
P49 Stripping -- Other
P51 Acid Leaching (other than for recovery/reuse)
P61 Solvent Extraction (other than recovery/reuse)
P99 Other Physical Treatment

RECOVERY/REUSE

R01 Reuse as Fuel -- Industrial Kiln
R02 Reuse as Fuel -- Industrial Furnace
R03 Reuse as Fuel -- Boiler
R04 Reuse as Fuel -- Fuel Blending
R09 Reuse as Fuel -- Other
R11 Solvents/Organics Recovery -- Batch Still
     Distillation
R12 Solvents/Organics Recovery -- Thin-Film
     Evaporation
R13 Solvents/Organics Recovery -- Fractionation
R14 Solvents/Organics Recovery -- Solvent
     Extraction
R19 Solvents/Organics Recovery -- Other
R21 Metals Recovery -- Electrolytic
R22 Metals Recovery -- Ion Exchange
R23 Metals Recovery -- Acid Leaching
R24 Metals Recovery -- Reverse Osmosis
R26 Metals Recovery -- Solvent Extraction
R29 Metals Recovery -- Other
R99 Other Reuse or Recovery

APPENDIX I
29 CFR PARTS 1910, 1915,
1917, 1918, 1926, AND 1928:
HAZARD COMMUNICATION;
FINAL RULE

Monday
August 24, 1987

federal register

Part III

Department of Labor
Occupational Safety and Health
Administration

29 CFR Parts 1910, 1915, 1917, 1918,
1926, and 1928
Hazard Communication; Final Rule

31852   Federal Register / Vol. 52, No. 163 / Monday, August 24, 1987 / Rules and Regulations

DEPARTMENT OF LABOR

Occupational Safety and Health Administration

29 CFR Parts 1910, 1915, 1917, 1918, 1926, and 1928

[Docket No. H–022D]

Hazard Communication

AGENCY: Occupational Safety and Health Administration (OSHA); Labor.

ACTION: Final rule.

SUMMARY: OSHA is revising its Hazard Communication Standard (HCS) (29 CFR 1910.1200), which currently applies to the manufacturing sector, to cover all employers with employees exposed to hazardous chemicals in their workplaces. Expansion of the scope of the HCS requires all non-manufacturing employers to establish hazard communication programs to transmit information on the hazards of chemicals to their employees by means of labels on containers, material safety data sheets, and training programs. This action will reduce the incidence of chemically-related occupational illnesses and injuries in non-manufacturing workplaces.

DATES: Effective September 23, 1987. The revised standard published today requires that chemical manufacturers, importers, and distributors ensure that material safety data sheets are provided with the next shipment of hazardous chemicals to non-manufacturing employers or distributors after September 23, 1987. All employers in the non-manufacturing sector are to be in compliance with all provisions of the standard by May 23, 1988.

FOR FURTHER INFORMATION CONTACT: Mr. James F. Foster, Office of Information and Consumer Affairs, Occupational Safety and Health Administration, 200 Constitution Avenue, NW., Room N3637, Washington, DC, 20210; telephone (202)523–8151.

SUPPLEMENTARY INFORMATION: References to the rulemaking record are made in the text of this preamble, and the following abbreviations have been used:

H–022, Ex.: Exhibit number in Docket H–022, which includes Dockets H–022A and H–022B.

Ex.: Exhibit number in Docket F'–022D for exhibits collected since the 1985 Court remand.

Tr.: Public hearing transcript page number.

Copies of the official list of entries in the record, as well as the exhibits

themselves, are available from the OSHA Docket Office, Dockets H–022 and H–022D, Occupational Safety and Health Administration, 200 Constitution Avenue, NW., Room N3670, Washington, DC, 20210; telephone (202)523–7894.

I. Background

A. History of OSHA's Hazard Communication Standard

When Congress passed the Occupational Safety and Health Act of 1970, 29 U.S.C. 651 *et seq.* (the Act), it included language in section 6(b)(7) stating that any occupational safety or health standard promulgated by the Secretary of Labor under section 6(b) rulemaking authority "shall prescribe the use of labels or other appropriate forms of warning as are necessary to insure that employees are apprised of all hazards to which they are exposed, relevant symptoms and appropriate emergency treatment, and proper conditions and precautions of safe use or exposure." Whenever OSHA has promulgated a substance-specific rule to address the hazards of a particular chemical, this Congressional directive has been followed. However, given the universe of chemicals present in American workplaces (as many as 575,000 hazardous chemical products), and the time-consuming nature of OSHA's rulemaking process, it soon became clear that little information would be available to employees if this substance-by-substance approach were the only one pursued. The Agency thus decided to address the issue of hazard information transmittal on a generic basis. OSHA's experience, as well as our rulemaking record to date, supports the view that when employees have access to, and understand, the nature of the chemical hazards they are exposed to during the course of their employment, they are better able to participate in their employers' protective programs, and take steps to protect themselves. In addition, providing employers with complete chemical hazard information enables them to better design and implement protective programs. Together these actions will result in more effective worker protection and the occurrence of fewer illnesses and injuries due to exposure to chemicals. *See, e.g.,* 48 FR 53282–84, 53321, 53323–24, 53327–29 (Nov. 25, 1983); 47 FR 12093–12101 (Mar. 19, 1982).

In 1974, OSHA established a Standards Advisory Committee on Hazardous Materials Labeling under section 7(b) of the Act to develop guidelines for the implementation of section 6(b)(7). On June 6, 1975, the

Committee submitted its final report to the Assistant Secretary for Occupational Safety and Health which recommended categorization and ranking of chemical hazards, as well as provisions for labels, material safety data sheets, and training programs for all workers.

The National Institute for Occupational Safety and Health (NIOSH) published a criteria document in 1974 which also recommended a standard to OSHA. The document, entitled "A Recommended Standard . . . An Identification System for Occupationally Hazardous Materials." included provisions for labels and material safety data sheets.

In 1976, Congressman Andrew Maguire from New Jersey and the Health Research Group petitioned OSHA to issue a standard to require the labeling of all workplace chemicals. The House of Representatives' Committee on Government Operations (1976 and 1977) recommended that OSHA enforce the health provisions of the Act by requiring manufacturers to disclose any toxic ingredients in their products, and by requiring all employers to disclose this information to workers.

On January 28, 1977, OSHA initiated the public participation phase of the rulemaking process on these issues by publishing an advance notice of proposed rulemaking (ANPR) on chemical labeling in the **Federal Register** (42 FR 5372). The ANPR requested comments and information on the need for such a standard, and the particular provisions that should be included. The Agency received eighty-one comments. Most supported the need for the rule, with opinions as to the specific approaches to be pursued varied significantly.

On January 16, 1981, OSHA published a notice of proposed rulemaking (NPRM) entitled "Hazards Identification" (46 FR 4412). The rule would have required manufacturing employers to assess the hazards in their workplaces using specified procedures, and to label specified containers. The requirements were quite different from the comprehensive approach previously recommended by the Standards Advisory Committee and NIOSH as they did not include provisions for material safety data sheet development or training.

OSHA withdrew the NPRM on February 12, 1981 (46 FR 12214) for further consideration of regulatory alternatives. A new NPRM was published on March 19, 1982, and was entitled "Hazard Communication" (47 FR 12092). It proposed to require producers of chemicals to evaluate them to determine their hazards, label containers, and provide material safety data sheets to manufacturing purchasers of their products. The standard also proposed that all employers in the manufacturing sector have a hazard communication program, label in-plant containers, maintain and provide access to material safety data sheets, and train workers. The proposal also invited comments on whether non-manufacturing employers should be subject to the rule.

Following a period for written comments, informal public hearings, and a post-hearing comment period, OSHA published the final Hazard Communication Standard on November 25, 1983 (48 FR 53280). The provisions of the final rule are very similar to those described above for the proposal, i.e., chemical manufacturers and importers are required to evaluate the hazards of the chemicals they produce or import, and all manufacturers are required to have hazard communication programs for their employees exposed to hazardous chemicals. This comprehensive standard was designed to reduce the hazards faced by manufacturing workers when they handle chemicals without adequate information on, among other things, the physical and health hazards of the chemicals, safe handling precautions, and emergency and first aid procedures. See, e.g., 48 FR 53321. OSHA found that inadequate communication regarding chemical hazards is a significant risk to workers. See, e.g., 48 FR 53321. Accord United Steelworkers of America v. Auchter, 763 F.2d 728, 735 (3d cir. 1985) (United Steelworkers I) ("[I]nadequate communication is itself a hazard, which the standard can eliminate or mitigate.").

OSHA decided to limit the scope of coverage of the HCS to the manufacturing sector based on an analysis of the chemical source illnesses and injuries occurring in each industrial sector. (See discussion at 48 FR 53284–85.) In particular, since the purpose of the standard is to reduce the occurrence of such incidents, OSHA determined that the rule should focus on those industrial sectors where they are recorded most frequently. The Agency found that over half of these incidents occur in manufacturing, although manufacturing accounts for only about 30 percent of total employment. Thus OSHA decided that the greatest need for transmittal of chemical hazard information is in the manufacturing sector. The Agency further recognized that since chemicals are developed and produced in the manufacturing sector, the hazard information would have to be

developed in the manufacturing sector first, regardless of the eventual coverage of the rule. OSHA believed that requiring the development of the chemical hazard information in manufacturing would lead to its increased availability in the other sectors without the standard specifically requiring the transmittal of hazard information to those sectors. The Agency acknowledged that hazardous chemicals are pervasive throughout industry and that chemical source injuries and illnesses have been recorded in all industry sectors. See, e.g., 46 FR 53282–87. See also United Steelworkers I, 763 F.2d at 737. The Agency planned to make a decision regarding the explicit coverage of the non-manufacturing sectors once the HCS was in effect, and a determination could be made as to whether the other industries were, in fact, obtaining the information they needed. OSHA believed that the Act gives the Secretary of Labor and the Agency the authority to regulate the most hazardous industry first under section 6(g), 29 U.S.C. 655(g), which states in part:

In determining the priority for establishing standards under this section, the Secretary shall give due regard to the urgency of the need for mandatory safety and health standards for particular industries, trades, crafts, occupations, businesses, workplaces or work environments.

B. Court Challenges

The HCS was challenged in the U.S. Court of Appeals for the Third Circuit (hereinafter referrred to as "the Court" or "the Third Circuit") on several grounds. The Court issued its decision on May 24, 1985 (United Steelworkers I, 763 F.2d 728 (3d Cir. 1985)). The standard was upheld in most respects, but three issues were remanded to the Agency for reconsideration. The decision was not appealed.

First, the Court concluded that the definition of trade secrets incorporated by OSHA included chemical identity information that was readily discoverable through reverse engineering and, therefore, was "broader than the protection afforded trade secrets by state law." The Court directed the Secretary of Labor to reconsider a trade secret definition which would not include chemical identity information that is readily discoverable through reverse engineering. Second, the Court held that trade secret access rule in the standard invalid insofar as it limited access to health professionals, but found the access rule otherwise valid. The Secretary was directed to adopt a rule permitting access by employees and

their collective bargaining representatives to trade secret chemical identities. OSHA complied with the Court orders regarding the two trade secret issues in a separate rule, published in final form on September 30, 1986 (51 FR 34590).

The third issue remanded to OSHA involved the scope of the standard's coverage. As noted, the HCS currently applies to employers and employees in the manufacturing sector. The Court rejected the Secretary's contention that section 6(g) gave him the flexibility to regulate the most hazardous sector first before commencing rulemaking for other sectors in which workers are exposed, to a lesser extent, to the same hazards. The Court agreed that section 6(g) "clearly permits the Secretary to set priorities for the use of the Agency's resources, and to promulgate standards sequentially." 763 F.2d at 738. The Court also acknowledged that "there is substantial evidence in the record that the manufacturing sector has the highest incidence rate of chemical exposures, which the Agency has authority to regulate." Id. at 737. However, the Court held that it is not enough merely to establish that the sector selected for coverage presents greater hazards than those that have been left for later rulemaking. Given the record evidence of high levels of exposure to hazardous chemicals in several job settings outside the manufacturing sector, the Secretary was required to explain "why coverage of workers outside the manufacturing sector would have seriously impeded the rulemaking process" or "why it is not feasible for the same standard to be applied in other sectors where workers are exposed to similar hazards." Id. at 738.

The Court was not persuaded that the HCS would provide protection to uncovered workers because chemical hazard warnings would be found on container labels and detailed information on material safety data sheets would become increasingly available in the unregulated sectors as a result of being required in manufacturing. Id. There was considerable record evidence that indicated that workers in the non-manufacturing industries are exposed to chemical hazards. The Court concluded that the Secretary had not stated why it would not be feasible to require employers in non-manufacturing industries to give workers material safety data sheets and training as required in the manufacturing sector. Id. The Court maintained that the Act required an explanation why the same information, that is, labels, material

31854   **Federal Register** / Vol. 52, No. 163 / Monday, August 24, 1987 / Rules and Regulations

safety data sheets, and training, is not needed for workers in other sectors similarly exposed to hazardous chemicals. *Id.* at 738–39. Therefore, as previously indicated, OSHA was directed by the Court to reconsider the application of the standard to employees in the non-manufacturing industries and to order its application to these other sectors unless the Secretary can state reasons why this application would not be feasible. It should be noted that in previous OSHA litigation, the Courts have defined "feasibility" in terms of OSHA rules as meaning "capable of being done." *American Textile Manufacturers Institute v. Donovan*, 452 U.S. 490, 508–509 (1980)(ATMI).

OSHA decided not to appeal this decision. As stated in the preamble to the final rule (48 FR 53286):

It should be emphasized that the Agency does not believe that employees in other industries are not exposed to hazardous chemicals, or that they should not be informed of those hazards. OSHA has merely exercised its discretion to establish rulemaking priorities, and chosen to first regulate those industries with the greatest demonstrated need.

OSHA was prepared to evaluate the HCS' effectiveness in getting information to downstream employers, and to extend the standard if necessary. In fact, the Agency initiated the process on March 4, 1985, prior to the Court decision, when the Assistant Secretary asked the National Advisory Committee on Occupational Safety and Health (NACOSH) to give OSHA its recommendation on the need and feasibility of expanding the scope of the HCS to other industries. On June 21, 1985, NACOSH adopted the following recommendation:

[NACOSH] strongly endorses the OSHA effort to promulgate a Hazard Communication Standard and selection of the manufacturing sector for its initial scope of coverage. It is the consensus recommendation of the Committee that the scope of the current Hazard Communication Standard should be expanded to cover all employees in all industries at as early a time as possible. Complete implementation may require phasing in gradually. The BLS [Bureau of Labor Statistics] incidence rates of occupational illnesses, and other appropriate factors, should be primary considerations in expanding the coverage. The Committee further recommends that OSHA establish a task force to address these issues.

Meanwhile, OSHA's review of the rulemaking record showed that while there was considerable evidence concerning the need for hazard communication in other industries, and general support for a finding that the HCS would be feasible for non-

manufacturing, there was a need for more direct evidence of the feasibility of expanded coverage, particularly in the area of economic feasibility. Accordingly, OSHA believed it was necessary and appropriate to initiate further rulemaking. OSHA commissioned a study of the economic impact of extending the HCS to the fifty major non-manufacturing industry groups within its jurisdiction, and issued an Advance Notice of Proposed Rulemaking (ANPR) seeking public comment on present hazard communication practices outside manufacturing, and the likely impact of extending the HCS to industries significantly different from the prototypical manufacturing worksites on which the original standard was based. 50 FR 48794 (Nov. 27, 1985). Over two hundred responses were received. Based on this newly acquired evidence and on the previous rulemaking record, OSHA was in the process of drafting a proposed rule which it expected to publish for notice and comment, followed by promulgation of a final rule in early 1988.

On January 27, 1987, however, the United Steelworkers of America, AFL–CIO–CLC and Public Citizen, Inc., petitioners in the 1985 challenge, filed a Motion For An Order Enforcing The Court's Judgment and Holding Respondent In Civil Contempt. Petitioners claimed that the Court's 1985 order had not authorized OSHA to embark on further fact gathering; that OSHA should have made a feasibility determination on the 1985 rulemaking record. Petitioners also argued that even if further fact gathering had been allowed by the Court's order, OSHA's pace was unduly slow.

In response, OSHA noted that the Court's 1985 order did not specify that OSHA should act on the then-existing record. OSHA believed that seeking further evidence on feasibility in non-manufacturing was appropriate in light of its statutory obligation to issue rules that are well grounded in a factual record. OSHA also asserted that, consistent with Supreme Court precedent, the Agency should be permitted to exercise its discretion in determining the appropriate rulemaking procedures for complying with the Court's remand order. Lastly, the Agency argued that its schedule to complete the rulemaking was reasonable and did not constitute undue delay.

On May 29, 1987, the Court issued a decision holding that the Court's 1985 remand order required consideration of the feasibility of an expanded standard without further rulemaking. *United*

Steelworkers of America, AFL–CIO–CLC v. Pendergrass, No. 83–3554 (3d Cir.) (*United Steelworkers II*). The Court declared that adequate notice had been provided to non-manufacturers during the original rulemaking that they might be covered by the HCS, *id.* slip op. at 7–10, 16–17, that the answers to the remaining questions OSHA may have had regarding feasibility were "self-evident" or "readily ascertainable" from the original record, *id.* at 15, 17, and that further fact finding was "unnecessary", *id.* at 15. The Court ordered the Agency to issue, within 60 days of its order, "a hazard communication standard applicable to all workers covered by the OSHA Act, including those which have not been covered in the hazard communication standard as presently written, or a statement of reasons why. on the basis of the present administrative record, a hazard communication standard is not feasible." *Id.* at 19. OSHA is responding to the Court order by issuing this final rule expanding the scope of the HCS' coverage to all workers within OSHA's jurisdiction.

OSHA continues to believe that it should have been permitted to follow the rulemaking procedures in the Act by issuing a notice of proposed rulemaking and developing a public record prior to promulgating a final rule. However, as discussed in the following section regarding feasibility, the Agency does not have sufficient evidence in the current record to indicate that the rule would be infeasible for any part of the non-manufacturing sector. OSHA recognizes that information submitted during a normal rulemaking process might have resulted in further changes to the provisions to better address feasibility or practicality concerns.

In light of the fact that there may be additional information regarding the feasibility or practicality of the rule as it applies to some non-manufacturing sectors, the Agency invites persons to provide such information and any recommendations for further rulemaking within sixty days of the date of publication of this final rule. OSHA will then evaluate these submissions and determine whether any additional rulemaking is required. Data or evidence related to feasibility should be addressed to: Directorate of Health Standards Programs, Occupational Safety and Health Administration, Attention: Hazard Communication, 200 Constitution Avenue, NW., Room N3718, Washington, DC, 20210.

C. Feasibility of the Standard

In the context of OSHA standard setting, feasibility constraints limit the extent to which standards can address health and safety concerns within the workplace. Section 6(b)(5) of the Act, 29 U.S.C 655(b)(5). Feasibility analysis involves an inquiry to determine whether a standard is both technologically and economically capable of being done. *ATMI*, 452 U.S. at 512–13 and 513 n.31 (1980). As the Third Circuit has indicated, "the Secretary was able to determine that the hazard communication standard could feasibly be applied in the manufacturing sector." *United Steelworkers II*, slip op. at 16. The Court further noted that OSHA had concluded in the final rule that importers and distributors could feasibly comply with the HCS based on the evidence in the record and that "this is equally true of all non-manufacturer user employers. Plainly, the ease with which the same information can be utilized by those employers can be easily determined from the information already in the record." *Id.* at 18. The Third Circuit has ordered expansion of the HCS to all workers *unless* OSHA can give reasons why the HCS is infeasible for particular industries, and has forbidden OSHA from gathering further evidence.

OSHA concludes that the original HCS rulemaking record (Docket H–022), does not contain credible evidence indicating the standard would be infeasible for any industrial sector. In fact, OSHA believes that the original record on the whole supports a finding that the performance-oriented HCS is feasible for all industries. In addition, the Agency's experience under the present HCS and other pertinent OSHA standards, the promulgation and implementation of State and local right-to-know laws, and evidence and data gathered by the Agency since the 1985 Court order (Docket H–022D), further supports OSHA's conclusion that non-manufacturing employers are "capable" of implementing the HCS for their employees potentially exposed to hazardous chemicals.

OSHA found that the HCS is technologically feasible for manufacturers, and believes it is clearly technologically feasible for non-manufacturers as well. Twelve of the OSHA-approved State plan States have already extended the rule to cover the non-manufacturing sector, and the requirements are being enforced in those States as workplace standards. This experience provides practical evidence of the technological feasibility of the requirements of the rule. The more

technical aspects of the standard—scientific evaluation of chemicals to determine their hazards and creation of material safety data sheets and warning labels—remain a burden on those producing or importing hazardous chemicals. The technical expertise needed to develop the chemical hazard information, and its associated costs, is subsumed within the current rule covering manufacturers, and it has been found feasible. All other requirements in the HCS, such as maintaining material safety data sheets, developing a written hazard communication program, and designing and implementing chemical hazard training, are conventional and common business practices that are administrative in nature, and no technological barriers prevent their development and implementation. OSHA has mandated such practices in some non-manufacturing workplaces since the early 1970's. *See, e.g.,* 29 CFR 1915.97 (requiring material safety data sheets and chemical hazard training for shipyard workers); 1917.22 (requiring marine terminal workers be instructed as to the chemical hazards presented by cargo); 1918.86 (requiring chemical hazard instruction for longshore workers); 1926.21 (requiring chemical hazard training for construction workers). *See, also,* H–022, Ex. 99 (journal article regarding usefulness of material safety data sheets, written by Dow Chemical Company representatives and published in December 1957).

OSHA also believes that the economic feasibility of extending the current HCS to the non-manufacturing sector is supported by the record. Simply put, economic feasibility is established by evidence that the standard will not threaten the regulated industry's "long-term profitability." *ATMI*, 452 U.S. at 531 n.55. Costs associated with expanding the standard to cover non-manufacturing workplaces will stem from the initial start-up costs and the less substantial recurring program implementation and upkeep costs for: maintaining material safety data sheets received from manufacturers, importers, distributors, and other employers; creating labels for in-house containers of hazardous chemicals; developing a written hazard communication program, including a list of hazardous chemicals present in the workplace; and developing and implementing chemical hazard training.

After careful analysis of the original HCS rulemaking record, OSHA concludes that, as a whole, it supports a finding that non-manufacturers are economically capable of providing

employees chemical hazard information in the manner prescribed by the HCS. As noted previously, development of the evidentiary record for the HCS began as early as 1974. In that year, NIOSH recommended that OSHA adopt a standard requiring *all* employers to implement a system of labels, placards and material safety data sheets in their workplaces to inform employees about the chemical hazards to which they may be exposed. (H–022, Ex. 4). The NIOSH recommended standard, like the HCS, included requirements that employers ensure that chemicals in the workplace are marked with hazard warnings and that material safety data sheets are "filed in the establishment" where they are "readily available for examination by workers". *Id.* at 3. This hazard identification and warning system was designed to additionally "help in the education of employees and provide the data necessary for employers to take proper action to safeguard their employees." *Id.* at 1. NIOSH concluded that such a chemical hazard communication program was appropriate for all employers. *See, also* comments of the Air Transport Association, H–022, Ex. 5–3 ("[T]he airlines have no general objection to the [NIOSH] Criteria . . . [except that it] should clearly delineate the responsibility of the manufacturer supplying the necessary data on the Material Safety Data Sheets.").

The 1975 report of the Standards Advisory Committee on Hazardous Materials Labeling (H–022, Ex. 3), recommended a "total system" approach to chemical hazard communication not unlike the comprehensive approach of the current HCS. The Advisory Committee, which included representatives of non-manufacturers, recommended labeling and placarding systems, the creation and availability of material safety data sheets, and employee education and training programs for all workers potentially exposed to hazardous chemicals. The Committee recognized that these practices "are not new and novel concepts" but "well established in many industries and professional associations as well as regulated by various governmental agencies and international agreements." *Id.* at 3. The Advisory Committee made "no distinction among employees in different sectors of the economy." *United Steelworkers II*, at 7.

As the Court has stated, *id.* at 8, the 1977 ANPR requested public comment from all interested persons on whether a chemical hazard communication standard should be promulgated by

31856   Federal Register / Vol. 52, No. 163 / Monday, August 24, 1987 / Rules and Regulations

OSHA. Comments on the Standards Advisory Committee's recommended standard were specifically requested. Although OSHA did not receive comment from employers in every industrial sector, those non-manufacturers that did respond supported a comprehensive hazard communication system for their workplaces. For example, Sea-Land Service, Inc. (H–022, Ex. 2A–6), supported requirements for container labels (consistent with transportation labels already in place), the availability of material safety data sheets to persons in the workplace, and individual training programs. Panhandle Eastern Pipe Line Company (H–022, Ex. 2A–7) and Truckline Gas Company (H–022, Ex. 2A–9) both "agree[d] that employees need information about the product with which they work" and that this could be accomplished by requiring suppliers of hazardous chemicals to label containers with the "degree and nature of the hazard" and by requiring user employers to "inform employees of the hazard." Those companies had already developed "a special manual of data for all chemicals, solvents and cleaners used in [their] operations and maintenance."

Wisconsin Electric Power Company (H–022, Ex. 2A–30), stated that given adequate labels and material safety data sheets from chemical manufacturers and suppliers, chemical users such as they "would be in a position to prepare their own Material Safety Data Sheets, hazard placard systems, proper labeling of auxiliary and secondary containers and training of personnel who may use or otherwise contact this material." Recognizing the need for "proper labeling, storage, handling and instructions in the use of hazardous materials," Wisconsin Electric Power Company had already "developed and put into effect a Hazardous Materials Control Program." Southern Gas Association (SGA) (H–022, Ex. 2A–75) also believed that given suppliers and manufacturers of hazardous materials should be required to provide proper labeling, warnings and other hazard information to all employers using these materials. SGA further suggested that OSHA promulgate a standard directing all employers "to establish required training for employees that may handle or otherwise be exposed to any hazardous materials." These comments and others filed in response to OSHA's 1977 ANPR indicate that many non-manufacturers consider maintaining labels received on chemical containers, making material safety data sheets received from suppliers available

to employees, and providing information and training to employees regarding the chemical hazards present in the workplace to be economically feasible. See, also H–022, Exs. 2A–2 (Schirmer Engineering Corporation); 2A–31 (Union Electric Company); 2A–32 (Texaco); 2A–36 (American Trucking Association, Inc.).

Moreover, comments received from non-manufacturers at later stages of the original rulemaking also indicate they are capable of implementing the performance-oriented HCS. In fact, there are comments which indicate that many of these requirements were already being implemented in the non-manufacturing sector.

For example, the Western Agricultural Chemicals Association indicated that its members provide material safety data sheets to anyone who requests them, including customers in the non-manufacturing sector (Tr. 2873). Their representative further stated that "[i]n the agricultural field, I would say most technical products have material safety data sheets. I would say maybe 75% to 80% of the inerts have them . . ." (Tr. 2881).

There was also testimony from employee representatives, including those in the non-manufacturing sector such as airline mechanics, that they requested and were able to obtain material safety data sheets from manufacturers for products in use in their facilities. Tr. 2819–21, 3131, 3828. One union testified that a joint employee-employer safety committee received every material safety data sheet it requested, and that the union then trained workers to be able to use the information. Tr. 2824–A.

Another non-manufacturing union representative, the International Brotherhood of Painters and Allied Trades, indicated that it shared collected material safety data sheets with employers who needed such information. "[T]o contractors who make requests of us for information, we do provide them material safety data sheets, write-ups on the chemicals and the products . . . We do everything—our union does everything they can as a service to our contractor members to provide them with the information they need to operate safely. . . ." Tr. 2101–2.

Other large companies with manufacturing as well as non-manufacturing establishments testified that information was made available throughout their corporations, and they provide information to all customers regardless of industry. For example, Atlantic Richfield Company testified that they have a company-wide material

safety data sheet policy and program. "[U]nder this program, a material safety data sheet is recognized as a basic source of information for practical health, safety and environmental information. The MSDS whether generated internally or obtained from a supplier is used to communicate relevant data within the company and to outside customers. It is the responsibility of our various operating companies to distribute copies of each MSDS to customers and company facilities for employee instruction and/ or information." Tr. 2439. Their company facilities include such non-manufacturing operations as petroleum production.

Similarly, Exxon, Inc. testified that it too provides material safety data sheets to all customers: "[W]e consider a material safety data sheet a matter of public information that's part of our literature, regularly available to anyone who requests it." Tr. 1708–09. See, also, Shell testimony at Tr. 1712 and 2500, and Uniroyal Chemicals at Tr. 1464.

Therefore, based on the recommendations of NIOSH, the Standards Advisory Committee and the comments received from non-manufacturers and their representatives participating in the lengthy rulemaking, OSHA concludes that the original record as a whole indicates that non-manufacturers are capable of complying with the HCS. As long as chemical suppliers provide adequate chemical hazard information in the form of labels and material safety data sheets to non-manufacturers using the chemicals, those user employers, like the manufacturers who use hazardous chemicals which they themselves did not manufacture or import, can develop hazard communication programs and provide employees information and training on the chemical hazards in the workplace.

In light of the evidence in the original rulemaking record, OSHA concludes that non-manufacturers can incorporate the HCS' administrative practices and provide chemical hazard information to their employees. OSHA believes all employers can ensure that containers of chemicals are maintained with proper hazard warnings just as an employer would maintain labels or marks on containers to ensure that employees comprehend their contents and intended uses. Likewise, all employers are able to acquire and maintain up-to-date material safety data sheets for hazardous chemicals just as they are able to acquire and maintain up-to-date cost information and performance specifications on those very same chemicals. OSHA also concludes that it is feasible for employers to inform and train their workers regarding the chemical hazards present in the workplace just as employers are capable of training their workers to perform their jobs in an efficient and speedy manner. These conclusions are further supported by the experience and evidence gathered by the Agency since promulgation of the HCS for manufacturers in 1983.

At this time, OSHA has no evidence indicating that the profitability of manufacturers generally, or even chemical manufacturers in SIC 28 (by far the most economically burdened by the HCS, *see* 48 FR 53333), has been threatened by complying with the HCS. Manufacturers have had the considerable costs of evaluating, collectively, hundreds of thousands of chemicals for their hazards and creating corresponding labels and material safety data sheets since November 1985, as well as the costs of implementing an in-plant program by May 1986. After thorough analysis, OSHA determined that the current HCS would not impose a substantial burden on manufacturers and that the HCS was economically feasible for them. *See* 48 FR 53333. Experience to date in implementation of the rule supports that finding. For example, if manufacturers were experiencing significant feasibility problems in complying with the rule, OSHA would have expected to receive numerous substantive comments regarding those problems in response to the 1985 ANPR questions addressing feasibility concerns. However, although some manufacturing employers objected to some requirements, substantive comments demonstrating infeasibility were not received, which appears to support OSHA's conclusion that compliance with the HCS was, and continues to be, economically feasible for manufacturers and indicates that the standard is also feasible for non-manufacturers. In fact, some manufacturers took the opportunity to state their continuing support for the rule and its requirements. *See, e.g.,* H–022D, Ex. 2–14, (The Chemical Manufacturers Association "strongly believes that the substantive provisions of the Hazard Communication Standard are sound as a matter of science and policy."); Ex. 2–67 (Economics Laboratory, Inc. "considers hazard communication worth the effort.")

Generally, the HCS costs to non-manufacturers would be a function of the number of hazardous chemicals in the workplace, and the number of employees exposed to hazardous

chemicals. If employees are not potentially exposed to hazardous chemicals in a particular work operation, the proposed standard does not apply. Also, to the extent that employers are voluntarily providing information, or providing information in order to comply with other regulations or laws, this should significantly reduce the burden of compliance with this rule. Approximately 32 States and several localities already have hazard communication/right-to-know laws covering non-manufacturing industries indicating that many others seeking to protect the safety and health of workers have concluded that industry can comply with these types of requirements. In fact, as evidenced in the original rulemaking record, many companies involved in interstate commerce would benefit from promulgation of a uniform Federal standard as it would preempt different and potentially conflicting State and local laws and lessen overall compliance burdens. 48 FR 53283. *See also, e.g.,* H–022D, Ex. 2–83 (The American Gas Association "believes that a Federal Standard, rather than a variety of differing state regulations, would best serve the needs of the natural gas industry, the employees in our industry, and the general public as well."); Ex. 2–108 (The National Constructors Association has found that "[i]t has been nearly impossible to establish uniform interstate policy" and "can clearly see the wisdom of having one workable/cost-effective government regulation that addresses hazard communication.")

Although the original HCS record contained no evidence to indicate the HCS would be economically infeasible for non-manufacturing, OSHA recognized that potential feasibility concerns could arise, for example, with small businesses, businesses with large employee turnover (such as retail stores and construction companies), and businesses with rapid turnover of hazardous chemicals in the workplace (such as warehouses and marine cargo operations). However, based on the original HCS rulemaking record, and additionally based on: (1) The apparent successful implementation of the present HCS by manufacturers; (2) the implementation of other Federal communication standards and of State plan States' laws by non-manfacturers; and, (3) on regulatory impact and regulatory flexibility analyses prepared by the Agency since the 1985 Court order and summarized in Section III of this document, OSHA concludes that the provisions in the current Hazard

Communication Standard are economically feasible for all of the non-manufacturing industries.

OSHA is also aware that many employers in the manufacturing sector have been able to satisfy some of their responsibilities under the HCS by using compliance materials obtained from various sources. Trade associations, for example, have frequently been instrumental in assisting their members in developing programs suitable for their type of industrial facility. This is particularly appropriate given the performance orientation of the HCS, and the flexibility employers are permitted to design appropriate compliance programs. Sample written programs and other written materials, as well as training programs regarding the requirements of the rule, have been developed and provided to association members and have facilitated compliance efforts. The ability of associations to accomplish this successfully demonstrates technical feasibility and enhances economic feasibility. Trade associations in states covering non-manufacturing workplaces under their right-to-know rules have also been able to develop materials to assist their members to comply. Materials developed for these State laws or for the manufacturing sector under the current HCS could be adapted for the non-manufacturing workplaces newly covered by the HCS.

There have also been a number of services provided by consultants in the private sector. These range from very specific items, such as computer programs to manage information, to a comprehensive compliance strategy, where a consultant will devise an entire program to enable a facility to comply. Such services will often minimize the burden of compliance by minimizing the time the facility staff must spend to develop and implement a program. The availability of such programs also provides support for the conclusion that the rule is feasible.

For large companies, the burden per facility will often be minimized by corporate development of a standardized program. It can be expected that most corporations with multiple facilities will use this approach (this has occurred in the manufacturing sector as well).

Therefore, OSHA concludes that similar resources will be available to employers in the non-manufacturing sectors, which further demonstrates that the rule is feasible for implementation in all sectors. In fact, given the pre-existing coverage of non-manufacturing under various state rules, and the extent of the

31858 Federal Register / Vol. 52, No. 163 / Monday, August 24, 1987 / Rules and Regulations

materials developed in response to the current HCS which would also be applicable in non-manufacturing, additional development of such materials should require considerably less effort and be easier for non-manufacturers to obtain.

Nevertheless, OSHA recognizes that the unique characteristics of some businesses render certain provisions of the current standard unnecessary or ineffective in communicating the hazards of chemicals to workers. The Agency has thus made some modifications to the standard to ensure that its provisions are practical and effective in communicating hazards to all workers. *Cf. ATMI*, 452 U.S. at 531 n.32 (OSHA may use cost-effectiveness analyses and choose the less costly of two equally effective standards). The inclusion of these "tailoring" provisions is consistent with the Agency's action in tailoring the original HCS to make it practical and cost-effective for all manufacturers. *See* 29 CFR 1910.1200(b) (3)–(5). Now that the coverage of the standard is being expanded to non-manufacturing employers as well, it is necessary to tailor the standard to the unique characteristics of these non-manufacturing employers. The tailoring provisions, explained in Section II of this preamble, are based on the original record in the HCS rulemaking, and also on Agency experience in implementing the current rule; State plan State experience in implementing expanded versions of the current rule; and comments submitted to the Agency in response to the ANPR published in November 1985. OSHA believes that the knowledge and experience gained during the past few years of implementation and enforcement of the current rule must be taken into consideration when crafting a rule to appropriately apply to the non-manufacturing sector.

The Agency's position is that all employees are entitled to information regarding the chemical hazards they are exposed to in the workplace, and that a uniform Federal hazard communication standard is the best method to ensure that information is provided. This position is consistent with the Act (protecting all employees to the extent feasible), as well as with the Court's decision upon review of the rule. Therefore, this final rule addresses communicating chemical hazards to all exposed employees.

It should be emphasized that in preparing a detailed regulatory impact analysis for the expansion of the scope of the HCS, OSHA has accumulated evidence to indicate that some

employees in every SIC code designation are exposed to hazardous chemicals, and that it is therefore not appropriate to exempt any particular industry sector. For example, OSHA has received suggestions that retail establishments be exempted since employee exposure to chemicals is believed to be unlikely in these types of facilities. However, there is testimony in the original rulemaking record from the United Food and Commercial Workers International Union (Tr. 3088–97) that demonstrates that workers in such facilities are exposed to hazardous chemicals, and therefore do need the protections afforded by coverage under the HCS:

> While supermarkets don't use hundreds of hazardous chemicals like some manufacturing industries, a large number of workers are exposed to the dozen or so they do use. Chemicals used include caustic and acid cleaning compounds, solvents, waxes, paints and disinfectants . . . Let me relate to you one case within our union where workers were overexposed to an unidentified substance. A group of supermarket workers began experiencing dizziness, upper respiratory tract irritation and headaches . . . Not until workers started to talk with one another did they start to suspect a possible link between their illness and a certain solvent that was used to remove old price labels from merchandise called Garvey XC-36.

See Tr. 3088–89. *See also* Tr. 414 and Tr. 1840–43. The testimony further relates other incidents, as well as the various activities the union had to pursue to obtain information for exposed workers—including chemical analysis of products to determine their contents. This illustrates the need for application of the standard in industries such as retail stores, as well as those industries where chemical exposures are more obvious. For additional testimony regarding the extent of chemical exposures in the non-manufacturing sector, *see, e.g.,* hospital workers: Tr. 411–14, 2738–41, and 3036 (". . . hospital workers are exposed to formaldehyde, ethylene oxide, cleaning agents which are often very caustic . . .") (Tr. 411); barbers and beauticians: Tr. 415–16 (". . . work around hair dyes . . . known to cause cancer . . .") longshore workers: Tr. 3143; utility workers: Tr. 417, 3078, 3130; workers in dry cleaners and laundries: Tr. 416, 4084–90 (" . . . [B]eyond the chlorinated solvents that your dry cleaners use, some cleaners and laundries also use dyes . . ."); farmworkers: Tr. 2260.

D. Construction Advisory Committee Recommendations

On June 23, 1987, the Construction Advisory Committee on Occupational

Safety and Health met to discuss a draft proposed standard prepared by OSHA to expand the scope of the HCS to the non-manufacturing industries. The draft proposed rule was very similar to the final standard being promulgated herein. OSHA has reviewed the recommendations of the Construction Advisory Committee, and incorporated a number of the suggested revisions into this document to tailor the rule for the construction industry. As with other industries which have similar concerns due to similar differences in work operations from the typical manufacturing establishment. Other recommendations called for more substantive changes to the HCS, affecting the obligations of chemical manufacturers and others, and OSHA does not believe they are supported by the record or appropriate to incorporate into this final rule without further opportunity for notice and comment from those affected. It is important to note, however, that despite the recommended changes there were no indications that members of the Construction Advisory Committee believe that it is infeasible to implement hazard communication programs in the construction industry. In fact, as OSHA has noted previously, the construction industry has been subject to training requirements concerning chemical hazards for many years (*see* 29 CFR 1926.21).

In preparing the draft proposed rule, and subsequently this final rule, OSHA did review the *Report on Occupational Health Standards for the Construction Industry* which was submitted by the Construction Advisory Committee to the Assistant Secretary on May 16, 1980. In that report, the Committee addressed recommendations for labels, material safety data sheets, and training—all of the major components of the HCS.

Of particular concern to the Committee at that time was that construction employers do not have access to the necessary information upon which to develop appropriate signs and labels or material safety data sheets, and therefore must depend upon suppliers for such information. "[C]onstruction employers may not always be aware of the hazard associated with a particular product or device if the items are not accompanied upon purchase by appropriate labels and data sheets. . . ." OSHA agrees that this lack of information has been a problem for all downstream users of chemicals, and thus developed the approach incorporated into the HCS—producers or importers of chemicals are responsible for evaluating the hazards and transmitting that information to downstream employers or users of the materials. Under the expanded rule, construction employers would be the recipients in this downstream flow of information.

The HCS did not exist at the time of the report, and the Committee thus recommended that a solution to the problem of lack of information "would be to modify and extend the existing OSHA standard for material safety data sheets which now applies only to ship repairing, shipbuilding, and ship breaking (29 CFR 1915, 1916 and 1917). The modified standard would require manufacturers or formulators of harmful materials or agents to supply material safety data sheets along with their products in such a fashion that they reach construction employers." Shipbuilding and ship repairing are in the manufacturing sector, and covered by the requirements of the 1983 final rule—ship breaking will be covered by these expanded provisions. Therefore, OSHA is doing what was recommended in 1980, *i.e.*, extending the existing OSHA standard for material safety data sheets to construction. The Advisory Committee concluded that although the hazard information may have been difficult for construction employers to acquire in the past, "such information was fundamental to the preparation of warning signs, labels, training programs, and other important job safety and health activities."

The Construction Advisory Committee is now recommending that the construction industry be regulated under a separate standard for Hazard Communication, rather than being treated as any other downstream employer who uses chemicals. The rationale is that construction sites are unique among industrial workplaces and should be addressed in a vertical standard specific to the industry. Although OSHA has found this argument persuasive for a few health standards, where there are fundamental differences in control strategies to achieve permissible exposures for a chemical in a fixed site facility versus the construction site, it does not appear to be appropriate in this situation which simply involves transmittal of information, that can be accomplished on any type of site. Arguments regarding transient workers, mobile work sites, etc. can appropriately be made for other non-manufacturing users of chemicals as well. The problems raised can be dealt with more effectively by modifying the provisions of the current rule to address them, rather than preparing completely separate standards for each industry.

It was interesting to note that although the Construction Advisory Committee was essentially maintaining that hazard communication in construction could be treated as a separate issue, many of the changes the members were recommending would often have required substantive changes in the requirements for the manufacturing sector. As noted above, the Committee expects to receive lables on containers and material safety data sheets from its suppliers. This is certainly consistent with OSHA's approach in the rule. But the Committee is also recommending that the labels on containers being shipped to construction contain additional information, and that the requirements for material safety data sheets be slightly different as well. They also recommended changes in hazard determination provisions, while maintaining that hazard determinations must be accomplished in the manufacturing sector. These recommendations serve to support OSHA's view that in an approach which requires a downstream flow of information, the relationship between the requirements for producers and downstream users are so inter-dependent that separation of them into two separate standards would be logically inconsistent. And furthermore, since the requirements for hazard determinations, labels, and material safety data sheets were based on an extensive rulemaking record, and are not industry-specific, it would not be appropriate to modify those requirements at this point.

Two separate standards would also require cross-referencing provisions from one rule to another to ensure proper information transmittal, a regulatory format which would be unnecessarily confusing to the regulated community. OSHA believes it is more effective to list, in one standard, the obligations of chemical producers, importers, and suppliers with those of the users so that employers using hazardous chemicals will be aware of the content and quality of the hazard information they are entitled to receive from their suppliers. Furthermore, it would not be appropriate to indicate requirements for chemical manufacturers and importers in a standard which purports to cover solely the construction industry, as would have to be done to accommodate all of the recommendations of the Committee. Therefore, construction employers are included with all other employers in this standard. However, OSHA will print the rule in full in 29 CFR Part 1926 (in § 1926.59) for ease of reference for

construction employers and employees. In addition, it will also be printed in 29 CFR Parts 1915, 1917, and 1918, for the use of maritime employers and employees (at new § 1915.99, 1917.28, and 1918.90, respectively), and will be referenced in Part 1928 covering agricultural employments.

E. Federal Community Right-to-Know Law

Expansion of OSHA's HCS will also have an impact on employers' obligations under another Federal law to inform State and local communities of the hazardous chemicals present in the workplace. On October 17, 1986, the President signed into law the Supefund Amendments and Reauthorization Act of 1986 ("SARA"). Part of the new law, Title III, the Emergency Planning and Community Right-to-Know Act of 1986, encourages and supports emergency planning efforts at the State and local level and provides citizens and local governments with information concerning potential chemical hazards present in their communities.

Two provisions in the new law, sections 311 and 312, mandate that employers required under the Occupational Safety and Health Act of 1970 and regulations under that Act to prepare or have available material safety data sheets for hazardous chemicals in their workplaces, must also submit chemical hazard information to State and local governments. Specifically, employers required by OSHA HCS to create or maintain material safety data sheets for employees must also submit to the State emergency response commissions, the local emergency planning committee and the local fire department: (1) A material safety data sheet for each hazardous chemical for which a data sheet is available (section 311); and (2) an emergency and hazardous chemical inventory form (section 312). The public may request material safety data sheets and inventory information from the local planning committee.

Because all manufacturing employers are currently subject to the OSHA HCS and required to create or maintain data sheets for the hazardous chemicals present in their workplaces, they must also comply with the community reporting requirements of the Emergency Planning and Community Right-to-Know Act. An expanded HCS covering non-manufacturers will require non-manufacturers to provide chemical hazard information not only to their employees but also to the surrounding communities.

31960 Federal Register / Vol. 52, No. 163 / Monday, August 24, 1987 / Rules and Regulations

On January 27, 1987, EPA proposed regulations to implement the community data sheet and inventory reporting requirements. A detailed explanation of the EPA proposal can be found at 52 FR 2836 (January 27, 1987). A final rule is expected to be published in the near future. OSHA has prepared a preliminary estimate of the costs of expansion of the EPA requirements into the non-manufacturing sector. This estimate is addressed further in the section of this preamble dealing with the regulatory impact analysis for the final rule.

EPA has established a toll-free hotline to answer questions concerning the requirements: Chemical Emergency Preparedness Program Hotline, 1–800/535–0202; in Washington, DC at 1–202/479–2449.

II. Summary and Explanation of the Issues and the Provisions of the Final Standard

This final rule is both an expansion and a revision of the current HCS. The regulatory text presented herein includes the unchanged provisions of the present rule, as well as those which OSHA is changing. This was done to ensure that readers can clearly follow where these changes would appear in the standard. As explained below, the substantive changes were found to be necessary and appropriate for a hazard communication standard covering all workers exposed to hazardous chemicals. OSHA is also making several corrections and minor technical amendments to the standard. OSHA finds prior public notice and comment for these minor amendments to the unnecessary because of their non-substantive nature. 5 U.S.C. 553(b); 29 CFR 1911.5.

The discussion which follows will address the changed provisions of the rule, as well as the issues related to these changes. A detailed summary and explanation of the current rule's provisions is only provided when necessary for the discussion of the modification. For a complete explanation of the existing provisions, please see the preamble to the current HCS (48 FR 53334–40). The current rule is codified at 29 CFR 1910.1200, and was published at 48 FR 53340–48. The modified trade secret provisions are discussed at 51 FR 34590.

This discussion is organized by paragraph of the standard, and is presented in the order these paragraphs appear in the HCS.

For ease of reference, OSHA will be printing the same rule in full in 29 CFR Part 1910 (in § 1910.1200) for general industry. 29 CFR Part 1926 (in § 1926.59)

for construction, and in 29 CFR Parts 1915, 1917, and 1918, for the use of the maritime industry (at new §§ 1915.99, 1917.28, and 1918.90, respectively).

(a) Purpose

All references to the manufacturing sector, SIC Codes 20 to 39, have been deleted to reflect the expansion of the scope to all employers and employees. It should be noted that these changes have been made throughout the provisions of the rule, wherever the HCS currently addresses employers and employees in the manufacturing sector rather than employers and employees in general. Despite the expansion of covered employers from manufacturers to all employers, however, OSHA retains in this final rule the distinction between chemical manufacturers and importers who produce or import hazardous chemicals, and downstream employers who merely use the chemicals. Only the former are to prepare the technical hazard information for labels and materials safety data sheets · accompanying hazardous chemicals, whereas all employers are to pass this information on to their workers potentially exposed to the chemicals through a comprehensive hazard communication program which includes individual training.

The original Hazard Communication Standard included, at 29 CFR 1910.1200(a)(2), a generally-worded statement concerning the Agency's position regarding the preemptive effect of the standard. This paragraph has been revised to more explicitly state the Agency's position regarding preemption based on the provisions of the Act and related legal actions. This final rule significantly expands the number of industrial groups to which the Federal standard applies, and thus it significantly expands the area in which state and local laws will be preempted. Section 18(a) of the Act, 29 U.S.C. 667(a), provides that a state may assert jurisdiction through any court or agency over "any occupational safety or health issue with respect to which no standard is in effect under section 6." Conversely, where OSHA has issued a standard, section 18 expressly preempts states from asserting jurisdiction through any court or agency over the issue addressed by that standard, unless a Federally-approved State plan is in effect. 29 U.S.C. 667(a) and (b); 29 CFR 1901.2.

The express preemption provisions of the Act apply to all state or local laws which relate to an issue covered by a Federal standard, without regard to whether the state law would conflict with, complement, or supplement the Federal standard, and without regard to

whether the state law appears to be "at least as effective as" the Federal standard. The "at least as effective as" test applies only to state standards adopted under an approved State plan. 29 U.S.C. 667(c)(2). In enacting OSHA, Congress rejected provisions which would have permitted states to enforce laws which were "not in conflict with" or "at least as effective as" Federal OSHA standards. *See* Senate Comm. on Labor and Public Welfare, 92d Cong., 1st Sess., Legislative History of the Occupational Safety and Health Act of 1970, at 58, 706 (Comm. Print 1971). Instead, Congress enacted section 18 providing that Federally-approved State plans are the exclusive alternative to preemption.

Since the promulgation of OSHA's original Hazard Communication Standard, a number of court decisions have dealt with the effect of express and implied Federal preemption upon state and local hazard communication or "right-to-know" laws. *United Steelworkers of America v. Auchter,* 763 F.2d 728, 733–36 (3d Cir. 1985) (Federal Hazard Communication Standard expressly preempts state hazard disclosure laws in manufacturing sector); *New Jersey State Chamber of Commerce v. Hughey,* 774 F.2d 587 (3d Cir. 1985) (provisions of New Jersey right-to-know law which pertain primarily to community or environmental safety and health are not expressly preempted; right-to-know laws subject to implied preemption if they make it impossible to comply with Federal law or pose an obstacle to objectives of the Federal Act); *Manufacturers Association of Tri-County v. Knepper,* 801 F.2d 130 (3d Cir. 1986) (similar holding in connection with Pennsylvania right-to-know law).

The revised paragraph (a)(2) specifically provides that both state and local laws pertaining to occupational hazard communication are preempted by the Federal standard. In the one court decision which has addressed the question, the United States Court of Appeals for the Sixth Circuit ruled that the Federal Hazard Communication Standard preempts local as well as state laws. *Ohio Manufacturers Association v. City of Akron,* 801 F.2d 824 (1986). The court noted that the text of § 1910.1200(a)(2) did not mention localities and referred only to preemption of "state" laws. *Id.* at 827, 831–832. Nevertheless, relying upon references to local as well as state laws in the preamble to the 1983 standard, the court correctly inferred that OSHA had intended to preempt all non-Federal occupational hazard communication

Federal Register / Vol. 52, No. 163 / Monday, August 24, 1987 / Rules and Regulations   31861

laws. *Id.* at 832. Therefore, in accordance with the Court decision, OSHA is making a technical amendment to paragraph (a)(2) so that it explicitly states that the HCS preempts local worker right-to-know laws.

The revised § 1910.1200(a)(2) not only defines hazard communication as an "issue" under the terms of the Act, but also enumerates the generic areas addressed by the standard for purposes of establishing the parameters of preemption. Thus any State or local government provision requiring the preparation of material safety data sheets, labeling of chemicals and identification of their hazards, development of written hazard communication programs including lists of hazardous chemicals present in the workplace, and development and implementation of worker chemical hazard training for the primary purpose of assuring worker safety and health, would be preempted by the HCS unless it was established under the authority of an OSHA-approved State plan.

(b) Scope and Application

Laboratories. With regard to the coverage of laboratories, specifically addressed in paragraph (b)(3), OSHA concludes that the current rule's provisions, requiring only that labels and material safety data sheets received with incoming chemicals be maintained and that the general training of paragraph (h) be provided, are feasible for non-manufacturing laboratories as well. *See, e.g.,* comments of the Massachusetts Institute of Technology, H–022D, Ex. 2–120 ("We agree that the Hazard Communication Standard's requirements for labs are adequate. . . . We expect our compliance costs to remain at the current level of spending because the majority of these are start-up costs and some activities have been absorbed and integrated within existing programs.") OSHA believes that these somewhat limited hazard communication requirements for manufacturing laboratories are also appropriate for non-manufacturing laboratories because both share the operating conditions that distinguish them from the typical industrial workplace: they commonly use small quantities of many different hazardous chemicals for short periods of time; the conditions and purposes of the use of the chemicals frequently change, often unpredictably; many substances are of unknown toxicity; and many workers are highly trained. *Compare* 48 FR 53287–89, *with* 51 FR 26663–64. OSHA concludes that the same HCS provisions tailored for manufacturing laboratories are appropriate for the protection of all

laboratory workers within OSHA's jurisdiction.

It should also be noted that OSHA is currently proceeding with a specific rulemaking to directly address "Occupational Exposure to Toxic Substances in Laboratories" (51 FR 26660; July 24, 1986). When that rule becomes final, its provisions may supplement the information transmittal requirements of the HCS by directly reducing hazardous chemical exposures in laboratories by requiring, among other things, safe work practices. As noted in that proposal, the final rule might modify the general information and training requirements in the HCS to incorporate other aspects of that standard. Any changes in the application of the HCS provisions to laboratories will be addressed in detail in the final rule for laboratories and will be based on that rulemaking record (Docket H–150).

Coverage determined by "exposure." The HCS covers situations where employees "may be exposed" to hazardous chemicals (paragraph (b)(2)), and such exposure is defined to include potential exposure as well as actual exposure. This is to ensure that employees receive information about all chemical hazards in their work areas, and that they are prepared to deal with any unexpected releases or emergency situations, as well as exposures during the normal course of employment. OSHA concluded that employees are entitled to information regarding the chemicals to which they are exposed in their work areas. It should be noted, however, that individual facilities and workplaces may have some employees who are covered since their work involves exposure to hazardous chemicals, and others who are not covered because their work does not. For example, in a retail department store, maintenance workers or workers in a graphic arts department may be covered since their jobs involve exposure to chemicals, but an accountant in the billing department would not be likely to experience exposure that would require coverage by the HCS.

There are a number of work situations where employees only handle sealed containers of chemicals, and under normal conditions of use would not open the containers and would not expect to experience any measurable exposure to the chemicals. Such work operations include, for example, warehousing, retail sales, marine cargo handling, and trucking terminals. It is reasonable to assume, however, that all such containers are subject to leakage and

breakage, and these employees are in fact potentially exposed by virtue of the presence of these hazardous chemicals in their workplaces. Because of this potential exposure, they need information to protect themselves from the hazards of these chemicals in the event such an emergency situation occurs.

However, OSHA has considered the extent of information necessary or appropriate in this type of operation, and the practicality of requiring such work operations to be subject to all of the provisions of the rule. The primary need is to ensure that these employees know how to acquire and use the hazard information available to them, and to handle an emergency exposure situation. As in laboratory operations, maintaining lists of chemicals where the chemicals present may change on short notice, sometimes on a daily basis, is not a useful requirement. Similarly, obtaining material safety data sheets for every chemical in a sealed container that passes through a facility—even if it is there less than a day in some situations—would result in a considerable amount of paperwork, with little discernable benefit for the employees involved. Therefore, OSHA has added a provision, paragraph (b)(4), to limit the duties of employers for those work operations where employees only handle sealed containers that are not intended to be opened under normal conditions of use. (Some States which have adopted right-to-know laws have also recognized the practical problems of coverage in this area, and have included provisions limiting coverage of workplaces where chemicals are handled in sealed containers. *See, e.g.,* Tennessee Hazardous Chemical Right to Know Law, Tennessee Code Annotated, 50–3–2001 through 50–3–20019.) In these situations, employers must not remove labels affixed to incoming containers of hazardous chemicals; must maintain and provide access to material safety data sheets that are received for hazardous chemicals while the chemicals are in the workplace, and obtain material safety data sheets when they are not received but an employee requests one; and must train employees in accordance with the provisions of the rule to ensure they are protected in the event of a spill or leak.

The employees in these operations will always have access to the label information, which will provide appropriate hazard warnings and be a visual reminder of the potential hazards if exposure occurs. Employees will also be trained regarding the general classes of chemical hazards faced and the means by which they can protect

31862 Federal Register / Vol. 52, No. 163 / Monday, August 24, 1987 / Rules and Regulations

themselves from these hazards when there is a spill or leak. The training must also address the availability and use of substance-specific information found on labels and material safety data sheets, where available. These requirements should provide employees handling only sealed containers of chemicals with the information they need.

This limited provision also addresses some of the concerns raised by representatives of industries with these types of workplaces. (*See, e.g.* Exs. 2–53, 2–75, 2–201, and 2–214). Although they generally were arguing that this type of operation warrants exclusion from the rule, OSHA does not agree that no protection under the HCS is required in these situations. As already described, a potential for exposure does exist, and therefore such employees must be appropriately covered. OSHA believes the limited coverage described will effectively protect employees while recognizing the constraints of the particular work operations involved with regard to the applicability of the current rule to these types of work.

Labeling exemptions. The HCS includes a number of labeling exemptions to ensure that OSHA does not provide duplicative coverage for products which are already labeled under the rules of another Federal agency. It should be reemphasized that these exemptions (in paragraph (b)(4) of the original rule; paragraph (b)(5) in this final rule) are only from the container labeling requirements under paragraph (f)—all other provisions of the rule are still in effect. A minor correction is being made, however, to these exemptions to indicate that medical or veterinary devices are labeled in accordance with the labeling requirements of the Food and Drug Administration (FDA) under authority of the Federal Food, Drug, and Cosmetic Act (21 U.S.C. 301 *et seq.*), those items are exempted from HCS labeling requirements. All other items regulated by FDA under that Act were listed in the HCS labeling exemption. Medical and veterinary devices were inadvertently omitted from the list of items that might be subject to FDA labeling requirements under the Federal Food, Drug, and Cosmetic Act, and they are exempted from HCS labels for the same reasons that the other items are exempt when subject to labeling under FDA. *See* 48 FR 53289. To ensure that all these FDA regulated items are treated in the same manner and that devices are exempted from HCS labeling if subject to FDA labeling, paragraph (b)(5)(ii) is amended by adding medical and veterinary devices.

Other exemptions. The HCS includes a number of specific, total exemptions from the requirements of the rule for certain types of chemicals. This rule adds three categories of exemptions: food, drugs, cosmetics, or alcoholic beverages in a retail establishment packaged for retail sale (paragraph (b)(6)(vi); consumer products (paragraph (b)(6)(vii)); and certain pharmaceuticals (paragraph (b)(6)(viii)).

Food, drugs, cosmetics, alcoholic beverages. The current HCS includes an exemption for food, drugs, or cosmetics brought into the workplace for employee consumption. These types of exposures are not related to an employee's work, and therefore do not need to be covered under the HCS.

The expansion of the HCS into the non-manufacturing sector will result in many of these types of products being present in workplaces (*e.g.,* liquor stores) where they are not intended for employee consumption, and where they normally would not result in employee exposure because they are packaged for sale to consumers. Although some of these products may meet the definition of a "hazardous chemical" (*e.g.,* vinegar is acetic acid), when packaged for retail sale they do not pose a hazard to workers that is any different than the hazards of such products in their homes. The label information required by other Federal agencies for foods, drugs, cosmetics, and alcoholic beverages should thus provide sufficient protection for workers, and OSHA has exempted these products from coverage under the rule. It should be noted that this is not an exemption for facilities of any particular industry, as all facilities may have other chemicals in use that would be covered by the HCS. In addition, since these products are exempted, employers which package them for retail sale would not have to furnish material safety data sheets to distributors receiving the products.

Consumer products. The current rule provides a labeling exemption for consumer products when they are labeled in accordance with the requirements of the Consumer Product Safety Commission (CPSC). CPSC requires consumer products which contain hazardous substances to be appropriately labeled. Examples of consumer products would include such items as oven cleaner, paint stripper, and adhesive, which may be found in various types of workplaces. In addition to the specific labeling exemption, OSHA has been interpreting the rule as not being applicable to consumer products when used as a consumer would use them. OSHA is now adding

this interpretation to the rule itself, paragraph (b)(6)(vi), stating that where such consumer products are used in the workplace in a a manner comparable to normal conditions of consumer use, resulting in a duration and frequency of exposure to employees which is no greater than exposures experienced by ordinary consumers, under such conditions the chemical would not have to be included in the employer's hazard communication program. This position is consistent with OSHA's reason for orginally limiting the exemption for hazardous consumer products used in the course of employment to only an exemption from HCS labeling, and not material safety data sheet and training requirements. "OSHA recognizes . . . that there may be situations where worker exposure is significantly greater than that of consumers, and that under these circumstances, substances which are safe for contemplated consumer use may pose unique hazards in the workplace." 48 FR 53289. However, to the extent that workers are exposed to the substances in a manner similar to that of the general public, there is no need for any HCS requirements.

One example of such a differentiation in exposure situations involves the use of abrasive cleaners in the workplace. Where these are used intermittently to clean a sink, much as they would be used at home, the cleaners would not be covered under the standard. But if they are used to clean out reactor vessels, thus resulting in a much greater level of exposure, they would be covered. Or if an employee cleans sinks all day long, thus resulting in more frequent exposures, the abrasive would also be included in the hazard communication program. Thus workplaces which only have chemicals which are consumer products used in the same way and as frequently as the general public would normally use them, would not have to have a hazard communication program.

It should be noted that OSHA intends to read this exemption narrowly. Where an employer is uncertain whether the duration and frequency of exposure to these products is comparable to consumer use, an employer should obtain or develop the material safety data sheet and make it available to employees.

In response to questions raised in the 1985 ANPR, OSHA received a few comments on the use of consumer products in the non-manufacturing sector. A number indicated that overexposure may occur from the use of such products, or that the frequency and duration of workplace exposure is typically greater than that experienced

APPENDIX I 313

by consumers (Exs. 2–59, 2–83, 2–100, 2–120, and 2–164). Others stated that the exposure was comparable to consumer use (Exs. 2–46 and 2–63). There were several that felt the label provided enough information, and no additional requirements were needed to protect employees (Exs. 2–75, 2–79, 2–99, 2–107, and 2–116), while others felt the employer should be required to request material safety data sheets because employees are not getting enough information (Exs. 2–109, 2–128, and 2–169). One suggested that the label note that a material safety data sheet is available on request (Ex. 2–100), while another contended that when a product is used by a professional, it is no longer a consumer product (Ex. 2–199). OSHA believes that the consumer product exemption in this final rule takes all of these concerns into consideration, and strikes a balance between the practical considerations of acquiring and maintaining material safety data sheets on CPSC regulated products which employees are exposed to at home as well as at work, and the worker's need for more hazard information than a CPSC label when exposures are greater or more frequent than typical public use of the chemical would generate.

A number of States adopting right-to-work laws have also developed consumer product exemptions. (See, e.g., Wisconsin "Employees' Right to Know Law"; Illinois "Toxic Substances Disclosure to Employees Act.") However, most of these rules have taken a broader approach to the consumer product exemption, generally eliminating coverage of such products unless exposure is "significantly greater" than consumer exposure during the "principal consumer use." OSHA considered and rejected such language for the consumer product exemption. It would be very difficult from an enforcement perspective to determine when exposure to a consumer product is "significantly greater" than consumer exposure. The key elements of concern to OSHA are as stated in the consumer product exemption included in this rule—that the consumer product be used in the same manner as a consumer would use it (and therefore as intended by the manufacturer when preparing the label information), and that the duration and frequency of exposure be essentially the same as would be experienced by a consumer (and thus the label warnings would provide adequate protection.) A broader exemption than this would not be appropriate to protect workers from occupational exposures that were not anticipated by the manufacturer when

the labels, and thus the protective measures, were developed.

Application to Office Products. A number of questions have been raised about the application of the rule to office products that may contain hazardous chemicals. It is OSHA's determination that office products such as pencils, pens, typewriter ribbons, and the like, are "articles" under the rule and therefore exempted, paragraph (b)(6)(iv). Employers are not therefore required to implement a program for such products. OSHA has also determined that intermittent, occasional use of a copying machine to make copies is not covered by the rule. The copying machine would also be considered an article for purposes of this standard. However, if a firm has a copying machine operator who is responsible for handling the chemicals associated with its use, or who operates the machine frequently, that individual would be entitled to information under the rule.

Medicine. The rule, paragraph (b)(6)(vii), also includes an exemption for drugs when they are solid, and are in final form for direct administration to the patient (i.e., pills or tablets). Employees handling such finished drug products would not be exposed to the chemicals involved, and would receive information other than that supplied on the container label under FDA requirements. (The State of North Carolina adopted a similar exemption in their Hazard Communication Standard, 13 NCAC s7C.101(a)(99)).

Wood dust. As OSHA has received a number of questions regarding the application of the wood and wood products exemption to wood dust, OSHA would like to reiterate its interpretation regarding the wood and wood product exemption in paragraph (b)(6)(iii) of this final rule. The wood and wood products exemption was included in the HCS for two reasons. First, the presence and identity of wood and wood products in the workplace is "unmistakable" and second, their hazards (i.e., flammability or combustibility) are well-known to workers. 48 FR 53289. Because wood and wood products, characteristic hazards are self-evident, regulations requiring formal notification were not thought to be necessary. Wood and wood products "are not expected to be hazardous for purposes of this standard." Id. at 53335. OSHA never intended, however, that wood dust be excluded from the standard's coverage under the wood and wood products exemption. Wood dust is not generally a wood "product," but is created as a byproduct during manufacturing

operations involving sawing, sanding, and shaping of wood. Wood dust does not share solid wood products' "self-evident" hazard characteristics that supported the exemption of wood products from the HCS' coverage. Except for the chemical additives present in the wood, products such as lumber, plywood, and paper are easily recognizable in the workplace and pose a risk of fire that is obvious and well-known to the employees working with them. The potential for exposure to wood dust within the workplace, especially with regard to respirable particles, is not self-evident, nor are its hazards through inhalation so well-known that hazard communication programs are unnecessary. "Wood dust" is a recognized health hazard, with exposure limits recommended by the American Conference of Governmental Industrial Hygienists (ACGIH) to control employee exposures to the substance. Under the provisions of the HCS, this means that wood dust is to be considered a hazardous chemical (paragraph (d)(3)(ii)), and therefore subject to the requirements of the rule including material safety data sheets and training.

(c) Definitions

The only changes to the definitions in the current HCS are those that need to be made to accomplish the expansion of the HCS.

The reference to SIC Codes 20 through 39 is being deleted from the definition of "chemical manufacturer" to be consistent with the extent scope of the rule. Any employer who produces a hazardous chemical for "use or distribution" is considered a "chemical manufacturer" under the HCS, and must prepare and provide the appropriate hazard information.

OSHA has modified the definition of "container" to exempt "engines, fuel tanks, or other operating systems in a vehicle." The Agency has received some questions regarding the need for labeling such parts of a vehicle in applying the rule to the manufacturing sector. Expansion into non-manufacturing will greatly increase the number of vehicles involved in work operations, and thus OSHA determined that this clarification will ensure that the Agency's position regarding this issue is clear—vehicles do not have to bear labels regarding hazardous chemicals used to operate them. This does not exempt such chemicals from coverage by the rule—it simply eliminates the need to label once they are placed into the vehicle.

The definition of "distributor" has also been changed to reflect the

314 FUNDAMENTALS OF HAZARDOUS MATERIALS INCIDENTS

extended scope of the rule. A "distributor" means "a business, other than a chemical manufacturer or importer, which supplies hazardous chemicals to other distributors or to employers." Among other things, distributors must transmit hazard information they receive from chemical manufacturers and importers to all their employer customers.

Under the current rule, OSHA defined "employee" as someone working in the manufacturing sector, and stated that those employees in manufacturing whose jobs did not involve routine potential exposure to hazardous chemicals to other distributors or to covered by the rule. Examples related to the manufacturing sector were provided. This was intended to limit the coverage primarily to those employees in the industry who were actually involved in production operations. However, since the scope of the entire standard is being expanded to cover employees in all types of work operations, the definition has been modified to clarify that workers who are exposed to hazardous chemicals as part of their assigned jobs would generally be covered under the rule, except for those who only encounter hazardous chemicals in non-routine, isolated instances. OSHA believes most office workers, and many other workers, are not exposed to the hazardous chemicals covered by the HCS in such a way that the rule would apply to those types of work operations. The rule, therefore, simply defines a covered "employee" as any "worker who is exposed to hazardous chemicals under normal operating conditions or in forseeable emergencies" and further states that "workers such as office workers or bank tellers who encounter hazardous chemicals only in non-routine, isolated instances are not covered." "Normal operating conditions" are those which employees encounter in performing their job duties in their assigned work areas. For example, if the receptionist in a facility receives and delivers a telephone message for someone in a different work area where hazardous chemicals are present, this does not mean that the receptionist would be covered under the rule by virtue of the one potential exposure from delivering the message. However, if performance of the receptionist's job entails walking through the production area every day, and thus being potentially exposed during the performance of regular duties, that job would be covered under the rule.

The definitions of "employer" and "importer" are also amended to indicate

that all employers are covered by the standard. In addition, the definition of "employer" is amended to indicate that the term includes contractors and subcontractors. This reflects the definition of employer used in OSHA's construction standards. Similarly, the definition of "workplace" has been modified to specifically include job sites and projects.

Hazard warning. While OSHA is not modifying the definition of "hazard warning" contained in the current rule, the Agency wishes to reiterate the intent to help employers better understand and comply with the requirements. "Hazard warning" means "any words, pictures, symbols, or combination thereof which convey the hazard(s) of the chemical(s) in the container(s)." "Appropriate hazard warnings" are to be put on container labels. (*See* final rule paragraphs (f)(1)(ii) and (f)(5)(ii)). Since the rule covers "physical" and "health" *hazards*, specific information regarding these would be required on a label to comply.

Many labels at the time the HCS was promulgated includes only precautionary statements, rather than providing necessary information about the specific hazards of the chemicals. Thus employees encountered statements such as "avoid inhalation" on virtually every chemical container, but were not provided with statements regarding what type or severity of effect inhalation could be expected to produce.

Therefore, OSHA's standard requires identity and hazard information on labels. Although employers can choose to provide additional statements, OSHA's requirements are limited to that required to convey the hazards to the workers. Under the OSHA scheme, other data regarding protective measures, first aid, etc., are to be included on the material safety data sheet or in training, rather than appearing on the label itself. This approach is in keeping with the Agency's evaluation of available data on effectiveness of labels which indicates that the more detail there is on a label, the less likely it is that employees will read and act on the information. The purpose of the label is to serve as an immediate visual warning of the chemical hazards in the workplace. (*See generally,* 48 FR 53300–03).

There have been misinterpretations of the requirements made based on statements in the preamble to the current rule concerning various labeling systems (*see* 48 FR 53301). This preamble discussion involves format of labels, and is not an unqualified

endorsement of any particular labeling system. It simply states that any format may be used, as long as the label includes the information regarding the chemical hazards required by the standard. It should be noted that it can be expected that some labels prepared in accordance with any of the available labeling systems can be expected to be found to be deficient. Again, the preamble discussion cited merely reemphasized that employers are not constrained to use any particular format or wording, but are constrained by the necessity to comply with the requirements of the rule concerning the information to be provided—the identity, the hazards, and for containers leaving the workplace, the name and address of the responsible party.

The terms "physical" and "health" hazards are already defined in the rule, and these are the specific hazards that are to be "conveyed" in an "appropriate" hazard warning. There are some situations where the specific target organ effect is not known. Where this is the case, a more general warning statement would be permitted. For example, if the only information available is an LC₅₀ test result, "harmful if inhaled" may be the only type of statement supported by the data and thus may be appropriate.

It will not necessarily be "appropriate" to warn on the label about every hazard listed in the MSDS. The data sheet is to address essentially everything that is known about the chemical. The selection of hazards to be highlighted on the label will involve some assessment of the weight of the evidence regarding each hazard reported on the data sheet. This does not mean, however, that only acute hazards are to be covered on the label, or that well-substantiated hazards can be omitted from the label because they appear on the data sheet.

It may be "appropriate" to provide less detailed information on the chemical hazards in an in-plant labeling system, where MSDSs and training are readily available, than on a label placed on a container leaving the workplace, where it may provide the only hazard information in certain situations and where there is no guarantee that the downstream employees handling or using the chemical will fully understand the less detailed label. This difference in appropriateness allows employers to establish standardized in-plant labeling systems, as long as training regarding the use of these systems is conducted, and MSDSs provide the required, detailed information.

Article. OSHA is not modifying the definition of "article" but would like to provide some clarification regarding the Agency's interpretation. Releases of very small quantities of chemicals are not considered to be covered by the rule. So if a few molecules or a trace amount are released, the item is still an article and therefore exempted. In an earlier discussion in this preamble, application of the rule to office products was discussed and it was stated that items such as pens or pencils are to be considered articles. Other examples would be: emissions from tires when in use; emissions from toner on pieces of paper; or emissions from newly varnished furniture.

Furthermore, it should be reiterated that the HCS is limited to hazardous chemicals "known to be present" (paragraph (b)(2)), and does not require any chemical analysis or testing to determine or verify such presence. *See* 48 FR 53334–35. Thus although one may assume that molecules are being emitted from an item, under the standard one does not "know" that a particular hazardous chemical is "present."

The article exemption applies solely to the ultimate end use—intermediate users which result in exposure are covered and require hazard information to be provided. The following are examples of items which would require information for intermediate use prior to being finally installed: encapsulated asbestos insulation where the normal installation involves hammering the material into openings, thus releasing the asbestos; tiles to be placed on a ship's hull which contain lead that is released during installation; and glass mercury switches to be installed in equipment, a percentage of which are expected to break during this installation process. In these cases, installation is the "normal condition of use" for the employees installing the items, and thus hazard information is required for these intermediate uses. Once installed, these items would be articles and thus exempted.

Although installation of an item may render the exemption temporarily void (until the item is installed, information must still be provided if there is a potential for exposure), OSHA does not believe that the possibility that exposure could occur when the item is repaired or worked on need be considered in the determination of when information must be transmitted downstream. Employers of employees performing repairs must provide the best information they have concerning the potential exposures. There would be no way to ensure, for example, that a material safety data sheet prepared for a lead pipe would be available to a worker repairing the pipe some years following installation. The employer would provide the employees with general information concerning the hazards of the operations they were performing in lieu of specific information on the pipe itself.

(d) Hazard Determination

OSHA is not modifying the current rule's hazard determination requirements. The burden of evaluating chemicals to determine whether they are hazardous remains on the chemical manufacturers and importers who produce or import them and on those user employers who choose not to rely on the evaluations made by their suppliers and instead evaluate the chemicals themselves. A detailed explanation of these provisions can be found at 48 FR 53296–99, 53335–36.

(e) Written Hazard Communication Program

Under the current rule, a written hazard communication program must be developed and implemented for each workplace. Since the current rule covers fixed manufacturing sites, it did not appear to be necessary to specifically state that the written program be available at the site. With expansion to non-manufacturing, particularly in the construction industry where a firm may have multiple sites, the standard must be tailored to specifically state that the intent is to maintain the written program at each site. Employees will then be able to access the information as required.

The current written hazard communication program requirements include a provision that requires manufacturing employers to provide hazard information to on-site contractor employers who have employees who may be exposed to the hazards generated by the manufacturer (current paragraph (e)(1)(iii)). The current standard does not address the reverse situation, *i.e.,* where a contractor employer brings hazardous materials on-site, and exposes the manufacturer's employees to them. Since the expanded rule will affect more worksites with work arrangements (*e.g.,* construction), and the need for an exchange of hazard information is obvious, OSHA has revised the requirements to tailor it to address the multi-employer workplace. (This was suggested in comments submitted in response to the ANPR. *See* Ex. 2–225, comments from the National Constructors Association. In addition, this situation has also been addressed in existing State right-to-know laws. *See,* *e.g.,* Alabama Act 85–658; Tennessee "Hazardous Chemical Right to Know Law.")

Under these provisions (paragraph (e)(2)), the employers must exchange material safety data sheets, as well as information about precautionary measures necessary to protect employees and an indication of the type of labeling system in use, where exposures may occur to another employer's employees. Each employer will then have the information necessary to inform and train their employees. This will help ensure that all employees have sufficient information to protect themselves in the workplace, regardless of which employer uses the hazardous chemical.

Consistent with the performance-orientation of the rule, the provisions do not specify how this coordination is to be accomplished. This is best left to the discretion of the parties involved. In many cases, it would probably be most efficient for the general contractor to coordinate the function. For example, the general contractor could keep and make available material safety data sheets in the office on the site.

It should be emphasized that the exchange of information is limited to those situations where exposures of other employers' employees may occur. Given the nature of multi-employer work sites in construction, there would be many situations where subcontractors responsible for various phases of the building project would not have employees present during other phases and thus no such exchange would be required. For example, if the electricians are not working near, or at the same time as, the paving contractor, then no interchange is required. But if a painting contractor's workers are using flammable solvents in an area where another subcontractor is welding pipes, this information exchange is vital to ensure proper protection of employees.

(f) Labels and Other Forms of Warning

A tailoring provision has been added concerning shipments which consist of solid metal. OSHA considers this change to be necessary since the problem addressed will occur more frequently in shipments to the nonmanufacturing sector than has been the case in the manufacturing sector. (Paragraph (f)(2)). Solid metal is often considered to be an "article" under the rule, and thus exempt. Where the metal is not an "article" since its downstream use results in hazardous chemical exposure to employees working with it, a provision has been added which allows shippers of this type of material

31866   **Federal Register** / Vol. 52, No. 163 / Monday, August 24, 1987 / Rules and Regulations

to send the label information once, similar to material safety data sheet transmittal, as long as the material is the same and it is being shipped to the same customer. In these situations, there should be no hazard to anyone handling the metal from the time it is produced in solid form, until the time someone works on it in a way that releases a chemical hazard. Since the label information transmitted would only reflect the chemical hazards released when it is later worked on, the label would not provide any hazard information that is needed by those handling the material in transit. It must be emphasized that this exception is only for the solid metal itself—any hazardous chemicals present in conjunction with the metal in such a form that employees may be exposed when handling the material (e.g., cutting fluids, lubricants, and greases), require labels with each shipment. This tailoring provision, therefore, does not diminish worker protection—workers get the hazard information they need.

(g) Material Safety Data Sheets

Under the hazard determination provisions, a requirement is included which indicates that there are situations where the percentage cut-off for mixtures would not apply—when the released chemical is particularly hazardous, or when it could exceed an established permissible exposure limit or Threshold Limit Value when released (paragraph (d)(5)(iv)). Although this is clearly a requirement of the rule, *see also* 48 FR 53336, the material safety data sheet provisions for disclosure of hazardous ingredient identities did not address that particular situation. Clearly it was OSHA's intent to have all hazardous ingredients of mixtures listed on a material safety data sheet, even those in very small concentrations, when the hazard determination provisions of paragraph (d) mandate that they are to be considered hazardous for purposes of the HCS. As noted in the HCS preamble discussion of the material safety data sheet provisions: "Employers must also list ingredients present in concentrations of less than one percent if there is evidence that the permissible exposure limit may be exceeded or if it could present a health hazard in those concentrations." *Id.* at 53337. This obvious oversight has been corrected by a minor amendment to the rule. Paragraph (g)(2)(i)(C)(2).

Another situation which raises practicality concerns because of the expansion of the scope of the rule involves employers who purchase hazardous chemicals from local retail distributors, rather than directly from the chemical manufacturer or importer,

or from wholesale distributors as is more commonly done in the manufacturing sector. Under the current HCS, distributors of hazardous chemicals must automatically provide commercial customers material safety data sheets (paragraph (g)(7)). Retail distributors, however, often sell to businesses and the general public and frequently have no way of knowing who a particular purchaser is. Under the current rule, retail distributors might have to give material safety data sheets to each customer to ensure that commercial customers get the information they need under the HCS. A specific statement regarding retail distributors is, therefore, included in paragraph (g)(7) to address this practical problem. Those retail distributors who sell hazardous chemicals to employers must provide a material safety data sheet upon request, and must post a sign or otherwise inform the employers that an MSDS is available. According to Schneider Hardware of Banksville, Inc., this is a reasonable approach (Ex. 2–179):

If OSHA does require commercial customers to get information through a retail outlet, I do not foresee any problems with that arrangement. The manufacturers could supply us with the information, as they are required to now for shipments to manufacturing plants, and we could make it available to customers upon request. We would merely keep the sheets in a file drawer and post a sign informing customers of their availability. We have less than 100 chemicals that would probably be affected, and keeping information on those would not require at most, one file drawer. It would not be burdensome.

The retail distributors likely affected are those selling building supplies, hardware, etc. Retail distributors will have to assess their product lines, and whether or not they have commercial accounts, to determine whether they must comply with this provision. It is clear that most other types of retail establishments (e.g., grocery stores, clothing stores, etc.) would not. With regard to the maintenance of material safety data sheets so that they are readily available to employees, whereas manufacturing facilities are generally fixed work sites with fixed locations for these materials, in some types of nonmanufacturing work operations, employees involved in servicing oil and gas wells may have a central office location, but then travel by truck to the wells to perform their work. These remote locations may not have any staff, or may not have a work facility. OSHA has added a provision to the MSDS requirements to allow MSDSs

to be kept at a central location in this type of situation, as along as the employer ensures that the employees can immediately obtain the information in an emergency, paragraph (g)(9). OSHA believes that this provision tailors the HCS so that it remains practical, yet effective, in getting workers the hazard information they need. This was also supported by a number of ANPR commenters (see, e.g., Exs. 2–83, 2–107, 2–114, 2–116, and 2–117).

The current rule, as well as the expanded standard, allows downstream employers to rely on upstream chemical manufacturers and importers to provide MSDSs. However, there is a duty for downstream users to request an MSDS when they don't receive one at the time of the first shipment. There have been some questions regarding how the downstream user will know a data sheet is required without doing a hazard evaluation. Such an evaluation is not necessary. If the label indicates a hazard, the employer will know he needs a data sheet and must request one if it is not received. If there are no hazards on the label, the downstream user can assume the product is not hazardous and a data sheet is not required.

(h) Employee Information and Training

OSHA is not making any modifications to the current rule's information and training provisions. These requirements remain performance-oriented and designed so that each employer will adequately address the hazards posed by chemicals in the workplace. An explanation of these provisions can be found at 48 FR 53310–12, 53337–38.

One question that does arise regarding training is whether it needs to be done specifically on each chemical, or whether employers can train regarding categories of hazards. Either method would be acceptable. See 48 FR 53312, 53338. If employees are exposed to a small number of chemicals, the employer may wish to discuss the particular hazards of each one. Where there are large numbers of chemicals, the training regarding hazards could be done on categories (e.g., flammable liquids; carcinogens), with employees being referred to substance-specific information on the labels and MSDSs. Similarly, the re-training occurs when the hazard changes, not just when a new chemical is introduced into the workplace. If the new chemical has hazards which employees have been trained about, no re-training occurs. If the chemical has a hazard they have not been trained about, re-training would be limited to that hazard.

(i) Trade Secrets

Paragraph (i)(11) of the current rule states that "[i]f, following the issuance of a citation and any protective order, the chemical manufacturer, importer, or employer continues to withhold the information, the matter is referrable to the Occupational Safety and Health Review Commission for enforcement of the citation. . . ." This provision was worded in such a manner that it left the impression that OSHA could refer the matter to the Review Commission. This is incorrect as a matter of law. An enforcement proceeding is referred to the Review Commission when a citation is issued by OSHA, and is subsequently contested by the employer receiving the citation. Therefore, OSHA has made a technical amendment to paragraph (i)(11) to reflect the applicable procedural law.

(j) Effective Dates

The expansion of the rule to cover all employers becomes effective nine months from the date of promulgation of the final standard. Since the chemical hazard information for labels and material safety data sheets has already been generated in the manufacturing sector, and in many cases has also been distributed in non-manufacturing due to State law requirements and voluntary transmittal by suppliers, one month should be sufficient time for chemical manufacturers, importers, and distributors to initiate provision of material safety data sheets to other distributors and to customers in the non-manufacturing sector. An additional eight months is being provided for non-manufacturers to complete preparation of a written hazard communication program for each facility and to conduct employee training. It should be noted that this eight month period for compliance only applies to those employers which are newly covered under the expanded provisions—employers in SIC Codes 20 through 39 are covered under the current HCS and are already required to be in compliance with the provisions of that rule. Those tailoring provisions that apply to manufacturing workplaces, such as the consumer product exemption, go into effect immediately for those facilities.

Appendices A and B

OSHA is not amending Appendix A's discussion of the health hazards posed by chemicals, or Appendix B's discussion of hazard determination. They remain applicable to all chemical manufacturers, importers, and

employers performing hazard determinations.

Appendix C

The reference sources listed in this non-mandatory appendix have been updated to reflect currently available sources.

Appendix D

The recent rulemaking on trade secrets added a new Appendix D regarding the evaluation of the validity of trade secret claims. 51 FR 34590. The full text of this appendix has been reprinted in this document as well.

III. Analyses of Regulatory Impact, Regulatory Flexibility, and Environmental Impact

The following is a summary of the regulatory impact and regulatory flexibility analysis prepared by OSHA for the revision of the Hazard Communication Standard which extends the scope of the existing standard to the nonmanufacturing sector. The full text of the document may be examined and copied in OSHA's Docket Office, 200 Constitution Avenue, NW., Room N3670, Washington, DC 20210; telephone (202) 523-7894.

Economic Analysis

As part of OSHA's efforts to gather information concerning the economic feasibility of extending the coverage of the HCS to include workplaces in the nonmanufacturing sector, the JACA Corporation performed a study examining the benefits, costs, and overall economic impact of such a revision. This report was used as the basis for the regulatory impact analysis prepared by OSHA.

The analysis reflects the extent to which employers in the nonmanufacturing sector are currently subject to state right-to-know laws and are voluntarily implementing their own hazard communication programs. The analysis also takes into account OSHA's existing policy regarding the use of consumer products and training requirements already imposed on employers by other OSHA standards. With respect to consumer products covered by the HCS, OSHA Instruction CPL 2-2.38A ("Inspection Procedures for the Hazard Communication Standard, 29 CFR 1910.1200") states:

A common sense approach must be employed whenever a product is used in a manner similar to which it could be used by a consumer, thus resulting in levels of exposure comparable to consumer exposure. The frequency and duration of use should be considered. For example, it may not be necessary to have a data sheet for a can of

cleanser used to clean the sink in an employee restroom. However, if such cleanser is used in large quantities to clean process equipment, it should be addressed in the Hazard Communication Program.

This policy has been incorporated into the revisions to the HCS, and was taken into account when evaluating data describing the number of hazardous chemicals in the various two-digit SIC groups that could be affected by extension of the HCS to the nonmanufacturing sector.

Assessing the net impact of the training provisions required identifying and deducting the costs of existing OSHA standards which already require employers to provide the types of information and training activities prescribed in the HCS. This was done for construction (§ 1926.21), shipbreaking (§ 1915.97), marine terminals (§ 1917.22), and longshoring (§ 1918.86). However, it was not possible to separately identify and deduct the existing training costs for substance-specific standards that currently apply to the nonmanufacturing sector. Thus, the compliance costs presented in this analysis are somewhat overstated.

In extending the rule for manufacturing to the nonmanufacturing sector, OSHA has made revisions to reflect unique aspects of some work operations. For example, the standard allows MSDSs to be maintained at central locations in circumstances where employees must travel between work operations during a worksheet, provided that the information can be obtained immediately in an emergency. This provision is expected to lower costs in SIC groups 07, 08, 09, 13, 46, 49, and 73. (*See* Table 1 for a description of the SICs.)

The standard also allows for limited coverage in those work situations where employees handle chemicals in sealed containers that are not opened under normal conditions of use, and thus have little potential for measurable exposures. Employers would be required to leave warning labels on containers, and make available any MSDSs received with the containers. Employers would also have to be trained in accordance with the standard, with particular emphasis on procedures to follow if there is a spill or leak of the hazardous chemicals in the normally sealed containers. Affected establishments would not have to make special efforts to obtain and keep MSDSs that are not received with the chemicals, and no written plan for complying with the HCS would be required. This provision is expected to

318 FUNDAMENTALS OF HAZARDOUS MATERIALS INCIDENTS

result in lower costs in SIC groups 42, 44, 45, 47, 51, and 52.

Thus the changes made to establish more appropriate provisions for unique work situations should result in lower costs than would be experienced if the HCS for manufacturing were extended to the nonmanufacturing sector without revision.

Table 1.—SIC Groups Covered in the OSHA Analysis

Division A. Agriculture, Forestry, and Fishing

Major Group 01. Agricultural production—crops
Major Group 02. Agricultural production—livestock
Major Group 07. Agricultural services
Major Group 08. Forestry
Major Group 09. Fishing, hunting, and trapping

Division B. Mining

Major Group 13. Oil and gas extraction

Division C. Construction

Major Group 15. Building construction—general contractors and operative builders
Major Group 16. Construction other than building construction—general contractors
Major Group 17. Construction—special trade contractors

Division E. Transportation, Communication, Electric, Gas, and Sanitary Services

Major Group 40. Railroad transportation
Major Group 41. Local and suburban transit and interurban highway passenger transportation
Major Group 42. Motor freight transportation and warehousing
Major Group 44. Water transportation
Major Group 45. Transportation by air
Major Group 46. Pipe lines, except natural gas
Major Group 47. Transportation services
Major Group 48. Communication
Major Group 49. Electric, gas, and sanitary services

Division F. Wholesale Trade

Major Group 50. Wholesale trade—durable goods
Major Group 51. Wholesale trade—nondurable goods

Division G. Retail Trade

Major Group 52. Building materials, hardware, garden supply, and mobile home dealers
Major Group 53. General merchandise stores
Major Group 54. Food stores
Major Group 55. Automotive dealers and gasoline service stations
Major Group 56. Apparel and accessory stores
Major Group 57. Furniture, home furnishing, and equipment stores
Major Group 58. Eating and drinking places
Major Group 59. Miscellaneous retail

Division H. Finance, Insurance, and Real Estate

Major Group 60. Banking
Major Group 61. Credit agencies other than banks

Major Group 62. Security and commodity brokers, dealers, exchanges, and services
Major Group 63. Insurance
Major Group 64. Insurance agents, brokers, and service
Major Group 65. Real estate
Major Group 66. Combinations of real estate, insurance, loans, law office
Major Group 67. Holding and other investment offices

Division I. Services

Major Group 70. Hotels, rooming houses, camps, and other lodging places
Major Group 72. Personal services
Major Group 73. Business services
Major Group 75. Automotive repair, services, and garages
Major Group 76. Miscellaneous repair services
Major Group 78. Motion pictures
Major Group 79. Amusement and recreation services, except motion pictures
Major Group 80. Health Services
Major Group 81. Legal Services
Major Group 82. Education Services
Major Group 83. Social Services
Major Group 84. Museums, art galleries, botanical and zoological gardens
Major Group 86. Membership organizations
Major Group 89. Miscellaneous services

The analysis of the benefits, costs, and economic impacts of extending the HCS to the nonmanufacturing sector are projected for 40 years. As indicated, the analysis reflects requirements of state right-to-know laws and voluntarily implemented hazard communication programs.

Risk Evaluation/Benefits Analysis

For this analysis OSHA estimated the percentage of workers exposed to hazardous chemicals. The percentage and numbers of exposed workers are shown in Table 2 [1] by SIC group. The analysis of risks and benefits proceeds from the current annual incidence of chemical-related injuries and illnesses in the nonmanufacturing sector. For workers in this sector, measures of acute chemical source injuries and illnesses included nonlost workday (NLWD) injuries (13,671) and LWD illnesses (38,249); and fatalities (102). Measures for chronic illnesses include: chronic illness cases (17,153), cancer cases (25,388), and cancer deaths (12,890). The cancer cases category includes cancer deaths. (Note that tables used in the computer models for this analysis may vary slightly from these figures due to rounding.)

The benefits of the standard result from its expected reduction of occupational injuries and illnesses that are chemically related. Specifically, OSHA projects that the standard will avert 20 percent of these injuries and illnesses. (Five percent of all cancer

¹ Tables 2 to 10 appear at the end of this article.

cases are assumed to be occupationally related; the 20 percent reduction is applied to this 5 percent of all cases among occupationally exposed workers in the nonmanufacturing sector.) However, the full reduction of chronic illnesses and cancers will not occur immediately; rather, the reduction for these cases is phased in over time. For chronic illnesses, the standard is expected to reduce 1 percent of the cases in the first year, 2 percent in the second year, and so on, until it reaches the full reduction of 20 percent. For cancer cases and cancer deaths, the standard is expected not to have an effect for the first 10 years, then it is expected to reduce 2 percent of the cases in the eleventh year, 4 percent in the twelfth year, and so on until it reaches the full reduction of 20 percent.

Benefits were monetized using two independent approaches. The first took into account medical costs and lost earnings incurred by each victim. This "human capital" approach resulted in first-year benefits of $56.3 million, and a 40 year present value of $6.66 billion (summarized in Table 3).

A second estimate of benefits was made using the "willingness-to-pay" approach. This approach resulted in first-year benefits of $568.7 million and a 40 year present value of $54.6 billion (Table 3).

To provide comparability with the estimates of compliance costs, benefits were attributed to the states with right-to-know laws in proportion to the share of hazard communication costs projected for firms in those states. Under the "human capital" approach the present value of the 40 year stream of benefits from the extension of the HCS, after deducting states with right-to-know-laws, is $3.80 billion (1985 dollars). Under the willingness-to-pay approach, the present value of the 40-year stream of benefits from extension of the HCS is $31.0 billion, after deducting the amount attributable to states with right-to-know laws.

The monetized benefits of hazard communication in the nonmanufacturing sector, whether monetized in terms of human capital or willingness to pay, are presented after discounting (at 10 percent). Such discounting does not convey the magnitude of the expected number of injuries, illnesses and deaths that should be averted by the extension of hazard communication to the nonmanufacturing sector. The actual number of NLWD cases, LWD cases, chronic illness cases, cancer cases, cancer deaths, and other fatalities that are expected to be averted in the first, twentieth, and fortieth years are presented in Table 4.

The numbers of cases presented in Table 4 are projections of cases that will be averted by the state right-to-know laws and the extension of the HCS. Approximately 43 percent of these cases will be averted as a result of the hazard communication (i.e., right-to-know) laws of the states. The remaining 57 percent uniquely relate to the extension of HCS and translate into the following: 148,400 cancer cases and 74,200 cancer deaths, 119,200 chronic disabling illnesses, 448,500 lost work day cases, 702,000 non-lost work day cases, and about 653 non-cancer fatalities avoided over the next 40 years. This estimate is believed to be conservative since OSHA assumed that only 5 percent of all cancers are occupationally related.

The original Regulatory Impact Analysis (RIA) for the HCS in manufacturing included estimates of benefits arising from the reduction of the incidence of chemical fires in the manufacturing sector. Using the RIA's methodology and newer data obtained from the U.S. Fire Administration's National Fire Incidence Reporting System, OSHA has determined that extension of the HCS to the nonmanufacturing sector would yield first-year benefits (i.e., the value of property damages and losses avoided) of $1.6 million (1985 dollars). For the twentieth and fortieth years, the estimates are $2.2 and $2.9 million, respectively. The present value of the 40-year stream of benefits is $20.3 million (using a 10 percent discount rate).

Extending the HCS to the nonmanufacturing sector will also yield benefits by eliminating the need for employers to comply with multiple state and local right-to-know laws with differing requirements. The estimated benefits for the first year amount to $39.6 million (1985 dollars). For the twentieth and fortieth years, the benefits are $69.5 and $125.5 million, respectively. The present value of the 40-year stream of benefits is $578 million (using a 10 percent discount rate).

Compliance Costs

Compliance costs were estimated for five items: preparation of a written hazard communication program; container labeling; provision of MSDSs; maintenance of MSDSs; and information and training.

Table 5 provides a summary of total regulatory costs, the costs attributable to state right-to-know laws and the costs attributable to the extension of the OSHA standard. Costs are presented for the first, twentieth, and fortieth year of

the standard, as well as in terms of total present value over forty years. Present values were calculated using a 10 percent discount rate. Table 6 presents the costs by provision.

The total cost attributable to hazard communication laws during the first year the expanded HCS is effective is $1.28 billion (1985 dollars). The first year cost associated with compliance with state right-to-know laws is $597.3 million and $687.3 million with the Federal HCS. The present value of the total HCS-related compliance costs over the 40 year period is $1.57 billion.

Recordkeeping activities are required in the maintenance of MSDSs. As shown in Table 6, the Year 1 costs for this function amount to $44.9 million (1985 dollars). The costs for the twentieth and fortieth years are $6.0 and $13.3 million. The present value of the costs over 40 years is $84.8 million.

Economic Impacts

In order to assess the potential economic impacts of expanding the hazard communication standard, OSHA studied the impact of the first year costs on typical establishments that have not implemented any of the provisions. No allowance was made for partial compliance. If establishments can pass through or absorb first year costs, it is assumed that they can afford the minimal recurring costs related to training new employees and the introduction of new hazards. Table 7 presents the average compliance costs, assuming no current compliance, for typical establishments in each SIC Code. Typical establishments in the preponderance for SICs (over 80 percent) would incur no compliance costs of less than $700 in the first year.

In only one of the SICs does the average total first year cost exceed $800 per establishment. The average first year cost per exposed employee in all SICs is less than $250, or less than $5.00 per worker per week.

Table 8 presents a comparison of the post-tax compliance costs to a typical firm's revenues and profits. A typical establishment's pre-tax compliance cost will be a negligible percentage (less than one-half of one percent) of the establishment's average annual revenue in over 96 percent of the SICs. The only exceptions, SIC 83 (Social Service) and SIC 86 (Membership Organizations), are primarily composed of nonprofit establishments that are characterized by relatively inelastic demand for their services. Given the magnitude of the compliance costs in relation to revenue, and the fact that the affected industry sectors are predominantly service providers, which are necessarily

characterized by localized markets, it appears likely that most firms will pass the compliance costs on to their customers. The post-tax compliance cost as a percent of profits is less than two percent in most (over 80 percent) of the SICs. Typical firms in these SICs should be able to absorb the costs even if they cannot pass them on to their customers. Given the small absolute magnitude of the compliance costs, and the fact that the analysis was conducted using first year compliance costs which are significantly higher than the recurring compliance costs for subsequent years, the expansion of the hazard communication standard should have little or no economic impact on typical firms.

Community Right-to-Know

The cost of extending the Superfund Amendments and Reauthorization Act (SARA) requirements for community right-to-know to the non-manufacturing sector was also estimated. Under Title III of SARA, establishments holding a given hazardous chemical in amounts greater than specified threshold quantities must report these chemicals and their quantities to State and local emergency planning committees and the local fire department. Cost estimates were based on EPA's projected phase-in threshold quantities of 10,000 pounds in the first two years, and 500 pounds in the third and subsequent years that the requirements apply to the non-manufacturing sector. The estimated costs for the first and second years are $8,614,300 and $3,524,000, respectively. Third and fourth year costs were estimated to be $63,492,800 and $32,736,300.

The economic impact of extending SARA to nonmanufacturing was also estimated by OSHA. The third year average total cost of SARA was combined with OSHA's recurring average total costs of the Hazard Communication Standard to estimate the impact. The analysis indicated that the economic impact per facility of extending SARA to nonmanufacturing is minor, and that costs incurred by affected establishments could be passed on to the consumer. OSHA believes that the extension of SARA to nonmanufacturing will not affect the feasibility of the Hazard Communication Standard.

Regulatory Flexibility

As is shown in Table 9, a majority of establishments in all of the potentially impacted SICs are small businesses with fewer than 20 employees. Thus, the average compliance costs for small firms

31870   **Federal Register** / Vol. 52, No. 163 / Monday, August 24, 1987 / Rules and Regulations

are very similar to those for typical firms. No disproportionate economic impact is foreseen for small firms.

Most establishments in the potentially affected SICs are service providers, which typically compete on the basis of many factors (e.g., location, specialized service, customer relations, etc.) in addition to price. Assuming all firms try to pass their compliance cost on to their customers, minor price differentials of less than one-half of one percent, shown in Table 10, are unlikely to adversely affect the overall competitive position of small entities.

As can be seen from Table 10, the cost differential between small and large firms in over 80 percent of the SICs is anticipated to be less than 0.2 percent of revenue. In SICs 83 and 86 the difference is about 2 percent. However, these SICs are dominated by non-profit firms which are less likely to be subject to price competition.

Environmental Impacts

At the time the current HCS was promulgated in the Federal Register (48 FR 53280), OSHA stated that the standard was unlikely to result in the occurrence of significant health or environmental impacts outside of the workplace. The extension of the HCS does not entail any change from the current HCS in terms of impacts outside the workplace. As concluded previously, the labeling of containers will not have a direct or significant impact on air or water quality, land or energy use, or solid waste disposal outside of the workplace. Similarly, the requirements for preparation of a written compliance plan, provision and maintenance of MSDSs, and provision of information and training should have no adverse environmental impact.

IV. Clearance of Information Collection Requirements

On March 31, 1983, the Office of Management and Budget (OMB) published a new 5 CFR Part 1320, implementing the information collection provisions of the Paperwork Reduction Act of 1980, 44 U.S.C. 3501 *et seq.* (48 FR 13666). Part 1320, which became effective on April 30, 1983, sets forth procedures for agencies to follow in

obtaining OMB clearance for information collection requirements. The sections of the Hazard Communication Standard which may create recordkeeping requirements are paragraphs (d) hazard determination; (e) written hazard communication program; (f) labels and other appropriate forms of warning; (g) material safety data sheets; (h) information and training; and (i) trade secrets.

In accordance with the provisions of the Paperwork Reduction Act and the regulations issued pursuant thereto, OSHA certifies that it has submitted the information collection requirements contained in its rule on hazard communication to OMB for review under section 3504(h) of that Act.

V. State Plan Applicability

The 25 States with their own OSHA-approved occupational safety and health plans must adopt a comparable standard within six months of the publication date of a final standard. These States include: Alaska, Arizona, California, Connecticut (for State and local government employees only), Hawaii, Indiana, Iowa, Kentucky, Maryland, Michigan, Minnesota, Nevada, New Mexico, New York (for State and local government employees only), North Carolina, Oregon, Puerto Rico, South Carolina, Tennessee, Utah, Vermont, Virginia, Virgin Islands, Washington, and Wyoming. Until such time as a State standard is promulgated, Federal OSHA will provide interim enforcement assistance, as appropriate. (Thirteen (13) of these States (Alaska, California, Iowa, Maryland, Michigan, Minnesota, New Mexico, North Carolina, Oregon, Tennessee, Vermont, Washington, and Wyoming) have already expanded the scope of their hazard communication standard/right-to-know law to cover private sector, non-manufacturing workplaces.)

Although a State HCS becomes effective in accordance with State promulgation provisions, and is enforceable upon promulgation, OSHA must also review and approve the standard to assure that it is "at least as effective" as the Federal standard. OSHA intends to closely scrutinize State standards submitted under current

or future State plans to assure not only equal or greater effectivenss, but also that any additional requirements do not conflict with, or adversely affect, the effectiveness of the national application of OSHA's standard. Because the HCS is "appliable to products" in that it permits the distribution and use of hazardous chemicals in commerce only if they are in labeled containers accompanied by material safety data sheets, OSHA must determine in its review whether any State plan standard provisions which differ from the Federal are "required by compelling local conditions and do not unduly burden interstate commerce." Section 18(c) of the Act, 29 U.S.C. 667(c).

VI. Authority, Signature, and the Final Rule

This document was prepared under the direction of John A. Pendergrass, Assistant Secretary of Labor for Occupational Safety and Health, U.S. Department of Labor, 200 Constitution Avenue, NW., Washington, DC 20210.

For the reasons set out in the preamble, and under the authority of section 41 of the Longshore and Harbor Workers' Compensation Act (33 U.S.C. 941), section 107 of the Contract Work Hours and Safety Standards Act (Construction Safety Act) (40 U.S.C. 333), sections 4, 6 and 8 of the Occupational Safety and Health Act of 1970 (29 U.S.C. 653, 655, 657), Secretary of Labor's Order No. 9–83 (48 FR 35736) and 29 CFR Part 1911, and 5 U.S.C. 553, the Occupational Safety and Health Administration hereby amends Parts 1910, 1915, 1917, 1918, 1926, and 1928 of Title 29 of the Code of Federal Regulations, as set forth below.

List of Subjects in 29 CFR Parts 1910, 1915, 1917, 1918, 1926, and 1928

Hazard communication, Occupational safety and health, Right-to-know, Labeling, Material safety data sheets; Employee training.

Signed at Washington, DC, this 16th day of August 1987.

John A. Pendergrass,

Assistant Secretary for Occupational Safety and Health.

TABLE 2.—WORKER EXPOSURE TO HAZARDOUS CHEMICALS

Industry	Total number of establishment	Total employment	Percent of workers exposed to hazardous chemicals	Number of exposed employees
SIC 01	31,739	504,025	70	352,818
SIC 02	10,994	126,039	70	88,227

TABLE 2.—WORKER EXPOSURE TO HAZARDOUS CHEMICALS—Continued

Industry	Total number of establishment	Total employment	Percent of workers exposed to hazardous chemicals	Number of exposed employees
SIC 07	65,704	459,479	70	321,635
SIC 08	2,117	20,223	70	14,156
SIC 09	3,886	13,549	20	2,710
SIC 13	31,572	591,714	70	414,200
SIC 15	166,012	1,137,853	70	796,497
SIC 16	44,702	791,892	70	554,324
SIC 17	320,208	2,406,916	70	1,684,841
SIC 40	18,539	324,206	40	129,682
SIC 41	15,539	285,578	20	57,116
SIC 42	99,805	1,323,495	20	264,699
SIC 44	8,346	178,013	70	124,609
SIC 45	8,691	490,395	40	196,158
SIC 46	959	18,405	60	11,043
SIC 47	30,783	267,113	40	106,845
SIC 48	22,910	1,321,116	5	66,056
SIC 49	15,571	890,586	40	356,234
SIC 50	300,972	3,357,168	10	335,717
SIC 51	191,745	2,295,451	25	573,863
SIC 52	66,756	662,051	50	331,026
SIC 53	29,818	2,230,449	5	111,522
SIC 54	137,393	2,696,839	20	539,368
SIC 55	173,902	1,850,359	60	1,110,215
SIC 56	99,022	1,004,666	5	50,233
SIC 57	93,338	714,264	5	35,713
SIC 58	309,650	5,479,633	25	1,369,908
SIC 59	261,694	2,133,614	20	426,723
SIC 60	24,949	1,681,408	5	84,070
SIC 61	43,408	733,201	5	36,660
SIC 62	17,995	346,214	5	17,311
SIC 63	30,139	1,190,103	5	59,505
SIC 64	96,706	536,223	5	26,811
SIC 65	191,400	1,077,550	5	53,878
SIC 66	2,937	13,752	5	688
SIC 67	15,792	138,488	5	6,924
SIC 70	44,697	1,273,343	25	318,336
SIC 72	158,272	1,068,670	50	534,335
SIC 73	284,684	4,092,820	50	2,046,410
SIC 75	121,431	713,798	50	356,899
SIC 76	57,900	316,365	60	189,819
SIC 78	15,338	216,806	30	65,042
SIC 79	58,064	757,287	20	151,457
SIC 80	365,758	6,167,908	60	3,700,745
SIC 81	119,861	670,317	5	33,516
SIC 82	23,280	1,174,052	10	117,405
SIC 83	66,380	1,182,651	5	59,133
SIC 84	1,592	39,021	25	9,755
SIC 86	83,774	724,283	5	36,214
SIC 89	117,155	1,200,885	5	60,044
Totals	4,503,879	58,890,236		18,391,096

Source: U.S. Department of Labor, OSHA, Office of Regulatory Analysis.

TABLE 3.—ESTIMATED BENEFITS OF HAZARD COMMUNICATION

[Millions of 1985 dollars]

Type of injury/illness	Benefits—Year			
	1	20	40	TPV
HUMAN CAPITAL APPROACH				
NLWD:				
Lost earnings	0.7	1.3	2.5	9.3
Medical costs	1.7	4.6	13.4	30.3
LWD:				
Lost earnings	15.2	28.3	57.1	209.3

31872   Federal Register / Vol. 52, No. 163 / Monday, August 24, 1987 / Rules and Regulations

TABLE 3.—ESTIMATED BENEFITS OF HAZARD COMMUNICATION—Continued

[Millions of 1985 dollars]

Type of injury/illness	Benefits—Year			
	1	20	40	TPV
Medical costs	10.9	29.2	86.5	192.2
Chronic:				
Lost earnings	20.5	722.8	1,365.8	2,967.5
Medical costs	2.8	143.4	404.1	582.8
Cancer:				
Lost earnings	0	651.6	1,309.6	1,735.2
Medical costs	0	298.9	906.4	875.8
Fatalities: Lost earnings	4.4	7.3	13.0	56.6
Total	56.3	1,887.3	4,158.3	6,659.1
WILLINGNESS-TO-PAY APPROACH				
NLWD	59.6	107.8	211.9	804.5
LWD	374.4	686.4	1,371.1	5,099.8
Chronic	61.7	2,173.7	4,121.6	8,924.3
Cancer	0	14,529.0	29,651.2	38,812.0
Fatalities	72.9	123.4	255.3	946.9
Total	568.7	17,620.7	35,581.2	54,587.4

Source: JACA Corporation Report.

TABLE 4.—INJURIES, ILLNESSES, AND FATALITIES AVERTED BY HAZARD COMMUNICATION IN THE NONMANUFACTURING SECTOR

Year	1	20	40	Cumulative total
FEDERAL AND STATE STANDARDS COMBINED				
NLWD	17,000	30,800	60,600	1,354,500
LWD	10,700	19,600	39,200	865,800
Chronic	150	6,200	11,800	230,100
Cancer cases	0	8,200	17,000	286,500
Cancer deaths	0	4,100	8,500	143,300
Noncancer deaths	0	20	80	1,260
IMPACT OF FEDERAL STANDARD ALONE				
NLWD	8,800	16,000	31,400	702,000
LWD	5,500	10,200	20,300	448,500
Chronic	78	3,200	6,100	119,200
Cancer cases	0	4,248	8,806	148,400
Cancer deaths	0	2,100	4,400	74,200
Noncancer deaths	0	10	41	653

Source: U.S. Department of Labor, OHSA, Office of Regulatory Analysis.

TABLE 5.—SUMMARY OF HAZARD COMMUNICATION COSTS

[Millions of 1985 dollars]

Year	Total	State	OHSA
1	1,284.5	597.3	687.2
20	214.5	101.3	113.2
40	384.0	184.0	200.0

TABLE 5.—SUMMARY OF HAZARD COMMUNICATION COSTS—Continued

[Millions of 1985 dollars]

Year	Total	State	OHSA
Total present value	2,926.4	1,356.3	1,570.1

Source: U.S. Department of Labor, OHSA, Office of Regulatory Analysis.

Federal Register / Vol. 52, No. 163 / Monday, August 24, 1987 / Rules and Regulations   31873

TABLE 6.—SUMMARY OF FEDERAL HCS COSTS BY PROVISION

[Millions of 1985 dollars]

Year	Main-tain MSDS's	Label-ing	Writ-ten	Train-ing	Provide MSDS's	Totals
1	44.9	12.8	137.4	472.9	19.3	687.2
20	6.0	20.3	5.7	78.7	2.5	113.2
40	13.3	35.2	9.4	136.5	5.6	200.0
TPV	84.8	170.9	170.9	1054.6	88.9	1570.1

Source: U.S. Department of Labor, OHSA, Office of Regulatory Analysis.

TABLE 7.—SUMMARY OF HCS COSTS PER ESTABLISHMENT NOT IN COMPLIANCE WITH HCS

[1985 dollars]

Industry	First year		Second year	
	Average costs per establish-ment	Average costs per exposed employee	Average costs per establish-ment	Average costs per exposed employee
SIC 01	502	45	32	3
SIC 02	475	59	23	3
SIC 07	490	100	28	6
SIC 08	358	54	26	4
SIC 09	304	242	6	5
SIC 13	497	36	72	5
SIC 15	150	31	12	3
SIC 16	225	16	34	3
SIC 17	169	32	14	3
SIC 40	603	86	51	7
SIC 41	285	76	11	3
SIC 42	273	98	12	4
SIC 44	442	30	55	4
SIC 45	892	40	72	3
SIC 46	461	40	55	5
SIC 47	398	115	15	4
SIC 48	319	50	15	2
SIC 49	798	35	64	3
SIC 50	472	238	14	7
SIC 51	700	234	32	11
SIC 52	335	68	20	4
SIC 53	372	50	27	4
SIC 54	323	82	18	5
SIC 55	437	68	31	5
SIC 56	265	149	6	3
SIC 57	288	190	6	4
SIC 58	337	76	17	4
SIC 59	321	184	7	4
SIC 60	410	61	21	3
SIC 61	217	76	9	3
SIC 62	312	79	18	5
SIC 63	250	46	16	3
SIC 64	236	155	5	3
SIC 65	306	186	8	5
SIC 66	238	181	5	4
SIC 67	415	167	12	5
SIC 70	408	57	37	5
SIC 72	500	148	16	5
SIC 73	444	62	43	6
SIC 75	381	130	14	5
SIC 76	325	99	15	5
SIC 78	351	83	26	6
SIC 79	346	117	20	7
SIC 80	581	57	57	6
SIC 81	242	153	7	5
SIC 82	287	46	10	2
SIC 83	337	132	11	4
SIC 84	608	99	39	6
SIC 86	273	149	6	3

31874   Federal Register / Vol. 52, No. 163 / Monday, August 24, 1987 / Rules and Regulations

TABLE 7.—SUMMARY OF HCS COSTS PER ESTABLISHMENT NOT IN COMPLIANCE WITH HCS—Continued

[1985 dollars]

Industry	First year		Second year	
	Average costs per establish-ment	Average costs per exposed employee	Average costs per establish-ment	Average costs per exposed employee
SIC 89	312	146	10	5

Source: U.S. Department of Labor, OSHA, Office of Regulatory Analysis.

TABLE 8.—ANALYSIS OF POST-TAX FIRST-YEAR COMPLIANCE COSTS

[1985 dollars]

Industry	Average annual revenue per establishment	Average cost as a percent of revenue per establishment	Average net income per establishment	Average post-tax cost per establishment	Post tax cost as a percent of net income per establishment
SIC 01	2,794,100	0.018	103,382	377	0.36
SIC 02	11,275,400	0.004	417,190	356	0.09
SIC 07	286,600	0.171	7,165	368	5.13
SIC 08	1,689,100	0.021	42,228	268	0.64
SIC 09	797,500	0.038	19,938	228	1.14
SIC 13	6,185,800	0.008	346,405	373	0.11
SIC 15	816,700	0.018	19,601	113	0.57
SIC 16	1,419,700	0.016	56,788	169	0.30
SIC 17	372,400	0.045	10,800	127	1.17
SIC 40	2,584,100	0.023	111,116	453	0.41
SIC 41	411,400	0.069	13,165	214	1.62
SIC 42	730,100	0.037	21,903	205	0.94
SIC 44	2,214,300	0.020	141,715	331	0.23
SIC 45	5,900,000	0.015	70,800	669	0.94
SIC 46	20,569,600	0.002	1,069,619	346	0.03
SIC 47	831,900	0.048	14,974	299	1.99
SIC 48	5,347,900	0.006	390,397	239	0.06
SIC 49	16,269,000	0.005	732,105	599	0.08
SIC 50	1,866,900	0.025	28,004	354	1.26
SIC 51	3,371,500	0.021	57,316	525	0.92
SIC 52	793,800	0.042	20,639	251	1.22
SIC 53	5,702,000	0.007	136,848	279	0.20
SIC 54	2,089,700	0.015	25,076	242	0.96
SIC 55	2,016,100	0.022	16,129	327	2.03
SIC 56	507,600	0.052	19,796	199	1.00
SIC 57	371,400	0.078	11,513	216	1.88
SIC 58	383,500	0.088	11,122	252	2.27
SIC 59	829,100	0.039	20,728	241	1.16
SIC 60	14,970,800	0.003	509,007	307	0.06
SIC 61	2,585,300	0.008	41,365	163	0.39
SIC 62	1,856,900	0.017	135,554	234	0.17
SIC 63	12,911,400	0.002	438,988	187	0.04
SIC 64	220,400	0.107	15,869	177	1.11
SIC 65	338,400	0.090	27,749	230	0.83
SIC 66	661,600	0.036	54,251	178	0.33
SIC 67	798,300	0.052	167,643	311	0.19
SIC 70	607,000	0.067	34,599	306	0.88
SIC 72	228,500	0.219	10,283	375	3.64
SIC 73	531,600	0.084	21,264	333	1.57
SIC 75	351,800	0.108	8,795	286	3.25
SIC 76	187,100	0.174	7,671	244	3.18
SIC 78	815,900	0.043	31,820	263	0.83
SIC 79	782,100	0.044	51,619	260	0.50
SIC 80	198,200	0.293	6,342	436	6.87
SIC 81	456,000	0.053	10,032	182	1.81
SIC 82	NA	0.168	NA	215	NA
SIC 83	NA	1.763	NA	252	NA
SIC 83	NA	1.763	NA	252	NA
SIC 84	NA	0.094	NA	456	NA
SIC 86	NA	1.007	NA	205	NA

Federal Register / Vol. 52, No. 163 / Monday, August 24, 1987 / Rules and Regulations   31875

TABLE 8.—ANALYSIS OF POST-TAX FIRST-YEAR COMPLIANCE COSTS—Continued

[1985 dollars]

Industry	Average annual revenue per establishment	Average cost as a percent of revenue per establishment	Average net income per establishment	Average post-tax cost per establishment	Post tax cost as a percent of net income per establishment
SIC 89	290,500	0.107	11,039	234	2.12

Source: U.S. Department of Labor, OSHA, Office of Regulatory Analysis.

TABLE 9.—ESTABLISHMENTS WITH FEWER THAN TWENTY EMPLOYEES

SIC code	Total number of establishments	Number of establishments with 1 to 19 employees	Percent of establishments with 1 to 19 employees
01	31,739	27,440	86
02	10,994	9,574	87
07	65,704	61,928	94
08	2,117	1,852	87
09	2,160	2,088	97
13	31,572	26,037	82
15	166,012	154,819	93
16	44,702	37,484	84
17	320,208	294,850	92
40	18,539	15,756	85
41	15,267	11,998	79
42	94,561	80,822	85
44	8,346	6,917	83
45	8,691	6,514	75
46	959	724	75
47	30,783	28,420	92
48	10,319	6,612	64
49	15,571	10,922	70
50	169,451	133,233	79

TABLE 9.—ESTABLISHMENTS WITH FEWER THAN TWENTY EMPLOYEES—Continued

SIC code	Total number of establishments	Number of establishments with 1 to 19 employees	Percent of establishments with 1 to 19 employees
51	191,745	166,562	87
52	66,756	60,097	90
53	14,909	8,963	60
54	137,393	114,738	84
55	173,902	152,920	88
56	28,181	23,874	85
57	23,582	20,474	87
58	309,650	241,282	78
59	244,849	227,803	93
60	12,475	6,318	51
61	12,912	9,561	74
62	4,380	3,079	70
63	10,998	7,263	66
64	17,577	15,608	89
65	32,714	28,099	86
66	524	488	93
67	2,790	2,239	80
70	44,697	34,693	78

TABLE 9.—ESTABLISHMENTS WITH FEWER THAN TWENTY EMPLOYEES—Continued

SIC code	Total number of establishments	Number of establishments with 1 to 19 employees	Percent of establishments with 1 to 19 employees
72	158,272	149,812	95
73	284,684	249,553	88
75	121,431	116,344	96
76	57,900	55,543	96
78	15,338	13,314	87
79	50,981	42,916	84
80	365,758	338,396	93
81	21,210	18,659	88
82	18,661	11,197	60
83	23,148	17,068	74
84	1,592	1,250	79
86	19,757	16,416	83
89	28,103	23,179	82

Source: U.S. Department of Labor, OSHA, Office of Regulatory Analysis.
ᵃ From Chapter 5 of the JACA Report [4].
ᵇ Column 2 divided by Column 1.

TABLE 10.—ANALYSIS OF IMPACT ON SMALLEST VERSUS LARGEST ESTABLISHMENTS

[Comparing average costs as a percent of revenue]

SIC code	Average cost as a percent of revenue per establishment 250+ employees	Average cost as a percent of revenue per establishment 1–19 employees	Difference in cost as a percent of revenue due to size of establishments
01	0.003	0.044	0.040
02	0.001	0.009	0.008
07	0.038	0.189	0.151
08	0.007	0.018	0.011
09	0.002	0.021	0.019
13	0.002	0.035	0.033
15	0.003	0.044	0.042
16	0.007	0.064	0.057
17	0.017	0.065	0.048
40	0.008	0.082	0.075
41	0.007	0.248	0.240
42	0.007	0.103	0.096
44	0.007	0.060	0.053
45	0.009	0.105	0.096
46	0.000	0.174	0.174
47	0.003	0.101	0.098
48	0.002	0.041	0.039

31876 Federal Register / Vol. 52. No. 163 / Monday. August 24. 1987 / Rules and Regulations

TABLE 10.—ANALYSIS OF IMPACT ON SMALLEST VERSUS LARGEST ESTABLISHMENTS—Continued

[Comparing average costs as a percent of revenue]

SIC code	Average cost as a percent of revenue per establishment 250 + employees	Average cost as a percent of revenue per establishment 1-19 employees	Difference in cost as a percent of revenue due to size of establishments
49	0.002	0.049	0.047
50	0.003	0.037	0.034
51	0.006	0.038	0.032
52	0.008	0.046	0.038
53	0.002	0.029	0.026
54	0.002	0.054	0.052
55	0.006	0.044	0.038
56	0.002	0.104	0.102
57	0.003	0.117	0.114
58	0.008	0.158	0.150
59	0.003	0.055	0.051
60	0.001	0.012	0.011
61	0.000	0.038	0.038
62	0.003	0.028	0.025
63	0.000	0.069	0.069
64	0.002	0.179	0.177
65	0.005	0.124	0.119
66	0.001	0.054	0.053
67	0.003	0.096	0.093
70	0.021	0.283	0.262
72	0.007	0.346	0.339
73	0.028	0.204	0.175
75	0.004	0.151	0.148
76	0.099	0.205	0.106
78	0.007	0.113	0.106
79	0.016	0.071	0.055
80	0.269	0.370	0.101
81	0.118	0.077	−0.041
82	0.025	0.915	0.890
83	0.428	2.293	1.865
84	0.033	0.259	0.226
86	0.035	2.109	2.074
89	0.008	0.210	0.202

Source: U.S. Department of Labor, OHSA, Office of Regulatory Analysis.

OSHA is amending Parts 1910, 1915, 1917, 1918, 1926, and 1928 of Title 29 of the Code of Federal Regulations as follows:

PART 1910—OCCUPATIONAL SAFETY AND HEALTH STANDARDS

1. The authority citation for Subpart Z of Part 1910 continues to read as follows:

Authority: Secs. 6, 8, Occupational Safety and Health Act (29 U.S.C. 655, 657); Secretary of Labor's Order No. 12–71 (36 FR 8754); 8–76 (41 FR 2509); or 9–83 (48 FR 35736) as applicable; and 29 CFR Part 1911.
Section 1910.1000 Tables Z-1. Z-2, Z-3 also issued under 5 U.S.C. 553.
Section 1910.1000 not issued under 29 CFR Part 1911, except for "Arsenic" and "Cotton Dust" listings in Table Z-1.
Section 1910.1001 not issued under Sec. 107 of Contract Work Hours and Safety Standards Act, 40 U.S.C. 333.

Section 1910.1002 not issued under 29 U.S.C. 655 or 29 CFR Part 1911; also issued under 5 U.S.C. 553.
Sections 1910.1003 through 1910.1018 also issued under 29 U.S.C. 653.
Section 1910.1025 also issued under 29 U.S.C. 653 and 5 U.S.C. 553.
Section 1910.1043 also issued under 5 U.S.C. 551 et seq.
Sections 1910.1045 and 1910.1047 also issued under 29 U.S.C. 653.
Sections 1910.1200, 1910.1499 and 1910.1500 also issued under 5 U.S.C. 553.

PART 1915—OCCUPATIONAL SAFETY AND HEALTH STANDARDS FOR SHIPYARD EMPLOYMENT

2. The authority citation for Part 1915 is revised to read as follows:

Authority: Sec. 41. Longshore and Harbor Workers' Compensation Act (33 U.S.C. 941); secs. 4, 6, 8, Occupational Safety and Health Act of 1970 (29 U.S.C. 653, 655, 657); Secretary of Labor's Order No. 12–71 (36 FR 8754), 8–76

(41 FR 25059), or 9–83 (48 FR 35736), as applicable; 29 CFR Part 1911.
Section 1915.99 also issued under 5 U.S.C. 553.

PART 1917—MARINE TERMINALS

3. The authority citation for Part 1917 is revised to read as follows:

Authority: Sec. 41. Longshore and Harbor Workers' Compensation Act (33 U.S.C. 941); secs. 4, 6, 8, Occupational Safety and Health Act of 1970 (29 U.S.C. 653, 655, 657); Secretary of Labor's Order No. 12–71 (36 FR 8754), 8–76 (41 FR 25059), or 9–83 (48 FR 35736), as applicable; 29 CFR Part 1911.
Section 1917.28 also issued under 5 U.S.C. 553.

PART 1918—SAFETY AND HEALTH REGULATIONS FOR LONGSHORING

4. The authority citation for Part 1918 is revised to read as follows:

Authority: Sec. 41. Longshore and Harbor Workers' Compensation Act (33 U.S.C. 941);

secs. 4, 6, 8, Occupational Safety and Health Act of 1970 (29 U.S.C. 653, 655, 657); Secretary of Labor's Order No. 12–71 (36 FR 8754), 8–76 (41 FR 25059), or 9–83 (48 FR 35736), as applicable.

Section 1918.90 also issued under 5 U.S.C. 553 and 29 CFR Part 1911.

PART 1926—SAFETY AND HEALTH REGULATIONS FOR CONSTRUCTION

5. The authority citation for Subpart D of Part 1926 is revised to read as follows:

Authority: Sec. 107, Contract Work Hours and Safety Standards Act (Construction Safety Act) (40 U.S.C. 333); secs. 4, 6, 8, Occupational Safety and Health Act of 1970 (29 U.S.C. 653, 655, 657); Secretary of Labor's Order No. 12–71 (36 FR 8754), 8–76 (41 FR 25059), or 9–83 (48 FR 35736), as applicable. Section 1926.59 also issued under 5 U.S.C. 553 and 29 CFR Part 1911.

PART 1928—OCCUPATIONAL SAFETY AND HEALTH STANDARDS FOR AGRICULTURE

6. The authority citation for Part 1928 is revised to read as follows:

Authority: Secs. 6 and 8, Occupational Safety and Health Act of 1970 (29 U.S.C. 655, 657); Secretary of Labor's Orders 12–71 (36 FR 8754), 8–76 (41 FR 25059), or 9–83 (48 FR 35736), as applicable; 29 CFR Part 1911. Section 1928.21 also issued under 5 U.S.C. 553.

PARTS 1910, 1915, 1917, 1918, 1926 and 1928—[AMENDED]

7. Parts 1910, 1915, 1917, 1918, and 1926 are amended by revising § 1910.1200 as set forth below, and by adding §§ 1915.99, 1917.28, 1918.90, and 1926.59 to contain the identical text of the revised § 1910.1200, including Apendices A, B, C, and D of 1910.1200:

§ _____ Hazard communication.

(a) *Purpose.* (1) The purpose of this section is to ensure that the hazards of all chemicals produced or imported are evaluated, and that information concerning their hazards is transmitted to employers and employees. This transmittal of information is to be accomplished by means of comprehensive hazard communication programs, which are to include container labeling and other forms of warning, material safety data sheets and employee training.

(2) This occupational safety and health standard is intended to address comprehensively the issue of evaluating the potential hazards of chemicals, and communicating information concerning hazards and appropriate protective measures to employees, and to preempt any legal requirements of a state, or

political subdivision of a state, pertaining to the subject. Evaluating the potential hazards of chemicals, and communicating information concerning hazards and appropriate protective measures to employees, may include, for example. but is not limited to, provisions for: developing and maintaining a written hazard communication program for the workplace, including lists of hazardous chemicals present; labeling of containers of chemicals in the workplace, as well as of containers of chemicals being shipped to other workplaces; preparation and distribution of material safety data sheets to employees and downstream employers; and development and implementation of employee training programs regarding hazards of chemicals and protective measures. Under section 18 of the Act, no state or political subdivision of a state may adopt or enforce, through any court or agency, any requirement relating to the issue addressed by this Federal standard, except pursuant to a Federally-approved state plan.

(b) *Scope and application.* (1) This section requires chemical manufacturers or importers to assess the hazards of chemicals which they produce or import, and all employers to provide information to their employees about the hazardous chemicals to which they are exposed, by means of a hazard communication program, labels and other forms of warning, material safety data sheets, and information and training. In addition, this section requires distributors to transmit the required information to employers.

(2) This section applies to any chemical which is known to be present in the workplace in such a manner that employees may be exposed under normal conditions of use or in a foreseeable emergency.

(3) This section applies to laboratories only as follows:

(i) Employers shall ensure that labels on incoming containers of hazardous chemicals are not removed or defaced;

(ii) Employers shall maintain any material safety data sheets that are received with incoming shipments of hazardous chemicals, and ensure that they are readily accessible to laboratory employees; and,

(iii) Employers shall ensure that laboratory employees are apprised of the hazards of the chemicals in their workplaces in accordance with paragraph (h) of this section.

(4) In work operations where employees only handle chemicals in sealed containers which are not opened under normal conditions of use (such as are found in marine cargo handling, warehousing, or retail sales), this

section applies to these operations only as follows:

(i) Employers shall ensure that labels on incoming containers of hazardous chemicals are not removed or defaced;

(ii) Employers shall maintain copies of any material safety data sheets that are received with incoming shipments of the sealed containers of hazardous chemicals, shall obtain a material safety data sheet for sealed containers of hazardous chemicals received without a material safety data sheet if an employee requests the material safety data sheet, and shall ensure that the material safety data sheets are readily accessible during each work shift to employees when they are in their work area(s);

(iii) Employers shall ensure that employees are provided with information and training in accordance with paragraph (h) of this section (except for the location and availability of the written hazard communication program under paragraph (h)(1)(iii)), to the extent necessary to protect them in the event of a spill or leak of a hazardous chemical from a sealed container.

(5) This section does not require labeling of the following chemicals:

(i) Any pesticide as such term is defined in the Federal Insecticide, Fungicide, and Rodenticide Act (7 U.S.C. 136 et seq.), when subject to the labeling requirements of that Act and labeling regulations issued under that Act by the Environmental Protection Agency;

(ii) Any food, food additive, color additive, drug, cosmetic, or medical or veterinary device, including materials intended for use as ingredients in such products (e.g. flavors and fragrances), as such terms are defined in the Federal Food, Drug, and Cosmetic Act (21 U.S.C. 301 et seq.) and regulations issued under that Act, when they are subject to the labeling requirements under that Act by the Food and Drug Administration;

(iii) Any distilled spirits (beverage alcohols), wine, or malt beverage intended for nonindustrial use, as such terms are defined in the Federal Alcohol Administration Act (27 U.S.C. 201 et seq.) and regulations issued under that Act, when subject to the labeling requirements of that Act and labeling regulations issued under that Act by the Bureau of Alcohol Tobacco, and Firearms; and,

(iv) Any consumer product or hazardous substance as those terms are defined in the Consumer Product Safety Act (15 U.S.C. 2051 et seq.) and Federal Hazardous Substances Act (15 U.S.C. 1261 et seq.) respectively, when subject to a consumer product safety standard or labeling requirement of those Acts, or regulations issued under those Acts by

31878 **Federal Register** / Vol. 52, No. 163 / Monday, August 24, 1987 / Rules and Regulations

the Consumer Product Safety Commission.

(6) This section does not apply to:

(i) Any hazardous waste as such term is defined by the Solid Waste Disposal Act, as amended by the Resource Conservation and Recovery Act of 1976, as amended (42 U.S.C. 6901 et seq.), when subject to regulations issued under that Act by the Environmental Protection Agency;

(ii) Tobacco or tobacco products;

(iii) Wood or wood products;

(iv) Articles;

(v) Food, drugs, cosmetics, or alcoholic beverages in a retail establishment which are packaged for sale to consumers;

(vi) Foods, drugs, or cosmetics intended for personal consumption by employees while in the workplace;

(vii) Any consumer product or hazardous substance, as those terms are defined in the Consumer Product Safety Act (15 U.S.C. 2051 et seq.) and Federal Hazardous Substances Act (15 U.S.C. 1261 et seq.) respectively, where the employer can demonstrate it is used in the workplace in the same manner as normal consumer use, and which use results in a duration and frequency of exposure which is not greater than exposures experienced by consumers; and,

(viii) Any drug, as that term is defined in the Federal Food, Drug, and Cosmetic Act (21 U.S.C. 301 et seq.), when it is in solid, final form for direct administration to the patient (i.e. tablets or pills).

(c) *Definitions.*

"Article" means a manufactured item: (i) Which is formed to a specific shape or design during manufacture; (ii) which has end use function(s) dependent in whole or in part upon its shape or design during end use; and (iii) which does not release, or otherwise result in exposure to, a hazardous chemical, under normal conditions of use.

"Assistant Secretary" means the Assistant Secretary of Labor for Occupational Safety and Health, U.S. Department of Labor, or designee.

"Chemical" means any element, chemical compound or mixture of elements and/or compounds.

"Chemical manufacturer" means an employer with a workplace where chemical(s) are produced for use or distribution.

"Chemical name" means the scientific designation of a chemical in accordance with the nomenclature system developed by the International Union of Pure and Applied Chemistry (IUPAC) or the Chemical Abstracts Service (CAS) rules of nomenclature, or a name which will clearly identify the chemical for the

purpose of conducting a hazard evaluation.

"Combustible liquid" means any liquid having a flashpoint at or above 100 °F (37.8 °C), but below 200 ° F (93.3 ° C), except any mixture having components with flashpoints of 200 °F (93.3 °C), or higher, the total volume of which make up 99 percent or more of the total volume of the mixture.

"Common name" means any designation or identification such as code name, code number, trade name, brand name or generic name used to identify a chemical other than by its chemical name.

"Compressed gas" means:

(i) A gas or mixture of gases having, in a container, an absolute pressure exceeding 40 psi at 70 °F (21.1 °C); or

(ii) a gas or mixture of gases having, in a container, an absolute pressure exceeding 104 psi at 130 °F (54.4 °C) regardless of the pressure at 70 °F (21.1 °C); or

(iii) A liquid having a vapor pressure exceeding 40 psi at 100 °F (37.8 °C) as determined by ASTM D–323–72.

"Container" means any bag, barrel, bottle, box, can, cylinder, drum, reaction vessel, storage tank, or the like that contains a hazardous chemical. For purposes of this section, pipes or piping systems, and engines, fuel tanks, or other operating systems in a vehicle, are not considered to be containers.

"Designated representative" means any individual or organization to whom an employee gives written authorization to exercise such employee's rights under this section. A recognized or certified collective bargaining agent shall be treated automatically as a designated representative without regard to written employee authorization.

"Director" means the Director, National Institute for Occupational Safety and Health, U.S. Department of Health and Human Services, or designee.

"Distributor" means a business, other than a chemical manufacturer or importer, which supplies hazardous chemicals to other distributors or to employers.

"Employee" means a worker who may be exposed to hazardous chemicals under normal operating conditions or in foreseeable emergencies. Workers such as office workers or bank tellers who encounter hazardous chemicals only in non-routine, isolated instances are not covered.

"Employer" means a person engaged in a business where chemicals are either used, distributed, or are produced for use or distribution, including a contractor or subcontractor.

"Explosive" means a chemical that causes a sudden, almost instantaneous release of pressure, gas, and heat when subjected to sudden shock, pressure, or high temperature.

"Exposure" or "exposed" means that an employee is subjected to a hazardous chemical in the course of employment through any route of entry (inhalation, ingestion, skin contact or absorption, etc.), and includes potential (e.g. accidental or possible) exposure.

"Flammable" means a chemical that falls into one of the following categories:

(i) "Aerosol, flammable" means an aerosol that, when tested by the method described in 16 CFR 1500.45, yields a flame projection exceeding 18 inches at full valve opening, or a flashback (a flame extending back to the valve) at any degree of valve opening;

(ii) "Gas, flammable" means:

(A) A gas that, at ambient temperature and pressure, forms a flammable mixture with air at a concentration of thirteen (13) percent by volume or less; or

(B) A gas that, at ambient temperature and pressure, forms flammable mixtures with air wider than twelve (12) percent by volume, regardless of the lower limit;

(iii) "Liquid, flammable" means any liquid having a flashpoint below 100 °F (37.8 °C), except any mixture having components with flashpoints of 100 °F (37.8 °C) or higher, the total of which make up 99 percent or more of the total volume of the mixture;

(iv) "Solid, flammable" means a solid, other than a blasting agent or explosive as defined in § 190.109(a), that is liable to cause fire through friction, absorption of moisture, spontaneous chemical change, or retained heat from manufacturing or processing, or which can be ignited readily and when ignited burns so vigorously and persistently as to create a serious hazard. A chemical shall be considered to be a flammable solid if, when tested by the method described in 16 CFR 1500.44, it ignites and burns with a self-sustained flame at a rate greater than one-tenth of an inch per second along its major axis.

"Flashpoint" means the minimum temperature at which a liquid gives off a vapor in sufficient concentration to ignite when tested as follows:

(i) Tagliabue Closed Tester (See American National Standard Method of Test for Flash Point by Tag Closed Tester, Z11.24–1979 (ASTM D 56–79)) for liquids with a viscosity of less than 45 Saybolt Universal Seconds (SUS) at 100 °F (37.8 °C), that do not contain suspended solids and do not have a

Federal Register / Vol. 52, No. 163 / Monday, August 24, 1987 / Rules and Regulations 31879

tendency to form a surface film under test; or

(ii) Pensky-Martens Closed Tester (See American National Standard Method of Test for Flash Point by Pensky-Martens Closed Tester, Z11.7-1979 (ASTM D 93-79)) for liquids with a viscosity equal to or greater than 45 SUS at 100 °F (37.8 °C), or that contain suspended solids, or that have a tendency to form a surface film under test; or

(iii) Setaflash Closed Tester (see American National Standard Method of Test for Flash Point by Setaflash Closed Tester (ASTMD 3278-78))

Organic peroxides, which undergo autoaccelerating thermal decomposition, are excluded from any of the flashpoint determination methods specified above.

"Foreseeable emergency" means any potential occurrence such as, but not limited to, equipment failure, rupture of containers, or failure of control equipment which could result in an uncontrolled release of a hazardous chemical into the workplace.

"Hazardous chemical" means any chemical which is a physical hazard or a health hazard.

"Hazard warning" means any words, pictures, symbols, or combination thereof appearing on a label or other appropriate form of warning which convey the hazard(s) of the chemical(s) in the container(s).

"Health hazard" means a chemical for which there is statistically significant evidence based on at least one study conducted in accordance with established scientific principles that acute or chronic health effects may occur in exposed employees. The term "health hazard" includes chemicals which are carcinogens, toxic or highly toxic agents, reproductive toxins, irritants, corrosives, sensitizers, hepatotoxins, nephrotoxins, neurotoxins, agents which act on the hematopoietic system, and agents which damage the lungs, skin, eyes, or mucous membranes. Appendix A provides further definitions and explanations of the scope of health hazards covered by this section, and Appendix B describes the criteria to be used to determine whether or not a chemical is to be considered hazardous for purposes of this standard.

"Identity" means any chemical or common name which is indicated on the material safety data sheet (MSDS) for the chemical. The identity used shall permit cross-references to be made among the required list of hazardous chemicals, the label and the MSDS.

"Immediate use" means that the hazardous chemical will be under the control of and used only by the person who transfers it from a labeled container and only within the work shift in which it is transferred.

"Importer" means the first business with employees within the Customs Territory of the United States which receives hazardous chemicals produced in other countries for the purpose of supplying them to distributors or employers within the United States.

"Label" means any written, printed, or graphic material, displayed on or affixed to containers of hazardous chemicals.

"Material safety data sheet (MSDS)" means written or printed material concerning a hazardous chemical which is prepared in accordance with paragraph (g) of this section.

"Mixture" means any combination of two or more chemicals if the combination is not, in whole or in part, the result of a chemical reaction.

"Organic peroxide" means an organic compound that contains the bivalent -O-O-structure and which may be considered to be a structural derivative of hydrogen peroxide where one or both of the hydrogen atoms has been replaced by an organic radical.

"Oxidizer" means a chemical other than a blasting agent or explosive as defined in § 1910.109(a), that initiates or promotes combustion in other materials, thereby causing fire either of itself or through the release of oxygen or other gases.

"Physical hazard" means a chemical for which there is scientifically valid evidence that it is a combustible liquid, a compressed gas, explosive, flammable, an organic peroxide, an oxidizer, pyrophoric, unstable (reactive) or water-reactive.

"Produce" means to manufacture, process, formulate, or repackage.

"Pyrophoric" means a chemical that will ignite spontaneously in air at a temperature of 130 °F (54.4 °C) or below.

"Responsible party" means someone who can provide additional information on the hazardous chemical and appropriate emergency procedures, if necessary.

"Specific chemical identity" means the chemical name, Chemical Abstracts Service (CAS) Registry Number, or any other information that reveals the precise chemical designation of the substance.

"Trade secret" means any confidential formula, pattern, process, device, information or compilation of information that is used in an employer's business, and that gives the employer an opportunity to obtain an advantage over competitors who do not know or use it. Appendix D sets out the

criteria to be used in evaluating trade secrets.

"Unstable (reactive)" means a chemical which in the pure state, or as produced or transported, will vigorously polymerize, decompose, condense, or will become self-reactive under conditions of shocks, pressure or temperature.

"Use" means to package, handle, react, or transfer.

"Water-reactive" means a chemical that reacts with water to release a gas that is either flammable or presents a health hazard.

"Work area" means a room or defined space in a workplace where hazardous chemicals are produced or used, and where employees are present.

"Workplace" means an establishment, job site, or project, at one geographical location containing one or more work areas.

(d) *Hazard determination.* (1) Chemical manufacturers and importers shall evaluate chemicals produced in their workplaces or imported by them to determine if they are hazardous. Employers are not required to evaluate chemicals unless they choose not to rely on the evaluation performed by the chemical manufacturer or importer for the chemical to satisfy this requirement.

(2) Chemical manufacturers, importers or employers evaluating chemicals shall identify and consider the available scientific evidence concerning such hazards. For health hazards, evidence which is statistically significant and which is based on at least one positive study conducted in accordance with established scientific principles is considered to be sufficient to establish a hazardous effect if the results of the study meet the definitions of health hazards in this section. Appendix A shall be consulted for the scope of health hazards covered, and Appendix B shall be consulted for the criteria to be followed with respect to the completeness of the evaluation, and the data to be reported.

(3) The chemical manufacturer, importer or employer evaluating chemicals shall treat the following sources as establishing that the chemicals listed in them are hazardous:

(i) 29 CFR Part 1910, Subpart Z, Toxic and Hazardous Substances, Occupational Safety and Health Administration (OSHA); or

(ii) *Threshold Limit Values for Chemical Substances and Physical Agents in the Work Environment,* American Conference of Governmental Industrial Hygienists (ACGIH) (latest edition)

31880   **Federal Register** / Vol. 52, No. 163 / Monday, August 24, 1987 / Rules and Regulations

The chemical manufacturer, importer, or employer is still responsible for evaluating the hazards associated with the chemicals in these source lists in accordance with the requirements of this standard.

(4) Chemical manufacturers, importers and employers evaluating chemicals shall treat the following sources as establishing that a chemical is a carcinogen or potential carcinogen for hazard communication purposes:

(i) National Toxicology Program (NTP), *Annual Report on Carcinogens* (latest edition);

(ii) International Agency for Research on Cancer (IARC) *Monographs* (latest editions); or

(iii) 29 CFR Part 1910, Subpart Z, Toxic and Hazardous Substances, Occupational Safety and Health Administration.

Note.—The *Registry of Toxic Effects of Chemical Substances* published by the National Institute for Occupational Safety and Health indicates whether a chemical has been found by NTP or IARC to be a potential carcinogen.

(5) The chemical manufacturer, importer or employer shall determine the hazards of mixtures of chemicals as follows:

(i) If a mixture has been tested as a whole to determine its hazards, the results of such testing shall be used to determine whether the mixture is hazardous;

(ii) If a mixture has not been tested as a whole to determine whether the mixture is a health hazard, the mixture shall be assumed to present the same health hazards as do the components which comprise one percent (by weight or volume) or greater of the mixture, except that the mixture shall be assumed to present a carcinogenic hazard if it contains a component in concentrations of 0.1 percent or greater which is considered to be a carcinogen under paragraph (d)(4) of this section;

(iii) If a mixture has not been tested as a whole to determine whether the mixture is a physical hazard, the chemical manufacturer, importer, or employer may use whatever scientifically valid data is available to evaluate the physical hazard potential of the mixture; and,

(iv) If the chemical manufacturer, importer, or employer has evidence to indicate that a component present in the mixture in concentrations of less than one percent (or in the case of carcinogens, less than 0.1 percent) could be released in concentrations which would exceed an established OSHA permissible exposure limit or ACGIH Threshold Limit Value, or could present

a health hazard to employees in those concentrations, the mixture shall be assumed to present the same hazard.

(6) Chemical manufacturers, importers, or employers evaluating chemicals shall describe in writing the procedures they use to determine the hazards of the chemical they evaluate. The written procedures are to be made available, upon request, to employees, their designated representatives, the Assistant Secretary and the Director. The written description may be incorporated into the written hazard communication program required under paragraph (e) of this section.

(e) *Written hazard communication program.* (1) Employers shall develop, implement, and maintain at the workplace, a written hazard communication program for their workplaces which at least describes how the criteria specified in paragraphs (f), (g), and (h) of this section for labels and other forms of warning, material safety data sheets, and employee information and training will be met, and which also includes the following:

(i) A list of the hazardous chemicals known to be present using an identity that is referenced on the appropriate material safety data sheet (the list may be compiled for the workplace as a whole or for individual work areas); and,

(ii) The methods the employer will use to inform employees of the hazards of non-routine tasks (for example, the cleaning of reactor vessels), and the hazards associated with chemicals contained in unlabeled pipes in their work areas.

(2) *Multi-employer workplaces.* Employers who produce, use, or store hazardous chemicals at a workplace in such a way that the employees of other employer(s) may be exposed (for example, employees of a construction contractor working on-site) shall additionally ensure that the hazard communication programs developed and implemented under this paragraph (e) include the following:

(i) The methods the employer will use to provide the other employer(s) with a copy of the material safety data sheet, or to make it available at a central location in the workplace, for each hazardous chemical the other employer(s)' employees may be exposed to while working;

(ii) The methods the employer will use to inform the other employer(s) of any precautionary measures that need to be taken to protect employees during the workplace's normal operating conditions and in foreseeable emergencies; and,

(iii) The methods the employer will use to inform the other employer(s) of the labeling system used in the workplace.

(3) The employer may rely on an existing hazard communication program to comply with these requirements, provided that it meets the criteria established in this paragraph (e).

(4) The employer shall make the written hazard communication program available, upon request, to employees, their designated representatives, the Assistant Secretary and the Director, in accordance with the requirements of 29 CFR 1910.20(e).

(f) *Labels and other forms of warning.* (1) The chemical manufacturer, importer, or distributor shall ensure that each container of hazardous chemicals leaving the workplace is labeled, tagged or marked with the following information:

(i) Identity of the hazardous chemical(s);

(ii) Appropriate hazard warnings; and

(iii) Name and address of the chemical manufacturer, importer, or other responsible party.

(2) For solid metal (such as a steel beam or a metal casting) that is not exempted as an article due to its downstream use, the required label may be transmitted to the customer at the time of the intial shipment, and need not be included with subsequent shipments to the same employer unless the information on the label changes. The label may be transmitted with the initial shipment itself, or with the material safety data sheet that is to be provided prior to or at the time of the first shipment. This exception to requiring labels on every container of hazardous chemicals is only for the solid metal itself and does not apply to hazardous chemicals used in conjunction with, or known to be present with, the metal and to which employees handling the metal may be exposed (for example, cutting fluids or lubricants).

(3) Chemical manufacturers, importers, or distributors shall ensure that each container of hazardous chemicals leaving the workplace is labeled, tagged, or marked in accordance with this section in a manner which does not conflict with the requirements of the Hazardous Materials Transportation Act (49 U.S.C. 1801 *et seq.*) and regulations issued under that Act by the Department of Transportation.

(4) If the hazardous chemical is regulated by OSHA in a substance-specific health standard, the chemical manufacturer, importer, distributor or employer shall ensure that the labels or

Federal Register / Vol. 52, No. 163 / Monday, August 24, 1987 / Rules and Regulations   31881

other forms of warning used are in accordance with the requirements of that standard.

(5) Except as provided in paragraphs (f)(6) and (f)(7) the employer shall ensure that each container of hazardous chemicals in the workplace is labeled, tagged or marked with the following information:

(i) Identity of the hazardous chemical(s) contained therein; and

(ii) Appropriate hazard warnings.

(6) The employer may use signs, placards, process sheets, batch tickets, operating procedures, or other such written materials in lieu of affixing labels to individual stationary process containers, as long as the alternative method identifies the containers to which it is applicable and conveys the information required by paragraph (f)(5) of this section to be on a label. The written materials shall be readily accessible to the employees in their work area throughout each work shift.

(7) The employer is not required to label portable containers into which hazardous chemicals are transferred from labeled containers, and which are intended only for the immediate use of the employee who performs the transfer.

(8) The employer shall not remove or deface existing labels on incoming containers of hazardous chemicals, unless the container is immediately marked with the required information.

(9) The employer shall ensure that labels or other forms of warning are legible, in English, and prominently displayed on the container, or readily available in the work area throughout each work shift. Employers having employees who speak other languages may add the information in their language to the material presented, as long as the information is presented in English as well.

(10) The chemical manufacturer, importer, distributor or employer need not affix new labels to comply with this section if existing labels already convey the required information.

(g) *Material safety data sheets.* (1) Chemical manufacturers and importers shall obtain or develop a material safety data sheet for each hazardous chemical they produce or import. Employers shall have a material safety data sheet for each hazardous chemical which they use.

(2) Each material safety data sheet shall be in English and shall contain at least the following information:

(i) The identity used on the label, and, except as provided for in paragraph (i) of this section on trade secrets:

(A) If the hazardous chemical is a single substance, its chemical and common name(s);

(B) If the hazardous chemical is a mixture which has been tested as a whole to determine its hazards, the chemical and common name(s) of the ingredients which contribute to these known hazards, and the common name(s) of the mixture itself; or,

(C) If the hazardous chemical is a mixture which has not been tested as a whole:

(1) The chemical and common name(s) of all ingredients which have been determined to be health hazards, and which comprise 1% or greater of the composition, except that chemicals identified as carcinogens under paragraph (d)(4) of this section shall be listed if the concentrations are 0.1% or greater; and,

(2) The chemical and common name(s) of all ingredients which have been determined to be health hazards, and which comprise less than 1% (0.1% for carcinogens) of the mixture, if there is evidence that the ingredient(s) could be released from the mixture in concentrations which would exceed an established OSHA permissible exposure limit or ACGIH Threshold Limit Value, or could present a health hazard to employees; and,

(3) The chemical and common name(s) of all ingredients which have been determined to present a physical hazard when present in the mixture;

(ii) Physical and chemical characteristics of the hazardous chemical (such as vapor pressure, flash point);

(iii) The physical hazards of the hazardous chemical, including the potential for fire, explosion, and reactivity;

(iv) The health hazards of the hazardous chemical, including signs and symptoms of exposure, and any medical conditions which are generally recognized as being aggravated by exposure to the chemical;

(v) The primary route(s) of entry;

(vi) The OSHA permissible exposure limit, ACGIH Threshold Limit Value, and any other exposure limit used or recommended by the chemical manufacturer, importer, or employer preparing the material safety data sheet, where available;

(vii) Whether the hazardous chemical is listed in the National Toxicology Program (NTP) *Annual Report on Carcinogens* (latest edition) or has been found to be a potential carcinogen in the International Agency for Research on Cancer (IARC) *Monographs* (latest editions), or by OSHA;

(viii) Any generally applicable precautions for safe handling and use which are known to the chemical manufacturer, importer or employer

preparing the material safety data sheet, including appropriate hygienic practices, protective measures during repair and maintenance of contaminated equipment, and procedures for clean-up of spills and leaks;

(ix) Any generally applicable control measures which are known to the chemical manufacturer, importer or employer preparing the material safety data sheet, such as appropriate engineering controls, work practices, or personal protective equipment;

(x) Emergency and first aid procedures;

(xi) The date of preparation of the material safety data sheet or the last change to it; and,

(xii) The name, address and telephone number of the chemical manufacturer, importer, employer or other responsible party preparing or distributing the material safety data sheet, who can provide additional information on the hazardous chemical and appropriate emergency procedures, if necessary.

(3) If not relevant information is found for any given category on the material safety data sheet, the chemical manufacturer, importer or employer preparing the material safety data sheet shall mark it to indicate that no applicable information was found.

(4) Where complex mixtures have similar hazards and contents (i.e. the chemical ingredients are essentially the same, but the specific composition varies from mixture to mixture), the chemical manufacturer, importer or employer may prepare one material safety data sheet to apply to all of these similar mixtures.

(5) The chemical manufacturer, importer or employer preparing the material safety data sheet shall ensure that the information recorded accurately reflects the scientific evidence used in making the hazard determination. If the chemical manufacturer, importer or employer preparing the material safety data sheet becomes newly aware of any significant information regarding the hazards of a chemical, or ways to protect against the hazards, this new information shall be added to the material safety data sheet within three months. If the chemical is not currently being produced or imported the chemical manufacturer or importer shall add the information to the material safety data sheet before the chemical is introduced into the workplace again.

(6) Chemical manufacturers or importers shall ensure that distributors and employers are provided an appropriate material safety data sheet with their intitial shipment, and with the first shipment after a material safety

31882 Federal Register / Vol. 52, No. 163 / Monday, August 24, 1987 / Rules and Regulations

data sheet is updated. The chemical manufacturer or importer shall either provide material safety data sheets with the shipped containers or send them to the employer prior to or at the time of the shipment. If the material safety data sheet is not provided with a shipment that has been labeled as a hazardous chemical, the employer shall obtain one from the chemical manufacturer, importer, or distributor as soon as possible.

(7) Distributors shall ensure that material safety data sheets, and updated information, are provided to other distributors and employers. Retail distributors which sell hazardous chemicals to commercial customers shall provide a material safety data sheet to such employers upon request, and shall post a sign or otherwise inform them that a material safety data sheet is available. Chemical manufacturers, importers, and distributors need not provide material safety data sheets to retail distributors which have informed them that the retail distributor does not sell the product to commercial customers or open the sealed container to use it in their own workplaces.

(8) The employer shall maintain copies of the required material safety data sheets for each hazardous chemical in the workplace, and shall ensure that they are readily accessible during each work shift to employees when they are in their work area(s).

(9) Where employees must travel between workplaces during a workshift, i.e., their work is carried out at more than one geographical location, the material safety data sheets may be kept at a central location at the primary workplace facility. In this situation, the employer shall ensure that employees can immediately obtain the required information in an emergency.

(10) Material safety data sheets may be kept in any form, including operating procedures, and may be designed to cover groups of hazardous chemicals in a work area where it may be more appropriate to address the hazards of a process rather than individual hazardous chemicals. However, the employer shall ensure that in all cases the required information is provided for each hazardous chemical, and is readily accessible during each work shift to employees when they are in in their work areas(s).

(11) Material safety data sheets shall also be made readily available, upon request, to designated representatives and to the Assistant Secretary, in accordance with the requirements of 29 CFR 1910.20 (e). The Director shall also be given access to material safety data sheets in the same manner.

(h) *Employee information and training.* Employers shall provide employees with information and training on hazardous chemicals in their work area at the time of their initial assignment, and whenever a new hazard is introduced into their work area.

(1) *Information.* Employees shall be informed of:

(i) The requirements of this section;

(ii) Any operations in their work area where hazardous chemicals are present; and,

(iii) The location and availability of the written hazard communication program, including the required list(s) of hazardous chemicals, and material safety data sheets required by this section.

(2) *Training.* Employee training shall include at least:

(i) Methods and observations that may be used to detect the presence or release of a hazardous chemical in the work area (such as monitoring conducted by the employer, continuous monitoring devices, visual appearance or odor of hazardous chemicals when being released, etc.);

(ii) The physical and health hazards of the chemicals in the work area;

(iii) The measures employees can take to protect themselves from these hazards, including specific procedures the employer has implemented to protect employees from exposure to hazardous chemicals, such as appropriate work practices, emergency procedures, and personal protective equipment to be used; and,

(iv) The details of the hazard communication program developed by the employer, including an explanation of the labeling system and the material safety data sheet, and how employees can obtain and use the appropriate hazard information.

(i) *Trade secrets.* (1) The chemical manufacturer, importer, or employer may withhold the specific chemical identity, including the chemical name and other specific identification of a hazardous chemical, from the material safety data sheet, provided that:

(i) The claim that the information withheld is a trade secret can be supported;

(ii) Information contained in the material safety data sheet concerning the properties and effects of the hazardous chemical is disclosed;

(iii) The material safety data sheet indicates that the specific chemical identity is being withheld as a trade secret; and,

(iv) The specific chemical identity is made available to health professionals, employees, and designated

representatives in accordance with the applicable provisions of this paragraph.

(2) Where a treating physician or nurse determines that a medical emergency exists and the specific chemical identity of a hazardous chemical is necessary for emergency or first-aid treatment, the chemical manufacturer, importer, or employer shall immediately disclose the specific chemical identity of a trade secret chemical to that treating physician or nurse, regardless of the existence of a written statement of need of a confidentiality agreement. The chemical manufacturer, importer, or employer may require a written statement of need and confidentiality agreement, in accordance with the provisions of paragraphs (i)(3) and (4) of this section, as soon as circumstances permit.

(3) In non-emergency situations, a chemical manufacturer, importer, or employer shall, upon request, disclose a specific chemical identity, otherwise permitted to be withheld under paragraph (i)(1) of this section, to a health professional (i.e. physician, industrial hygienist, toxicologist, epidemiologist, or occupational health nurse) providing medical or other occupational health services to exposed employee(s), and to employees or designated representatives, if:

(i) The request is in writing;

(ii) The request describes with reasonable detail one or more of the following occupational health needs for the information:

(A) To assess the hazards of the chemicals to which employees will be exposed;

(B) To conduct or assess sampling of the workplace atmosphere to determine employee exposure levels;

(C) To conduct pre-assignment or periodic medical surveillance of exposed employees;

(D) To provide medical treatment to exposed employees;

(E) To select or assess appropriate personal protective equipment for exposed employees;

(F) To design or assess engineering controls or other protective measures for exposed employees; and,

(G) To conduct studies to determine the health effects of exposure.

(iii) The request explains in detail why the disclosure of the specific chemical identity is essential and that, in lieu thereof, the disclosure of the following information to the health professional, employee, or designated representative, would not satisfy the purposes described in paragraph (i)(3)(ii) of this section:

(A) The properties and effects of the chemical;

(B) Measures for controlling workers' exposure to the chemical;

(C) Methods of monitoring and analyzing worker exposure to the chemical; and,

(D) Methods of diagnosing and treating harmful exposures to the chemical;

(iv) The request includes a description of the procedures to be used to maintain the confidentiality of the disclosed information; and,

(v) The health professional, and the employer or contractor of the services of the health professional (i.e. downstream employer, labor organization, or individual employee), employee, or designated representative, agree in a written confidentiality agreement that the health professional, employee, or designated representative, will not use the trade secret information for any purpose other than the health need(s) asserted and agree not to release the information under any circumstances other than to OSHA, as provided in paragraph (i)(6) of this section, except as authorized by the terms of the agreement or by the chemical manufacturer, importer, or employer.

(4) The confidentiality agreement authorized by paragraph (i)(3)(iv) of this section:

(i) May restrict the use of the information to the health purposes indicated in the written statement of need;

(ii) May provide for appropriate legal remedies in the event of a breach of the agreement, including stipulation of a reasonable pre-estimate of likely damages; and,

(iii) May not include requirements for the posting of a penalty bond.

(5) Nothing in this standard is meant to preclude the parties from pursuing non-contractual remedies to the extent permitted by law.

(6) If the health professional, employee, or designated representative receiving the trade secret information decides that there is a need to disclose it to OSHA, the chemical manufacturer, importer, or employer who provided the information shall be informed by the health professional, employee, or designated representative prior to, or at the same time as, such disclosure.

(7) If the chemical manufacturer, importer, or employer denies a written request for disclosure of a specific chemical identity, the denial must:

(i) Be provided to the health professional, employee, or designated representative, within thirty days of the request;

(ii) Be in writing;

(iii) Include evidence to support the claim that the specific chemical identity is a trade secret;

(iv) State the specific reasons why the request is being denied; and,

(v) Explain in detail how alternative information may satisfy the specific medical or occupational health need without revealing the specific chemical identity.

(8) The health professional, employee, or designated representative whose request for information is denied under paragraph (i)(3) of this section may refer the request and the written denial of the request to OSHA for consideration.

(9) When a health professional, employee, or designated representative refers the denial to OSHA under paragraph (i)(8) of this section, OSHA shall consider the evidence to determine if:

(i) The chemical manufacturer, importer, or employer has supported the claim that the specific chemical identity is a trade secret;

(ii) The health professional, employee, or designated representative has supported the claim that there is a medical or occupational health need for the information; and,

(iii) The health professional, employee, or designated representative has demonstrated adequate means to protect the confidentiality.

(10)(i) If OSHA determines that the specific chemical identity requested under paragraph (i)(3) of this section is not a *bona fide* trade secret, or that it is a trade secret, but the requesting health professional, employee, or designated representative has a legitimate medical or occupational health need for the information, has executed a written confidentiality agreement, and has shown adequate means to protect the confidentiality of the information, the chemical manufacturer, importer, or employer will be subject to citation by OSHA.

(ii) If a chemical manufacturer, importer, or employer demonstrates to OSHA that the execution of a confidentiality agreement would not provide sufficient protection against the potential harm from the unauthorized disclosure of a trade secret specific chemical identity, the Assistant Secretary may issue such orders or impose such additional limitations or conditions upon the disclosure of the requested chemical information as may be appropriate to assure that the occupational health services are provided without an undue risk of harm to the chemical manufacturer, importer, or employer.

(11) If a citation for a failure to release specific chemical identity information is

contested by the chemical manufacturer, importer, or employer, the matter will be adjudicated before the Occupational Safety and Health Review Commission in accordance with the Act's enforcement scheme and the applicable Commission rules of procedure. In accordance with the Commission rules, when a chemical manufacturer, importer, or employer continues to withhold the information during the contest, the Administrative Law Judge may review the citation and supporting documentation *in camera* or issue appropriate orders to protect the confidentiality or such matters.

(12) Notwithstanding the existence of a trade secret claim, a chemical manufacturer, importer, or employer shall, upon request, disclose to the Assistant Secretary any information which this section requires the chemical manufacturer, importer, or employer to make available. Where there is a trade secret claim, such claim shall be made no later than at the time the information is provided to the Assistant Secretary so that suitable determinations of trade secret status can be made and the necessary protections can be implemented.

(13) Nothing in this paragraph shall be construed as requiring the disclosure under any circumstances of process or percentage of mixture information which is a trade secret.

(j) *Effective dates.* (1) Chemical manufacturers, importers, and distributors shall ensure that material safety data sheets are provided with the next shipment of hazardous chemicals to employers after September 23, 1987.

(2) Employers in the non-manufacturing sector shall be in compliance with all provisions of this section by May 23, 1988. (Note: Employers in the manufacturing sector (SIC Codes 20 through 39) are already required to be in compliance with this section.)

Appendix A to § ―――― Health Hazard Definitions (*Mandatory*)

Although safety hazards related to the physical characteristics of a chemical can be objectively defined in terms of testing requirements (e.g. flammability), health hazard definitions are less precise and more subjective. Health hazards may cause measurable changes in the body—such as decreased pulmonary function. These changes are generally indicated by the occurrence of signs and symptoms in the exposed employees—such as shortness of breath, a non-measurable, subjective feeling. Employees exposed to such hazards must be apprised of both the change in body function and the signs and symptoms that may occur to signal that change.

31884   Federal Register / Vol. 52, No. 163 / Monday, August 24, 1987 / Rules and Regulations

The determination of occupational health hazards is complicated by the fact that many of the effects or signs and symptoms occur commonly in non-occupationally exposed populations, so that effects of exposure are difficult to separate from normally occurring illnesses. Occasionally, a substance causes an effect that is rarely seen in the population at large, such as angiosarcomas caused by vinyl chloride exposure, thus making it easier to ascertain that the occupational exposure was the primary causative factor. More often, however, the effects are common, such as lung cancer. The situation is further complicated by the fact that most chemicals have not been adequately tested to determine their health hazard potential, and data do not exist to substantiate these effects.

There have been many attempts to categorize effects and to define them in various ways. Generally, the terms "acute" and "chronic" are used to delineate between effects on the basis of severity or duration. "Acute" effects usually occur rapidly as a result of short-term exposures, and are of short duration. "Chronic" effects generally occur as a result of long-term exposure, and are of long duration.

The acute effects referred to most frequently are those defined by the American National Standards Institute (ANSI) standard for Precautionary Labeling of Hazardous Industrial Chemicals (Z129.1–1982)— irritation, corrosivity, sensitization and lethal dose. Although these are important health effects, they do not adequately cover the considerable range of acute effects which may occur as a result of occupational exposure, such as, for example, narcosis. Similarly, the term chronic effect is often used to cover only carcinogenicity, teratogenicity, and mutagenicity. These effects are obviously a concern in the workplace, but again, do not adequately cover the area of chronic effects, excluding, for example, blood dyscrasias (such as anemia), chronic bronchitis and liver atrophy.

The goal of defining precisely, in measurable terms, every possible health effect that may occur in the workplace as a result of chemical exposures cannot realistically be accomplished. This does not negate the need for employees to be informed of such effects and protected from them. Appendix B, which is also mandatory, outlines the principles and procedures of hazardous assessment.

For purposes of this section, any chemicals which meet any of the following definitions, as determined by the criteria set forth in Appendix B are health hazards:

1. *Carcinogen:* A chemical is considered to be a carcinogen if:

(a) It has been evaluated by the International Agency for Research on Cancer (IARC), and found to be a carcinogen or potential carcinogen; or

(b) It is listed as a carcinogen or potential carcinogen in the *Annual Report on Carcinogens* published by the National Toxicology Program (NTP) (latest edition); or,

(c) It is regulated by OSHA as a carcinogen.

2. *Corrosive:* A chemical that causes visible destruction of, or irreversible alterations in, living tissue by chemical action at the site of contact. For example, a chemical is considered to be corrosive if, when tested on the intact skin of albino rabbits by the method described by the U.S. Department of Transportation in Appendix A to 49 CFR Part 173, it destroys or changes irreversibly the structure of the tissue at the site of contact following an exposure period of four hours. This term shall not refer to action on inanimate surfaces.

3. *Highly toxic:* A chemical falling within any of the following categories:

(a) A chemical that has a median lethal dose (LD$_{50}$) of 50 milligrams or less per kilogram of body weight when administered orally to albino rats weighing between 200 and 300 grams each.

(b) A chemical that has a median lethal dose (LD$_{50}$) of 200 milligrams or less per kilogram of body weight when administered by continuous contact for 24 hours (or less if death occurs within 24 hours) with the bare skin of albino rabbits weighing between two and three kilograms each.

(c) A chemical that has a median lethal concentration (LC$_{50}$) in air of 200 parts per million by volume or less of gas or vapor, or 2 milligrams per liter or less of mist, fume, or dust, when administered by continuous inhalation for one hour (or less if death occurs within one hour) to albino rats weighing between 200 and 300 grams each.

4. *Irritant:* A chemical, which is not corrosive, but which causes a reversible inflammatory effect on living tissue by chemical action at the site of contact. A chemical is a skin irritant if, when tested on the intact skin of albino rabbits by the methods of 16 CFR 1500.41 for four hours exposure or by other appropriate techniques, it results in an empirical score of five or more. A chemical is an eye irritant if so determined under the procedure listed in 16 CFR 1500.42 or other appropriate techniques.

5. *Sensitizer:* A chemical that causes a substantial proportion of exposed people or animals to develop an allergic reaction in normal tissue after repeated exposure to the chemical.

6. *Toxic.* A chemical falling within any of the following categories:

(a) A chemical that has a median lethal dose (LD$_{50}$) of more than 50 milligrams per kilogram but not more than 500 milligrams per kilogram of body weight when administered orally to albino rats weighing between 200 and 300 grams each.

(b) A chemical that has a median lethal dose (LD$_{50}$) of more than 200 milligrams per kilogram but not more than 1,000 milligrams per kilogram of body weight when administered by continuous contact for 24 hours (or less if death occurs within 24 hours) with the bare skin of albino rabbits weighing between two and three kilograms each.

(c) A chemical that has a median lethal concentration (LC$_{50}$) in air of more than 200 parts per million but not more than 2,000 parts per million by volume of gas or vapor, or more than two milligrams per liter but not more than 20 milligrams per liter of mist,

fume, or dust, when administered by continuous inhalation for one hour (or less if death occurs within one hour) to albino rats weighing between 200 and 300 grams each.

7. *Target organ effects.* The following is a target organ categorization of effects which may occur, including examples of signs and symptoms and chemicals which have been found to cause such effects. These examples are presented to illustrate the range and diversity of effects and hazards found in the workplace, and the broad scope employers must consider in this area, but are not intended to be all-inclusive.

a. Hepatotoxins: Chemicals which produce liver damage

Signs & Symptoms: Jaundice; liver enlargement

Chemicals: Carbon tetrachloride; nitrosamines

b. Nephrotoxins: Chemicals which produce kidney damage

Signs & Symptoms: Edema; proteinuria

Chemicals: Halogenated hydrocarbons; uranium

c. Neurotoxins: Chemicals which produce their primary toxic effects on the nervous system

Signs & Symptoms: Narcosis; behavioral changes; decrease in motor functions

Chemicals: Mercury; carbon disulfide

d. Agents which act on the blood or hematopoietic system: Decrease hemoglobin function; deprive the body tissues of oxygen

Signs & Symptoms: Cyanosis; loss of consciousness

Chemicals: Carbon monoxide; cyanides

e. Agents which damage the lung: Chemicals which irritate or damage the pulmonary tissue

Signs & Symptoms: Cough; tightness in chest; shortness of breath

Chemicals: Silica; asbestos

f. Reproductive toxins: Chemicals which affect the reproductive capabilities including chromosomal damage (mutations) and effects on fetuses (teratogenesis)

Signs & Symptoms: Birth defects; sterility

Chemicals: Lead; DBCP

g. Cutaneous hazards: Chemicals which affect the dermal layer of the body

Signs & Symptoms: Defatting of the skin; rashes; irritation

Chemicals: Ketones; chlorinated compounds

h. Eye hazards: Chemicals which affect the eye or visual capacity

Signs & Symptoms: Conjunctivitis; corneal damage

Chemicals: Organic solvents; acids

Appendix B to § ———, Hazard Determination (Mandatory)

The quality of a hazard communication program is largely dependent upon the adequacy and accuracy of the hazard determination. The hazard determination requirement of this standard is performance-oriented. Chemical manufacturers, importers, and employers evaluating chemicals are not required to follow any specific methods for determining hazards, but they must be able to demonstrate that they have adequately ascertained the hazards of the chemicals produced or imported in accordance with the criteria set forth in this Appendix.

Hazard evaluation is a process which relies heavily on the professional judgment of the evaluator, particularly in the area of chronic hazards. The performance-orientation of the hazard determination does not diminish the duty of the chemical manufacturer, importer or employer to conduct a thorough evaluation, examining all relevant data and producing a scientifically defensible evaluation. For purposes of this standard, the following criteria shall be used in making hazard determinations that meet the requirements of this standard.

1. *Carcinogenicity*: As described in paragraph (d)(4) and Appendix A of this section, a determination by the National Toxicology Program, the International Agency for Research on Cancer, or OSHA that a chemical is a carcinogen or potential carcinogen will be considered conclusive evidence for purposes of this section.

2. *Human data*: Where available, epidemiological studies and case reports of adverse health effects shall be considered in the evaluation.

3. *Animal data*: Human evidence of health effects in exposed populations is generally not available for the majority of chemicals produced or used in the workplace. Therefore, the available results of toxicological testing in animal populations shall be used to predict the health effects that may be experienced by exposed workers. In particular, the definitions of certain acute hazards refer to specific animal testing results (see Appendix A).

4. *Adequacy and reporting of data*. The results of any studies which are designed and conducted according to established scientific principles, and which report statistically significant conclusions regarding the health effects of a chemical, shall be a sufficient basis for a hazard determination and reported on any material safety data sheet. The chemical manufacturer, importer, or employer may also report the results of other scientifically valid studies which tend to refute the findings of hazard.

Appendix C to §— Information Sources (Advisory)

The following is a list of available data sources which the chemical manufacturer, importer, distributor, or employer may wish to consult to evaluate the hazards of chemicals they produce or import:

—Any information in their own company files, such as toxicity testing results or illness experience of company employees.

—Any information obtained from the supplier of the chemical, such as material safety data sheets or product safety bulletins.

—Any pertinent information obtained from the following source list (latest editions should be used):

Condensed Chemical Dictionary
Van Nostrand Reinhold Co., 135 West 50th Street. New York, NY 10020.
The Merck Index: An Encyclopedia of Chemicals and Drugs
Merck and Company, Inc., 126 E. Lincoln Ave., Rahway, NJ 07065.

IARC Monographs on the Evaluation of the Carcinogenic Risk of Chemicals to Man
Geneva: World Health Organization, International Agency for Research on Cancer, 1972–Present. (Multivolume work). Summaries are available in supplement volumes. 49 Sheridan Street, Albany, NY 12210.
Industrial Hygiene and Toxicology, by F.A. Patty
John Wiley & Sons, Inc., New York, NY (Multivolume work).
Clinical Toxicology of Commercial Products
Gleason, Gosselin, and Hodge
Casarett and Doull's Toxicology; The Basic Science of Poisons
Doull, Klaassen, and Amdur, Macmillan Publishing Co., Inc., New York, NY.
Industrial Toxicology, by Alice Hamilton and Harriet L. Hardy
Publishing Sciences Group, Inc., Acton, MA.
Toxicology of the Eye, by W. Morton Grant
Charles C. Thomas, 301–327 East Lawrence Avenue, Springfield, IL.
Recognition of Health Hazards in Industry
William A. Burgess, John Wiley and Sons, 605 Third Avenue, New York, NY 10158.
Chemical Hazards of the Workplace
Nick H. Proctor and James P. Hughes, J.P. Lipincott Company, 6 Winchester Terrace, New York, NY 10022.
Handbook of Chemistry and Physics
Chemical Rubber Company, 18901 Cranwood Parkway, Cleveland. OH 44128.
Threshold Limit Values for Chemical Substances and Physical Agents in the Work Environment and Biological Exposure Indices with Intended Changes
American Conference of Governmental Industrial Hygienists (ACGIH). 6500 Glenway Avenue, Bldg. D–5, Cincinnati, OH 45211.
Information on the physical hazards of chemicals may be found in publications of the National Fire Protection Association, Boston, MA.

Note.—The following documents may be purchased from the Superintendent of Documents, U.S. Government Printing Office, Washington, DC 20402.

Occupational Health Guidelines
NIOSH/OSHA (NIOSH Pub. No. 81–123]
NIOSH Pocket Guide to Chemical Hazards
NIOSH Pub. No. 85–114
Registry of Toxic Effects of Chemical Substances
NIOSH Pub. No. 80–102
Miscellaneous Documents published by the National Institute for Occupational Safety and Health:
Criteria documents.
Special Hazard Reviews.
Occupational Hazard Assessments.
Current Intelligence Bulletins.
OSHA's General Industry Standards [29 CFR Part 1910]
NTP Annual Report on Carcinogens and Summary of the Annual Report on Carcinogens.
National Technical Information Service (NTIS], 5285 Port Royal Road, Springfield, VA 22161; (703) 487–4650.

BIBLIOGRAPHIC DATA BASES

Service provider	File name
Bibliographic Retrieval Services (BRS), 1200 Route 7, Latham, NY 12110.	Biosis Previews
	CA Search
	Medlars
	NTIS
	Hazardline
	American Chemcal Society Journal
	Excerpta Medica
	IRCS Medical Science Journal
Lockheed—DIALOG Information Service, Inc., 3460 Hillview Avenue, Palo Alto, CA 94304.	Biosis Prev. Files
	CAS Files
	CA Search Files
	CAB Abstracts
	Chemical Exposure
	Chemname
	Chemsis Files
	Chemzero
	Embase Files
	Environmental Bibliographies
	Enviroline
	Federal Research in Progress
	IRL Life Science Collection
	NTIS
	Occupational Safety and Health (NIOSH)
	Paper Chem
SDC—Orbit, SDC Information Service, 2500 Colorado Avenue, Santa Monica, CA 90406.	Chemdex, 2, 3
	NTIS
National Library of Medicine, Department of Health and Human Services, Public Health Service, National Institutes of Health, Bethesda, MD 20209.	Hazardous Substances Data Bank (NSDB)
	Medline files
	Toxline Files
	Cancerlit
	RTECS
	Chemline
Pergamon International Information Corp., 1340 Old Chain Bridge Rd., McLean, VA 22101.	Laboratory Hazard Bulletin
Questel, Inc., 1625 Eye Street, NW., Suite 818, Washington, DC 20006.	CIS/ILO
	Cancernet
Chemical Information System ICI (ICIS), Bureau of National Affairs, 1133 15th Street, NW., Suite 300, Washington, DC 20005.	Structure and Nomenclature Search System (SANSS)
	Acute Toxicity (RTECS)
	Clinical Toxicology of Commercial Products
	Oil and Hazardous Materials Technical Assistance Data System
	CCRIS
	CESARS
Occupational Health Services, 400 Plaza Drive, Secaucus, NJ 07094.	MSDS
	Hazardline

Appendix D to §———Definition of "Trade Secret" (Mandatory)

The following is a reprint of the *Restatement of Torts* section 757, comment *b* (1939):

b. Definition of trade secret. A trade secret may consist of any formula, pattern, device or compilation of information which is used in one's business, and which gives him an opportunity to obtain an advantage over competitors who do not know or use it. It may be a formula for a chemical compound, a process of manufacturing, treating or preserving materials, a pattern for a machine or other device, or a list of customers. It differs from other secret information in a business (see § 759 of the *Restatement of Torts* which is not included in this Appendix] in that it is not simply information as to single or ephemeral events in the conduct of the business, as, for example, the amount or

31886 **Federal Register** / Vol. 52, No. 163 / Monday, August 24, 1987 / Rules and Regulations

other terms of a secret bid for a contract or the salary of certain employees, or the security investments made or contemplated, or the date fixed for the announcement of a new policy or for bringing out a new model or the like. A trade secret is a process or device for continuous use in the operations of the business. Generally it relates to the production of goods, as, for example, a machine or formula for the production of an article. It may, however, relate to the sale of goods or to other operations in the business, such as a code for determining discounts, rebates or other concessions in a price list or catalogue, or a list of specialized customers, or a method of bookkeeping or other office management.

Secrecy. The subject matter of a trade secret must be secret. Matters of public knowledge or of general knowledge in an industry cannot be appropriated by one as his secret. Matters which are completely disclosed by the goods which one markets cannot be his secret. Substantially, a trade secret is known only in the particular business in which it is used. It is not requisite that only the proprietor of the business know it. He may, without losing his protection, communicate it to employees involved in its use. He may likewise communicate it to others pledged to secrecy. Others may also know of it independently, as, for example, when they have discovered the process or formula by independent invention and are keeping it secret. Nevertheless, a substantial element of secrecy must exist, so that, except by the use of improper means, there would be difficulty in acquiring the information. An exact definition of a trade secret is not possible. Some factors to be considered in determining whether given information is one's trade secret are: (1) The extent to which the information is known outside of his business; (2) the extent to which it is known by employees and others involved in his business; (3) the extent of measures taken by him to guard the secrecy of the information; (4) the value of the information to him and his competitors; (5) the amount of effort or money expended by him in developing the information; (6) the ease or difficulty with which the information could be properly acquired or duplicated by others.

Novelty and prior art. A trade secret may be a device or process which is patentable; but it need not be that. It may be a device or process which is clearly anticipated in the prior art or one which is merely a mechanical improvement that a good mechanic can make. Novelty and invention are not requisite for a trade secret as they are for patentability. These requirements are essential to patentability because a patent protects against unlicensed use of the patented device or process even by one who discovers it properly through independent research. The patent monopoly is a reward to the inventor. But such is not the case with a trade secret. Its protection is not based on a policy of rewarding or otherwise encouraging the development of secret processes or devices. The protection is merely against breach of faith and reprehensible means of learning another's secret. For this limited protection it is not appropriate to require also the kind of novelty and invention which is a requisite of patentability. The nature of the secret is, however, an important factor in determining the kind of relief that is appropriate against one who is subject to liability under the rule stated in this section. Thus, if the secret consists of a device or process which is a novel invention, one who acquires the secret wrongfully is ordinarily enjoined from further use of it and is required to account for the profits derived from his past use. If, on the other hand, the secret consists of mechanical improvements that a good mechanic can make without resort to the secret, the wrongdoer's liability may be limited to damages, and an injunction against future use of the improvements made with the aid of the secret may be inappropriate.

8. Section 1915.97 would be revised to read as follows:

§ 1915.97 Health and sanitation.

The provisions of this section shall apply to ship repairing, shipbuilding and shipbreaking, except where indicated otherwise.

(a) The employer shall provide all necessary controls, and the employees shall be protected by suitable personal protective equipment against the hazards identified under § 1915.99 of this part and those hazards for which specific precautions are required in Subparts B, C, and D of this part.

(b) The employer shall provide adequate washing facilities for employees engaged in the application of paints or coatings or in other operations where contaminants can, by ingestion or absorption, be detrimental to the health of the employees. The employer shall encourage good personal hygiene practices by informing the employees of the need for removing surface contaminants by thorough washing or hands and face prior to eating or smoking.

(c) The employer shall not permit employees to eat or smoke in areas undergoing surface preparation or preservation or where shipbreaking operations produce atmospheric contaminants.

(d) The employer shall not permit employees engaged in ship repair work on a vessel to work in the immediate vicinity of uncovered garbage and shall ensure that employees working beneath or on the outboard side of a vessel are not subject to contamination by drainage or waste from overboard discharges.

(e) No minor under 18 years of age shall be employed in shipbreaking or related employments.

9. Section 1928.21 would be amended by adding paragraph (a)(5) as follows:

§ 1928.21 Applicable standards in 29 CFR Part 1910.

(a) * * *

(5) Hazard communication— § 1910.1200.

* * * * *

[FR Doc. 87–19137 Filed 8–19–87; 8:45 am]
BILLING CODE 4510-26-M

APPENDIX J
29 CFR PART 1910:
HAZARDOUS WASTE OPERATIONS
AND EMERGENCY RESPONSE;
FINAL RULE

Monday
March 6, 1989

federal register

Part III

Department of Labor

Occupational Safety and Health
Administration

29 CFR Part 1910
Hazardous Waste Operations and
Emergency Response; Final Rule

9294   Federal Register / Vol. 54, No. 42 / Monday, March 6, 1989 / Rules and Regulations

DEPARTMENT OF LABOR

Occupational Safety and Health Administration

29 CFR Part 1910

[Docket No. S-760A]

Hazardous Waste Operations and Emergency Response

AGENCY: Occupational Safety and Health Administration; Labor.

ACTION: Final rule.

SUMMARY: The Occupational Safety and Health Administration (OSHA) is amending the OSHA standard for hazardous waste operations and emergency response found in 29 CFR 1910.120. This final rule will replace the existing interim final rule required by Congress in the Superfund Amendments and Reauthorization Act of 1986 (as amended) (SARA) (Pub. L. 99–499, 29 U.S.C. 655 note). When this final rule becomes effective one year from today, the interim final rule promulgated December 19, 1986 (51 FR 45654) will be revoked. The interim final rule remains in effect until then. The Notice of Proposed Rulemaking for this final rule was published in the **Federal Register** on August 10, 1987 (52 FR 29620).

This rule will regulate the safety and health of employees involved in clean-up operations at uncontrolled hazardous waste sites being cleaned-up under government mandate, in certain hazardous waste treatment, storage, and disposal (TSD) operations conducted under the Resource, Conservation and Recovery Act of 1976 as amended (RCRA) [42 U.S.C. 6901 et seq], and in any emergency response to incidents involving hazardous substances.

This standard provides for employee protection during initial site characterization and analysis, monitoring activities, materials handling activities, training, and emergency response.

DATES: This final rule will become effective March 6, 1990.

Paperwork authorization has been granted by the Office of Management and Budget (OMB) under control number 1218–0139.

ADDRESS: In compliance with 28 U.S.C. 2112(a), the Agency designates for receipt of petitions for review of the standard, the Associate Solicitor for Occupational Safety and Health, Office of the Solicitor, Room S-4004, U.S. Department of Labor, 200 Constitution Avenue NW., Washington, DC 20210.

FOR FURTHER INFORMATION CONTACT: Mr. James F. Foster, U.S. Department of

Labor, Occupational Safety and Health Administration, Division of Consumer Affairs, Room N–3647, 200 Constitution Avenue NW., Washington, DC 20210, 202–523–8151.

SUPPLEMENTARY INFORMATION:

I. Background

The U.S. Environmental Protection Agency estimates that approximately 57 million metric tons of hazardous waste are produced each year in the United States.[1] These wastes must be treated and stored or disposed in a manner that protects the environment from the adverse affects of the various constituents of those wastes.

In response to the need to protect the environment from the improper disposal of these hazardous wastes, Congress, over the years, has enacted several pieces of legislation intended to control the nation's hazardous waste problem. Federal laws passed in 1965[2] and 1970[3] initially addressed solid waste disposal. Several other pieces of legislation have been enacted by Congress that have ultimately led to the development of this rule and they are discussed below.

A. The Resource Conservation and Recovery Act of 1976

The first comprehensive, federal effort to deal with the solid waste problem in general, and hazardous waste specifically, came with the passage of the Resource Conservation and Recovery Act of 1976 (RCRA)[4]. The act provides for the development of federal and state programs for otherwise unregulated land disposal of waste materials and for the development of resource recovery programs. It regulates anyone engaged in the creation, transportation, treatment, and disposal of "hazardous wastes." It also regulates facilities for the disposal of all solid wastes and prohibits the use of open dumps for solid wastes in favor of requiring sanitary landfills.

There are, however, many hazardous waste disposal sites that were created prior to the passage of RCRA. These sites are often abandoned and contain unknown quantities of unknown wastes.

B. The Comprehensive Environmental Response, Compensation and Liability Act of 1980

In response to the need to clean-up and properly reclaim these pre-RCRA

sites, Congress enacted the Comprehensive Environmental Response, Compensation, and Liability Act of 1980 (CERCLA)[5] commonly known as "Superfund." Superfund established two related funds to be used for the immediate removal of hazardous substances released into the environment. Superfund is intended to establish a mechanism of response for the immediate clean-up of hazardous waste contamination from accidental spills and from chronic environmental damage such as is associated with abandoned hazardous waste disposal sites.

The treatment and disposal of hazardous wastes under RCRA and CERCLA creates a significant risk to the safety and health of employees who work in treatment and disposal operations. Exposure to hazardous wastes through skin contact, skin absorption, and inhalation pose the most significant risks to employees. Employee exposure to these risks occurs when employees respond to hazardous substance or waste emergencies, when they work with hazardous wastes during storage, treatment and disposal operations or when they participate in the clean-up of abandoned-waste sites.

This risk of exposure and the need for protecting employees exposed to hazardous wastes is addressed in the "Superfund Amendments and Reauthorization Act of 1986" (SARA).

C. Superfund Amendments and Reauthorization Act of 1986

On October 17, 1986, the President signed into law the "Superfund Amendments and Reauthorization Act of 1986" (SARA).[6] As part of SARA, in section 126 of Title I, Congress addressed the risk of injury to employees by providing that the Secretary of Labor ("Secretary") issue interim final worker protection regulations within 60 days after the date of enactment of SARA that would provide no less protection for workers engaged in hazardous waste operations than the protections contained in the U.S. Environmental Protection Agency's (EPA) "Health and Safety Requirements for Employees Engaged in Field Activities" manual (EPA Order 1440.2) dated 1981, and the existing OSHA standards under Subpart C of 29 CFR Part 1926. OSHA published those interim final regulations in the **Federal Register** on December 19, 1986 (51 FR 45654). A correction notice was published on May 4, 1987 (52 FR 16241).

---

[1] U.S. Environmental Protection Agency, *Everybody's Problem Hazardous Waste* at 1 (1980).
[2] Solid Waste Disposal Act, Pub. L. No. 89–272, 79 Stat 99.
[3] Resource Recovery Act, Pub. L. No. J1-512, 84 Stat 1427 and Pub. L. 93-14, 87 Stat II.
[4] 42 U.S.C. 6901 et seq.

[5] 42 U.S.C. 9601 et seq.
[6] Pub. L. 99-499.

Federal Register / Vol. 54, No. 42 / Monday, March 6, 1989 / Rules and Regulations   9295

With the exception of a few provisions that had delayed start-up dates, OSHA's interim final regulations became effective on December 19, 1986 in accordance with section 126(e) of SARA, and apply to all regulated workplaces until the final rule developed under sections 126 (a)–(d) becomes effective.

Section 126(a) of SARA provides that the Secretary shall " * * * pursuant to section 6 of the Occupational Safety and Health Act of 1970, promulgate standards for the health and safety of employees engaged in hazardous waste operations." These standards must be promulgated within one year after the date of enactment of SARA. This notice completes the development of those standards by issuing a final rule based upon the proposed regulations as indicated in sections 126(a) and 126(b) of SARA.

Pursuant to section 126(c) of SARA, the final regulations issued today are to take effect in one year. Section 126(c) also provides that the final regulations are to include each of the worker protection provisions listed in section 126(b), unless the Secretary determines that the evidence in the public record developed during this rulemaking and considered as a whole does not support inclusion of any such provision. A discussion of the public record for this rulemaking and the changes made to the proposed regulations issued August 10, 1987 follows.

This final rule has been adapted from the language of the proposed rule. Changes have been made to address more fully the provisions which Congress directed the Agency to cover and the comments made in the public record. OSHA utilized several sources for the proposal. These included the EPA manual entitled "Health and Safety Requirements for Employees Engaged in Field Activities" (1981), the language of OSHA's safety and health standards in Subpart C of 29 CFR Part 1926 and various documents issued either jointly or separately by the EPA, OSHA, the U.S. Coast Guard, and the National Institute for Occupational Safety and Health (NIOSH).

OSHA specifically used the joint OSHA/EPA/USCG/NIOSH manual entitled, *"Occupational Safety and Health Guidance Manual for Hazardous Waste Site Activities"* (Preamble Reference 6), as an outline in preparing the interim rule and the proposed rule. This manual was developed as a result of the collaborative efforts of professionals representing the four agencies. These professionals, who are knowledgeable in hazardous waste operations, worked with over 100

experts and organizations in the development of the criteria contained in this manual. The manual was published in October 1985 and is public information. The manual is a guidance document for managers responsible for occupational safety and health programs at inactive hazardous waste sites. The manual is intended for use by government officials at all levels and contractors involved in hazardous waste operations. The manual provides general guidance and is intended to be used as a preliminary basis for developing a specific health and safety program for hazardous waste operations. Further, the major subject areas listed in section 126(b) of SARA are nearly identical to the major chapters in the manual.

Based upon the extensive public comments and hearing testimony, OSHA has modified the proposal. The final rule takes into account the entire record. In addition, the language of this final rule clarifies some areas of confusion in the interim rule that OSHA has identified during the public comment period and since the promulgation of the interim final rule. The final rule also reorganizes some of the sections to clarify the standard.

D. Regulatory History

The Superfund Amendments and Reauthorization Act of 1986 (SARA) gave the Secretary of Labor 60 days to issue interim final regulations which would provide no less protection for workers employed by contractors than the protections contained in the Environmental Protection Agency Manual (1981) "Health and Safety Requirements for Employees Engaged in Field Activities" and existing standards under the Occupational Safety and Health Act of 1970 found in Subpart C of Part 1926 of the Code of Federal Regulations. Those interim final regulations were to take effect upon issuance and would apply until final regulations became effective (SARA, § 126(e)). OSHA issued its interim final regulations on December 19, 1986 (51 FR 45654).

SARA also instructed the Secretary of Labor to promulgate, within one year after the date of the enactment of section 126 of SARA and pursuant to section 6 of the Occupational Safety and Health Act of 1970, standards for the health and safety protection of employees engaged in hazardous waste operations (SARA, section 126(a)). On August 10, 1987 OSHA issued a Notice of Proposed Rulemaking and Public Hearings (52 FR 29620). That Notice set forth OSHA's proposed language for its

final rule and announced public hearings that would be held to gather further information to aid the agency in developing its permanent final rule.

Informal public hearings on the subject of this rulemaking were scheduled and held to afford interested parties the opportunity to comment on OSHA's proposals. The hearings were held October 13–16 and 20–21, 1987 in Washington, DC and October 27–28, 1987 in Seattle, Washington. The hearings originally scheduled for San Francisco, CA in the August 10, 1987 Notice of Proposed Rulemaking were rescheduled for Seattle, WA in an October 13, 1987 announcement (52 FR 37973).

Testimony from over 40 witnesses was presented at the hearings. Further, over 30 post hearing comments were submitted to the record of this rulemaking. In addition to the public hearings and the testimony received in response to those hearings, OSHA received over 125 written comments on its proposed language for a final rule.

II. Summary and Explanation of the Standard

Paragraph (a)—Scope, Application, and Definitions

1. *Scope.* OSHA proposed to define the scope of this final rule in paragraphs (a)(1) and (a)(2). "Scope" defines the specific worker populations to be covered by this rule.

The scope of this rulemaking has been an issue during the development and promulgation of the final rule. OSHA requested specific comment on whether the proposed rule was appropriate. Eastman Kodak's comment (10–36) states, "The preamble of the proposed standard at page 29622 requested 'specific comment on whether [OSHA's] interpretation of scope is too broad or too narrow.' The scope of applicability of the standard, especially with regard to ongoing operations at hazardous waste management facilities regulated under RCRA and/or corresponding state programs, appears to be appropriate."

While the language of the final rule is somewhat different from the language of the proposed rule, the four major areas of scope remain essentially the same. These four areas of scope include (1) clean-up operations at uncontrolled hazardous waste disposal sites that have been identified for clean-up by a governmental health or environmental agency, (2) routine operations at hazardous waste treatment, storage and disposal facilities or those portions of any facility regulated by 40 CFR Parts 264 and 265, (3) emergency response

9296   Federal Register / Vol. 54, No. 42 / Monday, March 6, 1989 / Rules and Regulations

operations at sites where hazardous substances have been or may be released, and (4) corrective actions at RCRA sites. In addition OSHA has clarified that the agency intends to cover voluntary clean-ups at government identified sites.

OSHA's proposal addressed the three specific populations of workers at the above operations. First, it was proposed to regulate those operations where employees are engaged in the clean-up of uncontrolled hazardous waste sites. These operations include those hazardous substance response operations under the Comprehensive Environmental Response, Compensation, and Liability Act of 1980 as amended (CERCLA), including initial investigations at CERCLA sites before the presence or absence of hazardous substances has been ascertained, those major corrective actions taken in clean-up operations under the Resource Conservation and Recovery Act of 1976 as amended (RCRA), and those hazardous waste operations at sites that have been designated for clean-up by state or local governmental authorities.

The second worker population proposed to be covered included those employees engaged in operations involving hazardous waste treatment, storage, and disposal (TSD) facilities regulated under 40 CFR Parts 264 and 265 pursuant to RCRA, except for small quantity generators and those employers with less than 90 days accumulation of hazardous wastes as defined in 40 CFR 262.34.

The third and final worker population proposed to be covered were those employees engaged in emergency response operations for releases or substantial threats of releases of hazardous substances, and post-emergency response operations to such releases at all workplaces.

In paragraph (a)(1)(i) of the final rule OSHA is regulating all government mandated clean-up operations at uncontrolled hazardous waste disposal sites. These operations were included in paragraphs (a)(1)(i) and (a)(1)(iii) of the proposal. For the purposes of this final rule, "Superfund" and other uncontrolled hazardous waste disposal sites include hazardous substance response operations at sites regulated under 40 CFR Part 300, Subpart F; RCRA closure activities conducted under 40 CFR Part 265, Subpart G; and those similar uncontrolled hazardous waste disposal sites that have been designated for clean-up by Federal, state or local governments.

OSHA intends and the change in language clarifies that all government mandated clean-ups are covered. These

include not only sites on the various "Superfund" lists, but also all other government mandated clean-ups as well. The changed language makes clear that such clean-ups are covered whether or not they are financed by the government. The language further clarifies that clean-ups mandated by any level of government are covered.

In paragraph (a)(1)(ii) of the final rule, OSHA is regulating corrective actions at RCRA facilities. This paragraph adopts the language proposed in paragraph (a)(1)(ii) of the proposal with one change. The term 'major' has been deleted as a modifier of "corrective action." Several commenters requested clarification of the term "major corrective action." International Technologies, a major hazardous waste clean-up contractor, requested in their comment (10-44), "Please clarify 'major' corrective actions conducted under RCRA.' What distinguishes 'major' corrective actions from other corrective actions?" The State of Indiana commented (10-23), "There is no definition of what constitutes a 'major corrective action' under RCRA." In addition, the term "major" is not used in EPA terminology.

"Corrective action" is a term unique to RCRA and has been defined for use with RCRA. OSHA's addition of the modifier "major" raised many definitional questions. Therefore OSHA, in the final rule, is deleting the word "major" to be consistent with EPA terminology and eliminate confusion. Rather than define "major corrective action," OSHA is amending the language of the proposal to include a phrase describing the level of corrective action that is to be regulated in the scope of this rule. OSHA will be regulating those corrective actions that potentially expose employees to a "safety or health hazard." OSHA is not concerned with those corrective actions that are intended to abate environmental risks without exposing employees to safety or health hazards. The phrase "safety or health hazard" in the introductory language is the phrase that OSHA has used to differentiate the type of releases that this standard regulates versus those release that may pose only environmental threats rather than safety or health threats to employees.

OSHA has decided to add a new paragraph (a)(1)(iii) to the final rule that would include within the scope of this rule those voluntary clean-up operations conducted at sites recognized by governmental bodies as uncontrolled hazardous waste disposal sites. All other voluntary clean-ups would be exempt from 29 CFR 1910.120. OSHA

does not have the statutory responsibility to identify hazardous waste sites. It will leave to agencies with that authority the responsibility to identify those sites. Those voluntary sites that are not recognized by the government as uncontrolled hazardous waste disposal sites would be exempt from 29 CFR 1910.120; however, they would still be regulated by the other OSHA general industry or construction industry standards applicable to the work being performed at the site.

OSHA did not propose to cover voluntary clean-ups of hazardous substances in its proposed rule. Many comments suggested this, however, the Agency has concluded that individuals involved in voluntary clean-ups may be exposed to the same safety and health risks at voluntary sites identified by the government whether or not the government is compelling action. However, it would be difficult to know whether or not sites not identified by the government are hazardous waste sites without a structured evaluation system for such potential sites.

OSHA raised an issue on the scope in the preamble to the proposal that generated several comments. On page 29622 of the preamble to the proposal, OSHA listed several TSD facilities that would not be covered by the final rule. The exemptions were taken from a list published by the U.S. EPA that are not directly regulated by U.S. EPA. However, the proposed standard's language did not grant these exemptions. Comments did not support the exemptions and OSHA did not believe that they were appropriate.

The particular exemption that generated the most comment exempted those TSD facilities which operate under a state hazardous waste program pursuant to RCRA section 3006. These state hazardous waste programs are recognized by U.S. EPA in a similar fashion to the OSHA state plan states under section 18 of the OSH Act. A number of commenters, such as the State of Indiana (10-23), objected to this type of exemption by OSHA as not being appropriate. They stated OSHA jurisdiction should not be impacted by U.S. EPA state agreements, but only those state agreements provided in the OSH Act. OSHA agrees with these commenters and therefore OSHA jurisdiction will be delegated to only those states which OSHA has formal agreements with under the OSH Act. However, it should be noted that the U.S. EPA jurisdictions under SARA section 126 may make use of their state agreements.

Other commenters, EXXON (10–33) and CONOCO (10–32), suggested that OSHA incorporate the exemptions on page 29622 as a separate paragraph in the final rule.

Typical TSD facilities range from the hazardous waste generator with a hazardous waste storage area to the large, complex hazardous waste disposal facility. EPA estimates that approximately 80 percent of all generators also treat, store, or dispose of their hazardous wastes and thereby qualify as a TSD facility. Over 30,000 TSD facilities notified EPA in 1980 that they would qualify for regulation under section 3004 of RCRA.

OSHA continues to regulate RCRA TSD facilities in paragraph (a)(1)(iv) of the final rule as it was proposed in the regulatory language of the proposal. The list of exemptions on page 29622 will not be incorporated into the final rule. OSHA believes that such a list would create too great a gap in the protection of workers. For example, with respect to workers at TSD facilities operating under a state hazardous waste program pursuant to RCRA section 3006, OSHA agrees with a comment made by the State of Indiana (10–23) that it is possible that the workers in those 42 authorized states identified by Indiana could be without the protections mandated by Congress.

In paragraph (a)(1)(v) OSHA would continue to regulate emergency response operations for releases of, or substantial threats of releases of, hazardous substances without regard to the location of the operation as proposed in paragraph (a)(2) of the proposal. Such emergency response operations are not limited to those responses at uncontrolled hazardous waste disposal sites or RCRA TSD facilities. With respect to transportation incidents, responders to the scene are covered but operators (i.e., truck drivers and train crews) are not covered unless they become actively involved in the response action.

OSHA is making major revisions to proposed paragraph (1). These revisions have been made in response to comments concerning OSHA's involvement in regulating emergency response at every site involving hazardous substance release or potential release. Some of the comments were in favor of OSHA's continued involvement with emergency response (i.e., American Chemical Society, 10–44) and others were opposed to continued involvement (i.e., ECOLAB, 10–64). Others supported OSHA involvement in emergency response activities at uncontrolled hazardous waste sites and certain RCRA facilities but opposed the

agency's involvement with non-waste clean-up or non-RCRA facilities (i.e., The Chlorine Institute, 10–24). Yet others called for two separate areas in the rule; one for hazardous waste operations, and one for emergency response (i.e., Allied Signal, 10–38). Others opposed coverage of emergency response to petroleum spills (CONOCO, Ex. 10–32).

OSHA after reviewing all the comments, continues to believe that it is the clear intent of Congress that any employees participating in an emergency response to the release or potential for release of hazardous substance be covered by this rulemaking. This Congressional intent applies to all such emergency responses including those both off and on hazardous waste sites.

The statutory language indicates that all emergency responses where the threat of hazardous substance spills exist are to be covered.

Section 126(b)(11) of SARA specifically provides that "requirements for emergency response" are to be included and is not limited to hazardous waste sites.

In addition, section 126(d)(4) states:

Training of Emergency Response Personnel.—Such training standards shall set forth requirements for the training of workers who are responsible for responding to *hazardous emergency situations* who may be exposed to toxic substances in carrying out their responsibilities. (emphasis added)

This is very broad language that is not limited to hazardous waste operations or hazardous wastes or substances on CERCLA or RCRA sites. It covers all "hazardous emergency situations" for all "toxic substances" which would clearly cover all types of emergency response for chemical spills including chemical tanker spills and the like. It should also be noted that once a tank truck spills a toxic chemical in an emergency it creates a hazardous waste in the very real sense.

Further, the grant provision of the statute clearly indicates that grants can be made to train workers for emergency response at any location, not just on hazardous waste sites.

Section 126(g)(1) states:

Grant Purposes.—Grants for the training and education of workers who are or may be engaged in activities related to hazardous waste removal or containment or *emergency response* may be made under this section. (emphasis added)

Other statutory sections also indicate the legislative intent to cover all emergency responses where hazardous chemical spills are possible.

In addition to the statutory language, the documents cited by Congress as the

minimum guides for OSHA to use in developing this rule refer to all emergency responses. The EPA manual and the OSHA construction standards referred to in the statute require preparations and planning for emergencies generally, not just for hazardous waste site emergencies.

In addition the legislative history indicates that Congress intended Section 126 to cover emergency response to all situations where spills of hazardous chemicals were a possibility and not just emergency response on hazardous waste sites. For example, Senator Hatch stated:

This amendment will address the concerns that have been raised that the Department of Labor issue standards for employees engaged in hazardous waste operations, as well as emergency response. (9/24/88 Cong. Rec. pg. S–12031)

As discussed elsewhere in this preamble OSHA believes there is a clear need for training and other provisions to protect workers engaged in all emergency responses when there is the possibility of hazardous substance spills. This is needed whether or not the emergency occurs on a hazardous waste site. The agency believes that the hazards are the same in these cases.

Finally, other parts of SARA, in particular Title III, address emergency response actions and planning by communities and local government employers outside of the hazardous waste clean-up operation. The Congressional concerns on toxic emergencies also discussed in *Task Force on Toxic Emergencies*, Environmental and Energy Study Conference Special Report, September 18, 1986. This report stresses the need for training of emergency response personnel as well as emergency response planning and related areas. This was part of the legislative research which led to the passage of section 126 of SARA.

OSHA's final rule rulemaking divides emergency response into three separate areas. First, OSHA is regulating emergency response by employees at uncontrolled hazardous waste sites in paragraph (l) of the final rule. This paragraph contains the requirements that were in paragraphs (l)(1) and (l)(2) of the proposal and the interim rule. These regulations applied to the "on-site" operations of the interim rule. Second, OSHA is regulating emergency response at RCRA facilities in paragraph (p)(8). This paragraph contains the requirements that were in paragraphs (l)(1) and (l)(3) of the proposal and interim rule. These regulations applied to the "off-site"

9298   Federal Register / Vol. 54, No. 42 / Monday, March 6, 1989 / Rules and Regulations

operations of the interim rule. Third, OSHA is regulating emergency response to hazardous substance releases by employees not covered by paragraphs (l) and (p)(8) in paragraph (q). Paragraph (q) contains the requirements proposed in paragraphs (l)(1), (l)(3), (l)(4), and (l)(5) of the proposal and interim rule. These regulations were directed toward emergency response teams, industrial fire brigades, and hazardous materials teams.

In its proposal OSHA covered emergency response to releases of hazardous substances. The agency did not propose to limit emergency response to uncontrolled hazardous waste sites but decided instead to propose to cover all emergency response whether it was done at uncontrolled hazardous waste sites or anywhere else, including petrochemical and similar manufacturing facilities.

OSHA's decision to propose coverage of all emergency response was based upon the high risk associated with emergency response by untrained and unprotected employees and the need for proper training and equipment to be provided for emergency response to hazardous substance releases. In testimony during the public hearings on this rulemaking, Mr. William Bunner stated, "The highest-risk incidents are the persons who respond to spills and accidental releases of hazardous chemicals; and those personnel, particularly public first responders, have had the least protection in terms of chemical emergency response safety and health plans, training and equipment." (Tr. pgs. 24–25). Mr. Bunner goes on to state, "The real strength of 29 CFR 1910.120 is that it not only provides for a more consistent and thorough approach to protecting workers involved in hazardous waste operations, but also for personnel who face extremely high risk to life and health that's associated with chemical emergency response." (Tr. pg. 25).

Another witness, Mr. Ray Simpson, one of OSHA's expert witnesses on fire suppression, fire inspection, and training, testified, "I like to support any concept that advocates properly equipping, training, and supporting emergency responders. When I talk about an emergency responder, I'm not talking simply about fire fighters although that's basically my expertise. I'm talking about the emergency medical technicians, the people who handle the victims. I'm talking about the police officer who, many times, is first on the scene before any of us get there: the many who really must make, in some situations, the initial decision about

what's going to happen. I have learned over these many years that the two greatest dangers that face us as emergency responders are ignorance or non-awareness of what we're facing and the lack of plan or any procedure that will take us to the end that we're trying to accomplish." (Tr. pgs. 89–90).

Margaret Seminario, Associate Director, Department of Occupational Safety, Health and Social Security of the American Federation of Labor-Congress of Industrial Organizations (AFL–CIO) also testified at OSHA's public hearings on the issue of emergency response. Ms. Seminario discussed the participation of the AFL–CIO in hearings before the House Subcommittee on Employment and Housing of the Government Operations Committee and the Safety and Health Subcommittee of Education and Labor. Ms. Seminario stated, "Those hearings dealt with the issues of the problems for hazardous waste workers in both Superfund operations and RCRA operations, but they also got into an issue that had really not been explored very fully: the problems facing emergency response workers, particularly for the AFL–CIO firefighters. The members of our firefighter's union were the ones who were called in when there were spills, leaks and other accidental releases. These weren't hazardous waste sites, per se, at the time. They became hazardous waste sites and were defined as such after the fact; but they were routinely called in without information, without adequate protection to deal with these problems. That was an issue, as I said, that was fully explored in those hearings and it was the reason that, when we moved from those hearings into a legislative opportunity in Superfund, we looked beyond the language that we had come up with in the 1980 law which dealt only with hazardous waste and expanded it to include emergency response operations." (Tr. pgs. 345–346)

Further, OSHA still believes that Congress intended this rule to have such coverage. This is indicated by the language of SARA as well as the legislative history.

As OSHA stated in the preamble to the proposed rule, "The language of section 126(a) mandates safety and health standards for the protection of employees engaged "in hazardous waste operations." The term "hazardous waste operation" is not limited in the legislation and a response to spills of hazardous substances on the highway or from a railway tank car in order to control and contain the hazardous substance (which has become a waste

once it is not contained) is in the common sense meaning a hazardous waste operation."

"This interpretation is reinforced by the fact that SARA is a free-standing statutory provision and not an amendment to CERCLA. The clear Congressional intent then is to provide protection to employees whenever they deal with hazardous wastes."

In addition section 126(d)(4) discussing training for emergency response personnel utilizes the very broad term "hazardous emergency situation." Section 126(g)(1) indicates that training grants may be given independently for emergency response training separate from hazardous waste removal training. Section 126(b)(11) also indicates emergency response is an independent concept separate from hazardous waste removal operation. For those and other reasons OSHA believes section 126 is intended to cover emergency response to hazardous substances whether on an uncontrolled hazardous waste disposal site, a RCRA site or elsewhere. However, the clarified language in these sections makes it clear the only employers whose employees have the reasonable possibility of engaging in emergency response are covered. Emergency response employees who respond or will respond to incidents involving hazardous substances are covered by this final rule to the extent that they are exposed to hazardous substances. State and local government employees in states that have agreements with OSHA under section 18 of the OSH Act must be regulated by state regulations at least as effective as these to protect public employees. Those state regulations must be issued within six months of the date of promulgation of this final rule.

However, some commenters have commented that OSHA has exceeded the intent of Congress with the scope of the proposed rule. Many of these commenters stated that OSHA's coverage of emergency response at sites other than specific cleanup or TSD facilities was too broad and unwarranted. AMOCO's comment (10–26) is representative of some of the comments made on this issue. In their comment AMOCO stated that, "Section 126(a) of SARA is the directive to OSHA to promulgate standards 'for the health and safety protection of employees engaged in hazardous waste operations.' There is no reference whatsoever in this directive to promulgate standard (sic) with respect to emergency response activities outside of hazardous waste operations." However, other comments received from the petrochemical

Federal Register / Vol. 54, No. 42 / Monday, March 6, 1989 / Rules and Regulations   9299

industry support, on a limited basis, OSHA's decision to cover emergency response with the scope of the standard. CONOCO's comment (10–32) is representative of this point of view. CONOCO states, "Conoco's primary concern with the proposed rule centers on the extremely broad scope of employee coverage under this standard and compared to Congress' intent to cover 'hazardous waste operations and emergency response.' We believe that Congress intended section 126 of SARA to cover employees engaged in hazardous waste operations and emergency response to these operations on a full-time basis." While this comment would seem to support OSHA's coverage of employees engaged in emergency response, that support is limited to those employees engaged in response on a full-time basis at hazardous waste operations.

Based upon public testimony and written comments received into the record of this rulemaking, OSHA has concluded, that because of the high risk associated with emergency response to the releases of hazardous substances and the number of these incidents occurring, that coverage of workers conducting such emergency response activities is both appropriate and necessary.

OSHA believes that the scope of this final rule carries out the intent of Congress and is consistent with good occupational safety and health policy. Employees performing clean-up operations under CERCLA, RCRA (corrective actions) and state or local government designated sites—generally those employees likely to have the highest exposures to hazardous substances over a longer period—would be covered by virtually all the provisions of this final rule. Employees exposed to hazardous wastes in routine RCRA hazardous waste operations, who are regularly exposed to hazardous wastes but in a more controlled environment, would be covered by the more limited requirements of paragraphs (p) and (q). Emergency response workers, exposed usually for short periods to often unknown but possibly high levels of hazardous substances, would be regulated by paragraph (q).

2. *Application.* OSHA proposed to define the application of this final rule in paragraph (a)(3) of OSHA's Notice of Proposed Rulemaking (NPRM) published on August 10, 1987 (52 FR 29620). "*Application*" establishes which regulations within this rule apply to the specific worker populations to be protected by this rule.

In paragraph (a)(3)(i) OSHA proposed that the employer would have to comply

with the standards in 29 CFR Parts 1910 and 1926, as well as with the requirements specifically covered in the proposed rule. If there were a conflict or overlap between standards, it was proposed that the more protective provisions would apply. Since this rule does not cover all of the hazards present at hazardous waste operations, other OSHA standards in Parts 1910 and 1926 should apply to ensure employee safety and health. Other OSHA standards regulate many other hazards, and OSHA wants to make clear that the other standards continue to apply. Also, OSHA proposed that hazardous waste operators who are not within the scope of this standard should continue to be regulated by the Parts 1910 and 1926 standards. OSHA is keeping those provisions in the final rule for the reasons stated with minor editorial changes for clarification.

In paragraph (a)(3)(ii), OSHA proposed that all paragraphs of section 1910.120 except paragraph (o) would apply to hazardous wastes operations at CERCLA sites, at major corrective action at RCRA sites, and at sites designated for clean-up by state and local governments. Paragraph (o) of the proposal addressed certain operations conducted under the Resource Conservation and Recovery Act of 1976 (RCRA).

OSHA recognizes that the hazards presented to employees engaged in clean-up operations involving uncontrolled hazardous wastes are far greater than those presented to employees engaged in the routine, day-to-day operations of an EPA licensed TSD facility.

OSHA has made two editorial changes in its proposed language in paragraph (a)(3)(ii) without changing the intent of the paragraph. First, rather than referring to each of the types of sites individually, OSHA is making reference to the scope paragraphs (a)(1)(i) through (a)(1)(iii) to identify the sites that this application paragraph addresses. The sites to be addressed remain the same as proposed. Second, because the codification of paragraphs has changed in this final rule due to changes made to the proposal, paragraph (p) of § 1910.120 rather than paragraph (o) will apply specifically to hazardous waste operations at RCRA sites which are involved in treatment, storage, disposal and handling of hazardous waste. The new requirements of paragraph (p) are discussed later in the preamble.

In paragraph (a)(3)(iii), OSHA proposed that the requirements set forth in paragraph (o) of section 1910.120 would apply specifically to the

hazardous waste operations at RCRA sites which are involved in treatment, storage, disposal and handling of hazardous waste. The proposal contained a limited exclusion from these regulations for certain small quantity generators and less than 90-day accumulators, such as dry cleaners and gas stations, which come within the purview of RCRA, but are not hazardous waste operations in the normal meaning of the term. The exclusion was available to these operations depending upon the employer's decision to provide or not provide emergency response by employees to releases of, or substantial threats of releases of, hazardous substances.

OSHA proposed to exempt small quantity generators and less than 90 day accumulators from all parts of the rule if they did not provide emergency response by their employees to releases of, or substantial threats of releases of, hazardous substances. OSHA further proposed to exempt small quantity generators and less than 90 day accumulators from all parts of the rule except paragraph (l) if they did provide emergency response by their employees to releases of, or substantial threats of, releases of, hazardous substances.

OSHA recognized that many small quantity generators are smaller businesses with limited employee populations. Since most of these establishments rely on the emergency response services of local fire and rescue departments, OSHA is providing a complete exemption from these proposed standards when the employer can show that employees are not required to engage in emergency response, but are directed in the case of emergency spills of hazardous substances to maintain a safe distance and to call local fire or other emergency response organizations. In cases where such establishments do provide emergency response by employees, and thereby expose employees to hazardous substances, OSHA proposed that such employers meet the emergency response requirements of paragraph (l) of this proposed rule. OSHA concludes its proposal is supported by the record.

Without these exemptions, these regulations could be interpreted to cover gas stations, dry cleaners, and other small businesses which temporarily store small quantities of a hazardous waste. These businesses are not engaged in hazardous waste operations as that term is conceived of normally. In addition, it is not believed that Congress intended such businesses to be covered. They do not present the relatively high

9300   **Federal Register** / Vol. 54, No. 42 / Monday, March 6, 1989 / Rules and Regulations

exposure to a number of hazardous health risks to employees that hazardous waste sites typically do.

In paragraph (a)(3)(iv) OSHA proposed that the requirements set forth in paragraph (l) of this section would specifically apply to the work conducted by emergency response personnel when they respond to hazardous substance emergency incidents. Emergency response personnel include firefighters, EMS personnel, and police as well as other employees.

The regulation of employees providing emergency response has been discussed under the "Scope" portion of this preamble discussion. Further discussion of OSHA's changes to the emergency response portion of this rulemaking is contained in the discussion of paragraph (l) of the proposal.

OSHA also requested comment on what other operations should be and are intended by Congress to be covered, and whether specific operations should be excluded because of low exposures. For example, municipal or other sanitary landfills that handle domestic wastes would not normally be regulated by this rule. Similarly, waste paper or scrap metal operations would not normally be regulated because of the type of wastes they handle. However, both types of operations would be regulated if they have clean-ups for or handle hazardous wastes meeting the scope provisions of the standard.

Also, employees at hazardous waste disposal sites who will not be exposed to, or do not have the potential to be exposed to, hazardous substances are not covered by this rule. The provisions of these regulations are designed to protect employees who have or may have exposures, and would not be needed for those employees who do not. Operations with no exposure to hazardous substances, i.e., road building for site access, construction of or the setting up of temporary facilities in the clean zone, or the closure of a RCRA site involving the building of a clay cap over hazard wastes, are considered to be construction activities covered by the standards in 29 CFR Part 1926.

As a result of the comments received during the public comment phase of this rulemaking, OSHA has made some changes to the degree of regulation for emergency response workers. However, the scope of this standard continues to cover such workers and paragraph (a)(2)(iv) identifies the new paragraphs within the rule that reflect the changes and identifies the standards that apply to emergency response operations.

3. *Definitions.* In paragraph (a)(4), *Definitions,* OSHA proposed to identify and define the various terms used in this

rulemaking that may cause confusion. However, the following new definitions have been added as a result of comments made in the record: "published exposure level," and "uncontrolled hazardous waste site". Except for the definition of "established exposure level" which has been amended to define "published exposure level," no definitions have been removed from the proposal.

The term "established permissible exposure limit" was incorporated as part of the determination of whether medical surveillance was required. There were a number of comments on this definition.

One commenter, Four Seasons Industrial Services (10–5), believed that the definition should be broadened. Four Seasons stated, "The routes of chemical exposure are through inhalation, skin absorption, and ingestion. All of these have to be considered when dealing with permissible exposure limits. Your definition as written does not include ingestion." Other commenters were concerned that OSHA included the NIOSH Recommended Exposure Limits in the definition of PELs. The E. I. DuPont de Nemours, Co. (10–28) comments summarize many of those received against the NIOSH levels. DuPont stated, "NIOSH limits have not undergone public review and comment as have national consensus standards and regulatory exposure standards. Therefore, they should not be included in the definition of 'established exposure limits'." CONOCO (10–32) agreed when they stated, "These NIOSH limits are not subject to peer review or public comment as are OSHA's PELs. We encourage OSHA to remove the references to these NIOSH limits. We strongly urge OSHA to *only* use PELs given in 29 CFR 1910, Subpart Z."

Dr. James Melius, Division of Environment Health Assessment, State of New York Department of Health Medical Surveillance, and one of OSHA's expert witnesses in occupational medicine, stated in his testimony (Tr. pg. 115) that, "I think it's important that the OSHA standards be supplemented by information from NIOSH and ACGIH, both of which cover a larger number of chemicals or toxic substances and both of which include, or at least consider, more up to date information on the toxicity of these substances."

In addition, during the public hearings, OSHA, through its panel of staff members specifically asked Captain Richard A. Lemen, Director of the Division of Standards Development and Technology Transfer within NIOSH, about the peer review process of NIOSH

RELs. Mr. Thomas Seymour of OSHA directed the following question to Captain Lemen (Tr. pg. 195).

[Mr. Seymour:] We have received some feedback in our record about the order of the hierarchy that we have used for permissible exposure limits. There have been statements made that the RELs are not peer-reviewed or developed. I wonder if you might describe to us how the RELs have been developed by the National Institute for Occupational Safety and Health.

Captain Lemen's response (Tr. pgs. 195–197) to Mr. Seymour was:

Okay. In response to the peer review question, I would say that probably the RELs were more peer reviewed than most recommendations. The first process of developing an REL at National Institute for Occupational Safety and Health is to propose such a recommendation be developed to the Director of NIOSH through a position paper. That is then reviewed by senior NIOSH staff. Once that approval is given, a criteria manager then develops the recommended exposure standard in the form of a criteria document.

Once the criteria document has gone through several layers of review within NIOSH, it is then sent out for external review. On the average, we usually send out to more than 40 experts in the field that document for their external comments. We receive those comments back. We incorporate those comments where appropriate. We then submit the document through the same internal review within NIOSH. It goes through every division within the Institute and it finally winds up in the senior review staff at the directors' level at our headquarters in Atlanta.

A meeting is then held with the Director of NIOSH at which time the criteria manager and the senior scientists within the Institute [review] all of the reviewer—external and internal reviewer—comments and a decision is made by the Director of NIOSH whether or not to publish a recommended exposure limit. If the decision is to publish, then we do so and submit that criteria document to the Director of OSHA.

So it then goes through a very extensive peer review process compared to, for example, an article submitted to peer-review journal which is written by a researcher and generally sent out to three to five individuals to review before it's placed in a peer review journal.

So the extensiveness of our review process, I think, is much greater than that for a peer-preview journal article.

In light of these comments OSHA has concluded that the NIOSH REL's have undergone the necessary peer review to be included in the standard's hierarchy of limits.

The term "established exposure levels" was defined in the proposal to indicate the levels which would trigger medical surveillance of the exposed employees. The term included not only OSHA established PELs, but also

Federal Register / Vol. 54, No. 42 / Monday, March 6, 1989 / Rules and Regulations   9301

exposure limits suggested by NIOSH and ACGIH. After review of these and other comments, OSHA concludes that it is appropriate to go beyond the OSHA established PELs in triggering medical surveillance. First, medical surveillance is appropriate for workers exposed to toxic chemicals other than those covered by the PEL's. Second, because of the broadly-worded language in section 126(b)(3), which requires medical surveillance for workers engaged in hazardous waste operations "which would expose them to toxic substances." Some of these "toxic substances" are not included in the OSHA PELs. When OSHA completes its rulemaking on the air contamination proposal (PEL's project), there will be fewer toxic substances not covered by PEL's. But in light of Congressional language and the large number of hazardous chemicals present in an uncontrolled hazardous waste site, OSHA concludes that this definition is appropriate to protect employee safety and health.

The term "permissible exposure limits" was defined in the proposal as the inhalation or dermal permissible exposure limit specified in 29 CFR Part 1910, Subpart Z. As a result of the comments received in the record, OSHA has amended its definition that ignored the health limits specified in Subpart G, for "permissible exposure limits."

OSHA has amended the definition for "permissible exposure limits" to include a reference to Subpart G of Part 1910. It now includes both Subpart Z health hazards and those requirements in Subpart G of Part 1910.

First, OSHA has changed the term "established exposure levels" to the term "published exposure level" to reduce confusion. Second, the term "published exposure level" is defined as the exposure limits published in the "NIOSH Recommendations for Occupational Health Standards" dated 1986, incorporated by reference, or if none is specified, the exposure limits published in the standards specified by the American Conference of Governmental Industrial Hygienists in their publication "Threshold Limit Values and Biological Exposure Indices for 1987-88" dated 1987, incorporated by reference. Third, the provisions of (f)(2) on medical surveillance have been changed to cover overexposures to both PEL's and, if none, then over-exposure to published exposure limits. OSHA concludes that with these changes the definitions are clear, comprehensive and carry out both statutory directives and appropriate medical criteria in determining whether medical

surveillance is required. Some commenters stated a broader guide is necessary for respirator use and that is discussed under paragraph (g).

OSHA requested comment on the appropriateness of its definitions of hazardous waste, health hazard and hazardous substance and whether they were consistent with EPA and DOT practice. Several comments were received on these issues. One set of comments criticized OSHA's incorporation of petroleum and petroleum products in its definition of hazardous substances.

A typical comment was made by EXXON (10–33). In their comments EXXON presented the following discussion:

Perhaps the most fundamental misinterpretation contained in this rule is the inclusion of petroleum and petroleum products in the definition of hazardous substance. As discussed in Comment II.A.4.v, below at pages 11 to 14 [internal EXXON comment references], Congress, the Environmental Protection Agency (EPA), and the Department of Transportation (DOT) have uniformly recognized the inappropriateness of characterizing petroleum as a hazardous substance. There is no indication in SARA Section 126 that Congress intended to change the petroleum exclusion or to subject petroleum releases to emergency response regulation.

EXXON further stated:

It is EXXON's understanding that a situation is not an emergency response subject to the requirements of paragraph (1) unless there is a release of a "hazardous substance." Therefore it is essential that the definition of "hazardous substance" be accurate and correct.

The proposed definition of "hazardous substance" references the Department of Transportation's definition of "hazardous materials" under 49 CFR 171.8. By so doing, petroleum and petroleum products have been included as hazardous substances; and, related spills may be subject to the burdensome requirements for emergency response operations.

Congress, in the very CERCLA sections cited in the proposed definition of "hazardous substance," has recognized that petroleum and petroleum products are excluded from the federal definition of "hazardous substance." EPA regulations under CERCLA have incorporated this congressional directive. See 40 CFR Part 302 and discussion at 50 FR 13456, 13460 (April 4, 1985). DOT has specifically recognized this Federal petroleum exclusion and incorporated the exclusion in its definition of "hazardous substance." See 52 FR 24474 (July 1, 1987). As such, the proposed OSHA definition is inconsistent with the CERCLA, EPA and DOT definitions of "hazardous substance."

The proposed definition of "hazardous waste" includes the EPA RCRA definition of hazardous waste and the DOT definitions at 49 CFR 171.8. The cited DOT regulation defines both hazardous substances and

hazardous wastes. As noted above, the DOT definition of hazardous substance at 49 CFR 171.8 should properly be incorporated in the proposed OSHA definition of hazardous substance. It is not a waste definition. Therefore, the proposed definition of hazardous waste should be limited to waste materials; and, the DOT definition of hazardous substance should be clearly excluded.

OSHA does not agree with these arguments. Section 126 of SARA is directed to protecting workers from the hazards of all hazardous waste spills. Petroleum products create significant health and safety hazards. Many comments supported OSHA's incorporation of petroleum and petroleum products.

During the questioning of Dr. Kenneth H. Chase, M.D., President of the Washington Occupational Health Associates, Inc., Mr. Chappell Pierce of the OSHA panel asked Dr. Chase the following question (Tr. pg. 551):"Do you feel that medical monitoring for these types of products [petroleum products] is appropriate?"

Dr. Chase responded,"Petroleum products is just too broad a term for me to answer that in a general way. Certain petroleum derivatives are more toxic than others. Some have acute toxicity; others subacute toxicity; and others, the concern is more about chronic toxicity that is most difficult to detect."

During the hearings, OSHA asked many of the individuals who testified if petroleum and petroleum products should be included in the definition of hazardous substances.

Representative of the responses made to this question was the testimony of the Prince Georges County Fire Department; the International Association of Fire Fighters, AFL–CIO; NIOSH; and the Seattle, Washington Fire Department.

Mr. Gregory Noll, the Hazardous Materials Coordinator for the Prince George's County Maryland Fire Department, testified on the issue (Tr. pg. 448). Mr. Thomas Seymour of the OSHA panel addressed Mr. Noll by stating: "I notice in your testimony, on page 3, that you indicate that at least 50 percent of your responses are involved with flammable liquids or gas emergencies. The definition that OSHA is using in this rulemaking for hazardous substances dealing with and covering flammable liquids and gases you find, then, appropriate?"

Mr. Noll responded, "I think realistically, from the perspective of fire service, we've been successfully handling flammable liquid and gas emergencies for a number of years. Today, with HAZMATs being the buzz word, certainly those categories of

9302   Federal Register / Vol. 54, No. 42 / Monday, March 6, 1989 / Rules and Regulations

commodities have been thrown into the hazardous materials field.

"We now regard them in the hazardous materials field from a practical perspective."

Mr. Thomas Seymour of the OSHA panel asked Mr. Richard Duffy of the International Association of Fire Fighters (Tr. pg. 110), "Mr. Duffy, we have had some previous commenters who have advocated that petroleum and petroleum products be excluded from the scope of the standard.

The example that you just gave about the propane tank inside the building exploding and killing fire fighters, what is your opinion about whether we should exclude petroleum products from this standard?"

Mr. Duffy responded: "I don't know how we would classify them. I would object to that. I mean, I don't know how to better qualify—I could talk to you for days about incidents involving petroleum products. I don't see any reason to exclude them any more than excluding the oxidizers or any group. I mean, you could pick lots of products and ask to exclude them. And I'm sure a lot of the lobbying entities can establish reasons for it. But I can't see any in terms for fire fighters."

Mr. Charles Gordon of the Department of Labor's Office of the Solicitor and a member of the OSHA panel asked Captain Richard A. Lemen, Director of the Division of Standards Development and Technology Transfer of, NIOSH the following question (Tr. pg. 200–201): "In the case of spills of petroleum or petroleum products in either an emergency response situation or in a hazardous waste dump were there are petroleum products as one of the major contaminants, is it appropriate for all the provisions of the OSHA standard or the recommendations to apply in those circumstances?"

Captain Lemen responded, "We believe it is appropriate and they should apply in those circumstances, as well."

Mr. Seymour also asked Deputy Chief Roger Ramsey of the Seattle Fire Department (Tr. pg. 142): "I gather from what you have said you said that the definition we have, including the DOT hazardous material definition for hazardous substance and materials is appropriate, and that we should not exclude petroleum products from the coverage of this standard?"

Deputy Chief Ramsey responded, "Absolutely not."

Many spills and emergency response to these spills involve petroleum products. These spills present both health and safety risks. Training is necessary to protect employees who respond to petroleum spills as with

other spills. In fact, these are usually the same employees.

OSHA concludes that it is crucial to cover responses to petroleum spills as well as all other spills because petroleum products constitute a substantial threat to employees responding to accidental releases of these substances. Many petroleum products present health hazards as well as fire and explosion hazards. In addition they often contain fractions which present high health hazards. For example, many contain benezene, a carcinogen to which employees may be exposed.

Therefore, OSHA is not amending its definition for "hazardous substance" to include the petroleum exclusion referenced by some of the commenters.

The other definitions are discussed in the preamble to the proposal for this rulemaking. There were no major comments. OSHA concludes that those definitions are appropriate for the reasons stated in the proposal preamble.

Paragraph (b)—Safety and Health Program

Paragraph (b) of the proposal has been reorganized for clarity as a result of the public comment. Basic requirements remain the same. Specific changes are discussed below. This paragraph basically requires that a written safety and health program cover safety and health organization and specific work practices to assure employee safety and health. OSHA has concluded that it is crucial for employee safety and health to have a written safety and health program that would force the systematic identification of site hazards and identify employee response to those hazards. The written plan is necessary to communicate hazards to employees for their awareness and protection. (See preamble discussion at 52 FR 29624.)

OSHA received many comments supporting the requirement for a written safety and health program (i.e, State of Wyoming, 10–9; James T. Dufour, 10–78; International Association of Fire Fighters Local 291, 10–12); other commenters have made suggestions for changes to the proposed language.

OSHA concludes that for the reasons stated a written program is necessary. The following discussion covers specific changes.

OSHA has included a non-mandatory note at the beginning of new paragraph (b) that explains the acceptability of safety and health programs developed and implemented to meet other Federal, state, or local regulations in meeting the requirements of this paragraph. Some commenters believed that OSHA's

requirements for a safety and health program were somewhat duplicative of the contingency plans and emergency response plans required by the E.P.A. for its permit requirements (i.e., Tennessee Valley Authority, 10–43; National Paint and Coating Association, 10–72; Johnson Wax, 10–84). OSHA will permit existing programs that have been designed to meet other government or corporate requirements. For example, contingency plans developed under 40 CFR 265.50 are acceptable in meeting this requirement if they are supplemented with the provisions established by the OSHA standard. OSHA does not intend to require the duplication of efforts made to meet other governmental regulations. Therefore, any plan containing all of the elements required for the OSHA plan will be acceptable in meeting this requirement without the need for developing a separate OSHA plan.

In paragraph (b)(1) of the final rule OSHA is using the language proposed in paragraphs (b)(1)(i), (b)(2), and (b)(3) of the proposal and subdivided it into paragraphs (b)(1)(i), (b)(1)(ii), (b)(1)(iii), and (b)(1)(iv). Paragraph (b)(1)(i) contains the first two sentences of the proposal along with two new sentences that clarify what the safety and health program shall include. OSHA has included the new sentences and the new note to this paragraph to provide further guidance to employers who may need assistance in developing their safety and health program.

In paragraph (b)(1)(ii) of the final rule OSHA is using the last sentence and the list of chapters proposed in paragraph (b)(1)(i) and subparagraphs (A) through (C). There are no changes made to the language as proposed other than a recodification of the paragraphs.

In paragraph (b)(1)(iii) of the final rule OSHA is using the exact language proposed in paragraph (b)(2). The proposed language has been moved to this paragraph because it contains a requirement that is of a general nature.

In paragraph (b)(1)(iv) of the final rule OSHA is using the language proposed in paragraph (b)(3)(i) with one exception. A new phrase would require the employer to inform contractors and sub-contractors of the site emergency response procedures in addition to the proposed information. One commenter, CDM Federal Programs Corporation (10–83), suggested revised language to the proposal that would assure that the contractors and subcontractors received the site specific safety and health plan as well as the safety and health programs. OSHA agrees with the suggestion of the commenter and that

Federal Register / Vol. 54, No. 42 / Monday, March 6, 1989 / Rules and Regulations   9303

the new language accomplishes the recommended change suggested by CDM Federal Programs.

In paragraph (b)(1)(v) of the final rule OSHA is using the exact language of proposed paragraph (b)(3)(ii).

In paragraphs (b)(2), (b)(3), and (b)(4) of the final rule OSHA is using the exact language of paragraphs (b)(1)(ii), (b)(1)(iii), and (b)(1)(iv) of the proposal. One commenter, James T. Dufour (10–78), while supporting the use of safety and health plans as an appropriate communication tool for identifying site hazards, suggested that OSHA should require a more comprehensive review and control of the plan to assure its professional quality. OSHA believes that the language of paragraph (b)(4)(iv) would provide for this type of oversight and control. Therefore, the only change to paragraphs (b)(1)(ii) through (b)(1)(iv) is a recodification of the paragraphs.

Paragraph (c)—Site Characterization and Analysis.

The employer needs to know the hazards faced by employees in order to develop and implement effective control measures. Site characterization provides the information needed to identify site hazards and to select employee protection methods. The more accurate, detailed, and comprehensive the information available about a site, the more the protective measures can be tailored to the actual hazards that the employees may encounter. Congress clearly intended that such a requirement be included. Section 126(b)(1) of SARA provides that the proposal include "requirements for a formal hazard analysis of the site * * *."

It is important to recognize that site characterization is a continuous process. At each phase of site characterization, information is obtained and evaluated to define the potential hazards of the site. This assessment is to be used to develop a safety and health plan for the next phase of work. In addition to the formal information gathering that takes place during the phases of site characterization described above, all site personnel should be constantly alert for new information about site conditions.

In paragraph (c) of the final rule OSHA has used most of the language in paragraph (c) of the proposal. New headnotes have been added to the major paragraphs to make reading the requirements easier.

In paragraphs (c)(1) through (c)(4) of the final rule, OSHA has used the language of paragraphs (c)(1) through (c)(3) of the proposal. The reason for the one additional paragraph in the final rule is that OSHA has numbered the

initial unnumbered paragraph in the proposal, and renumbered the rest. This is an editorial change and does not change any of the proposed requirements.

In paragraph (c)(5) of the final rule, OSHA is using the language of paragraph (c)(4) of the proposal with one change. Paragraph (c)(4)(ii) of the proposal has been revised as paragraph (c)(5)(ii). The new requirement still requires the use of a five minute escape self-contained breathing apparatus, however, its need is now based upon two conditions. In the proposal, all employees had to have access to an ESCBA during initial site entry. Two commenters, the State of Wyoming (10–9) and CDM Federal Programs Corporation (10–83), suggested that OSHA revise this requirement to recognize that the use of ESCBAs should be determined by the nature of the health hazards and the nature of the work to be performed. OSHA agrees that all employees who cannot be exposed to site conditions where possible health hazards may occur should not be required to carry ESCBAs. Therefore OSHA has amended its proposal as follows. Two conditions will now limit the employee population that must be provided access to ESCBA. They are (1) if positive-pressure self-contained breathing apparatus is not used as part of the entry personal protective equipment; and (2) if respiratory protection is warranted by the potential hazards identified during the preliminary site evaluation. Workers in populations where these two conditions are not met need not be provided with ESCBA.

Paragraphs (c)(5)(i), (c)(5)(iii), and (c)(5)(iv) contain the exact language as proposed in paragraphs (c)(4)(i), (c)(4)(iii), and (c)(4)(iv).

In paragraph (c)(6) of the final rule; OSHA is using the language from paragraph (c)(5) of the proposal with some changes.

In paragraph (c)(6)(ii) of the final rule, OSHA has required that direct reading instruments be used where available. In the proposal OSHA had required only that appropriate OSHA had required that appropriate equipment be used. The agency believes that direct reading instruments, where they are available for specific chemical hazards, will provide a more expeditious assessment of the hazards when there is not enough time during a specific work cycle to send samples out to a laboratory for analysis. In some situations, employees may be present at a particular job site for only a brief time. Certain sampling techniques, other than direct reading instruments, may require a longer time for analysis than the employee's actual

exposure time on the job. Therefore, OSHA is amending its proposal by recognizing direct reading instruments as an alternative to standard testing procedures. OSHA has added the phrase "appropriate direct reading test equipment" in place of "appropriate equipment."

OSHA has also added a paragraph (c)(6)(iv) that would require that an ongoing air monitoring program be implemented in accordance with paragraph (h) of the final rule after site characterization has determined that the site is safe for start-up of operations. This is not a new requirement since it uses the same language as that proposed in paragraph (c)(8) of the proposal. OSHA has moved the paragraph from its position in the proposal to paragraph (c)(6)(iv) of the final rule because it is related to the subject matter of paragraph (c)(6). OSHA considers this to be an editorial change because there is no change in the proposed language.

Paragraphs (c)(6)(i) and (c)(6)(iii) continue to use the language of proposed paragraphs (c)(5)(i) and (c)(5)(iii).

In paragraph (c)(7) of the final rule OSHA is using the language of proposed paragraph (c)(6) with one change. In the note which describes risks to be considered, OSHA has amended paragraph (a) by changing the language to reflect the exposure limits to be used in the final rule. Direct reference to Permissible Exposure Limits (PELs), Threshold Limit Values (TLVs), or Recommended Exposure Limits (RELs) has been deleted and a reference is made to permissible exposure limits and published exposure levels as defined in the final rule. No substantive change is made since those terms incorporate PELs, TLVs, and RELs by definition.

Paragraph (d)—Site Control

In paragraph (d) of the final rule OSHA is using the language of paragraph (d) of the proposal. Minor editorial changes have been made for clarity without changing the proposed requirements. The need for requirements for site control is discussed at 52 FR 29625 in the preamble to our proposal. There were few substantive comments. OSHA concludes that these provisions are necessary as discussed in the proposal.

Paragraph (e)—Training

The proposed rule included specific provisions for initial and routine training of employees before they would be permitted to engage in hazardous waste operations that could expose them to

9304   Federal Register / Vol. 54, No. 42 / Monday, March 6, 1989 / Rules and Regulations

safety and health hazards. Section 126(b)(2) of SARA requires initial and recurrent training to be included in the final rule. The intent of the final training provisions is to provide employees with the knowledge and skills necessary to perform hazardous waste clean-up operations with minimal risk to their safety and health.

The proposed requirements for training in paragraph (e) addressed the needs of employees who will be working at CERCLA sites, certain RCRA sites, and sites designated or identified for clean-up by state or local governments.

The proposed provisions included a minimum of 40 hours of initial instruction off the site, and a minimum of three days of actual field experience under the direct supervision of a trained and experienced supervisor, at the time of job assignment. Congress has specifically imposed these hour and day requirements under section 126(d) of SARA for the proposed final standard. The proposed requirement represented a one-time effort by the employer for each employee covered by this standard. Employees would not need to be retrained for 40 hours at each site at which they work. Employees who had received the required training at one site could use that training to meet the proposed requirement at other sites even if it involved a different employer, provided the previous training addressed the hazards at the new site.

There are often many hazards at a waste site. The employee must be trained to recognize the hazards and appropriate work practices to minimize those hazards. The employee must also be well trained in the use of respirators and other forms of personal protective equipment. Without training, that equipment may not be used effectively and may not provide adequate protection. An extensive training program is necessary to assure that employees can use personal protective equipment effectively.

Managers and supervisors at regulated facilities, who are directly responsible for the site's operations, must have the same training as that of site employees and additional time for specialized training on managing hazardous waste operations. Since these managers and supervisors are responsible for directing others, it is necessary to enhance their ability to provide guidance and to make informed decisions. Section 126(d)(2) of SARA provides that there shall be eight hours of additional training for supervisors and managers.

The provisions also proposed that employees be retrained on an annual basis on relevant matters such as review

of health hazards and the use of personal protective equipment. Employees at hazardous waste operations may face serious health and safety risks. Reminders are needed of this and of work practices necessary to avoid hazards. Personal protective equipment provides much of this protection. If there is no retraining in the use, care and maintenance of personal protective equipment, such equipment is unlikely to be properly utilized to provide adequate protection.

In all areas of training, whether it be for general site employees, supervisors at the site, or for the use of specific equipment, the level of training provided must be consistent with the worker's job function and responsibilities. Refresher training must be provided to reemphasize the initial training and to update employees on any new policies or procedures.

Section 126(d)(3) of SARA requires that OSHA provide for certification that an employee has received the training required by the standard. Section 126(d)(1) provides that OSHA not require training for employees who have already received equivalent training. The final standard has provisions to meet this directive.

OSHA requested comment as to whether its proposed training requirements were appropriate for hazardous waste operations. OSHA's proposed training requirements in paragraph (e) were limited to hazardous waste operations that involve the clean-up of uncontrolled hazardous waste disposal sites. Of all the issues raised by OSHA in its proposal, training was one that received a substantial amount of comment. Important comments directed to the paragraph (e) training requirements follow. Comments addressing the training of emergency response workers will be discussed later in this preamble under the appropriate paragraphs.

In paragraph (e)(1) of the final rule OSHA has combined the introductory paragraph of proposed paragraph (e) with the language proposed in paragraph (e)(5). The introductory paragraph of the proposal was designated paragraph (e)(1)(i) and proposed paragraph (e)(5) has been designated (e)(1)(ii). OSHA considers this an editorial change which groups two general requirements under a single paragraph titled "General."

In paragraph (e)(2) of the final rule OSHA is using the language of paragraph (e)(1) in the proposal. Some minor changes are made to reflect the renumbering of the paragraph without changing any of the proposed requirements.

In paragraph (e)(3) of the final rule OSHA is revising the proposed language of paragraph (e)(2) of the proposal. Several comments addressed the proposed 40-hour training requirement for all employees who work on hazardous waste sites (i.e., Wassau Insurance Company, 10–8; International Technologies, 10–44; Cooperweld Steel, 10–41; James T. Dufour, 10–78). Some of the commenters believed that 40 hours of training for some employees at this type of site was excessive. For example, it was argued that 40 hours of training was excessive for general laborers who may be installing perimeter fencing around an unopened site and who are not exposed to any hazards. This type of employee normally will not be wearing the type of protective equipment or be performing the type of tasks normally associated with removal of hazardous wastes. On the other hand, employees who will be "digging in the dirt" after the site has been opened in order to remove hazardous waste may need additional training because of the types of equipment they will be using and the types of hazards to which they will be exposed.

Wassau Insurance commented, "I feel the 40 hour minimum training requirement is excessive for many employees who will never be required to work above level D protection." The commenter continues, "The excessive training requirements of the current proposal add a significant burden to employers in situations where only low levels of protection are required (e.g., level D and level C situations)."

OSHA has revised its proposal for 40 hours of training for all employees engaged in hazardous waste operations at uncontrolled hazardous waste sites. For general site workers, OSHA is retaining the 40-hour, three-day on-the-job training requirement. OSHA has concluded that this level of training is necessary to protect general site workers because they are engaged in difficult work in areas with safety and health hazards. Moreover, OSHA believes the Congressional language is quite clear on this matter.

However, for certain types of other workers, OSHA has concluded that less training may be appropriate. For example, those workers who visit sites only on occasion and then under the supervision of experienced site workers are required to have 24 hours of training and one-day of on-the-job training. OSHA has also concluded that this same level of training would be appropriate for those general site workers who work in areas which have been monitored and fully characterized

Federal Register / Vol. 54, No. 42 / Monday, March 6, 1989 / Rules and Regulations 9305

indicating that exposures are under both permissible exposure limits and published exposure limits and that respirators are not necessary.

In paragraph (e)(4) of the final rule OSHA is using the language proposed in paragraph (e)(3).

In paragraph (e)(5) of the final rule OSHA is using the language proposed in paragraph (e)(4) with the addition of a new sentence. Some commenters thought that the proposed language for the qualification of trainers was too broad and ambiguous. The State of Indiana (10–23) offered a representative comment: "Knowledge or training equivalent to (redundant phrase removed) a level of training higher than the level that they are presenting is no assurance that an employee is capable of providing adequate training to other employees."

Another commenter, the International Union of Operating Engineers (10–58), stated, "We believe it irresponsible to summarily state that trainers must be 'qualified,' without defining the term other than to suggest that one who knows more than the person he trains may be a qualified trainer."

Subsequent to the receipt of post-hearing briefs, Congress amended section 126(d) of SARA to require the Secretary of Labor to develop requirements for the certification of training programs offered to employees and employers who must meet the training requirements of this standard. OSHA will soon be publishing a Notice of Proposed Rulemaking to carry out this Congressional direction. The requirements of that rulemaking will expand on the provisions stated in this rulemaking.

In order to provide interim guidance to employees and employers in determining the competency of trainers and their qualifications, OSHA has added two sentences to the proposed language. These sentences require the use and demonstration of training, credentials and experience to show competency as a trainer.

In paragraph (e)(6) of the final rule OSHA is using the language of proposed paragraph (e)(6) with one minor change. In addition to permitting certification to be given by the classroom instructor, OSHA will also recognize certifications given by the head or supervisory instructor of the training facility. This change recognizes the fact that some training certificates are signed by the head instructor upon recommendation of the classroom instructor, rather than by the individual classroom instructor.

In paragraph (e)(7) of the final rule OSHA is using the exact language of proposed paragraph (e)(7).

In paragraph (e)(8) of the final rule OSHA is using the language of proposed paragraph (e)(8) with the addition of an example of the type of refresher training that OSHA would consider acceptable. OSHA considers, and has now suggested, that critiques of prior emergency response performance can serve as a means of refresher training. Critiques of performance during an emergency response can give employees a training experience in which they have actual knowledge of the acceptable or nonacceptable actions taken during the response. Such critiques can also provide employees with the experience they may need to perform in a more appropriate manner during their next response. The proposed requirement for annual refresher training has not been changed.

In paragraph (e)(9) of the final rule OSHA is using the exact language of paragraph (e)(9) in the proposal.

Paragraph (f)—Medical Surveillance

The proposed rule included specific provisions for baseline, periodic and termination medical examinations. Section 126(b)(3) of SARA provides that this rule include requirements for medical examinations of workers engaged in hazardous waste operations. In addition, the EPA manual referred to in section 126(e) of SARA has more detailed requirements for initial or baseline, periodic and termination medical examinations. The clear Congressional direction is to provide a comprehensive medical surveillance program for employees engaged in hazardous waste operations where it is medically prudent.

In paragraphs (f)(1) and (f)(2) OSHA is making some changes for clarity. In addition, OSHA is using the new term "permissible exposure limits or published exposure levels" instead of the term "established exposure levels." The reasoning for this change has been discussed under the paragraph of this preamble addressing definitions.

OSHA would like to clarify an issue concerning who is covered by medical surveillance under paragraph (f)(2) that has cause confusion since the promulgation of the interim final rule. After reviewing the record of comments addressing medical surveillance, it seems that several commenters, in particular from the fire service (i.e., 10–1, 10–3, 10–4, 10–12, 10–32 10–79), believe that all firefighters must have the medical surveillance protections of paragraph (f) since they may wear respirators 30 days or more a year. Firefighters responding to structural fires will typically wear self-contained breathing apparatus when they enter

burning structures or other hazardous locations and they may make such responses 30 days or more a year. OSHA is not requiring all firefighters who wear respirators 30 days or more a year to have medical surveillance. Paragraph (f) applies only to individuals within the scope of paragraph (a)(1)(i) through (a)(1)(iii) and (a)(2)(ii). Typical firefighters from local fire departments do not fall within this scope. These firefighters are normally covered by the requirements of paragraph (q) as specified in paragraph (a)(2)(iv). Paragraph (q) does not contain requirements for medical surveillance of firefighters unless they are members of an organized and designated hazardous materials response team, are hazardous materials specialists, or have been injured due to an overexposure to health hazards during an emergency incident involving hazardous substances as established in paragraphs (e)(9) (i) and (ii) of the final rule.

In paragraph (f)(3) of the final OSHA is using the language proposed in paragraph (f)(2) with some changes. In new paragraph (f)(3)(i)(B), OSHA is adding the phrase "unless the attending physician believes a longer interval is appropriate" to the proposed language of paragraph (f)(2)(i)(B). Several commenters (State of Wyoming, 10–9; American Society of Safety Engineers, 10–29; Union Carbide Corporation, 10–58) suggested that an annual medical examination may be excessive for some employees, particularly when an attending physician can make a recommendation for a less frequent schedule. The American Society of Safety Engineers (10–29) stated, "This reviewer concurs in the approach that OSHA has outlined in this comment area that the practical health benefit of annual medical examination for hazardous waste operation workers is indeed uncertain. This is a broad area that requires input from the attending physician, the employee and the employer. It is recommended that annual medical examination *not* be required rigidly, that this be a flexible time frequency."

Wyoming (10–9) stated, "Periodic occupational health physical examination on an annual basis may not be warranted under all conditions." They go on to state, "It seems reasonable that a good occupational health program requiring physical examination would be based upon documented personal exposure levels and a medical physician's recommendation rather than on an arbitrary administrative decision to require personnel to undergo annual

9306 **Federal Register** / Vol. 54, No. 42 / Monday, March 6, 1989 / Rules and Regulations

periodic physicals if they fit into the categories under § 1910.120 (f)(1)(i) and (f)(1)(ii)."

Union Carbide (10–56) said, "The frequency of medical examinations and consultations in this proposed rule has been redefined and the proposed change clarifies the issue of medical surveillance but retains the annual requirement for 'all employees who wear a respirator * * *.' This frequency of examination is arbitrary. There is not medically-supportable rational for this annual requirement."

There were also comments in support of OSHA's annual physical examination requirement. The Occupational Health Nurses (10–30) stated, "AAOHN supports pre-exposure, annual, and exit examinations with provision of additional exams if over-exposure or signs or symptoms develop." Lockheed (10–45) responded to OSHA's question on whether examinations should be performed yearly, or at other intervals by stating, "Medical exams should be performed at least yearly."

GSX Chemical Services, Inc. (10–63) stated, "(12) Paragraph (f) describes medical surveillance requirements. The general program described by OSHA for pre-employment, annual, post-exposure, and termination medical examinations is excellent."

BP America, Inc. (10–85) stated, "The need for medical surveillance of workers who would be covered under the provisions of the proposed regulation is appropriate and is supported." They further state, "The proposed requirement to examine workers exposed in emergency situations, but not continue periodic surveillance simply because of the single episode, per se, is logical, and is strongly supported. Having such employees continue under periodic medical surveillance on the basis of the findings of the medical examination is, of course, appropriate."

Because of variations in employee exposures due to work schedules, annual physicals may not be medically necessary. OSHA concludes that annual medical examinations may not always be appropriate. Accordingly the standard is amended to permit the physician to reduce the frequency to not less than bi-annually if the physician believes it is appropriate. The physician may also increase the frequency if it is medically appropriate.

OSHA has also replaced the term "established exposure limits" with the phrase "permissible exposure limits or published exposure levels" in new paragraph (f)(3)(i)(D) since the terms have been redefined as previously explained.

The rest of the language in new paragraph (f)(3) remains as it was proposed in paragraph (f)(2).

In paragraph (f)(4)(i) of the final rule OSHA is using the exact language proposed in paragraph (f)(3)(i).

In paragraph (f)(4)(ii) of the final OSHA is using the language of proposed paragraph (f)(3)(ii) with one change. OSHA is still requiring that the content of medical examination and consultations be determined by the attending physician. However, OSHA has added language that would direct the employee, employer, and physician to Appendix D for guidelines in developing the examination.

Several commenters requested guidance on the content of the medical examinations required by the proposal. The Okolona Fire District (10–1) commented, "As, written the current document is rather vague." They continued, ". . . the document should give guidance on what the physical examination should entail." The American Association of Occupational Health Nurses (10–30), suggested, "At least minimum content of the physical examination should be specified. An "exam" may be no more than visual inspection of an individual's eyes, ears and throat and have no relevance to the exposure situation."

Other commenters supported OSHA's proposal for the employer and the physician to determine examination protocols. Eastman Kodak (10–38) commented, "We support OSHA's position that the physician is best able to determine an appropriate medical surveillance protocol. As noted by OSHA, employees may be exposed to differing substances and may be required to use differing levels of personal protective equipment, such as respirators. In view of the particular circumstances presented, the physician is in the best position to formulate and follow an appropriate medical protocol. OSHA should not include a detailed protocol for medical surveillance." Lockheed (10–45) responded to OSHA's issue on protocols, "No. As with training, differences in amounts, kinds and combinations of exposures in different working situations require that protocol for medical surveillance be left to the discretion of the attending physician."

Dr. James Melius testified, "I'd like to direct most of my testimony to discussions of medical surveillance programs for hazardous waste and emergency response workers. I'd like to begin be saying that programs for both of these sets of workers are extremely important." (Tr. pg. 107) He goes on to say, "The medical surveillance program

for the workers, therefore, should start with initially assessing their ability to work at the site and their capability for conducting that work. It should include an assessment that focuses through a medical history and initial physical examination on their cardiovascular and respiratory system, also looking for signs of other major medical problems. Selective testing may also be useful in these instance, including pulmonary function testing, chest x-rays and electrocardiograms. However, the workers may differ in their benefits from this testing depending on their age and other risk factors." (Tr. pgs. 110–111)

OSHA believes both sides of the argument can be addressed by placing recommended criteria for medical examination protocols in the Appendix to this section. Some commenters have suggested protocols that OSHA considered for placement in the Appendix. The St. Petersburg Fire Department (10–4) suggested, "A full physical examination: height, weight, eyesight, pulse, blood pressure, respiratory, skin examination, neurological examination, heart and lungs, medical history, and any other aspects determined by the physician. Also included are: Pulmonary function test, chest X-ray, urine analysis, SMA 18 blood test, and hearing examination." The chapter on medical surveillance found in the OSHA/NIOSH/EPA/Coast Guard manual in Appendix F also provides guidance. OSHA also believes that the language of Appendix F will provide guidance for developing the examination protocol.

In paragraph (f)(5) of the final rule OSHA is using the language of paragraph (f)(4) in the proposal with one change. OSHA has added a recommendation that a physician licensed in occupational medicine be used to supervise or administer the examination. Several commenters suggested that the use of such a physician would assure a more complete occupation-oriented examination than one offered by a physician licensed in another field.

Representative of these comments was the suggestion of the American Association of Occupational Health Nurses (10–30). The AAOHN (10–30) stated, "The nature of the potential exposures in hazardous waste operations requires specialized knowledge in toxicology—knowledge of signs and symptoms and effects of exposure to various substances—not common in basic health professional curricula. This is information that both occupational health nurses and physicians may have via advance

Federal Register / Vol. 54, No. 42 / Monday, March 6, 1989 / Rules and Regulations   9307

education degrees or continuing education, certification and experience." The AAOHN recommended that OSHA change its proposed language to require the examination to be performed "by a registered professional nurse or licensed physician with training and expertise in evaluating exposures to hazardous substances."

In recognition of AAOHN's comments, OSHA has added the recommendation for the use of a physician from the field of occupational health. The language of the final rule, while it does not preclude the use of occupational nurses, does not specifically call for the use of an occupational nurse. The final language requires that the examination be conducted under the supervision of a licensed physician and that would certainly allow the use of occupational nurses if the attending physician permits.

In paragraphs (f)(6), (f)(7) and (f)(8) of the final rule OSHA is using the exact language proposed in paragraphs (f)(5), (f)(6) and (f)(7)

Paragraph (g)—Engineering controls, work practices, and personal protective equipment for employee protection

OSHA is using the same opening paragraph for paragraph (g) that was in the opening paragraph for paragraph (g) in the proposal.

In paragraph (g)(1)(i) of the final rule OSHA is using the language of paragraph (g)(1)(i) of the proposal.

In paragraphs (g)(1)(ii) and (g)(1)(iii) of the final rule OSHA is using the exact language of paragraphs (g)(1)(ii) and (g)(1)(iii) of the proposal, except that the reference to Subpart G is deleted. A new paragraph (g)(1)(iv) is added to cross reference the requirements of Subpart G for clarity.

In paragraph (g)(2) of the final rule OSHA is using the language proposed in paragraph (g)(2) with some editorial modifications.

In paragraphs (g)(3), (g)(4) and (g)(5) of the final rule OSHA is using the language of paragraphs (g)(3), (g)(4) and (g)(5) in the proposal with minor editorial corrections to be consistent with the terms and language of the final rule.

Paragraph (h)—Monitoring

In paragraph (h)(1) of the final rule OSHA has combined the proposed language in the opening paragraph and paragraph (h)(1) of the proposal with a clarification. The new paragraphs are designated (h)(1)(i) and (h)(1)(ii).

In paragraph (h)(1)(i), OSHA has modified its proposed language by adding the phrase, "where it is not obvious that an exposure does or does

not exist." OSHA is adding this phrase to clarify that monitoring is not necessary where the site environment or safety precautions taken by the employer prevent employee exposure to hazardous levels of chemical exposure. OSHA is only requiring monitoring where there may be a question as to an employee's exposure. When there is a question then the employer should monitor. Where there is no question of exposure, then monitoring is not necessary. For example, if it is obvious through site characterization and analysis that there are no exposures at the worksite, monitoring need not be performed unless worksite conditions or work practices change to the extent that workers could be potentially exposed to hazardous concentrations of chemical exposure. If an employer decides that employees should wear level B protection in an area where exposure will most probably be below the PEL's, then during initial entry monitoring will not be necessary because the employees are more than adequately protected.

In paragraphs (h)(2) and (h)(3) of the final rule, OSHA is using the language proposed in paragraphs (h)(2) and (h)(3) except for two changes. First, OSHA is adding language to clarify that monitoring should be used to determine exposure above permissible exposure limits which are not immediately dangerous to life or health. Second, OSHA is deleting proposed subparagraph (h)(3)(v) because it is too general in nature and the previous four subparagraphs adequately cover the hazard.

In paragraph (h)(4) OSHA is using the exact language proposed in paragraph (h)(4) with one addition. If employees with the highest exposure are overexposed, then representative samples of other employees who may be overexposed must be taken to determine if controls or PPE are needed.

Paragraph (i)—Informational programs

In paragraph (i) of the final rule OSHA is using the language of paragraph (i) of the proposal. Minor editorial changes have been made for clarity without changing the proposed requirements. The need for requirements for informational programs is discussed at 52 FR 29628 in the preamble to our proposal. There were few substantive comments. OSHA concludes that these provisions are necessary as discussed in the proposal.

Paragraph (j)—Handling drums and containers

In paragraph (j) of the final rule OSHA is using the language proposed in paragraph (j). Minor editorial changes

have been made for clarity without changing the proposed requirements. The need for requirements for handling drums and containers is discussed at 52 FR 29629 in the preamble to our proposal. There were few substantive comments. OSHA concludes that these provisions are necessary as discussed in the proposal.

Paragraph (k)—Decontamination

In paragraph (k) of the final rule OSHA is using the language of paragraph (k) in the proposal. However, the agency has reorganized the paragraph and provided headnotes to make the reading of the paragraph easier. The need for requirements for decontamination is discussed at 52 FR 29629 in the preamble to our proposal. There were few substantive comments. OSHA concludes that these provisions are necessary as discussed in the proposal.

Paragraph (l)—Emergency response by employees at uncontrolled hazardous waste sites

In paragraph (l)(1) OSHA is using the exact language from proposed paragraph (l)(1)(i).

In paragraphs (l)(2)(i) through (l)(2)(xi) OSHA is using the exact text from paragraph (l)(1)(ii)(A) through (l)(1)(ii)(K).

In paragraph (l)(3) OSHA is using the language of proposed paragraph (l)(2)(i)(A) with some modification. The modifications are considered editorial and are made because of OSHA's reorganization of the overall proposed paragraph (l). In paragraph (l)(3) OSHA will require that employees performing emergency response at uncontrolled hazardous waste sites be trained in accordance with paragraph (e) of this section. This requirement is the same as proposed in the final part of proposed paragraph (l)(2)(i)(A). The portion of proposed paragraph (l)(2)(i)(A) that addresses training at RCRA sites is moved to the discussion of training in paragraph (p) of this rulemaking because of OSHA's reorganization of this paragraph.

The language proposed in paragraph (l)(2)(i)(B) has been moved to paragraph (e)(9) of this final rule. This move is considered editorial since it does not change any duties imposed on the employer, it only reflects the reorganization of proposed paragraph (l).

In paragraphs (l)(4)(i) through (l)(4)(vii) OSHA is using the exact language from paragraphs (l)(2)(ii)(A) through (l)(2)(ii)(G).

9308   Federal Register / Vol. 54, No. 42 / Monday, March 6, 1989 / Rules and Regulations

In summary, paragraphs (l)(1) through (l)(4) of the final rule use the language of paragraphs (l)(1) and (l)(2) of the proposal with some modifications due to the reorganization of the emergency response requirements of the proposal.

Paragraph (m)—Illumination

In paragraph (m) and Table H–120.2 of the final rule OSHA is using the language of paragraph (m) and Table H–102.1 of the proposal with one minor change. OSHA has combined the language of the opening paragraph and paragraph (m)(1) of the proposal into one paragraph designated paragraph (m). Minor editorial changes have been made for clarity without changing the proposed requirements. OSHA has combined the language of the opening paragraph and paragraph (m)(1) of the proposal into one paragraph designated paragraph (m). The need for requirements for illumination is discussed at 52 FR 29631 in the preamble to our proposal. There were few substantive comments. OSHA concludes that these provisions are necessary as discussed in the proposal.

Paragraph (n)—Sanitation at temporary workplaces

In paragraph (n) of the final rule OSHA is using the language of paragraph (n) in the proposal with some minor editorial changes. The opening paragraph of proposed paragraph (n) has been deleted because it is not a requirement, and Table H–102.3 has been renumbered Table H–102.3. Minor editorial changes have been made for clarity without changing the proposed requirements. The need for requirements for illumination is discussed at 52 FR 29631 in the preamble to our proposal. There were few substantive comments. OSHA concludes that these provisions are necessary as discussed in the proposal.

Paragraph (o)—New technology programs

In paragraph (o) of the final rule OSHA is using the language of proposed paragraph (p). This change is necessary due to the reorganization of the emergency response requirements and the moving of proposed paragraph (o), *Certain Operations Conducted Under the Resource Conservation and Recovery Act of 1976 (RCRA)*. Proposed paragraph (o) has been moved to paragraph (p) of the final rule.

In paragraph (o)(1) of the final rule OSHA is using the exact language that was proposed in paragraph (p)(1).

In paragraph (o)(2) of the final rule OSHA has used the language of paragraph (p)(2) with some changes.

OSHA has revised the paragraph to include some additional examples of acceptable means of suppression. The agency has also added additional information to provide guidance to the employer in making evaluations of products and new technologies. These changes are considered to be editorial since the requirement of the proposal has not changed.

Paragraph (p)—Certain operations conducted under the Resource Conservation and Recovery Act of 1976 (RCRA)

In paragraph (p) of the final rule OSHA is using the language proposed in paragraph (o) with some changes.

OSHA has revised the opening paragraph of the proposal to include large quantity generators of hazardous waste that store those wastes less than 90 days within the scope of this paragraph.

In paragraphs (p)(1), (p)(2), (p)(3), and (p)(4) of the final rule OSHA has used the proposed language of paragraphs (o)(1), (o)(2), (o)(3), and (o)(4) with some minor editorial changes. The proposed requirements for each individual paragraph remain the same.

OSHA is adding two new paragraphs. (p)(5) and (p)(6), to address new technology programs and material handling programs respectively. In paragraph (p)(5) OSHA requires the employee to develop and implement procedures for using new technologies and equipment. Congress, in the SARA legislation, directed OSHA to address new technology programs in its rule. The language of the proposal limited new technology programs to uncontrolled hazardous waste sites. OSHA is adding this paragraph to complete Congress's directive and to address these programs at RCRA TSD facilities.

In paragraph (p)(6) OSHA is requiring employers to develop and implement a material handling program for the same reasons as stated above.

In paragraph (p)(7) OSHA is using the language from paragraph (o)(5) of the proposal with some changes. In paragraph (p)(7)(i) OSHA is using the language of paragraph (o)(5)(i) with one change. OSHA has moved a requirement to paragraph (p)(7)(i) from the last sentence of proposed paragraph (o)(5)(ii) that requires employees with a certificate indicating that they have successfully completed the training required in the paragraph. OSHA believes that the issuance of this certificate will make it easier for employers to determine if new employees have completed the necessary training and are .eady for employment.

In paragraph (p)(7)(ii) of the final rule OSHA is using the language from paragraph (o)(5)(ii) of the proposal with two exceptions. First, the last sentence of proposed paragraph (o)(5)(ii) has been moved to paragraph (p)(7)(i) of the final rule as discussed above. Second, the requirement for eight hours of annual refresher training is added to this paragraph. OSHA has added this requirement to this paragraph because the new format of the final rule now addresses training for new employees and current employees separately. In the proposal there was no distinction between the two groups of employees.

In paragraph (p)(7)(iii) OSHA has added a new paragraph addressing the training of trainers who will be providing the required training to employees. OSHA received many comments on trainers' qualifications. The proposed language for RCRA facilities did not address these qualifications. Therefore OSHA is now requiring that trainers be properly trained and qualified to conduct the type of training that they are expected to provide.

In paragraph (p)(8) of the final OSHA is addressing emergency response at RCRA facilities. Paragraph (p)(8) addresses the subject matter proposed in paragraph (1) of the proposal as that paragraph applied to RCRA TSD facilities. Most of the language used in this paragraph has been taken from proposed paragraphs (l)(1).

In paragraph (p)(8)(i) of the final rule OSHA has used some of the language from paragraph (l)(1)(i) of the proposal. The basic requirement for the development and implementation of a written emergency action plan that addresses site procedures for handling emergency response is the same in the final rule as it was in the proposal. OSHA will still permit an exemption from this paragraph if the employer totally evacuates the facility at the time of the emergency and has an emergency action plan meeting the requirements of 29 CFR 1910.38(a). OSHA considers the changes made in this paragraph to be editorial since the proposed obligations of the employer remain the same.

In paragraph (p)(8)(ii) of the final rule OSHA has used the language of proposed paragraph (l)(1)(ii). This paragraph contains the minimum elements that must be addressed in the employers emergency response plan. The basic elements of the required plan remain the same as proposed.

As stated before, training and certification of training were among the many issues discussed during the rulemaking for this final rule. Several commenters indicated that there was a need for more specific training criteria for the courses to be offered and the quality of the instructors presenting the courses. In light of those comments, OSHA has added a new paragraph (p)(8)(iii) that addresses emergency response training on RCRA TSD facilities. The language that is used in the final rule was developed from that suggested in the comments made to the record of this proceeding.

Basically OSHA is requiring that all employees who are expected to perform emergency response at RCRA TSD facilities be trained in how to safely perform emergency response duties prior to being called upon to perform those duties [See paragraph (p)(8)(iii)(A).] Examples of the types of training to be provided have been given. Exemptions are provided in Exception #1 and Exception #2 when employee exposure is reduced through pre-emergency planning that includes development of employee awareness of hazards. OSHA is also requiring that employees who have attended and successfully completed the training that is required in paragraph (p)(8) be certified as having done so. Employers would also have to certify the continued competency of employees on an annual basis [See paragraph (p)(8)(iii)(C)].

In paragraph (p)(8)(iv) of the final rule OSHA is addressing the procedures to be used for handling emergency incidents. The language in the final rule has been taken from paragraph (1)(2)(ii) and the requirements remain the same as proposed.

Paragraph (q)—Emergency response to hazardous substance releases not previously covered

In paragraph (q) OSHA is covering those emergency response situations that occur at locations other than uncontrolled hazardous waste sites and RCRA TSD facilities. The typical site covered by this paragraph would be a transportation accident where hazardous substances are or have the potential for leaking into the environment. Other sites covered by this paragraph would include hazardous substance releases at chemical manufacturing facilities such as the release that occurred at the Union Carbide plants in Buphol, India, and Institute, WV.

A typical scenario where this paragraph would be applicable would be the emergency response to a derailed tank car containing a hazardous substance that has begun to leak its contents into the atmosphere. The emergency response to this type of accident would usually include the first

responders (i.e., witnesses, police, employees on the train), the first dispatched-responsers (i.e., the first due rescue and fire apparatus), any multiple alarm dispatches (i.e., additional fire and rescue apparatus, HAZMAT teams, state fire marshal, Coast Guard or Federal E.P.A. national response teams), and the clean-up crew (i.e., initial response employees of the site owner who clean-up the release). Employees of outside clean-up contractors would be covered by paragraphs (b) through (p).

As the clean-up scenario proceeds towards completion, the various employees on the scene will need different levels of training and protective equipment required in this paragraph.

In paragraph (q)(1) of the final rule OSHA is using the language taken from paragraph (1)(1)(i) with some minor editorial changes. OSHA wants to emphasize that employers who will evacuate their employees from the workplace when an emergency occurs and who do not permit any of their employees to assist in handling the emergency are exempt from the requirements of this paragraph if they provide an emergency action plan in accordance with § 1910.38(a).

In paragraph (q)(2) of the final rule OSHA is using the exact language of paragraph (1)(1)(ii).

In paragraph (q)(3) of the final rule OSHA is using the language proposed in paragraph (1)(3)(ii) with the following changes. In paragraph (q)(3)(i) OSHA has used the language proposed in paragraph (1)(3)(ii)(A) with some change. OSHA has deleted the requirement that the senior official responding to an hazardous substance emergency establish the Incident Command System (ICS). As a result of other requirements in this final rule, the Incident Command System should already be established prior to an emergency. The senior official responding to an incident scene should only need to take charge of the incident and begin to implement the preplanned ICS.

In paragraph (q)(3)(iv) OSHA has used the proposed language of paragraph (1)(3)(ii)(D) with a change. The proposed language required all employees engaged in emergency response and exposed to hazardous substances in any way to wear positive pressure self-contained breathing apparatus while engaged in emergency response. The final rule will require only those employees engaged in emergency response and exposed to hazardous substances "presenting an inhalation hazard or potential inhalation hazard" to wear positive pressure self-contained breathing apparatus. OSHA has made

this change since several comments suggested that some individuals engaged in emergency response may be exposed to hazardous substances that do not pose an inhalation hazard and, therefore, would negate the need for respiratory protection. Such protection would become a burden to those employees engaged in operations not requiring the use of such equipment.

In paragraph (q)(3)(vi) of the final rule OSHA has used the language of paragraph (1)(3)(i)(F) with the following change. In the proposal OSHA called for "qualified basic life support" personnel to be present at the site. In some emergency medical service (EMS) systems the term "basic-life support (BLS)" identifies a unique group of trained individuals who have received an established level of specialized training. Typically emergency medical response begins at the first-responder level, and progresses through basic-first aid and basic-life support to advanced-life support (ALS). The amount of training and expertise increases as individuals progress through the system. As a result of several comments, OSHA has decided to reduce the level of training required for a minimum stand-by capability at a hazardous waste sites. Employees trained and qualified in basic first aid have the basic skills such as initial patient assessment, maintenance of airway, control of bleeding, immobilization of fractures, and possibly cardiopulmonary resuscitation (CPR) to control injuries until a higher level responder arrives. If response time for BLS or ALS is long enough that it is necessary for this level of training to be at the site in case of an emergency, this rule does not prohibit the stationing of this level at the site. However, OSHA believes that if BLS or ALS service is available within a reasonable time, a qualified basic first aider can provide the necessary interim care.

The rest of the language in paragraph (q)(3) contains the language that was proposed in paragraph (1)(3)(ii) without change.

In paragraph (q)(4) of the final rule OSHA has used the language from paragraph (1)(3)(i)(C) with some minor editorial changes to reflect the changes made to other paragraphs in this rule. The basic requirement for the use and training of skilled support personnel remains the same as it was proposed.

In paragraph (q)(5) of the final rule OSHA has used the language from paragraph (1)(3)(i)(B) with one major change. OSHA has eliminated the requirement for 24 hours of training for specialist employees and has replaced it

354 FUNDAMENTALS OF HAZARDOUS MATERIALS INCIDENTS

9310   Federal Register / Vol. 54, No. 42 / Monday, March 6, 1989 / Rules and Regulations

with a requirement for annual training or demonstration of competency in their area of specialization. The required minimum hours of training was deleted because some employees may need more or may need less than 24 hours for their area of specialization. Specialized employees are by definition individuals specialized in their area of expertise and should only require whatever level of training is necessary to maintain their level of competency. OSHA considers the other changes made to the language of this paragraph to be editorial.

In paragraph (q)(6) of the final rule OSHA addresses the training requirements for employees who will be responding to hazardous materials incidents. In paragraph (q)(6) (i), (ii), (iii), and (iv) OSHA has provided tiered training criteria for those employees who may be designated as members of an emergency response team. The various levels of response and the required competency levels are based upon recognized levels of response being discussed in the hazardous materials response industry as recommended in several of the comments made during this rulemaking. To illustrate OSHA's tiered approach to training, the following scenario describes a possible emergency response call.

A state trooper is on routine patrol along a highway passing through a residential and light industrial area of a large metropolitan city. Ahead in his path of travel, the trooper notices a multi-vehicle accident involving a large overturned tank truck. Immediately the trooper uses his radio to contact his dispatcher to report the accident. After letting the dispatcher know the location and type of accident, the trooper places his vehicle across the travel lanes of the highway approaching the accident site to stop traffic. While he is doing this the dispatcher is alerting the fire and rescue companies in the immediate area and dispatching an established number of fire and rescue vehicles. The trooper then surveys the accident scene from his vehicle trying to identify the type of cargo on the overturned truck. Seeing three different U.S. DOT placards on the vehicle the trooper makes note of the four digit numbers and checks his DOT Emergency Response Guide for a summary of actions to be taken for the chemicals identified on the placards. After determining his next on-site responsibility, he recontacts his dispatcher with the additional information and secures the scene. He stays away from the immediate accident site and does not become involved in rescue or site mitigation.

While the trooper has been securing the scene, the fire and rescue units dispatched after his first radio call begin to arrive on the scene with the additional information from the trooper's second call. The officer-in-charge (OIC) of the fire/rescue response stops his vehicles in a safe location and contacts the state trooper. After determining the type of accident and vehicles involved, the OIC takes control of the scene and directs his crews to take a predetermined defensive action in controlling a leak that has begun on the tanker. The OIC then contacts the dispatcher and reports his assessment of the accident scene including the fact that the tanker is now leaking. He requests the dispatcher to send him the closest hazardous materials response team. He also asks for representatives from the shipper of the liquid and the liquid's manufacturer.

In the meantime, firefighters have established a perimeter defense of the accident scene using fire hose lines and proper personal protective equipment. They begin to evacuate surrounding homes and businesses as indicated in the Emergency Response Guide in case the leaking tanker should explode. They construct dikes and diversion pits to contain water and chemical run-off from the fire hose lines. Rescue personnel, including emergency medical technicians, have made a preliminary assessment of the accident scene and have determined whether any individuals in the spill area are trapped in their vehicles or need immediate assistance. They report their observations to the OIC.

A decision is made by the OIC, based upon the reports of the police officer, the emergency response crew, and the data on the DOT placards, that no rescue attempts can be made safely until such time as the leaking liquid is positively identified and controlled by the HAZMAT team. The proper local authorities are notified under the requirements of SARA Title III.

As firefighters continue to provide defensive protection of the scene and as emergency medical technicians establish a triage area for the treatment of injured passengers, the HAZMAT team arrives and begins to take control of the accident scene. Hazardous materials technicians and specialists assess the scene and plan their attack on the leaking tanker.

After equiping themselves properly, the HAZMAT team makes a final, pre-attack evaluation of the scene, including a scan of the area with appropriate monitoring equipment, and reports its findings to the fire and rescue personnel.

Based upon the results of the pre-attack evaluation and a determination by HAZMAT team members using monitoring equipment that the spill area is non-hazardous, rescue personnel now enter the area of the accident to provide emergency medical treatment to injured passengers and to extricate those passengers who may have been trapped in their vehicles. The HAZMAT team proceeds to the point of release and secures the leak.

After all the injured have been cared for and after the leak has been stopped, the firefighters and HAZMAT team begin to clean-up the accident scene in accordance with pre-planned procedures.

All four levels of hazardous materials response have played a role in this scenario. The state trooper, the first on the scene, is the first responder awareness level. The first responding fire and rescue companies who provided the defensive attack are the first responder operations level. The responding HAZMAT team had both hazardous materials technicians and hazardous materials specialists. In this scenario the state trooper would have to have a sufficient amount of training, the first responding fire/rescue companies would need eight hours of training, and the HAZMAT team would need 24 hours of training. The tiered training schedule is based upon the duties and responsibilities of the individuals involved in the various levels of response illustrated in the scenario.

In paragraph (q)(7) of the final rule OSHA is addressing the competency of the trainers who will be providing the training necessary for those employees responding to hazardous materials incidents. As discussed before, several commenters were concerned that OSHA's proposal for the qualifications of trainers was too weak.

In paragraph (q)(8) of the final rule OSHA is addressing refresher training for those employees who have been trained in accordance with paragraph (q)(6). In paragraph (l)(3)(i)(A) of the proposal OSHA addressed the training of employees who perform emergency response at non-hazardous waste clean-up sites. OSHA is using this proposed language in paragraphs (q)(8)(i) and (q)(8)(ii) because the language of the proposal was intended to cover the type of emergency response now regulated by paragraph (q).

In paragraph (q)(9) of the final rule OSHA is using the language of paragraph (l)(4)(ii) of the proposal with some editorial change. The basic requirement that employees who are members of an organized or designated

HAZMAT team and hazardous materials specialists receive a baseline physical examination in accordance with paragraph (f) of this section remains the same as proposed.

In paragraph (q)(10) of the final rule OSHA is using the proposed language of paragraph (l)(4)(iii).

In paragraph (q)(11) of the final rule OSHA is using the exact language as proposed in paragraph (l)(5). In paragraph (l)(5) OSHA regulated post-emergency clean-up and the language used in that paragraph has caused some confusion. Rather than change the basic requirement, OSHA is offering the following clarification of the intent of paragraph (q)(11):

Post-emergency response can be performed by two basic groups of employees: employees of the site, or employees from off of the site. Post-emergency clean-up begins when the individual in charge of the initial emergency response declares the site to be under control and ready for clean-up. For the purposes of this rule, paragraph (q)(11) will apply to those employees who come from other employers located off-of-the-site to perform post-emergency clean-up. Employees of the employer at the site where the release occurred, and who perform post-emergency clean-up, are considered, under this rule, to be part of the initial emergency response and not subject to paragraph (q)(11). The reason for this distinction is that employees at the site are more familiar with the types of emergencies that may occur and the types of clean-up operations that may have to take place. The more hazardous exposure to employees occurs when outside contractors or other off-site employees are brought into a strange environment and are expected to clean-up the residue from a release. With this clarification, OSHA concludes that no change to the proposed language is necessary.

III. Summary of the Preliminary Regulatory Impact and Regulatory Flexibility Analysis and Environmental Impact Assessment

Introduction

Executive Order 12291 (46 FR 13197, February 19, 1981) requires that a regulatory impact analysis be conducted for any rule having major economic consequences for the national economy, individual industries, geographical regions, or levels of government. In addition, the Regulatory Flexibility Act of 1980 (Pub. L. 96–353, 94 Stat. 1164 (5 U.S.C. 601 et seq.)) requires the Occupational Safety and Health Administration (OSHA) to determine

whether a regulation will have a significant economic impact on a substantial number of small entities, and the National Environmental Policy Act (NEPA) of 1969 (42 U.S.C. 4321, et seq.) requires the agency to assess the environmental consequences of regulatory actions.

In order to comply with these requirements, OSHA has prepared a Regulatory Impact and Regulatory Flexibility Analysis (RIA) for the hazardous waste operations and emergency response standard. This analysis includes a profile of the industries that will be affected, the estimated number of employees who are at risk from occupational exposures to hazardous wastes, technological feasibility, costs, benefits, and an overall economic impact of the standard. The RIA is available in the OSHA Docket Office.

Data Sources

The primary sources of information used for this analysis are: an April 1987 report by the Eastern Research Group (ERG) entitled, "Preparation of Data To Support a Regulatory Analysis and Environmental Assessment of the Proposed Standard for Working at Hazardous Waste Sites;" and the comments supplied in response to the Notice of Proposed Rulemaking, the comments made during the public hearings, and the post-hearing comments and submissions. The information contained in the ERG report was gathered from the Environmental Protection Agency sources, industry sources, experts in the area of hazardous waste management, etc. Consequently, OSHA believes that it has given due notice to all responsive parties and that the data used are the best available data for this final Regulatory Impact Analysis (RIA).

Industry Profile

The standard will affect about 20,000 uncontrolled hazardous waste sites, about 4,000 hazardous waste operations conducted under the Resource Conservation and Recovery Act (RCRA) of 1976, about 13,600 spills of hazardous materials that occur annually outside a fixed facility, and about 11,000 spills of hazardous material that occur annually inside a fixed facility. The firms that will be affected by this standard are as follows: about 100 contractors that perform hazardous waste site clean-ups, about 50 engineering or technical services firms that perform hazardous waste preliminary assessments or site investigations and remedial investigations or feasibility studies for hazardous waste site cleanups, about

300 RCRA-regulated commercial treatment, storage and disposal facilities; about 3,700 RCRA-regulated facilities that are operated by a hazardous waste generator; about 19,000 state and local police departments; about 28,000 fire departments; about 750 private hazardous materials (HAZMAT) response teams; and about 22,000 manufactures that use in-hours personnel to respond to emergency spills of hazardous materials within the facility.

Population at Risk

As many as 1.758 million employees, may be at risk from exposure to hazardous waste or to hazardous materials during an emergency response to a hazardous material spill. Of these employees, about 14,000 work at uncontrolled hazardous waste site cleanups, 52,700 at RCRA-regulated facilities, 563,200 are police officers, 944,500 are firefighters, 7,500 are private HAZMAT members, and 176,000 are members of industrial fire brigades that provide in-plant emergency responses to hazardous material spills. Most of these employees, however, do not work fulltime around hazardous waste. In fact, most police officers will not face a hazardous material emergency response and most fire fighters and industrial fire brigade personnel, who are at risk, are annually exposed to hazardous materials for only a few hours.

Feasibility

The standard does not require the use of any large-scale capital equipment that is not currently used in normal work operations. In addition, each provision requires equipment and work practices that are currently available. Thus, OSHA has determined that the standard is technologically feasible.

Benefits

This standard will protect 1.757 million employees and firefighters from health and safety hazards caused by their exposure to hazardous wastes. The benefits of this standard are quantified in Chapter 3 of the Final Regulatory Analysis (FRA). The FRA indicates that this standard will prevent 20 cancer deaths per year and from 6 to 20 deaths per year from cardiovascular, neurological, renal and liver disorders. The standard will also prevent 1,925 injuries per year involving 18,700 lost work days. The FRA also estimates that 6 fatalities that are not illness related will be prevented. This last figure is likely to be an underestimate. Individual incidents which are discussed in

356 FUNDAMENTALS OF HAZARDOUS MATERIALS INCIDENTS

Chapter 3 and which may have been prevented by following the standard have sometimes led to more than 6 deaths. Also, the FRA does not take into account the benefits to the surrounding, non-worker community derived from the better handing of hazardous waste and emergency response incidents by the more qualified, properly trained and equiped response teams that are likely to result from compliance with this standard.

Chapter 3 of the FRA also presents risk rates. For example, the 17 excess cancer deaths per 1000 exposed hazardous waste workers for an occupational lifetime of exposures is likely to be reduced by 75 per cent.

OSHA concludes therefore, that this standard will substantially reduce the significant risk of material impairment of health which results from exposure to hazardous waste either at hazardous waste operations or from emergency response.

However, section 126 of SARA gives OSHA clear statutory directions to issue this standard and is reasonably explicit about what type of provisions should be included. Section 126 is also a free standing provision and not an amendment to the OSH Act. Accordingly, it evidences a legislative intent to issue these regulations without the specific need to quantify benefits and reach significant risk conclusions.

Cost of Compliance

OSHA used current work practices as its baseline for estimating the cost of full compliance with the standard. This estimated cost does not include any cost that is currently being incurred by employers as part of their work practices because those work practices, and therefore those costs, would continue whether or not the final standard were promulgated.

OSHA estimated that the total annualized incremental cost of full compliance with the standard will be about $153.422 million, of which $27.966 million will be spent by contractors on government-mandated clean-ups of uncontrolled hazardous waste sites, $18.372 million will be spent by RCRA-regulated facility cleanups and operations, $17.332 million will be spent by police departments, $50.553 million will be spent by fire departments, $4.226 million will be spent by private HAZMAT teams, and $29.179 million will be spent by industrial fire brigades. The provision with the largest annual cost of compliance is the employee training provision ($92.978 million), followed by the medical surveillance provision ($11.293 million), the use of escape self-contained breathing

apparatus ($9.507 million), and the written plan to minimize employee exposure to hazardous materials during postemergency cleanups of hazardous materials spills ($8.381 million).

Economic Impacts

Most of the incremental cost of compliance will be paid by the government or the private firm responsible for the hazardous waste cleanup. OSHA calculated that it is economically feasible for every affected industry or group to comply with the standard. There may be an impact upon some labor markets as a consequence of the provision that only sufficiently experience employees, or employees certified to have received the necessary training at an appropriate training facility, will be allowed to work on hazardous waste sites. This provision will effectively curtail the current practice of using local subcontractors to provide short-term employees for hazardous waste site cleanups and limit the number of employees eligible to work at hazardous waste sites. This in turn, may increase future wage rates and the cost of hazardous waste site cleanups.

Regulatory Flexibility Analysis

Pursuant to the Regulatory Flexibility Act of 1980, the Assistant Secretary has assessed the expected impacts of the standard on small entities. Based on the available information, OSHA determined that the standard may have some impact upon some small entities. The cost of adequately training an employee off-site prior to working at a hazardous waste site cleanup will substantially reduce the use of subcontractor labor on a one-time basis. Thus, some local subcontractors face a potential reduction in hazardous waste site cleanup work. The majority of this subcontracted work will probably be performed by those subcontractors who concentrate upon this type of work. Subcontractors who have performed cleanup work but who do not elect to train employees needed to qualify for future work will probably be excluded from working in this market.

In addition, there could be an economic impact upon some small local fire departments depending upon the amount of financial resources available to them for additional training. With the allowance for different amounts of training hours depending upon the expected extent of involvement with hazardous materials spills, OSHA believes that this economic impact will not significantly affect a substantive number of local fire departments.

Environmental Impact Assessment— Finding of No Significant Impact

OSHA reviewed the final standard and concluded that no significant environmental impacts are likely to result from its promulgation. In OSHA's December 19, 1986, interim final rule for the protection of workers engaged in hazardous waste and emergency response operations, information was solicited from the public on various issues, including possible environmental impacts of the regulation. On the basis of the review detailed below, and in accordance with the requirements of the National Environmental Policy Act (NEPA) of 1969 (42 U.S.C. 4321 et seq.), the Council on Environmental Quality (CEQ) NEPA regulations (40 CFR Part 1500 et seq.), and the Department of Labor's implementing regulations for NEPA compliance (29 CFR Part 11), the Assistant Secretary determined that the standard will not have a significant impact on the external environment.

In most OSHA regulatory actions, two environments may be affected: (1) The workplace environment, and (2) the general human environment external to the workplace, including impacts on air and water pollution, solid waste, and energy and land use. The hazardous waste standard, however, is unique in that it focuses on the external environment because during these operations, the workplace and the external environment are usually one and the same. The standard is also unusual in that it is the first regulation since the passage of the Occupational Safety and Health Act of 1970 (the Act) to be mandated specifically by Congress under section 126 of the Superfund Amendments and Reauthorization Act (SARA). As indicated in the earlier sections of this Notice, the provisions of section 126 detail those protections that OSHA must include for workers at hazardous waste and emergency response operations. For example, section 126 requires that provisions for site analysis, training, and medical surveillance, among others, be included in the standard. In addition, there is a wide range of OSHA, EPA, and other standards that already apply to some activities that occur at hazardous waste sites and during emergency response operations. For example, there are existing OSHA standards that cover construction activities, onsite machinery and equipment, selection and use of personal protective equipment, handling of toxic and explosive materials, and general environmental and safety issues such as walking-working surfaces, noise, and illumination. Moreover, the final

Federal Register / Vol. 54, No. 42 / Monday, March 6, 1989 / Rules and Regulations   9313

standard, in many instances, either reflects OSHA regulations, procedures adopted by other federal agencies (e.g. , EPA), or practices that are commonly used by those knowledgeable in hazardous waste and emergency response operations. To the extent that existing standards, rules, or standard operating procedures are incorporated into this rule, no significant change in the environment is anticipated.

Potential Positive Environmental Effects

While OSHA does not anticipate any significant environmental effects as a result of this standard, there is a potential for some beneficial impacts. In general, as the work practices and procedures requirements of the standard reduce the incidence of employee injury, an indirect result should be a reduction in the likelihood of environmental releases of hazardous materials. (Virtually all provisions of the standard can be categorized in this manner, because once they are implemented, they will have a positive influence on worker safety.) As these requirements also provide guidance for routine reactions to situations encountered in emergencies, they may help to reduce the severity of such emergencies. Additional potentially positive impacts might be categorized as follows: (1) Direct benefits associated with reduced incidences in, or the severity of, the release of hazardous materials, and (2) indirect benefits associated with the improved flow of information and increased worker awareness of hazardous materials or with improved worker preparedness (either for normal site operations or for unexpected accidents). The following discussion highlights those provisions with potentially beneficial environmental effects.

Monitoring (h). The requirements of this provision will increase the amount of monitoring for airborne hazardous substances at uncontrolled hazardous waste sites. In some cases, hazardous materials will be detected, and steps will be taken to more quickly control release to the atmosphere, thereby providing an environmental benefit.

Handling drums and containers (j). A number of specific requirements of this paragraph will result in potentially positive environmental impacts. Relevant subsections include: Inspecting drums and containers; making salvage drums or absorbents available; initiating a spill containment program; emptying unsound drums and containers; requiring ground penetrating radar; and decontaminating equipment. These are discussed briefly in the following sections.

Inspection of drums/containers before moving (j)(1)(iii). This section requires that drums and containers be inspected for their integrity prior to handling and moving. Under current practices at hazardous waste cleanup sites, drums and containers are often handled with mechanized equipment (e.g., a barrel grapple on a backhoe arm) before being inspected, if unsound drums rupture or leak, any solid contaminated by the rupture or leak is removed for disposal upon completion of drum handling operations. This provision will, through worker awareness, increase the probability of averting ruptures and leakage. In addition, any hazardous materials in containers that cannot be moved without rupturing will have to be transferred to safe containers (as required in paragraph (j)(1)(ix)), with obvious positive environmental effect. These procedures will reduce the volume of contaminated soil requiring disposal and will also lower the possibility that leachate or runoff will carry contaminants offsite. This requirement does not have an impact on emergency response actions because the routines outlined are already standard procedure.

Availability of salvage drums/absorbents (j)(1)(vii). This provision specifies that salvage drums or containers as well as suitable amounts of proper absorbent be kept available for use in areas where spills, leaks, or ruptures might occur. This requirement will result in increased availability of salvage drums and spill absorbents at uncontrolled hazardous waste sites and in emergency response situations where spills are imminent, thereby reducing the environmental consequences related to spills of hazardous materials. In those instances where salvage drums/absorbents would have been inadequate without this requirement, there is a potential benefit to the environment.

Implement a spill containment program (j)(1)(viii). The purpose of this provision is to develop a program to be implemented, in the event of a major spill, that would contain and isolate hazardous materials being transferred into containers and drums. To the extent that this program is implemented, there will be a potential for reducing the negative environmental effects that occur as a result of spills, leakage, etc. This requirement will reduce the environmental impact of potential spills at cleanup sites.

Empty unsound drum/containers (j)(1)(ix). Unsound containers often rupture during handling operations. This provision requires that drums and containers that cannot be moved

without spillage, leakage, or rupture be emptied into a sound container. This requirement will reduce the incidence of drum and container rupture and will provide concomitant environmental benefits.

Use of a ground penetrating system to estimate depth and location of containers (j)(1)(x). At present, when preliminary investigations at hazardous waste sites indicate that buried drums or containers may be present, ground penetrating systems are frequently used to determine the depth and location of the drums. The requirements of this provision will very likely cause an increase in the use of these systems, thereby reducing the number of instances in which buried containers would go undetected or where undetected containers would be accidentally ruptured during excavation activities. Where it applies, the requirement will help prevent accidental ruptures and spills, improve the thoroughness of remedial actions, and benefit the site environment.

Develop Decontamination Procedures (k). The requirement to clean and decontaminate equipment, personnel, and personal protective equipment will prevent the migration of hazardous substances offsite, thereby benefiting the surrounding environment. It will also eliminate or minimize the contamination of personnel. Decontamination is already standard practice at most cleanup sites.

Inform Contractors of Existing Hazards (b)(1)(iv). Under this provision, contractors are to be informed of any "fire, explosion, health or other safety hazards" that are present. By ensuring that contractors know the location and nature of site hazards, this requirement will reduce the possibility that contractor activities will result in inadvertent releases or spills of hazardous materials.

Gather Information Before Site Entry (c)(4). Among the various requirements for site evaluation are those for information to be gathered regarding the (a) pathways for hazardous substance dispersion, and (b) status and capability of emergency response teams. These procedural requirements will result in an increased ability to predict and prevent movement offsite of hazardous materials, will mitigate emergency situations quickly and effectively, and will reduce the possibility or severity of contaminant release. As the requirements of the section mirror current practices, compliance will be accomplished with little difficulty.

Provide Worker Training (e). The training requirement will assure that site

9314   Federal Register / Vol. 54, No. 42 / Monday, March 6, 1989 / Rules and Regulations

activities will be carried out by qualified personnel, with the knowledge and ability to fulfill their job functions in a safe and responsible manner. To the extent that this occurs, there will be a potential benefit to the environment (in emergency-response situations, similar benefits accrue from emergency response training and RCRA-regulated facility employee training.) For example, worker training will result in a more careful handling of materials accompanied by a reduction in the potential for inadvertent spills, improper disposal, etc. In emergency situations this training will assure a more efficient and effective cleanup of hazardous materials or a quicker response to avert further hazardous material releases.

Informational Programs (i). These provisions include requirements for a site safety and health plan, pre-entry briefings, and site inspections. These requirements will not directly affect the existing environment; their purpose is to provide workers with the information necessary to carry out their activities safely. To the extent that this occurs, there will be a potential benefit to the environment. For example, implementing comprehensive site plans will reduce the incidence of accident releases of hazardous materials. Similarly, requiring pre-entry briefings will reduce the likelihood of employees unknowingly encountering contaminants or allowing their improper release or disposal.

Emergency Response Plan (l) and (r). The development and implementation of a response plan for on-site and off-site emergencies will provide for greater worker preparedness. In emergencies, workers will be able to respond more quickly and effectively, thereby benefitting the environment.

Potentially Negative Impacts

In some situations, there may be a potential for negative effects on the environment as a result of the standard. Any potential negative impacts, however, are not expected to be significant. To illustrate this, negative impacts may occur if there is an increase in the time required to implement specific cleanup and spill response activities, or to implement safe work practices or procedures required by the standard. Any such effects are likely to be negligible, however since response teams already have established operating procedures similar to those in OSHA's standard.

Another potential negative impact may result from the requirement that salvage drums and absorbents be readily available. This may increase the number of repacked hazardous waste

drums and the amount of spent absorbent used, which could add to the amount of material that would require safe disposal. Similarly, the requirements for implementation of proper decontamination procedures for all equipment, personal protective gear, and personnel at hazardous waste emergencies, cleanup sites, and RCRA sites may result in an increase in the frequency and use of decontamination materials. This, in turn, could generate a larger volume of spent decontamination fluids which would then require proper handling and disposal. Again, any such impact should be negligible since decontamination is largely standard procedure for most hazardous waste operations. A possible exception may be during activities that take place in the early stages of site evaluation before cleanup, or at spill response, where decontamination procedures are not yet standardized.

Conclusion

To the extent that the work practices and procedures are implemented, increased worker awareness and preparedness will result in a safer and more healthful work environment, which may indirectly benefit the environment. Any negative impacts that may occur as a result of the implementation of these work practices or procedures are expected to be negligible. Based on this assessment and the information presented earlier in the preamble, OSHA concludes that no significant environmental changes are anticipated as a result of the standard.

IV. International Trade

OSHA has evaluated the potential impact that this final standard would have upon international trade. OSHA has determined that the final standard would have a minimal potential impact upon the prices of products, so that there would be no effective change in the level of exported or imported products.

V. OMB Approval Under the Paperwork Reduction Act

This section contains a collection of information pertaining to the preparation of a written safety and health plan site characterization and analysis, site control, training, medical surveillance, emergency controls, work practices, PPE, monitoring, informational programs, handling drums and containers, decontamination, emergency response planning, and emergency response drills. OMB has reviewed these collections and has approved them under approval number 1218-0139.

VI. Public Reporting Burden

Public reporting burden for the collection of information identified in paragraph IV above is estimated to average 3.7 hours per response, including the time for reviewing instruction, searching existing data sources, gathering and maintaining the data needed, and completing and reviewing the collection of information. Send comments regarding this burden estimate or any other suggestions for reducing this burden to the Director, Directorate of Safety Standards Programs, OSHA Room N-3605, U.S. Department of Labor, Washington, DC 20210; and to the Office of Information and Regulatory Affairs, Office of Management and Budget, Washington, DC 20503.

VII. State Plan States

This Federal Register document amends an interim final rule (section 1910.120, "Hazardous Waste Operations and Emergency Response") in Subpart H of 29 CFR Part 1910, OSHA's general industry standards on hazardous materials. The 25 states with their own OSHA approved occupational safety and health plans must develop a comparable standard applicable to both the private and public (state and local government employees) sectors within six months of the publication date of this permanent final rule or show OSHA why there is no need for action, e.g., because an existing state standard covering this area is already "at least as effective" as the new Federal standard. These states are Alaska, Arizona, California (for state and local government employees only), Connecticut (for state and local government employees only) Hawaii, Indiana, Iowa, Kentucky, Maryland, Michigan, Minnesota, Nevada, New Mexico, New York (for state and local government employees only), North Carolina, Oregon, Puerto Rico, South Carolina, Tennessee, Utah, Vermont, Virginia, Virgin Islands, Washington, and Wyoming. After the effective date of this final rule, until such time as a state standard is promulgated, Federal OSHA will provide interim enforcement assistance, as appropriate, in these states.

VIII. Federal and State Coverage of the Public Sector and Volunteers

Federal OSHA is specifically precluded by section 3(5) of the Occupational Safety and Health Act from covering employees of any State or political subdivision thereof. However, States that elect to have their own occupational safety and health program under a plan approved and monitored by OSHA under section 18(b) of the Act are required to extend their coverage to these employees (see section VII of this preamble for a list of these states). Thus, a State hazardous waste operations standard that is either identical to or at least as effective as this Federal OSHA standard will apply to public sector as well as private sector employees in these States. Public sector employees in States without State plans will be protected from exposure to hazardous waste under Title I, section 126(f) of the Superfund Amendments and Reauthorization Act of 1986 (SARA), administered by the U.S. Environmental Protection Agency (EPA). This section requires EPA to promulgate, within 90 days of the promulgation date of this Federal OSHA standard, an identical standard that applies to employees of State and local governments in each State which does not have an OSHA-approved State plan.

OSHA's hazardous waste operations standard and the identical or equivalent standards which will be promulgated by States with OSHA-approved State plans apply under certain circumstances to volunteer firefighters and other volunteers engaged in emergency response operations or hazardous waste operations within the scope of these standards (see paragraphs (a) (1) and (2) of this standard). In many communities, fire and other emergency response services are provided by volunteer companies. In some cases, these companies are established as independent, private sector entities. In others, they are considered a component of State or local government (see 29 CFR 1975.5 for factors to consider in determining whether or not an entity is a public agency). A volunteer working for a public or private entity in a State with an OSHA-approved State plan must be considered an employee under State law in order to be covered by the State's hazardous waste operations and emergency response standard—for example, because of an employer-employee relationship or because of pay, retirement benefits, health insurance coverage, workers' compensation benefits, etc. This determination is made by each State as part of its standards promulgation process. In a State without an OSHA-approved State plan, a private entity fire company with one or more paid employees would be covered under this Federal standard (29 CFR 1975.4).

IX. Federalism

This final regulation has been reviewed in accordance with Executive Order 12612 (52 FR 41685; October 30,

1987) regarding Federalism. Executive Order 12612 requires that agencies, to the extent possible, refrain from limiting state policy options, consult with states prior to taking any actions that would restrict state policy options, and take such actions only when there is clear constitutional authority and the presence of a problem of national scope. The Executive Order provides for preemption of state law only if there is a clear Congressional intent for the Agency to do so. Any such preemption is to be limited to the extent possible.

During the development of this rule, OSHA has, to the extent possible, refrained from limiting state policy options by developing a rule that permits flexibility on the part of the States through the use of performance language. We have also consulted with the States, in particular those states with approved state OSHA plans, during the public hearings and comment period called for in the notice of proposed rulemaking for this rule. We will continue to work with the States that have state occupational safety and health plans approved under section 18 of the OSHA Act to encourage those states to develop their own policies to achieve program objectives and continue to work with appropriate state officials as they present their state standards for approval.

This rulemaking is directed by Congress under the Superfund Amendments and Reauthorization Act of 1986 (SARA). The Constitutional authority and Congressional intent for Federal action in the area of worker protection standards for employees engaged in hazardous waste operations is mandated clearly in section 126 of SARA. Congress therefore has identified the protection of employees engaged in hazardous waste operations and emergency response as a problem of national scope through the enactment of SARA.

Section 18 of the Occupational Safety and Health Act (OSH Act), permits any state to develop its own independent state occupational safety and health program. Any state may develop and submit to OSHA, for approval and use, a state occupational safety and health program that provides, among other things, worker protection "at least as effective as" that protection provided under the Federal program.

With respect to Section 4 of Executive Order 12612, Section 18 of the OSH Act also expresses Congress' clear intent to preempt state laws relating to issues with respect to which Federal OSHA has promulgated occupational safety or health standards. Under the OSH Act, a

state can avoid preemption only if it submits, and obtains Federal OSHA approval of, a plan for the development of such standards and their enforcement as mentioned above. Occupational safety and health standards developed by such approved Plan-States must, among other things, be as least as effective in providing safe and healthful employment and places of employment as the Federal standards.

OSHA has used its regulatory preemption of State law to the minimum level necessary to achieve the objectives of the OSH Act and section 126 of SARA.

Section 126 of SARA, under paragraph (f), requires that the U.S. Environmental Protection Agency (EPA) provide those state and local government workers who are not covered by the protections of approved OSHA state plans with protection that is identical to that provided under the Federal OSHA standards. Non-state and local government employees would be regulated by the Federal OSHA standard. State and local government workers, employed in 25 non-OSHA state plan states, would not normally be covered by standards promulgated under Federal OSHA or approved state OSHA programs. OSHA has worked with EPA in the development of this final rule to assure that the protections provided to all state and local government employees is consistent with that provided by the Federal OSHA standard and the OSH Act. EPA as the regulatory authority for the non-OSHA state plan states will address their actions with respect to worker protection policies that have federalism implications in their rulemaking.

This final rule is written so that employees engaged in hazardous waste operations and related emergency response operations in every state, including those state and local government employees in states regulated by EPA, would be protected by general, performance oriented standards. To the extent that there are state or regional peculiarities caused by the types of hazardous waste operations, including the types of related emergency response provided, states with occupational safety and health plans approved by OSHA under section 18 of the OSH Act would be able to develop their own state standards to address any special problems. This would assure the compatibility of state or local emergency response plans developed independently by state or local emergency planning committees under Title III of SARA with Federal

9316 Federal Register / Vol. 54, No. 42 / Monday, March 6, 1989 / Rules and Regulations

worker protection standards issued by OSHA and EPA.

And, under the OSH Act, if a state develops its own OSHA approved state program, it could make additional requirements in its standards. States that will be covered by regulations issued by EPA under paragraph 126(f) of SARA will be provided the same option. Moreover, the performance nature of this final rule, of and by itself, allows for flexibility by states and owners or operators of hazardous wastes sites or providers of emergency response to provide as much safety as possible using varying methods consonant with the conditions in each state.

In summary, there is a clear national problem, identified by Congress, related to occupational safety and health in hazardous waste operations and related emergency response. While the individual states, if all acted collectively, might be able to deal with the safety problems involved, most have not elected to do so in the seventeen years since the enactment of the OSH Act. Those states which have elected to participate under section 18 of the OSH Act, would not be preempted by this final regulation and would be able to address special, local conditions within the framework provided by this performance oriented standard while ensuring that their standards are at least as effective as the Federal standard. State comments were invited on the proposal and those that were submitted to the record were fully considered prior to promulgation of this Final Rule.

The agency certifies that this document has been assessed in light of the principles, criteria, and requirements stated in sections 2 through 5 of Executive Order 12821. There are no provisions of this rulemaking that are inconsistent with the principles, criteria, and requirements stated in sections 2 through 5 of Executive Order 12821. States which have approved state occupational safety and health plans may incur additional costs associated with standards development and enforcement as a result of this rulemaking. Funding for these approved state plan programs is available from OSHA under section 18 of the OSH Ac' This rulemaking would not change the State's ability to discharge traditional State governmental functions or other aspects of State sovereignty.

An outline of § 1910.120 is included for the convenience of the reader as follows:

Table of Contents

Hazardous Waste Operations and Emergency Response

§ 1910.120 *Hazardous waste operations and emergency response.*

(a) Scope, application, and definitions.
(1) Scope.
(2) Application.
(3) Definitions.
(b) Safety and health program.
(1) General.
(2) Organizational structure chapter of the site program.
(3) Comprehensive workplan chapter of the site program.
(4) Site-specific safety and health plan.
(i) General.
(ii) Elements.
(iii) Pre-entry briefing.
(iv) Effectiveness of site safety and health plan.
(c) Site characterization and analysis.
(1) General.
(2) Preliminary evaluation.
(3) Hazard identification.
(4) Required information.
(5) Personal protective equipment.
(6) Monitoring.
(7) Risk identification.
(8) Employee notification.
(d) Site control.
(1) General.
(2) Site control program.
(3) Elements of the site control program.
(e) Training.
(1) General.
(2) Elements to be covered.
(3) Frequency of training.
(4) Management and supervisor training.
(5) Qualifications for trainers.
(6) Training certification.
(7) Emergency response.
(8) Refresher training.
(9) Equivalent training.
(f) Medical surveillance.
(1) General.
(2) Employees covered.
(3) Frequency of medical examinations and consultations.
(4) Content of medical examinations and consultations.
(5) Examination by a physician and costs.
(6) Information provided to the physician.
(7) Physician's written opinion.
(8) Recordkeeping.
(g) Engineering controls, work practices, and personal protective equipment for employee protection.
(1) Engineering controls, work practices and PPE for substances regulated in Subparts G and Z.
(2) Engineering controls, work practices, and PPE for substances not regulated in Subparts G and Z.
(3) Personal protective equipment selection.
(4) Totally-encapsulating chemical protective suits.
(5) Personal protective equipment (PPE) program.

(h) Monitoring.
(1) General.
(2) Initial entry.
(3) Periodic monitoring.
(4) Monitoring of high-risk employees.
(i) Informational programs.
(j) Handling drums and containers.
(1) General.
(2) Opening drums and containers.
(3) Material handling equipment.
(4) Radioactive wastes.
(5) Shock sensitive wastes.
(6) Laboratory waste packs.
(7) Sampling of drum and container contents.
(8) Shipping and transport.
(9) Tank and vault procedures.
(k) Decontamination.
(1) General.
(2) Decontamination procedures.
(3) Location.
(4) Equipment and solvents.
(5) Personal protective clothing and equipment.
(6) Unauthorized employees.
(7) Commercial laundries or cleaning establishments.
(8) Showers and change rooms.
(l) Emergency response by employees at uncontrolled hazardous waste sites.
(1 Emergency response plan.
(2) Elements of an emergency response plan.
(3) Procedures for handling emergency incidents.
(m) Illumination.
(n) Sanitation at temporary workplaces.
(1) Potable water.
(2) Nonpotable water.
(3) Toilets facilities.
(4) Food handling.
(5) Temporary sleeping quarters.
(6) Washing facilities.
(7) Showers and change rooms.
(o) New technology programs.
(p) Certain Operations Conducted Under the Resource Conservation and Recovery Act of 1976 (RCRA).
(1) Safety and health program.
(2) Hazard communication program.
(3) Medical surveillance program.
(4) Decontamination program.
(5) New technology program.
(6) Material handling program.
(7) Training program.
(i) New employees.
(ii) Current employees.
(iii) Trainers.
(8) Emergency response program.
(i) Emergency response plan.
(ii) Elements of an emergency response plan.
(iii) Training.
(iv) Procedures for handling emergency incidents.
(q) Emergency response to hazardous substance releases by employees not previously covered.
(1) Emergency response plan.
(2) Elements of an emergency response plan.
(3) Procedures for handling emergency response.

(4) Skilled support personnel.
(5) Specialist employees.
(6) Training.
  (i) First responder awareness level.
  (ii) First responder operations level.
  (iii) Hazardous materials technician.
  (iv) Hazardous materials specialist.
  (v) On scene commander.
(7) Trainers.
(8) Refresher training.
(9) Medical surveillance and consultation.
(10) Chemical protective clothing.
(11) Post-emergency response operations.

Appendices to § 1910.120—Hazardous Waste Operations and Emergency Response

Appendix A—Personnel Protective Equipment Test Methods.
Appendix B—General Description and Discussion of the Levels of Protection and Protective Gear.
Appendix C—Compliance Guidelines.
Appendix D—References.

List of Subjects in 29 CFR Part 1910

Containers, Drums, Emergency response, Flammable and combustible liquids, Hazardous materials, Hazardous substances, Hazardous wastes, Incorporation by reference, Materials handling and storage, Personal protective equipment, Storage areas, Training, Waste disposal.

Authority

This document has been prepared under the direction of John A. Pendergrass, Assistant Secretary of Labor for Occupational Safety and Health, U.S. Department of Labor, 200 Constitution Avenue NW., Washington, DC 20210. Pursuant to section 126 of the Superfund Amendments and Reauthorization Act of 1986 as amended (Pub. L. 99–499, 100 Stat. 1690 as amended by Pub. L. 100–202, section 101(f), 101 Stat. 1329–198, 29 U.S.C. 655 note), sections 6 and 8 of the Occupational Safety and Health Act of 1970 (29 U.S.C. 655, 657), section 4 of the Administrative Procedures Act (5 U.S.C. 553), 29 CFR Part 1911 and Secretary of Labor's Order 9–83 (48 FR 35736), it is proposed to amend 29 CFR Part 1910 by revising § 1910.120, Hazardous Waste Operations and Emergency Response, as set forth below.

Signed at Washington, DC this 28th day of February 1989.

John A. Pendergrass,
Assistant Secretary of Labor.

PART 1910—OCCUPATIONAL SAFETY AND HEALTH STANDARDS

1. The authority citation for Subpart H

of Part 1910 is amended by adding the following paragraph:

Authority: * * *
  Section 1910.120 issued under the authority of section 126 of the Superfund Amendments and Reauthorization Act of 1986 as amended (29 U.S.C. 655 note), sections 6 and 8 of the Occupational Safety and Health Act of 1970 (29 U.S.C. 655, 657), section 4 of the Administrative Procedure Act (5 U.S.C. 553), 29 CFR Part 1911 and Secretary of Labor's Order 9–83 (48 FR 35736).

2. Section 1910.120 of Title 29 of the Code of Federal Regulations is revised to read as follows:

§ 1910.120   Hazardous waste operations and emergency response.

(a) *Scope, application, and definitions*—(1) *Scope.* This section covers the following operations, unless the employer can demonstrate that the operation does not involve employee exposure or the reasonable possibility for employee exposure to safety or health hazards:
  (i) Clean-up operations required by a governmental body, whether Federal, state, local or other involving hazardous substances that are conducted at uncontrolled hazardous waste sites (including, but not limited to, the EPA's National Priority Site List (NPL), state priority site lists, sites recommended for the EPA NPL, and initial investigations of government identified sites which are conducted before the presence or absence of hazardous substances has been ascertained);
  (ii) Corrective actions involving clean-up operations at sites covered by the Resource Conservation and Recovery Act of 1976 (RCRA) as amended (42 U.S.C. 6901 *et seq.*);
  (iii) Voluntary clean-up operations at sites recognized by Federal, state, local or other governmental bodies as uncontrolled hazardous waste sites;
  (iv) Operations involving hazardous wastes that are conducted at treatment, storage, and disposal (TSD) facilities regulated by 40 CFR Parts 264 and 265 pursuant to RCRA; or by agencies under agreement with U.S.E.P.A. to implement RCRA regulations; and
  (v) Emergency response operations for releases of, or substantial threats of releases of, hazardous substances without regard to the location of the hazard.

(2) *Application.* (i) All requirements of Part 1910 and Part 1926 of Title 29 of the Code of Federal Regulations apply

pursuant to their terms to hazardous waste and emergency response operations whether covered by this section or not. If there is a conflict or overlap, the provision more protective of employee safety and health shall apply without regard to 29 CFR 1910.5(c)(1).
  (ii) Hazardous substance clean-up operations within the scope of paragraphs (a)(1)(i) through (a)(1)(iii) of this section must comply with all paragraphs of this section except paragraphs (p) and (q).
  (iii) Operations within the scope of paragraph (a)(1)(iv) of this section must comply only with the requirements of paragraph (p) of this section.

Exceptions: For large quantity generators of hazardous waste who store those wastes less than 90 days and for small quantity generators of hazardous wastes, who have emergency response teams that respond to releases of, or substantial threats of releases of, hazardous substances, for their RCRA workplaces only paragraph (p)(8) of this section is applicable. Such generators of hazardous wastes who do not have emergency response teams that respond to releases of, or substantial threats of releases of, hazardous substances are exempt from the requirements of this section.

  (iv) Emergency response operations for releases of, or substantial threats of releases of, hazardous substances which are not covered by paragraphs (a)(1)(i) through (a)(1)(iv) of this section must only comply with the requirements of paragraph (q) of this section.

(3) *Definitions*—"*Buddy system*" means a system of organizing employees into work groups in such a manner that each employee of the work group is designated to be observed by at least one other employee in the work group. The purpose of the buddy system is to provide rapid assistance to employees in the event of an emergency.

"*Clean-up operation*" means an operation where hazardous substances are removed, contained, incinerated, neutralized, stabilized, cleared-up, or in any other manner processed or handled with the ultimate goal of making the site safer for people or the environment.

"*Decontamination*" means the removal of hazardous substances from employees and their equipment to the extent necessary to preclude the occurrence of foreseeable adverse health affects.

"*Emergency response*" or "*responding to emergencies*" means a response effort by employees from outside the

9318 Federal Register / Vol. 54, No. 42 / Monday, March 6, 1989 / Rules and Regulations

immediate release area or by other designated responders (i.e., mutual-aid groups, local fire departments, etc.) to an occurrence which results, or is likely to result, in an uncontrolled release of a hazardous substance. Responses to incidental releases of hazardous substances where the substance can be absorbed, neutralized, or otherwise controlled at the time of release by employees in the immediate release area, or by maintenance personnel are not considered to be emergency responses within the scope of this standard. Responses to releases of hazardous substances where there is no potential safety or health hazard (i.e., fire, explosion, or chemical exposure) are not considered to be emergency responses.

"Facility" means (A) any building, structure, installation, equipment, pipe or pipeline (including any pipe into a sewer or publicly owned treatment works), well, pit, pond, lagoon, impoundment, ditch, storage container, motor vehicle, rolling stock, or aircraft, or (B) any site or area where a hazardous substance has been deposited, stored, disposed of, or placed, or otherwise come to be located; but does not include any consumer product in consumer use or any water-borne vessel.

(3) *"Hazardous materials response (HAZMAT) team"* means an organized group of employees, designated by the employer, who are expected to perform work to handle and control actual or potential leaks or spills of hazardous substances requiring possible close approach to the substance. The team members perform responses to releases or potential releases of hazardous substances for the purpose of control or stabilization of the incident. A HAZMAT team is not a fire brigade nor is a typical fire brigade a HAZMAT team. A HAZMAT team, however, may be a separate component of a fire brigade or fire department.

"Hazardous substance" means any substance designated or listed under paragraphs (A) through (D) of this definition, exposure to which results or may result in adverse affects on the health or safety of employees:

(A) Any substance defined under section 101(14) of CERCLA;

(B) Any biological agent and other disease-causing agent as defined in section 101(33) of CERCLA;

(C) Any substance listed by the U.S. Department of Transportation as hazardous materials under 49 CFR 172.101 and appendices; and

(D) Hazardous waste as herein defined.

"Hazardous waste" means—

(A) A waste or combination of wastes as defined in 40 CFR 261.3, or

(B) Those substances defined as hazardous wastes in 49 CFR 171.8.

"Hazardous waste operation" means any operation conducted within the scope of this standard.

"Hazardous waste site" or "Site" means any facility or location within the scope of this standard at which hazardous waste operations take place.

"Health hazard" means a chemical, mixture of chemicals or a pathogen for which there is statistically significant evidence based on at least one study conducted in accordance with established scientific principles that acute or chronic health effects may occur in exposed employees. The term "health hazard" includes chemicals which are carcinogens, toxic or highly toxic agents, reproductive toxins, irritants, corrosives, sensitizers, heptaotoxins, nephrotoxins, neurotoxins, agents which act on the hematopoietic system, and agents which damage the lungs, skin, eyes, or mucous membranes. It also includes stress due to temperature extremes. Further definition of the terms used above can be found in Appendix A to 29 CFR 1910.1200.

"IDLH" or "Immediately dangerous to life or health" means an atmospheric concentration of any toxic, corrosive or asphyxiant substance that poses an immediate threat to life or would cause irreversible or delayed adverse health effects or would interfere with an individual's ability to escape from a dangerous atmosphere.

"Oxygen deficiency" means that concentration of oxygen by volume below which atmosphere supplying respiratory protection must be provided. It exists in atmospheres where the percentage of oxygen by volume is less than 19.5 percent oxygen.

"Permissible exposure limit" means the exposure, inhalation or dermal permissible exposure limit specified in 29 CFR Part 1910, Subparts G and Z.

"Published exposure level" means the exposure limits published in "NIOSH Recommendations for Occupational Health Standards" dated 1986 incorporated by reference, or if none is specified, the exposure limits published in the standards specified by the American Conference of Governmental Industrial Hygienists in their publication "Threshold Limit Values and Biological Exposure Indices for 1987–88" dated 1987 incorporated by reference.

"Post emergency response" means that portion of an emergency response performed after the immediate threat of a release has been stabilized or

eliminated and clean-up of the site has begun. If post emergency response is performed by an employer's own employees who were part of the initial emergency response, it is considered to be part of the initial response and not post emergency response. However, if a group of an employer's own employees, separate from the group providing initial response, performs the clean-up operation, then the separate group of employees would be considered to be performing post-emergency response and subject to paragraph (g)(11) of this section.

"Qualified person" means a person with specific training, knowledge and experience in the area for which the person has the responsibility and the authority to control.

"Site safety and health supervisor (or official)" means the individual located on a hazardous waste site who is responsible to the employer and has the authority and knowledge necessary to implement the site safety and health plan and verify compliance with applicable safety and health requirements.

"Small quantity qenerator" means a generator of hazardous wastes who in any calendar month generates no more than 1,000 kilograms (2,205 pounds) of hazardous waste in that month.

"Uncontrolled hazardous waste site" means an area where an accumulation of hazardous waste creates a threat to the health and safety of individuals or the environment or both. Some sites are found on public lands, such as those created by former municipal, county or state landfills where illegal or poorly managed waste disposal has taken place. Other sites are found on private property, often belonging to generators or former generators of hazardous waste. Examples of such sites include, but are not limited to, surface impoundments, landfills, dumps, and tank or drum farms. Normal operations at TSD sites are not covered by this definition.

(b) *Safety and health program.*
Note to (b): Safety and health programs developed and implemented to meet other Federal, state, or local regulations are considered acceptable in meeting this requirement if they cover or are modified to cover the topics required in this paragraph. An additional or separate safety and health program is not required by this paragraph.

(1) *General.* (i) Employers shall develop and implement a written safety and health program for their employees involved in hazardous waste operations. The program shall be designed to identify, evaluate, and control safety and health hazards, and provide for

Federal Register / Vol. 54, No. 42 / Monday, March 6, 1989 / Rules and Regulations 9319

emergency response for hazardous waste operations.

(ii) The written safety and health program shall incorporate the following:

(A) An organizational structure;

(B) A comprehensive workplan;

(C) A site-specific safety and health plan which need not repeat the employer's standard operating procedures required in paragraph (b)(1)(ii)(F) of this section;

(D) The safety and health training program;

(E) The medical surveillance program;

(F) The employer's standard operating procedures for safety and health; and

(G) Any necessary interface between general program and site specific activities.

(iii) *Site excavation.* Site excavations created during initial site preparation or during hazardous waste operations shall be shored or sloped as appropriate to prevent accidental collapse in accordance with Subpart P of 29 CFR Part 1926.

(iv) *Contractors and sub-contractors.* An employer who retains contractor or sub-contractor services for work in hazardous waste operations shall inform those contractors, sub-contractors, or their representatives of the site emergency response procedures and any potential fire, explosion, health, safety or other hazards of the hazardous waste operation that have been identified by the employer, including those identified in the employer's information program.

(v) *Program availability.* The written safety and health program shall be made available to any contractor or subcontractor or their representative who will be involved with the hazardous waste operation; to employees; to employee designated representatives; to OSHA personnel, and to personnel of other Federal, state, or local agencies with regulatory authority over the site.

(2) *Organizational structure part of the site program.*—(i) The organizational structure part of the program shall establish the specific chain of command and specify the overall responsibilities of supervisors and employees. It shall include, at a minimum, the following elements:

(A) A general supervisor who has the responsibility and authority to direct all hazardous waste operations.

(B) A site safety and health supervisor who has the responsibility and authority to develop and implement the safety and health plan and verify compliance.

(C) All other personnel needed for hazardous waste site operations and emergency response and their general functions and responsibilities.

(D) The lines of authority, responsibility, and communication.

(ii) The organizational structure shall be reviewed and updated as necessary to reflect the current status of waste site operations.

(3) *Comprehensive workplan part of the site program.* The comprehensive workplan part of the program shall address the tasks and objectives of the site operations and the logistics and resources required to reach those tasks and objectives.

(i) The comprehensive workplan shall address anticipated clean-up activities as well as normal operating procedures which need not repeat the employer's procedures available elsewhere.

(ii) The comprehensive workplan shall define work tasks and objectives and identify the methods for accomplishing those tasks and objectives.

(iii) The comprehensive workplan shall establish personnel requirements for implementing the plan.

(iv) The comprehensive workplan shall provide for the implementation of the training required in paragraph (e) of this section.

(v) The comprehensive workplan shall provide for the implementation of the required informational programs required in paragraph (i) of this section.

(vi) The comprehensive workplan shall provide for the implementation of the medical surveillance program described in paragraph (f) of this section.

(4) *Site-specific safety and health plan part of the program.*—(i) *General.* The site safety and health plan, which must be kept on site, shall address the safety and health hazards of each phase of site operation and include the requirements and procedures for employee protection.

(ii) *Elements.* The site safety and health plan, as a minimum, shall address the following:

(A) A safety and health risk or hazard analysis for each site task and operation found in the workplan.

(B) Employee training assignments to assure compliance with paragraph (e) of this section.

(C) Personal protective equipment to be used by employees for each of the site tasks and operations being conducted as required by the personal protective equipment program in paragraph (g)(5) of this section.

(D) Medical surveillance requirements in accordance with the program in paragraph (f) of this section.

(E) Frequency and types of air monitoring, personnel monitoring, and environmental sampling techniques and instrumentation to be used, including methods of maintenance and calibration of monitoring and sampling equipment to be used.

(F) Site control measures in accordance with the site control program required in paragraph (d) of this section.

(G) Decontamination procedures in accordance with paragraph (k) of this section.

(H) An emergency response plan meeting the requirements of paragraph (l) of this section for safe and effective responses to emergencies, including the necessary PPE and other equipment.

(I) Confined space entry procedures.

(J) A spill containment program meeting the requirements of paragraph (j) of this section.

(iii) *Pre-entry briefing.* The site specific safety and health plan shall provide for pre-entry briefings to be held prior to initiating any site activity, and at such other times as necessary to ensure that employees are apprised of the site safety and health plan and that this plan is being followed. The information and data obtained from site characterization and analysis work required in paragraph (c) of this section shall be used to prepare and update the site safety and health plan.

(iv) *Effectiveness of site safety and health plan.* Inspections shall be conducted by the site safety and health supervisor or, in the absence of that individual, another individual who is knowledgeable in occupational safety and health, acting on behalf of the employer as necessary to determine the effectiveness of the site safety and health plan. Any deficiencies in the effectiveness of the site safety and health plan shall be corrected by the employer.

(c) *Site characterization and analysis*—(1) *General.* Hazardous waste sites shall be evaluated in accordance with this paragraph to identify specific site hazards and to determine the appropriate safety and health control procedures needed to protect employees from the identified hazards.

(2) *Preliminary evaluation.* A preliminary evaluation of a site's characteristics shall be performed prior to site entry by a qualified person in order to aid in the selection of appropriate employee protection methods prior to site entry. Immediately after initial site entry, a more detailed evaluation of the site's specific characteristics shall be performed by a qualified person in order to further identify existing site hazards and to further aid in the selection of the appropriate engineering controls and personal protective equipment for the tasks to be performed.

(3) *Hazard identification.* All suspected conditions that may r͡

ɔ320   **Federal Register** / Vol. 54, No. 42 / Monday, March 6, 1989 / Rules and Regulations

inhalation or skin absorption hazards that are immediately dangerous to life or health (IDLH), or other conditions that may cause death or serious harm, shall be identified during the preliminary survey and evaluated during the detailed survey. Examples of such hazards include, but are not limited to, confined space entry, potentially explosive or flammable situations, visible vapor clouds, or areas where biological indicators such as dead animals or vegetation are located.

(4) *Required information.* The following information to the extent available shall be obtained by the employer prior to allowing employees to enter a site:

(i) Location and approximate size of the site.

(ii) Description of the response activity and/or the job task to be performed.

(iii) Duration of the planned employee activity.

(iv) Site topography and accessibility by air and roads.

(v) Safety and health hazards expected at the site.

(vi) Pathways for hazardous substance dispersion.

(vii) Present status and capabilities of emergency response teams that would provide assistance to hazardous waste clean-up site employees at the time of an emergency.

(viii) Hazardous substances and health hazards involved or expected at the site, and their chemical and physical properties.

(5) *Personal protective equipment.* Personal protective equipment (PPE) shall be provided and used during initial site entry in accordance with the following requirements:

(i) Based upon the results of the preliminary site evaluation, an ensemble of PPE shall be selected and used during initial site entry which will provide protection to a level of exposure below permissible exposure limits and published exposure levels for known or suspected hazardous substances and health hazards, and which will provide protection against other known and suspected hazards identified during the preliminary site evaluation. If there is no permissible exposure limit or published exposure level, the employer may use other published studies and information as a guide to appropriate personal protective equipment.

(ii) If positive-pressure self-contained breathing apparatus is not used as part of the entry ensemble, and if respiratory protection is warranted by the potential hazards identified during the preliminary site evaluation, an escape self-contained breathing apparatus of at least five minute's duration shall be carried by employees during initial site entry.

(iii) If the preliminary site evaluation does not produce sufficient information to identify the hazards or suspected hazards of the site, an ensemble providing protection equivalent to Level B PPE shall be provided as minimum protection, and direct reading instruments shall be used as appropriate for identifying IDLH conditions. (See Appendix B for a description of Level B hazards and the recommendations for Level B protective equipment.)

(iv) Once the hazards of the site have been identified, the appropriate PPE shall be selected and used in accordance with paragraph (g) of this section.

(6) *Monitoring.* The following monitoring shall be conducted during initial site entry when the site evaluation produces information that shows the potential for ionizing radiation or IDLH conditions, or when the site information is not sufficient reasonably to eliminate these possible conditions:

(i) Monitoring with direct reading instruments for hazardous levels of ionizing radiation.

(ii) Monitoring the air with appropriate direct reading test equipment (i.e., combustible gas meters, detector tubes) for IDLH and other conditions that may cause death or serious harm (combustible or explosive atmospheres, oxygen deficiency, toxic substances).

(iii) Visually observing for signs of actual or potential IDLH or other dangerous conditions.

(iv) An ongoing air monitoring program in accordance with paragraph (h) of this section shall be implemented after site characterization has determined the site is safe for the start-up of operations.

(7) *Risk identification.* Once the presence and concentrations of specific hazardous substances and health hazards have been established, the risks associated with these substances shall be identified. Employees who will be working on the site shall be informed of any risks that have been identified. In situations covered by the Hazard Communication Standard, 29 CFR 1910.1200, training required by that standard need not be duplicated.

Note to (c)(7).—Risks to consider include, but are not limited to:

(a) Exposures exceeding the permissible exposure limits and published exposure levels.

(b) IDLH concentrations.

(c) Potential skin absorption and irritation sources.

(d) Potential eye irritation sources.

(e) Explosion sensitivity and flammability ranges.

(f) Oxygen deficiency.

(8) *Employee notification.* Any information concerning the chemical, physical, and toxicologic properties of each substance known or expected to be present on site that is available to the employer and relevant to the duties an employee is expected to perform shall be made available to the affected employees prior to the commencement of their work activities. The employer may utilize information developed for the hazard communication standard for this purpose.

(d) *Site control*—(1) *General.* Appropriate site control procedures shall be implemented to control employee exposure to hazardous substances before clean-up work begins.

(2) *Site control program.* A site control program for protecting employees which is part of the employer's site safety and health program required in paragraph (b) of this section shall be developed during the planning stages of a hazardous waste clean-up operation and modified as necessary as new information becomes available.

(3) *Elements of the site control program.* The site control program shall, as a minimum, include: A site map; site work zones; the use of a "buddy system"; site communications including alerting means for emergencies; the standard operating procedures or safe work practices; and, identification of the nearest medical assistance. Where these requirements are covered elsewhere they need not be repeated.

(e) *Training*—(1) *General.* (i) All employees working on site (such as but not limited to equipment operators, general laborers and others) exposed to hazardous substances, health hazards, or safety hazards and their supervisors and management responsible for the site shall receive training meeting the requirements of this paragraph before they are permitted to engage in hazardous waste operations that could expose them to hazardous substances, safety, or health hazards, and they shall receive review training as specified in this paragraph.

(ii) Employees shall not be permitted to participate in or supervise field activities until they have been trained to a level required by their job function and responsibility.

(2) *Elements to be covered.* The training shall thoroughly cover the following:

(i) Names of personnel and alternates responsible for site safety and health;

Federal Register / Vol. 54, No. 42 / Monday, March 6, 1989 / Rules and Regulations 9321

(ii) Safety, health and other hazards present on the the site;

(iii) Use of personal protective equipment;

(iv) Work practices by which the employee can minimize risks from hazards;

(v) Safe use of engineering controls and equipment on the site;

(vi) Medical surveillance requirements, including recognition of symptoms and signs which might indicate overexposure to hazards; and

(vii) The contents of paragraphs (G) through (J) of the site safety and health plan set forth in paragraph (b)(4)(ii) of this section.

(3) Initial training. (i) General site workers (such as equipment operators, general laborers and supervisory personnel) engaged in hazardous substance removal or other activities which expose or potentially expose workers to hazardous substances and health hazards shall receive a minimum of 40 hours of instruction off the site, and a minimum of three days actual field experience under the direct supervision of a trained, experienced supervisor.

(ii) Workers on site only occasionally for a specific limited task (such as, but not limited to, ground water monitoring, land surveying, or geo-physical surveying) and who are unlikely to be exposed over permissible exposure limits and published exposure limits shall receive a minimum of 24 hours of instruction off the site, and the minimum of one day actual field experience under the direct supervision of a trained, experienced supervisor.

(iii) Workers regularly on site who work in areas which have been monitored and fully characterized indicating that exposures are under permissible exposure limits and published exposure limits where respirators are not necessary, and the characterization indicates that there are no health hazards or the possibility of an emergency developing, shall receive a minimum of 24 hours of instruction off the site and the minimum of one day actual field experience under the direct supervision of a trained, experienced supervisor.

(iv) Workers with 24 hours of training who are covered by paragraphs (a)(3)(ii) and (a)(3)(iii) of this section, and who become general site workers or who are required to wear respirators, shall have the additional 16 hours and two days of training necessary to total the training specified in paragraph (e)(3)(i).

(4) Management and supervisor training. On-site management and supervisors directly responsible for, or who supervise employees engaged in,

hazardous waste operations shall receive 40 hours initial training, and three days of supervised field experience (the training may be reduced to 24 hours and one day if the only area of their responsibility is employees covered by paragraphs (e)(3)(ii) and (e)(3)(iii)) and at least eight additional hours of specialized training at the time of job assignment on such topics as, but not limited to, the employer's safety and health program and the associated employee training program, personal protective equipment program, spill containment program, and health hazard monitoring procedure and techniques.

(5) Qualifications for trainers. Trainers shall be qualified to instruct employees about the subject matter that is being presented in training. Such trainers shall have satisfactorily completed a training program for teaching the subjects they are expected to teach, or they shall have the academic credentials and instructional experience necessary for teaching the subjects. Instructors shall demonstrate competent instructional skills and knowledge of the applicable subject matter.

(6) Training certification. Employees and supervisors that have received and successfully completed the training and field experience specified in paragraphs (e)(1) through (e)(4) of this section shall be certified by their instructor or the head instructor and trained supervisor as having successfully completed the necessary training. A written certificate shall be given to each person so certified. Any person who has not been so certified or who does not meet the requirements of paragraph (e)(9) of this section shall be prohibited from engaging in hazardous waste operations.

(7) Emergency response. Employees who are engaged in responding to hazardous emergency situations at hazardous waste clean-up sites that may expose them to hazardous substances shall be trained in how to respond to such expected emergencies.

(8) Refresher training. Employees specified in paragraph (e)(1) of this section, and managers and supervisors specified in paragraph (e)(4) of this section, shall receive eight hours of refresher training annually on the items specified in paragraph (e)(2) and/or (e)(4) of this section, any critique of incidents that have occurred in the past year that can serve as training examples of related work, and other relevant topics.

(9) Equivalent training. Employers who can show by documentation or certification that an employee's work experience and/or training has resulted in training equivalent to that training

required in paragraphs (e)(1) through (e)(4) of this section shall not be required to provide the initial training requirements of those paragraphs to such employees. However, certified employees new to a site shall receive appropriate, site specific training before site entry and have appropriate supervised field experience at the new site. Equivalent training includes any academic training or the training that existing employees might have already received from actual hazardous waste site work experience.

(f) Medical surveillance—(1) General. Employers engaged in operations specified in paragraphs (a)(1)(i) through (a)(1)(iv) of this section and not covered by (a)(2)(iii) exceptions and employers of employees specified in paragraph (q)(9) shall institute a medical surveillance program in accordance with this paragraph.

(2) Employees covered. The medical surveillance program shall be instituted by the employer for the following employees:

(i) All employees who are or may be exposed to hazardous substances or health hazards at or above the permissible exposure limit or, if there is no permissible exposure limit, above the published exposure levels for these substances, without regard to the use of respirators, for 30 days or more a year;

(ii) All employees who wear a respirator for 30 days or more a year or as required by § 1910.134;

(iii) All employees who are injured due to overexposure from an emergency incident involving hazardous substances or health hazards; or

(iv) Members of HAZMAT teams.

(3) Frequency of medical examinations and consultations. Medical examinations and consultations shall be made available by the employer to each employee covered under paragraph (f)(2) of this section on the following schedules:

(i) For employees covered under paragraphs (f)(2)(i), (f)(2)(ii), and (f)(2)(iv):

(A) Prior to assignment;

(B) At least once every twelve months for each employee covered unless the attending physician believes a longer interval (not greater than biennially) is appropriate;

(C) At termination of employment or reassignment to an area where the employee would not be covered if the employee has not had an examination within the last six months;

(D) As soon as possible upon notification by an employee that the employee has developed signs or symptoms indicating possible

9322   Federal Register / Vol. 54, No. 42 / Monday, March 6, 1989 / Rules and Regulations

overexposure to hazardous substances or health hazards, or that the employee has been injured or exposed above the permissible exposure limits or published exposure levels in an emergency situation;

(E) At more frequent times, if the examining physician determines that an increased frequency of examination is medically necessary.

(ii) For employees covered under paragraph (f)(2)(iii) and for all employees including those of employers covered by paragraph (a)(1)(v) who may have been injured, received a health impairment, developed signs or symptoms which may have resulted from exposure to hazardous substances resulting from an emergency incident, or exposed during an emergency incident to hazardous substances at concentrations above the permissible exposure limits or the published exposure levels without the necessary personal protective equipment being used:

(A) As soon as possible following the emergency incident or development of signs or symptoms;

(B) At additional times, if the examining physician determines that follow-up examinations or consultations are medically necessary.

(4) *Content of medical examinations and consultations.* (i) Medical examinations required by paragraph (f)(3) of this section shall include a medical and work history (or updated history if one is in the employee's file) with special emphasis on symptoms related to the handling of hazardous substances and health hazards, and to fitness for duty including the ability to wear any required PPE under conditions (i.e., temperature extremes) that may be expected at the work site.

(ii) The content of medical examinations or consultations made available to employees pursuant to paragraph (f) shall be determined by the attending physician. The guidelines in the *Occupational Safety and Health Guidance Manual for Hazardous Waste Site Activities* (See Appendix D, Reference #10) should be consulted.

(5) *Examination by a physician and costs.* All medical examinations and procedures shall be performed by or under the supervision of a licensed physician, preferably one knowledgeable in occupational medicine, and shall be provided without cost to the employee, without loss of pay, and at a reasonable time and place.

(6) *Information provided to the physician.* The employer shall provide one copy of this standard and its appendices to the attending physician.

and in addition the following for each employee:

(i) A description of the employee's duties as they relate to the employee's exposures.

(ii) The employee's exposure levels or anticipated exposure levels.

(iii) A description of any personal protective equipment used or to be used.

(iv) Information from previous medical examinations of the employee which is not readily available to the examining physician.

(v) Information required by § 1910.134.

(7) *Physician's written opinion.* (i) The employer shall obtain and furnish the employee with a copy of a written opinion from the attending physician containing the following:

(A) The physician's opinion as to whether the employee has any detected medical conditions which would place the employee at increased risk of material impairment of the employee's health from work in hazardous waste operations or emergency response, or from respirator use.

(B) The physician's recommended limitations upon the employee's assigned work.

(C) The results of the medical examination and tests if requested by the employee.

(D) A statement that the employee has been informed by the physician of the results of the medical examination and any medical conditions which require further examination or treatment.

(ii) The written opinion obtained by the employer shall not reveal specific findings or diagnoses unrelated to occupational exposure.

(8) *Recordkeeping.* (i) An accurate record of the medical surveillance required by paragraph (f) of this section shall be retained. This record shall be retained for the period specified and meet the criteria of 29 CFR 1910.20.

(ii) The record required in paragraph (f)(8)(i) of this section shall include at least the following information:

(A) The name and social security number of the employee;

(B) Physician's written opinions, recommended limitations, and results of examinations and tests;

(C) Any employee medical complaints related to exposure to hazardous substances;

(D) A copy of the information provided to the examining physician by the employer, with the exception of the standard and its appendices.

(g) *Engineering controls, work practices, and personal protective equipment for employee protection.* Engineering controls, work practices, personal protective equipment, or a combination of these shall be

implemented in accordance with this paragraph to protect employees from exposure to hazardous substances and safety and health hazards.

(1) *Engineering controls, work practices and PPE for substances regulated in Subparts G and Z.* (i) Engineering controls and work practices shall be instituted to reduce and maintain employee exposure to or below the permissible exposure limits for substances regulated by 29 CFR 1910, to the extent required by Subpart Z, except to the extent that such controls and practices are not feasible.

Note to (g)(1)(i): Engineering controls which may be feasible include the use of pressurized cabs or control booths on equipment, and/or the use of remotely operated material handling equipment. Work practices which may be feasible are removing all non-essential employees from potential exposure during opening of drums, wetting down dusty operations and locating employees upwind of possible hazards.

(ii) Whenever engineering controls and work practices are not feasible, PPE shall be used to reduce and maintain employee exposures to or below the permissible exposure limits or dose limits for substances regulated by 29 CFR Part 1910, Subpart Z.

(iii) The employer shall not implement a schedule of employee rotation as a means of compliance with permissible exposure limits or dose limits except when there is no other feasible way of complying with the airborne or dermal dose limits for ionizing radiation.

(iv) The provisions of 29 CFR, Subpart G, shall be followed.

(2) *Engineering controls, work practices, and PPE for substances not regulated in Subparts G and Z.* An appropriate combination of engineering controls, work practices and personal protective equipment shall be used to reduce and maintain employee exposure to or below published exposure levels for hazardous substances and health hazards not regulated by 29 CFR Part 1910, Subparts G and Z. The employer may use the published literature and MSDS as a guide in making the employer's determination as to what level of protection the employer believes is appropriate for hazardous substances and health hazards for which there is no permissible exposure limit or published exposure limit.

(3) *Personal protective equipment selection.* (i) Personal protective equipment (PPE) shall be selected and used which will protect employees from the hazards and potential hazards they are likely to encounter as identified during the site characterization and analysis.

(ii) Personal protective equipment selection shall be based on an evaluation of the performance characteristics of the PPE relative to the requirements and limitations of the site, the task-specific conditions and duration, and the hazards and potential hazards identified at the site.

(iii) Positive pressure self-contained breathing apparatus, or positive pressure air-line respirators equipped with an escape air supply, shall be used when chemical exposure levels present will create a substantial possibility of immediate death, immediate serious illness or injury, or impair the ability to escape.

(iv) Totally-encapsulating chemical protective suits (protection equivalent to Level A protection as recommended in Appendix B) shall be used in conditions where skin absorption of a hazardous substance may result in a substantial possibility of immediate death, immediate serious illness or injury, or impair the ability to escape.

(v) The level of protection provided by PPE selection shall be increased when additional information on site conditions indicates that increased protection is necessary to reduce employee exposures below permissible exposure limits and published exposure levels for hazardous substances and health hazards. (See Appendix B for guidance on selecting PPE ensembles.)

Note to (g)(3): The level of employee protection provided may be decreased when additional information or site conditions show that decreased protection will not result in hazardous exposures to employees.

(vi) Personal protective equipment shall be selected and used to meet the requirements of 29 CFR Part 1910, Subpart I, and additional requirements specified in this section.

(4) *Totally-encapsulating chemical protective suits.* (i) Totally-encapsulating suits shall protect employees from the particular hazards which are identified during site characterization and analysis.

(ii) Totally-encapsulating suits shall be capable of maintaining positive air pressure. (See Appendix A for a test method which may be used to evaluate this requirement.)

(iii) Totally-encapsulating suits shall be capable of preventing inward test gas leakage of more than 0.5 percent. (See Appendix A for a test method which may be used to evaluate this requirement.)

(5) *Personal protective equipment (PPE) program.* A written personal protective equipment program, which is part of the employer's safety and health program required in paragraph (b) of

this section or required in paragraph (p)(1) of this section and which is also a part of the site-specific safety and health plan shall be established. The PPE program shall address the elements listed below. When elements, such as donning and doffing procedures, are provided by the manufacturer of a piece of equipment and are attached to the plan, they need not be rewritten into the plan as long as they adequately address the procedure or element.

(i) PPE selection based upon site hazards,

(ii) PPE use and limitations of the equipment,

(iii) Work mission duration,

(iv) PPE maintenance and storage,

(v) PPE decontamination and disposal,

(vi) PPE training and proper fitting,

(vii) PPE donning and doffing procedures,

(viii) PPE inspection procedures prior to, during, and after use,

(ix) Evaluation of the effectiveness of the PPE program, and

(x) Limitations during temperature extremes, heat stress, and other appropriate medical considerations.

(h) *Monitoring*—(1) *General.* (i) Monitoring shall be performed in accordance with this paragraph where there may be a question of employee exposure to hazardous concentrations of hazardous substances in order to assure proper selection of engineering controls, work practices and personal protective equipment so that employees are not exposed to levels which exceed permissible exposure limits or published exposure levels for hazardous substances.

(ii) Air monitoring shall be used to identify and quantify airborne levels of hazardous substances and safety and health hazards in order to determine the appropriate level of employee protection needed on site.

(2) *Initial entry.* Upon initial entry, representative air monitoring shall be conducted to identify any IDLH condition, exposure over permissible exposure limits or published exposure levels, exposure over a radioactive material's dose limits or other dangerous condition such as the presence of flammable atmospheres or oxygen-deficient environments.

(3) *Periodic monitoring.* Periodic monitoring shall be conducted when the possibility of an IDLH condition or flammable atmosphere has developed or when there is indication that exposures may have risen over permissible exposure limits or published exposure levels since prior monitoring. Situations where it shall be considered whether the possibility that exposures have risen are as follows:

(i) When work begins on a different portion of the site.

(ii) When contaminants other than those previously identified are being handled.

(iii) When a different type of operation is initiated (e.g., drum opening as opposed to exploratory well drilling).

(iv) When employees are handling leaking drums or containers or working in areas with obvious liquid contamination (e.g., a spill or lagoon).

(4) *Monitoring of high-risk employees.* After the actual clean-up phase of any hazardous waste operation commences; for example, when soil, surface water or containers are moved or disturbed; the employer shall monitor those employees likely to have the highest exposures to hazardous substances and health hazards likely to be present above permissible exposure limits or published exposure levels by using personal sampling frequently enough to characterize employee exposures. If the employees likely to have the highest exposure are over permissible exposure limits or published exposure limits, then monitoring shall continue to determine all employees likely to be above those limits. The employer may utilize a representative sampling approach by documenting that the employees and chemicals chosen for monitoring are based on the criteria stated above.

Note to (h): It is not required to monitor employees engaged in site characterization operations covered by paragraph (c) of this section.

(i) *Informational programs.* Employers shall develop and implement a program, which is part of the employer's safety and health program required in paragraph (b) of this section, to inform employees, contractors, and subcontractors (or their representative) actually engaged in hazardous waste operations of the nature, level and degree of exposure likely as a result of participation in such hazardous waste operations. Employees, contractors and subcontractors working outside of the operations part of a site are not covered by this standard.

(j) *Handling drums and containers*— (1) *General.* (i) Hazardous substances and contaminated soils, liquids, and other residues shall be handled, transported, labeled, and disposed of in accordance with this paragraph.

(ii) Drums and containers used during the clean-up shall meet the appropriate DOT, OSHA, and EPA regulations for the wastes that they contain.

(iii) When practical, drums and containers shall be inspected and their integrity shall be assured prior to being

9324 **Federal Register** / Vol. 54, No. 42 / Monday, March 6, 1989 / Rules and Regulations

moved. Drums or containers that cannot be inspected before being moved because of storage conditions (i.e., buried beneath the earth, stacked behind other drums, stacked several tiers high in a pile, etc.) shall be moved to an accessible location and inspected prior to further handling.

(iv) Unlabelled drums and containers shall be considered to contain hazardous substances and handled accordingly until the contents are positively identified and labeled.

(v) Site operations shall be organized to minimize the amount of drum or container movement.

(vi) Prior to movement of drums or containers, all employees exposed to the transfer operation shall be warned of the potential hazards associated with the contents of the drums or containers.

(vii) U.S. Department of Transportation specified salvage drums or containers and suitable quantities of proper absorbent shall be kept available and used in areas where spills, leaks, or ruptures may occur.

(viii) Where major spills may occur, a spill containment program, which is part of the employer's safety and health program required in paragraph (b) of this section, shall be implemented to contain and isolate the entire volume of the hazardous substance being transferred.

(ix) Drums and containers that cannot be moved without rupture, leakage, or spillage shall be emptied into a sound container using a device classified for the material being transferred.

(x) A ground-penetrating system or other type of detection system or device shall be used to estimate the location and depth of buried drums or containers.

(xi) Soil or covering material shall be removed with caution to prevent drum or container rupture.

(xii) Fire extinguishing equipment meeting the requirements of 29 CFR Part 1910, Subpart L, shall be on hand and ready for use to control incipient fires.

(2) *Opening drums and containers.* The following procedures shall be followed in areas where drums or containers are being opened:

(i) Where an airline respirator system is used, connections to the source of air supply shall be protected from contamination and the entire system shall be protected from physical damage.

(ii) Employees not actually involved in opening drums or containers shall be kept a safe distance from the drums or containers being opened.

(iii) If employees must work near or adjacent to drums or containers being opened, a suitable shield that does not interfere with the work operation shall

be placed between the employee and the drums or containers being opened to protect the employee in case of accidental explosion.

(iv) Controls for drum or container opening equipment, monitoring equipment, and fire suppression equipment shall be located behind the explosion-resistant barrier.

(v) When there is a reasonable possibility of flammable atmospheres being present, material handling equipment and hand tools shall be of the type to prevent sources of ignition.

(vi) Drums and containers shall be opened in such a manner that excess interior pressure will be safely relieved. If pressure can not be relieved from a remote location, appropriate shielding shall be placed between the employee and the drums or containers to reduce the risk of employee injury.

(vii) Employees shall not stand upon or work from drums or containers.

(3) *Material handling equipment.* Material handling equipment used to transfer drums and containers shall be selected, positioned and operated to minimize sources of ignition related to the equipment from igniting vapors released from ruptured drums or containers.

(4) *Radioactive wastes.* Drums and containers containing radioactive wastes shall not be handled until such time as their hazard to employees is properly assessed.

(5) *Shock sensitive wastes.* As a minimum, the following special precautions shall be taken when drums and containers containing or suspected of containing shock-sensitive wastes are handled:

(i) All non-essential employees shall be evacuated from the area of transfer.

(ii) Material handling equipment shall be provided with explosive containment devices or protective shields to protect equipment operators from exploding containers.

(iii) An employee alarm system capable of being perceived above surrounding light and noise conditions shall be used to signal the commencement and completion of explosive waste handling activities.

(iv) Continuous communications (i.e., portable radios, hand signals, telephones, as appropriate) shall be maintained between the employee-in-charge of the immediate handling area and both the site safety and health supervisor and the command post until such time as the handling operation is completed. Communication equipment or methods that could cause shock sensitive materials to explode shall not be used.

(v) Drums and containers under pressure, as evidenced by bulging or swelling, shall not be moved until such time as the cause for excess pressure is determined and appropriate containment procedures have been implemented to protect employees from explosive relief of the drum.

(vi) Drums and containers containing packaged laboratory wastes shall be considered to contain shock-sensitive or explosive materials until they have been characterized.

Caution: Shipping of shock sensitive wastes may be prohibited under U.S. Department of Transportation regulations. Employers and their shippers should refer to 49 CFR 173.21 and 173.50.

(6) *Laboratory waste packs.* In addition to the requirements of paragraph (j)(5) of this section, the following precautions shall be taken, as a minimum, in handling laboratory waste packs (lab packs):

(i) Lab packs shall be opened only when necessary and then only by an individual knowledgeable in the inspection, classification, and segregation of the containers within the pack according to the hazards of the wastes.

(ii) If crystalline material is noted on any container, the contents shall be handled as a shock-sensitive waste until the contents are identified.

(7) *Sampling of drum and container contents.* Sampling of containers and drums shall be done in accordance with a sampling procedure which is part of the site safety and health plan developed for and available to employees and others at the specific worksite.

(8) *Shipping and transport.* (i) Drums and containers shall be identified and classified prior to packaging for shipment.

(ii) Drum or container staging areas shall be kept to the minimum number necessary to identify and classify materials safely and prepare them for transport.

(iii) Staging areas shall be provided with adequate access and egress routes.

(iv) Bulking of hazardous wastes shall be permitted only after a thorough characterization of the materials has been completed.

(9) *Tank and vault procedures.* (i) Tanks and vaults containing hazardous substances shall be handled in a manner similar to that for drums and containers, taking into consideration the size of the tank or vault.

(ii) Appropriate tank or vault entry procedures as described in the employer's safety and health plan shall

APPENDIX J 369

be followed whenever employees must enter a tank or vault.

(k) *Decontamination*—(1) *General.* Procedures for all phases of decontamination shall be developed and implemented in accordance with this paragraph.

(2) *Decontamination procedures.* (i) A decontamination procedure shall be developed, communicated to employees and implemented before any employees or equipment may enter areas on site where potential for exposure to hazardous substances exists.

(ii) Standard operating procedures shall be developed to minimize employee contact with hazardous substances or with equipment that has contacted hazardous substances.

(iii) All employees leaving a contaminated area shall be appropriately decontaminated; all contaminated clothing and equipment leaving a contaminated area shall be appropriately disposed of or decontaminated.

(iv) Decontamination procedures shall be monitored by the site safety and health supervisor to determine their effectiveness. When such procedures are found to be ineffective, appropriate steps shall be taken to correct any deficiencies.

(3) *Location.* Decontamination shall be performed in geographical areas that will minimize the exposure of uncontaminated employees or equipment to contaminated employees or equipment.

(4) *Equipment and solvents.* All equipment and solvents used for decontamination shall be decontaminated or disposed of properly.

(5) *Personal protective clothing and equipment.* (i) Protective clothing and equipment shall be decontaminated, cleaned, laundered, maintained or replaced as needed to maintain their effectiveness.

(ii) Employees whose non-impermeable clothing becomes wetted with hazardous substances shall immediately remove that clothing and proceed to shower. The clothing shall be disposed of or decontaminated before it is removed from the work zone.

(6) *Unauthorized employees.* Unauthorized employees shall not remove protective clothing or equipment from change rooms.

(7) *Commercial laundries or cleaning establishments.* Commercial laundries or cleaning establishments that decontaminate protective clothing or equipment shall be informed of the potentially harmful effects of exposures to hazardous substances.

(8) *Showers and change rooms.* Where the decontamination procedure

indicates a need for regular showers and change rooms outside of a contaminated area, they shall be provided and meet the requirements of 29 CFR 1910.141. If temperature conditions prevent the effective use of water, then other effective means for cleansing shall be provided and used.

(l) *Emergency response by employees at uncontrolled hazardous waste sites*—
(1) *Emergency response plan.* (i) An emergency response plan shall be developed and implemented by all employers within the scope of this section to handle anticipated emergencies prior to the commencement of hazardous waste operations. The plan shall be in writing and available for inspection and copying by employees, their representatives, OSHA personnel and other governmental agencies with relevant responsibilities.

(ii) Employers who will evacuate their employees from the workplace when an emergency occurs, and who do not permit any of their employees to assist in handling the emergency, are exempt from the requirements of this paragraph if they provide an emergency action plan complying with section 1910.38(a) of this part.

(2) *Elements of an emergency response plan.* The employer shall develop an emergency response plan for emergencies which shall address, as a minimum, the following:

(i) Pre-emergency planning.
(ii) Personnel roles, lines of authority, and communication.
(iii) Emergency recognition and prevention.
(iv) Safe distances and places of refuge.
(v) Site security and control.
(vi) Evacuation routes and procedures.
(vii) Decontamination procedures which are not covered by the site safety and health plan.
(viii) Emergency medical treatment and first aid.
(ix) Emergency alerting and response procedures.
(x) Critique of response and follow-up.
(xi) PPE and emergency equipment.
(3) *Procedures for handling emergency incidents.* (i) In addition to the elements for the emergency response plan required in paragraph (1)(2) of this section, the following elements shall be included for emergency response plans:
(A) Site topography, layout, and prevailing weather conditions.
(B) Procedures for reporting incidents to local, state, and federal governmental agencies.

(ii) The emergency response plan shall be a separate section of the Site Safety and Health Plan.

(iii) The emergency response plan shall be compatible and integrated with the disaster, fire and/or emergency response plans of local, state, and federal agencies.

(iv) The emergency response plan shall be rehearsed regularly as part of the overall training program for site operations.

(v) The site emergency response plan shall be reviewed periodically and, as necessary, be amended to keep it current with new or changing site conditions or information.

(vi) An employee alarm system shall be installed in accordance with 29 CFR 1910.165 to notify employees of an emergency situation; to stop work activities if necessary; to lower background noise in order to speed communication; and to begin emergency procedures.

(vii) Based upon the information available at time of the emergency, the employer shall evaluate the incident and the site response capabilities and proceed with the appropriate steps to implement the site emergency response plan.

(m) *Illumination.* Areas accessible to employees shall be lighted to not less than the minimum illumination intensities listed in the following Table H–120.1 while any work is in progress:

TABLE H–120.1.—Minimum Illumination Intensities in Foot-Candles

Foot-candles	Area or operations
5	General site areas.
3	Excavation and waste areas, access-ways, active storage areas, loading platforms, refueling, and field main-tenance areas.
5	Indoors: Warehouses, corridors, hall-ways, and exitways.
5	Tunnels, shafts, and general under-ground work areas. (Exception: Min-imum of 10 foot-candles is required at tunnel and shaft heading during drilling mucking, and scaling. Mine Safety and Health Administration approved cap lights shall be ac-ceptable for use in the tunnel head-ing.)
10	General shops (e.g., mechanical and electrical equipment rooms, active storerooms, barracks or living quar-ters, locker or dressing rooms, dining areas, and indoor toilets and workrooms.)
30	First aid stations, infirmaries, and of-fices.

(n) *Sanitation at temporary workplaces.*—(1) *Potable water.* (i) An adequate supply of potable water shall be provided on the site.

(ii) Portable containers used to dispense drinking water shall be

370 FUNDAMENTALS OF HAZARDOUS MATERIALS INCIDENTS

9326   **Federal Register** / Vol. 54, No. 42 / Monday, March 6, 1989 / Rules and Regulations

capable of being tightly closed, and equipped with a tap. Water shall not be dipped from containers.

(iii) Any container used to distribute drinking water shall be clearly marked as to the nature of its contents and not used for any other purpose.

(iv) Where single service cups (to be used but once) are supplied, both a sanitary container for the unused cups and a receptacle for disposing of the used cups shall be provided.

(2) *Nonpotable water.* (i) Outlets for nonpotable water, such as water for firefighting purposes, shall be identified to indicate clearly that the water is unsafe and is not to be used for drinking, washing, or cooking purposes.

(ii) There shall be no cross-connection, open or potential, between a system furnishing potable water and a system furnishing nonpotable water.

(3) *Toilet facilities.* (i) Toilets shall be provided for employees according to the following Table H–120.2.

TABLE H–120.2.—TOILET FACILITIES

Number of employees	Minimum number of facilities
20 or fewer	One.
More than 20, fewer than 200.	One toilet seat and one urinal per 40 employees.
More than 200	One toilet seat and one urinal per 50 employees.

(ii) Under temporary field conditions, provisions shall be made to assure that at least one toilet facility is available.

(iii) Hazardous waste sites not provided with a sanitary sewer shall be provided with the following toilet facilities unless prohibited by local codes:

(A) Chemical toilets;
(B) Recirculating toilets;
(C) Combustion toilets; or
(D) Flush toilets.

(iv) The requirements of this paragraph for sanitation facilities shall not apply to mobile crews having transportation readily available to nearby toilet facilities.

(v) Doors entering toilet facilities shall be provided with entrance locks controlled from inside the facility.

(4) *Food handling.* All food service facilities and operations for employees shall meet the applicable laws, ordinances, and regulations of the jurisdictions in which they are located.

(5) *Temporary sleeping quarters.* When temporary sleeping quarters are provided, they shall be heated, ventilated, and lighted.

(6) *Washing facilities.* The employer shall provide adequate washing

facilities for employees engaged in operations where hazardous substances may be harmful to employees. Such facilities shall be in near proximity to the worksite; in areas where exposures are below permissible exposure limits and published exposure levels and which are under the controls of the employer; and shall be so equipped as to enable employees to remove hazardous substances from themselves.

(7) *Showers and change rooms.* When hazardous waste clean-up or removal operations commence on a site and the duration of the work will require six months or greater time to complete, the employer shall provide showers and change rooms for all employees exposed to hazardous substances and health hazards involved in hazardous waste clean-up or removal operations.

(i) Showers shall be provided and shall meet the requirements of 29 CFR 1910.141(d)(3).

(ii) Change rooms shall be provided and shall meet the requirements of 29 CFR 1910.141(e). Change rooms shall consist of two separate change areas separated by the shower area required in paragraph (n)(7)(i) of this section. One change area, with an exit leading off the worksite, shall provide employees with a clean area where they can remove, store, and put on street clothing. The second area, with an exit to the worksite, shall provide employees with an area where they can put on, remove and store work clothing and personal protective equipment.

(iii) Showers and change rooms shall be located in areas where exposures are below the permissible exposure limits and published exposure levels. If this cannot be accomplished, then a ventilation system shall be provided that will supply air that is below the permissible exposure limits and published exposure levels.

(iv) Employers shall assure that employees shower at the end of their work shift and when leaving the hazardous waste site.

(o) *New technology programs.* (1) The employer shall develop and implement procedures for the introduction of effective new technologies and equipment developed for the improved protection of employees working with hazardous waste clean-up operations, and the same shall be implemented as part of the site safety and health program to assure that employee protection is being maintained.

(2) New technologies, equipment or control measures available to the industry, such as the use of foams, absorbents, adsorbents, neutralizers, or other means to suppress the level of air contaminates while excavating the site

or for spill control, shall be evaluated by employers or their representatives. Such an evaluation shall be done to determine the effectiveness of the new methods, materials, or equipment before implementing their use on a large scale for enhancing employee protection. Information and data from manufacturers or suppliers may be used as part of the employer's evaluation effort. Such evaluations shall be made available to OSHA upon request.

(p) *Certain Operations Conducted Under the Resource Conservation and Recovery Act of 1976 (RCRA).* Employers conducting operations at treatment, storage, and disposal (TSD) facilities specified in paragraph (a)(1)(iv) of this section not exempted by paragraph (a)(2)(iii) of this section shall provide and implement the programs specified in this paragraph.

(1) *Safety and health program.* The employer shall develop and implement a written safety and health program for employees involved in hazardous waste operations that shall be available for inspection by employees, their representatives and OSHA personnel. The program shall be designed to identify, evaluate and control safety and health hazards in their facilities for the purpose of employee protection, to provide for emergency response meeting the requirements of paragraph (p)(8) of this section and to address as appropriate site analysis, engineering controls, maximum exposure limits, hazardous waste handling procedures and uses of new technologies.

(2) *Hazard communication program.* The employer shall implement a hazard communication program meeting the requirements of 29 CFR 1910.1200 as part of the employer's safety and program.

Note to 1910.120.—The exemption for hazardous waste provided in § 1910.1200 is applicable to this section.

(3) *Medical surveillance program.* The employer shall develop and implement a medical surveillance program meeting the requirements of paragraph (f) of this section.

(4) *Decontamination program.* The employer shall develop and implement a decontamination procedure meeting the requirements of paragraph (k) of this section.

(5) *New technology program.* The employer shall develop and implement procedures meeting the requirements of paragraph (o) of this section for introducing new and innovative equipment into the workplace.

(6) *Material handling program.* Where employees will be handling drums or containers, the employer shall develop

and implement procedures meeting the requirements of paragraphs (j)(1) (ii) through (viii) and (xi) of this section, as well as (j)(3) and (j)(8) of this section prior to starting such work.

(7) *Training program*—(i) *New employees.* The employer shall develop and implement a training program, which is part of the employer's safety and health program, for employees involved with hazardous waste operations to enable employees to perform their assigned duties and functions in a safe and healthful manner so as not to endanger themselves or other employees. The initial training shall be for 24 hours and refresher training shall be for eight hours annually. Employees who have received the initial training required by this paragraph shall be given a written certificate attesting that they have successfully completed the necessary training.

(ii) *Current employees.* Employers who can show by an employee's previous work experience and/or training that the employee has had training equivalent to the initial training required by this paragraph, shall be considered as meeting the initial training requirements of this paragraph as to that employee. Equivalent training includes the training that existing employees might have already received from actual site work experience. Current employees shall receive eight hours of refresher training annually.

(iii) *Trainers.* Trainers who teach initial training shall have satisfactorily completed a training course for teaching the subjects they are expected to teach or they shall have the academic credentials and instruction experience necessary to demonstrate a good command of the subject matter of the courses and competent instructional skills.

(8) *Emergency response program*—(i) *Emergency response plan.* An emergency response plan shall be developed and implemented by all employers. Such plans need not duplicate any of the subjects fully addressed in the employer's contingency planning required by permits, such as those issued by the U.S. Environmental Protection Agency, provided that the contingency plan is made part of the emergency response plan. The emergency response plan shall be a written portion of the employers safety and health program required in paragraph (p)(1) of this section. Employers who will evacuate their employees from the worksite location when an emergency occurs and who do not permit any of their employees to assist in handling the emergency are

exempt from the requirements of paragraph (p)(8) if they provide an emergency action plan complying with § 1910.38(a) of this part.

(ii) *Elements of an emergency response plan.* The employer shall develop an emergency response plan for emergencies which shall address, as a minimum, the following areas to the extent that they are not addressed in any specific program required in this paragraph:

(A) Pre-emergency planning and coordination with outside parties.

(B) Personnel roles, lines of authority. and communication.

(C) Emergency recognition and prevention.

(D) Safe distances and places of refuge.

(E) Site security and control.

(F) Evacuation routes and procedures.

(G) Decontamination procedures.

(H) Emergency medical treatment and first aid.

(I) Emergency alerting and response procedures.

(J) Critique of response and follow-up.

(K) PPE and emergency equipment.

(iii) *Training.* (A) Training for emergency response employees shall be completed before they are called upon to perform in real emergencies. Such training shall include the elements of the emergency response plan, standard operating procedures the employer has established for the job, the personal protective equipment to be worn and procedures for handling emergency incidents.

Exception #1: An employer need not train all employees to the degree specified if the employer divides the work force in a manner such that a sufficient number of employees who have responsibility to control emergencies have the training specified, and all other employees, who may first respond to an emergency incident, have sufficient awareness training to recognize that an emergency response situation exists and that they are instructed in that case to summon the fully trained employees and not attempt control activities for which they are not trained.

Exception #2: An employer need not train all employees to the degree specified if arrangements have been made in advance for an outside fully-trained emergency response team to respond in a reasonable period and all employees, who may come to the incident first, have sufficient awareness training to recognize that an emergency response situation exists and they have been instructed to call the designated outside fully-trained emergency response team for assistance.

(B) Employee members of TSD facility emergency response organizations shall be trained to a level of competence in the recognition of health and safety

hazards to protect themselves and other employees. This would include training in the methods used to minimize the risk from safety and health hazards; in the safe use of control equipment; in the selection and use of appropriate personal protective equipment; in the safe operating procedures to be used at the incident scene; in the techniques of coordination with other employees to minimize risks; in the appropriate response to over exposure from health hazards or injury to themselves and other employees; and in the recognition of subsequent symptoms which may result from over exposures.

(C) The employer shall certify that each covered employee has attended and successfully completed the training required in paragraph (p)(8)(iii) of this section, or shall certify the employee's competency at least yearly. The method used to demonstrate competency for certification of training shall be recorded and maintained by the employer.

(iv) *Procedures for handling emergency incidents.* (A) In addition to the elements for the emergency response plan required in paragraph (p)(8)(ii) of this section, the following elements shall be included for emergency response plans to the extent that they do not repeat any information already contained in the emergency response plan:

(1) Site topography, layout, and prevailing weather conditions.

(2) Procedures for reporting incidents to local, state, and federal governmental agencies.

(B) The emergency response plan shall be compatible and integrated with the disaster, fire and/or emergency response plans of local, state, and federal agencies.

(C) The emergency response plan shall be rehearsed regularly as part of the overall training program for site operations.

(D) The site emergency response plan shall be reviewed periodically and, as necessary, be amended to keep it current with new or changing site conditions or information.

(E) An employee alarm system shall be installed in accordance with 29 CFR 1910.165 to notify employees of an emergency situation; to stop work activities if necessary; to lower background noise in order to speed communication; and to begin emergency procedures.

(F) Based upon the information available at time of the emergency, the employer shall evaluate the incident and the site response capabilities and proceed with the appropriate steps to

372 FUNDAMENTALS OF HAZARDOUS MATERIALS INCIDENTS

implement the site emergency response plan.

(q) *Emergency response to hazardous substance releases.* This paragraph covers employers whose employees are engaged in emergency response no matter where it occurs except that it does not cover employees engaged in operations specified in paragraphs (a)(1)(i) through (a)(1)(iv) of this section. Those emergency response organizations who have developed and implemented programs equivalent to this paragraph for handling releases of hazardous substances pursuant to section 303 of the Superfund Amendments and Reauthorization Act of 1986 (Emergency Planning and Community Right-to-Know Act of 1986, 42 U.S.C. 11003) shall be deemed to have met the requirements of this paragraph.

(1) *Emergency response plan.* An emergency response plan shall be developed and implemented to handle anticipated emergencies prior to the commencement of emergency response operations. The plan shall be in writing and available for inspection and copying by employees, their representatives and OSHA personnel. Employers who will evacuate their employees from the workplace when an emergency occurs, and who do not permit any of their employees to assist in handling the emergency, are exempt from the requirements of this paragraph if they provide an emergency action plan in accordance with § 1910.38(a) of this part.

(2) *Elements of an emergency response plan.* The employer shall develop an emergency response plan for emergencies which shall address, as a minimum, the following to the extent that they are not addressed elsewhere:

(i) Pre-emergency planning and coordination with outside parties.

(ii) Personnel roles, lines of authority, training, and communication.

(iii) Emergency recognition and prevention.

(iv) Safe distances and places of refuge.

(v) Site security and control.

(vi) Evacuation routes and procedures.

(vii) Decontamination.

(viii) Emergency medical treatment and first aid.

(ix) Emergency alerting and response procedures.

(x) Critique of response and follow-up.

(xi) PPE and emergency equipment.

(xii) Emergency response organizations may use the local emergency response plan or the state emergency response plan or both, as part of their emergency response plan to avoid duplication. Those items of the emergency response plan that are being

properly addressed by the SARA Title III plans may be substituted into their emergency plan or otherwise kept together for the employer and employee's use.

(3) *Procedures for handling emergency response.* (i) The senior emergency response official responding to an emergency shall become the individual in charge of a site-specific Incident Command System (ICS). All emergency responders and their communications shall be coordinated and controlled through the individual in charge of the ICS assisted by the senior official present for each employer.

Note to (q)(3)(i).—The "senior official" at an emergency response is the most senior official on the site who has the responsibility for controlling the operations at the site. Initially it is the senior officer on the first-due piece of responding emergency apparatus to arrive on the incident scene. As more senior officers arrive (i.e., battalion chief, fire chief, state law enforcement official, site coordinator, etc.) the position is passed up the line of authority which has been previously established.

(ii) The individual in charge of the ICS shall identify, to the extent possible, all hazardous substances or conditions present and shall address as appropriate site analysis, use of engineering controls, maximum exposure limits, hazardous substance handling procedures, and use of any new technologies.

(iii) Based on the hazardous substances and/or conditions present, the individual in charge of the ICS shall implement appropriate emergency operations, and assure that the personal protective equipment worn is appropriate for the hazards to be encountered. However, personal protective equipment shall meet, at a minimum, the criteria contained in 29 CFR 1910.156(e) when worn while performing fire fighting operations beyond the incipient stage for any incident or site.

(iv) Employees engaged in emergency response and exposed to hazardous substances presenting an inhalation hazard or potential inhalation hazard shall wear positive pressure self-contained breathing apparatus while engaged in emergency response, until such time that the individual in charge of the ICS determines through the use of air monitoring that a decreased level of respiratory protection will not result in hazardous exposures to employees.

(v) The individual in charge of the ICS shall limit the number of emergency response personnel at the emergency site, in those areas of potential or actual exposure to incident or site hazards, to those who are actively performing

emergency operations. However, operations in hazardous areas shall be performed using the buddy system in groups of two or more.

(vi) Back-up personnel shall stand by with equipment ready to provide assistance or rescue. Advance first aid support personnel, as a minimum, shall also stand by with medical equipment and transportation capability.

(vii) The individual in charge of the ICS shall designate a safety official, who is knowledgable in the operations being implemented at the emergency response site, with specific responsibility to identify and evaluate hazards and to provide direction with respect to the safety of operations for the emergency at hand.

(viii) When activities are judged by the safety official to be an IDLH condition and/or to involve an imminent danger condition, the safety official shall have the authority to alter, suspend, or terminate those activities. The safety official shall immediately inform the individual in charge of the ICS of any actions needed to be taken to correct these hazards at an emergency scene.

(ix) After emergency operations have terminated, the individual in charge of the ICS shall implement appropriate decontamination procedures.

(x) When deemed necessary for meeting the tasks at hand, approved self-contained compressed air breathing apparatus may be used with approved cylinders from other approved self-contained compressed air breathing apparatus provided that such cylinders are of the same capacity and pressure rating. All compressed air cylinders used with self-contained breathing apparatus shall meet U.S. Department of Transportation and National Institute for Occupational Safety and Health criteria.

(4) *Skilled support personnel.* Personnel, not necessarily an employer's own employees, who are skilled in the operation of certain equipment, such as mechanized earth moving or digging equipment or crane and hoisting equipment, and who are needed temporarily to perform immediate emergency support work that cannot reasonably be performed in a timely fashion by an employer's own employees, and who will be or may be exposed to the hazards at an emergency response scene, are not required to meet the training required in this paragraph for the employer's regular employees. However, these personnel shall be given an initial briefing at the site prior to their participation in any emergency response. The initial briefing shall

include instruction in the wearing of appropriate personal protective equipment, what chemical hazards are involved, and what duties are to be performed. All other appropriate safety and health precautions provided to the employer's own employees shall be used to assure the safety and health of these personnel.

(5) *Specialist employees.* Employees who, in the course of their regular job duties, work with and are trained in the hazards of specific hazardous substances, and who will be called upon to provide technical advice or assistance at a hazardous substance release incident to the individual in charge, shall receive training or demonstrate competency in the area of their specialization annually.

(6) *Training.* Training shall be based on the duties and function to be performed by each responder of an emergency response organization. The skill and knowledge levels required for all new responders, those hired after the effective date of this standard, shall be conveyed to them through training before they are permitted to take part in actual emergency operations on an incident. Employees who participate, or are expected to participate, in emergency response, shall be given training in accordance with the following paragraphs:

(i) *First responder awareness level.* First responders at the awareness level are individuals who are likely to witness or discover a hazardous substance release and who have been trained to initiate an emergency response sequence by notifying the proper authorities of the release. They would take no further action beyond notifying the authorities of the release. First responders at the awareness level shall have sufficient training or have had sufficient experience to objectively demonstrate competency in the following areas:

(A) An understanding of what hazardous materials are, and the risks associated with them in an incident.

(B) An understanding of the potential outcomes associated with an emergency created when hazardous materials are present.

(C) The ability to recognize the presence of hazardous materials in an emergency.

(D) The ability to identify hazardous materials, if possible.

(E) An understanding of the role of the first responder awareness individual in the employer's emergency response plan including site security and control and the U.S. Department of Transportation's Emergency Response Guidebook.

(F) The ability to realize the need for additional resources, and to make appropriate notifications to the communication center.

(ii) *First responder operations level.* First responders at the operations level are individuals who respond to releases or potential releases of hazardous substances as part of the initial response to the site for the purpose of protecting nearby persons, property, or the environment from the effects of the release. They are trained to respond in a defensive fashion without actually trying to stop the release. Their function is to contain the release from a safe distance, keep it from spreading, and prevent exposures. First responders at the operational level shall have received at least eight hours of training or have had sufficient experience to objectively demonstrate competency in the following areas in addition to those listed for the awareness level and the employer shall so certify:

(A) Knowledge of the basic hazard and risk assessment techniques.

(B) Know how to select and use proper personal protective equipment provided to the first responder operational level.

(C) An understanding of basic hazardous materials terms.

(D) Know how to perform basic control, containment and/or confinement operations within the capabilities of the resources and personal protective equipment available with their unit.

(E) Know how to implement basic decontamination procedures.

(F) An understanding of the relevant standard operating procedures and termination procedures.

(iii) *Hazardous materials technician.* Hazardous materials technicians are individuals who respond to releases or potential releases for the purpose of stopping the release. They assume a more aggressive role than a first responder at the operations level in that they will approach the point of release in order to plug, patch or otherwise stop the release of a hazardous substance. Hazardous materials technicians shall have received at least 24 hours of training equal to the first responder operations level and in addition have competency in the following areas and the employer shall so certify:

(A) Know how to implement the employer's emergency response plan.

(B) Know the classification, identification and verification of known and unknown materials by using field survey instruments and equipment.

(C) Be able to function within an assigned role in the Incident Command System.

(D) Know how to select and use proper specialized chemical personal protective equipment provided to the hazardous materials technician.

(E) Understand hazard and risk assessment techniques.

(F) Be able to perform advance control, containment, and/or confinement operations within the capabilities of the resources and personal protective equipment available with the unit.

(G) Understand and implement decontamination procedures.

(H) Understand termination procedures.

(I) Understand basic chemical and toxicological terminology and behavior.

(iv) *Hazardous materials specialist.* Hazardous materials specialists are individuals who respond with and provide support to hazardous materials technicians. Their duties parallel those of the hazardous materials technician, however, those duties require a more directed or specific knowledge of the various substances they may be called upon to contain. The hazardous materials specialist would also act as the site liaison with Federal, state, local and other government authorities in regards to site activities. Hazardous materials specialists shall have received at least 24 hours of training equal to the technician level and in addition have competency in the following areas and the employer shall so certify:

(A) Know how to implement the local emergency response plan.

(B) Understand classification, identification and verification of known and unknown materials by using advanced survey instruments and equipment.

(C) Know of the state emergency response plan.

(D) Be able to select and use proper specialized chemical personal protective equipment provided to the hazardous materials specialist.

(E) Understand in-depth hazard and risk assessment techniques.

(F) Be able to perform specialized control, containment, and/or confinement operations within the capabilities of the resources and personal protective equipment available.

(G) Be able to determine and implement decontamination procedures.

(H) Have the ability to develop a site safety and control plan.

(I) Understand chemical, radiological and toxicological terminology and behavior.

(v) *On scene incident commander.* Incident commanders, who will assume control of the incident scene beyond the

374 FUNDAMENTALS OF HAZARDOUS MATERIALS INCIDENTS

9330 Federal Register / Vol. 54, No. 42 / Monday, March 6, 1989 / Rules and Regulations

first responder awareness level, shall receive at least 24 hours of training equal to the first responder operations level and in addition have competency in the following areas and the employer shall so certify:

(A) Know and be able to implement the employer's incident command system.

(B) Know how to implement the employer's emergency response plan.

(C) Know and understand the hazards and risks associated with employees working in chemical protective clothing.

(D) Know how to implement the local emergency response plan.

(E) Know of the state emergency response plan and of the Federal Regional Response Team.

(F) Know and understand the importance of decontamination procedures.

(7) *Trainers.* Trainers who teach any of the above training subjects shall have satisfactorily completed a training course for teaching the subjects they are expected to teach, such as the courses offered by the U.S. Fire Academy, or they shall have the training and/or academic credentials and instructional experience necessary to demonstrate competent instructional skills and a good command of the subject matter of the courses they are to teach.

(8) *Refresher training.* (i) Those employees who are trained in accordance with paragraph (q)(6) of this section shall receive annual refresher training of sufficient content and duration to maintain their competencies, or shall demonstrate competency in those areas at least yearly.

(ii) A statement shall be made of the training or competency, and if a statement of competency is made, the employer shall keep a record of the methodology used to demonstrate competency.

(9) *Medical surveillance and consultation.* (i) Members of an organized and designated HAZMAT team and hazardous materials specialists shall receive a baseline physical examination and be provided with medical surveillance as required in paragraph (f) of this section.

(ii) Any emergency response employees who exhibits signs or symptoms which may have resulted from exposure to hazardous substances during the course of an emergency incident, either immediately or subsequently, shall be provided with medical consultation as required in paragraph (f)(3)(ii) of this section.

(10) *Chemical protective clothing.* Chemical protective clothing and equipment to be used by organized and designated HAZMAT team members, or

to be used by hazardous materials specialists, shall meet the requirements of paragraphs (g) (3) through (5) of this section.

(11) *Post-emergency response operations.* Upon completion of the emergency response, if it is determined that it is necessary to remove hazardous substances, health hazards, and materials contaminated with them (such as contaminated soil or other elements of the natural environment) from the site of the incident, the employer conducting the clean-up shall comply with one of the following:

(i) Meet all of the requirements of paragraphs (b) through (o) of this section; or

(ii) Where the clean-up is done on plant property using plant or workplace employees, such employees shall have completed the training requirements of the following: 29 CFR 1910.38(a); 1910.134; 1910.1200, and other appropriate safety and health training made necessary by the tasks that they are expected to be performed such as personal protective equipment and decontamination procedures. All equipment to be used in the performance of the clean-up work shall be in serviceable condition and shall have been inspected prior to use.

APPENDICES TO § 1910.120—HAZARDOUS WASTE OPERATIONS AND EMERGENCY RESPONSE

Note: The following appendices serve as non-mandatory guidelines to assist employees and employers in complying with the appropriate requirements of this section. However paragraph 1910.120(g) makes mandatory in certain circumstances the use of Level A and Level B PPE protection.

Appendix A—Personal Protective Equipment Test Methods

This appendix sets forth the non-mandatory examples of tests which may be used to evaluate compliance with § 1910.120 (g)(4) (ii) and (iii). Other tests and other challenge agents may be used to evaluate compliance.

A. Totally-encapsulating chemical protective suit pressure test

1.0—Scope

1.1 This practice measures the ability of a gas tight totally-encapsulating chemical protective suit material, seams, and closures to maintain a fixed positive pressure. The results of this practice allow the gas tight integrity of a totally-encapsulating chemical protective suit to be evaluated.

1.2 Resistance of the suit materials to permeation, penetration, and degradation by specific hazardous substances is not determined by this test method.

2.0—Definition of terms

2.1 *"Totally-encapsulated chemical protective suit (TECP suit)"* means a full body garment which is constructed of protective clothing materials; covers the wearer's torso, head, arms, legs and

respirator; may cover the wearer's hands and feet with tightly attached gloves and boots; completely encloses the wearer and respirator by itself or in combination with the wearer's gloves and boots.

2.2 *"Protective clothing material"* means any material or combination of materials used in an item of clothing for the purpose of isolating parts of the body from direct contact with a potentially hazardous liquid or gaseous chemicals.

2.3 *"Gas tight"* means, for the purpose of this test method, the limited flow of a gas under pressure from the inside of a TECP suit to atmosphere at a prescribed pressure and time interval.

3.0—Summary of test method

3.1 The TECP suit is visually inspected and modified for the test. The test apparatus is attached to the suit to permit inflation to the pre-test suit expansion pressure for removal of suit wrinkles and creases. The pressure is lowered to the test pressure and monitored for three minutes. If the pressure drop is excessive, the TECP suit fails the test and is removed from service. The test is repeated after leak location and repair.

4.0—Required Supplies

4.1 Source of compressed air.

4.2 Test apparatus for suit testing, including a pressure measurement device with a sensitivity of at least ¼ inch water gauge.

4.3 Vent valve closure plugs or sealing tape.

4.4 Soapy water solution and soft brush.

4.5 Stop watch or appropriate timing device.

5.0—Safety Precautions

5.1 Care shall be taken to provide the correct pressure safety devices required for the source of compressed air used.

6.0—Test Procedure

6.1 Prior to each test, the tester shall perform a visual inspection of the suit. Check the suit for seam integrity by visually examining the seams and gently pulling on the seams. Ensure that all air supply lines, fittings, visor, zippers, and valves are secure and show no signs of deterioration.

6.1.1 Seal off the vent valves along with any other normal inlet or exhaust points (such as umbilical air line fittings or face piece opening) with tape or other appropriate means (caps, plugs, fixture, etc.). Care should be exercised in the sealing process not to damage any of the suit components.

6.1.2 Close all closure assemblies.

6.1.3 Prepare the suit for inflation by providing an improvised connection point on the suit for connecting an airline. Attach the pressure test apparatus to the suit to permit suit inflation from a compressed air source equipped with a pressure indicating regulator. The leak tightness of the pressure test apparatus should be tested before and after each test by closing off the end of the tubing attached to the suit and assuring a pressure of three inches water gauge for three minutes can be maintained. If a component is removed for the test, that component shall be replaced and a second test conducted with another component removed to permit a complete test of the ensemble.

6.1.4 The pre-test expansion pressure (A) and the suit test pressure (B) shall be supplied by the suit manufacturer, but in no case shall they be less than: (A)=three inches water gauge; and (B)=two inches water gauge. The ending suit pressure (C) shall be no less than 80 percent of the test pressure (B); i.e., the pressure drop shall not exceed 20 percent of the test pressure (B).

6.1.5 Inflate the suit until the pressure inside is equal to pressure (A), the pre-test expansion suit pressure. Allow at least one minute to fill out the wrinkles in the suit. Release sufficient air to reduce the suit pressure to pressure (B), the suit test pressure. Begin timing. At the end of three minutes, record the suit pressure as pressure (C), the ending suit pressure. The difference between the suit test pressure and the ending suit test pressure (B–C) shall be defined as the suit pressure drop.

6.1.6 If the suit pressure drop is more than 20 percent of the suit test pressure (B) during the three-minute test period, the suit fails the test and shall be removed from service.

7.0—Retest Procedure

7.1 If the suit fails the test check for leaks by inflating the suit to pressure (A) and brushing or wiping the entire suit (including seams, closures, lens gaskets, glove-to-sleeve joints, etc.) with a mild soap and water solution. Observe the suit for the formation of soap bubbles, which is an indication of a leak. Repair all identified leaks.

7.2 Retest the TECP suit as outlined in Test procedure 6.0.

8.0—Report

8.1 Each TECP suit tested by this practice shall have the following information recorded:

8.1.1 Unique identification number, identifying brand name, date of purchase, material of construction, and unique fit features, e.g., special breathing apparatus.

8.1.2 The actual values for test pressures (A), (B), and (C) shall be recorded along with the specific observation times. If the ending pressure (C) is less than 80 percent of the test pressure (B), the suit shall be identified as failing the test. When possible, the specific leak location shall be identified in the test records. Retest pressure data shall be recorded as an additional test.

8.1.3 The source of the test apparatus used shall be identified and the sensitivity of the pressure gauge shall be recorded.

8.1.4 Records shall be kept for each pressure test even if repairs are being made at the test location.

Caution

Visually inspect all parts of the suit to be sure they are positioned correctly and secured tightly before putting the suit back into service. Special care should be taken to examine each exhaust valve to make sure it is not blocked.

Care should also be exercised to assure that the inside and outside of the suit is completely dry before it is put into storage.

B. Totally-encapsulating chemical protective suit qualitative leak test

1.0—Scope

1.1 This practice semi-qualitatively tests gas tight totally-encapsulating chemical protective suit integrity by detecting inward leakage of ammonia vapor. Since no modifications are made to the suit to carry

out this test, the results from this practice provide a realistic test for the integrity of the entire suit.

1.2 Resistance of the suit materials to permeation, penetration, and degradation is not determined by this test method. ASTM test methods are available to test suit materials for these characteristics and the tests are usually conducted by the manufacturers of the suits.

2.0—Definition of terms

2.1 "Totally-encapsulating chemical protective suit (TECP suit) means a full body garment which is constructed of protective clothing materials; covers the wearer's torso, head, arms, legs and respirator; may cover the wearer's hands and feet with tightly attached gloves and boots; completely encloses the wearer and respirator by itself or in combination with the wearer's gloves, and boots.

2.2 "Protective clothing material" means any material or combination of materials used in an item of clothing for the purpose of isolating parts of the body from direct contact with a potentially hazardous liquid or gaseous chemicals.

2.3 "Gas tight" means, for the purpose of this test method, the limited flow of a gas under pressure from the inside of a TECP suit to atmosphere at a prescribed pressure and time interval.

2.4 "Intrusion Coefficient" means a number expressing the level of protection provided by a gas tight totally-encapsulating chemical protective suit. The intrusion coefficient is calculated by dividing the test room challenge agent concentration by the concentration of challenge agent found inside the suit. The accuracy of the intrusion coefficient is dependent on the challenge agent monitoring methods. The larger the intrusion coefficient the greater the protection provided by the TECP suit.

3.0—Summary of recommended practice

3.1 The volume of concentrated aqueous ammonia solution (ammonia hydroxide NH_4OH) required to generate the test atmosphere is determined using the directions outlined in 6.1. The suit is donned by a person wearing the appropriate respiratory equipment (either a positive pressure self-contained breathing apparatus or a positive pressure supplied air respirator) and worn inside the enclosed test room. The concentrated aqueous ammonia solution is taken by the suited individual into the test room and poured into an open plastic pan. A two-minute evaporation period is observed before the test room concentration is measured, using a high range ammonia length of stain detector tube. When the ammonia vapor reaches a concentration of between 1000 and 1200 ppm, the suited individual starts a standardized exercise protocol to stress and flex the suit. After this protocol is completed, the test room concentration is measured again. The test protocol exit the test room and his stand-by person measures the ammonia concentration inside the suit using a low range ammonia length of stain detector tube or other more sensitive ammonia detector. A stand-by person is required to observe the test individual during the test procedure; aid the person in donning and doffing the TECP suit; and monitor the

suit interior. The intrusion coefficient of the suit can be calculated by dividing the average test area concentration by the interior suit concentration. A colorimetric ammonia indicator strip of bromophenol blue or equivalent is placed on the inside of the suit face piece lens so that the suited individual is able to detect a color change and know if the suit has a significant leak. If a color change is observed the individual shall leave the test room immediately.

4.0—Required supplies

4.1 A supply of concentrated aqueous (58 percent ammonium hydroxide by weight).

4.2 A supply of bromophenol/blue indicating paper or equivalent, sensitive to 5–10 ppm ammonia over a two-minute period of exposure. [pH 3.0 (yellow) to pH 4.6 (blue)]

4.3 A supply of high range (0.5–10 volume percent) and low range (5–700 ppm) detector tubes for ammonia and the corresponding sampling pump. More sensitive ammonia detectors can be substituted for the low range detector tubes to improve the sensitivity of this practice.

4.4 A shallow plastic pan (PVC) at least 12":14":1" and a half pint plastic container (PVC) with tightly closing lid.

4.5 A graduated cylinder or other volumetric measuring device of at least 50 milliliters in volume with an accuracy of at least ± 1 milliliter.

5.0—Safety precautions

5.1 Concentrated aqueous ammonium hydroxide, NH_4OH, is a corrosive volatile liquid requiring eye, skin, and respiratory protection. The person conducting the test shall review the MSDS for aqueous ammonia.

5.2 Since the established permissible exposure limit for ammonia is 50 ppm, only persons wearing a positive pressure self-contained breathing apparatus or a positive pressure supplied air respirator shall be in the chamber. Normally only the person wearing the totally-encapsulating suit will be inside the chamber. A stand-by person shall have a positive pressure self-contained breathing apparatus, or a positive pressure supplied air respirator available to enter the test area should the suited individual need assistance.

5.3 A method to monitor the suited individual must be used during this test. Visual contact is the simplest but other methods using communication devices are acceptable.

5.4 The test room shall be large enough to allow the exercise protocol to be carried out and then to be ventilated to allow for easy exhaust of the ammonia test atmosphere after the test(s) are completed.

5.5 Individuals shall be medically screened for the use of respiratory protection and checked for allergies to ammonia before participating in this test procedure.

6.0—Test procedure

6.1.1 Measure the test area to the nearest foot and calculate its volume in cubic feet. Multiply the test area volume by 0.2 milliliters of concentrated aqueous ammonia solution per cubic foot of test area volume to determine the approximate volume of

9332   Federal Register / Vol. 54, No. 42 / Monday, March 6, 1989 / Rules and Regulations

concentrated aqueous ammonia required to generate 1000 ppm in the test area.

6.1.2 Measure this volume from the supply of concentrated aqueous ammonia and place it into a closed plastic container.

6.1.3 Place the container, several high range ammonia detector tubes, and the pump in the clean test pan and locate it near the test area entry door so that the suited individual has easy access to these supplies.

6.2.1 In a non-contaminated atmosphere, open a pre-sealed ammonia indicator strip and fasten one end of the strip to the inside of the suit face shield lens where it can be seen by the wearer. Moisten the indicator strip with distilled water. Care shall be taken not to contaminate the detector part of the indicator paper by touching it. A small piece of masking tape or equivalent should be used to attach the indicator strip to the interior of the suit face shield.

6.2.2 If problems are encountered with this method of attachment, the indicator strip can be attached to the outside of the respirator face piece lens being used during the test.

6.3 Don the respiratory protective device normally used with the suit, and then don the TECP suit to be tested. Check to be sure all openings which are intended to be sealed (zippers, gloves, etc.) are completely sealed. DO NOT, however, plug off any venting valves.

6.4 Step into the enclosed test room such as a closet, bathroom, or test booth, equipped with an exhaust fan. No air should be exhausted from the chamber during the test because this will dilute the ammonia challenge concentrations.

6.5 Open the container with the pre-measured volume of concentrated aqueous ammonia within the enclosed test room, and pour the liquid into the empty plastic test pan. Wait two minutes to allow for adequate volatilization of the concentrated aqueous ammonia. A small mixing fan can be used near the evaporation pan to increase the evaporation rate of the ammonia solution.

6.6 After two minutes a determination of the ammonia concentration within the chamber should be made using the high range colorimetric detector tube. A concentration of 1000 ppm ammonia or greater shall be generated before the exercises are started.

6.7 To test the integrity of the suit the following four minute exercise protocol should be followed:

6.7.1 Raising the arms above the head with at least 15 raising motions completed in one minute.

6.7.2 Walking in place for one minute with at least 15 raising motions of each leg in a one-minute period.

6.7.3 Touching the toes with a least 10 complete motions of the arms from above the head to touching of the toes in a one-minute period.

6.7.4 Knee bends with at least 10 complete standing and squatting motions in a one-minute period.

6.8 If at any time during the test the colorimetric indicating paper should change colors, the test should be stopped and section 6.10 and 6.12 initiated (See ¶4.2).

6.9 After completion of the test exercise, the test area concentration should be measured again using the high range colorimetric detector tube.

6.10 Exit the test area.

6.11 The opening created by the suit zipper or other appropriate suit penetration should be used to determine the ammonia concentration in the suit with the low range length of stain detector tube or other ammonia monitor. The internal TECP suit air should be sampled far enough from the enclosed test area to prevent a false ammonia reading.

6.12 After completion of the measurement of the suit interior ammonia concentration the test is concluded and the suit is doffed and the respirator removed.

6.13 The ventilating fan for the test room should be turned on and allowed to run for enough time to remove the ammonia gas. The fan shall be vented to the outside of the building.

6.14 Any detectable ammonia in the suit interior (five ppm ammonia (NH_3) or more for the length of stain detector tube) indicates that the suit has failed the test. When other ammonia detectors are used a lower level of detection is possible, and it should be specified as the pass/fail criteria.

6.15 By following this test method, an intrusion coefficient of approximately 200 or more can be measured with the suit in a completely operational condition. If the intrusion coefficient is 200 or more, then the suit is suitable for emergency response and field use.

7.0—Retest procedures

7.1 If the suit fails this test, check for leaks by following the pressure test in test A above.

7.2 Retest the TECP suit as outlined in test procedure 6.0.

8.0—Report

8.1 Each gas tight totally-encapsulating chemical protective suit tested by this practice shall have the following information recorded.

8.1.1 Unique identification number, identifying brand name, date of purchase, material of construction, and unique suit features; e.g., special breathing apparatus.

8.1.2 General description of test room used for test.

8.1.3 Brand name and purchase date of ammonia detector strips and color change data.

8.1.4 Brand name, sampling range, and expiration date of the length of stain ammonia detector tubes. The brand name and model of the sampling pump should also be recorded. If another type of ammonia detector is used, it should be identified along with its minimum detection limit for ammonia.

8.1.5 Actual test results shall list the two test area concentrations, their average, the interior suit concentration, and the calculated intrusion coefficient. Retest data shall be recorded as an additional test.

8.2 The evaluation of the data shall be specified as "suit passed" or "suit failed," and the date of the test. Any detectable ammonia (five ppm or greater for the length of stain detector tube) in the suit interior indicates the suit has failed this test. When other ammonia detectors are used, a lower level of detection is possible and it should be specified as the pass fail criteria.

Caution

Visually inspect all parts of the suit to be sure they are positioned correctly and secured tightly before putting the suit back into service. Special care should be taken to examine each exhaust valve to make sure it is not blocked.

Care should also be exercised to assure that the inside and outside of the suit is completely dry before it is put into storage.

Appendix B—General Description and Discussion of the Levels of Protection and Protective Gear

This appendix sets forth information about personal protective equipment (PPE) which may be used to assist employers in complying with the PPE requirements of this section.

As required by the standard, PPE must be selected which will protect employees from the specific hazards which they are likely to encounter during their work on-site.

Selection of the appropriate PPE is a complex process which should take into consideration a variety of factors. Key factors involved in this process are identification of the hazards, or suspected hazards; their routes of potential hazard to employees (inhalation, skin absorption, ingestion, and eye or skin contact); and the performance of the PPE *materials* (and seams) in providing a barrier to these hazards. The amount of protection provided by PPE is material-hazard specific. That is, protective equipment materials will protect well against some hazardous substances and poorly, or not at all, against others. In many instances, protective equipment materials cannot be found which will provide continuous protection from the particular hazardous substance. In these cases the breakthrough time of the protective material should exceed the work durations, or the exposure after breakthrough may not pose a hazardous level.

Other factors in this selection process to be considered are matching the PPE to the employee's work requirements and task-specific conditions. The durability of PPE materials, such as tear strength and seam strength, should be considered in relation to the employee's tasks. The effects of PPE in relation to heat stress and task duration are a factor in selecting and using PPE. In some cases layers of PPE may be necessary to provide sufficient protection, or to protect expensive PPE inner garments, suits or equipment.

The more that is known about the hazards at the site, the easier the job of PPE selection becomes. As more information about the hazards and conditions at the site becomes available, the site supervisor can make decisions to up-grade or down-grade the level of PPE protection to match the tasks at hand.

The following are guidelines which an employer can use to begin the selection of the appropriate PPE. As noted above, the site information may suggest the use of combinations of PPE selected from the different protection levels (i.e., A, B, C, or D) as being more suitable to the hazards of the

Federal Register / Vol. 54, No. 42 / Monday, March 6, 1989 / Rules and Regulations   9333

work. It should be cautioned that the listing below does not fully address the performance of the specific PPE material in relation to the specific hazards at the job site, and that PPE selection, evaluation and re-selection is an ongoing process until sufficient information about the hazards and PPE performance is obtained.

Part A. Personal protective equipment is divided into four categories based on the degree of protection afforded. (See Part B of this appendix for further explanation of Levels A, B, C, and D hazards.)

I. *Level A*—To be selected when the greatest level of skin, respiratory, and eye protection is required.

The following constitute Level A equipment; it may be used as appropriate:

1. Positive pressure, full face-piece self-contained breathing apparatus (SCBA), or positive pressure supplied air respirator with escape SCBA, approved by the National Institute for Occupational Safety and Health (NIOSH).
2. Totally-encapsulating chemical-protective suit.
3. Coveralls. [1]
4. Long underwear. [1]
5. Gloves, outer, chemical-resistant.
6. Gloves, inner, chemical-resistant.
7. Boots, chemical-resistant, steel toe and shank.
8. Hard hat (under suit). [1]
9. Disposable protective suit, gloves and boots (depending on suit construction, may be worn over totally-encapsulating suit).

II. *Level B*—The highest level of respiratory protection is necessary but a lesser level of skin protection is needed.

The following constitute Level B equipment; it may be used as appropriate.

1. Positive pressure, full-facepiece self-contained breathing apparatus (SCBA), or positive pressure supplied air respirator with escape SCBA (NIOSH approved).
2. Hooded chemical-resistant clothing (overalls and long-sleeved jacket; coveralls; one or two-piece chemical-splash suit; disposable chemical-resistant overalls).
3. Coveralls. [1]
4. Gloves, outer, chemical-resistant.
5. Gloves, inner, chemical-resistant.
6. Boots, outer, chemical-resistant steel toe and shank.
7. Boot-covers, outer, chemical-resistant (disposable). [1]
8. Hard hat. [1]
9. [Reserved]
10. Face shield. [1]

III. *Level C*—The concentration(s) and type(s) of airborne substance(s) is known and the criteria for using air purifying respirators are met.

The following constitute Level C equipment; it may be used as appropriate.

1. Full-face or half-mask, air purifying respirators (NIOSH approved).
2. Hooded chemical-resistant clothing (overalls; two-piece chemical-splash suit; disposable chemical-resistant overalls).
3. Coveralls. [1]
4. Gloves, outer, chemical-resistant.

---

[1] Optional, as applicable.

5. Gloves, inner, chemical-resistant.
6. Boots (outer), chemical-resistant steel toe and shank. [1]
7. Boot-covers, outer, chemical-resistant (disposable) [1].
8. Hard hat. [1]
9. Escape mask. [1]
10. Face shield. [1]

IV. *Level D*—A work uniform affording minimal protection, used for nuisance contamination only.

The following constitute Level D equipment; it may be used as appropriate:

1. Coveralls.
2. Gloves. [1]
3. Boots/shoes, chemical-resistant steel toe and shank.
4. Boots, outer, chemical-resistant (disposable). [1]
5. Safety glasses or chemical splash goggles*.
6. Hard hat. [1]
7. Escape mask. [1]
8. Face shield. [1]

Part B. The types of hazards for which levels A, B, C, and D protection are appropriate are described below:

I. *Level A*—Level A protection should be used when:

1. The hazardous substance has been identified and requires the highest level of protection for skin, eyes, and the respiratory system based on either the measured (or potential for) high concentration of atmospheric vapors, gases, or particulates; or the site operations and work functions involve a high potential for splash, immersion, or exposure to unexpected vapors, gases, or particulates of materials that are harmful to skin or capable of being absorbed through the skin;
2. Substances with a high degree of hazard to the skin are known or suspected to be present, and skin contact is possible; or
3. Operations are being conducted in confined, poorly ventilated areas, and the absence of conditions requiring Level A have not yet been determined.

II. *Level B*—Level B protection should be used when:

1. The type and atmospheric concentration of substances have been identified and require a high level of respiratory protection, but less skin protection;
2. The atmosphere contains less than 19.5 percent oxygen; or
3. The presence of incompletely identified vapors or gases is indicated by a direct-reading organic vapor detection instrument, but vapors and gases are not suspected of containing high levels of chemicals harmful to skin or capable of being absorbed through the skin.

Note: This involves atmospheres with IDLH concentrations of specific substances that present severe inhalation hazards and that do not represent a severe skin hazard; or that do not meet the criteria for use of air-purifying respirators.

III. *Level C*—Level C protection should be used when:

1. The atmospheric contaminants, liquid splashes, or other direct contact will not

adversely affect or be absorbed through any exposed skin;
2. The types of air contaminants have been identified, concentrations measured and an air-purifying respirator is available that can remove the contaminants; and
3. All criteria for the use of air-purifying respirators are met.

IV. *Level D*—Level D protection should be used when:

1. The atmosphere contains no known hazard; or
2. Work functions preclude splashes, immersion, or the potential for unexpected inhalation of or contact with hazardous levels of any chemicals.

Note: As stated before, combinations of personal protective equipment other than those described for Levels A, B, C, and D protection may be more appropriate and may be used to provide the proper level of protection.

As an aid in selecting suitable chemical protective clothing, it should be noted that the National Fire Protection Association is developing standards on chemical protective clothing. These standards are currently undergoing public review and include:

NFPA 1991—Standard on Vapor-Protective Suits for Hazardous Chemical Emergencies (EPA Level A Protective Clothing)

NFPA 1991—Standard on Liquid Splash-Protective Suits for Hazardous Chemical Emergencies (EPA Level B Protective Clothing)

NFPA 1993—Standard on Liquid Splash-Protective Suits for Non-emergency, Non-flammable Hazardous Chemical Situations (EPA Level B Protective Clothing)

These standards would apply documentation and performance requirements to the manufacture of chemical protective suits. Chemical protective suits meeting these requirements would be labelled as compliant with the appropriate standard. When these standards are adopted by the National Fire Protection Association, it is recommended that chemical protective suits which meet these standards be used.

Appendix C—Compliance Guidelines

1. *Occupational Safety and Health Program.* Each hazardous waste site clean-up effort will require an occupational safety and health program headed by the site coordinator or the employer's representative. The purpose of the program will be the protection of employees at the site and will be an extension of the employer's overall safety and health program. The program will need to be developed before work begins on the site and implemented as work proceeds as stated in paragraph (b). The program is to facilitate coordination and communication of safety and health issues among personnel responsible for the various activities which will take place at the site. It will provide the overall means for planning and implementing the needed safety and health training and job orientation of employees who will be working at the site. The program will provide the means for identifying and controlling worksite hazards and the means for

9334   Federal Register / Vol. 54, No. 42 / Monday, March 6, 1989 / Rules and Regulations

monitoring program effectiveness. The program will need to cover the responsibilities and authority of the site coordinator or the employer's manager on the site for the safety and health of employees at the site, and the relationships with contractors or support services as to what each employer's safety and health responsibilities are for their employees on the site. Each contractor on the site needs to have its own safety and health program so structured that it will smoothly interface with the program of the site coordinator or principal contractor.

Also those employers involved with treating, storing or disposal of hazardous waste as covered in paragraph (p) must have implemented a safety and health program for their employees. This program is to include the hazard communication program required in paragraph (p)(1) and the training required in paragraphs (p)(7) and (p)(8) as parts of the employers comprehensive overall safety and health program. This program is to be in writing.

Each site or workplace safety and health program will need to include the following: (1) Policy statements of the line of authority and accountability for implementing the program, the objectives of the program and the role of the site safety and health supervisor or manager and staff; (2) means or methods for the development of procedures for identifying and controlling workplace hazards at the site; (3) means or methods for the development and communication to employees of the various plans, work rules, standard operating procedures and practices that pertain to individual employees and supervisors; (4) means for the training of supervisors and employees to develop the needed skills and knowledge to perform their work in a safe and healthful manner; (5) means to anticipate and prepare for emergency situations; and (6) means for obtaining information feedback to aid in evaluating the program and for improving the effectiveness of the program. The management and employees should be trying continually to improve the effectiveness of the program thereby enhancing the protection being afforded those working on the site.

Accidents on the site or workplace should be investigated to provide information on how such occurrences can be avoided in the future. When injuries or illnesses occur on the site or workplace, they will need to be investigated to determine what needs to be done to prevent this incident from occurring again. Such information will need to be used as feedback on the effectiveness of the program and the information turned into positive steps to prevent any reoccurrence. Receipt of employee suggestions or complaints relating to safety and health issues involved with site or workplace activities is also a feedback mechanism that can be used effectively to improve the program and may serve in part as an evaluative tool(s).

For the development and implementation of the program to be the most effective, professional safety and health personnel should be used. Certified Safety Professionals, Board Certified Industrial Hygienists or Registered Professional Safety

Engineers are good examples of professional stature for safety and health managers who will administer the employer's program.

2. *Training.* The training programs for employees subject to the requirements of paragraph (e) of this standard should address: the safety and health hazards employees should expect to find on hazardous waste clean-up sites; what control measures or techniques are effective for those hazards; what monitoring procedures are effective in characterizing exposure levels; what makes an effective employer's safety and health program; what a site safety and health plan should include; hands on training with personal protective equipment and clothing they may be expected to use; the contents of the OSHA standard relevant to the employee's duties and function; and, employee's responsibilities under OSHA and other regulations. Supervisors will need training in their responsibilities under the safety and health program and its subject areas such as the spill containment program, the personal protective equipment program, the medical surveillance program, the emergency response plan and other areas.

The training programs for employees subject to the requirements of paragraph (p) of this standard should address: the employers safety and health program elements impacting employees; the hazard communication program; the medical surveillance program; the hazards and the controls for such hazards that employees need to know for their job duties and functions. All require annual refresher training.

The training programs for employees covered by the requirements of paragraph (q) of this standard should address those competencies required for the various levels of response such as: the hazards associated with hazardous substances; hazard identification and awareness; notification of appropriate persons; the need for and use of personal protective equipment including respirators; the decontamination procedures to be used; preplanning activities for hazardous substance incidents including the emergency reponse plan; company standard operating procedures for hazardous substance emergency responses; the use of the incident command system and other subjects. Hands-on training should be stressed whenever possible. Critiques done after an incident which include an evaluation of what worked and what did not and how could the incident be better handled the next time may be counted as training time.

For hazardous materials specialists (usually members of hazardous materials teams), the training should address the care, use and/or testing of chemical protective clothing including totally encapsulating suits, the medical surveillance program, the standard operating procedures for the hazardous materials team including the use of plugging and patching equipment and other subject areas.

Officers and leaders who may be expected to be in charge at an incident should be fully knowledgeable of their company's incident command system. They should know where and how to obtain additional assistance and be familiar with the local district's emergency

response plan and the state emergency response plan.

Specialist employees such as technical experts, medical experts or environmental experts that work with hazardous materials in their regular jobs, who may be sent to the incident scene by the shipper, manufacturer or governmental agency to advise and assist the person in charge of the incident should have training on an annual basis. Their training should include the care and use of personal protective equipment including respirators; knowledge of the incident command system and how they are to relate to it; and those areas needed to keep them current in their respective field as it relates to safety and health involving specific hazardous substances.

Those skilled support personnel, such as employees who work for public works departments or equipment operators who operate bulldozers, sand trucks, backhoes, etc., who may be called to the incident scene to provide emergency support assistance, should have at least a safety and health briefing before entering the area of potential or actual exposure. These skilled support personnel, who have not been a part of the emergency response plan and do not meet the training requirements, should be made aware of the hazards they face and should be provided all necessary protective clothing and equipment required for their tasks.

3. *Decontamination.* Decontamination procedures should be tailored to the specific hazards of the site, and may vary in complexity and number of steps, depending on the level of hazard and the employee's exposure to the hazard. Decontamination procedures and PPE decontamination methods will vary depending upon the specific substance, since one procedure or method may not work for all substances. Evaluation of decontamination methods and procedures should be performed, as necessary, to assure that employees are not exposed to hazards by re-using PPE. References in Appendix F may be used for guidance in establishing an effective decontamination program. In addition, the U.S. Coast Guard's Manual, "Policy Guidance for Response to Hazardous Chemical Releases," U.S. Department of Transportation, Washington, DC (COMDTINST M16465.30) is a good reference for establishing an effective decontamination program.

4. *Emergency response plans.* States, along with designated districts within the states, will be developing or have developed local emergency response plans. These state and district plans should be utilized in the emergency response plans called for in the standard. Each employer should assure that its emergency response plan is compatible with the local plan. The major reference being used to aid in developing the state and local district plans is the *Hazardous Materials Emergency Planning Guide,* NRT-1. The current Emergency Response Guidebook from the U.S. Department of Transportation, CMA's CHEMTREC and the Fire Service Emergency Management Handbook may also be used as resources.

Federal Register / Vol. 54, No. 42 / Monday, March 6, 1989 / Rules and Regulations 9335

Employers involved with treatment, storage, and disposal facilities for hazardous waste, which have the required contingency plan called for by their permit, would not need to duplicate the same planning elements. Those items of the emergency response plan that are properly addressed in the contingency plan may be substituted into the emergency response plan required in 1910.120 or otherwise kept together for employer and employee use.

5. *Personal protective equipment programs.* The purpose of personal protective clothing and equipment (PPE) is to shield or isolate individuals from the chemical, physical, and biologic hazards that may be encountered at a hazardous substance site.

As discussed in Appendix B, no single combination of protective equipment and clothing is capable of protecting against all hazards. Thus PPE should be used in conjunction with other protective methods and its effectiveness evaluated periodically.

The use of PPE can itself create significant worker hazards, such as heat stress, physical and psychological stress, and impaired vision, mobility, and communication. For any given situation, equipment and clothing should be selected that provide an adequate level of protection. However, over-protection, as well as under-protection, can be hazardous and should be avoided where possible.

Two basic objectives of any PPE program should be to protect the wearer from safety and health hazards, and to prevent injury to the wearer from incorrect use and/or malfunction of the PPE. To accomplish these goals, a comprehensive PPE program should include hazard identification, medical monitoring, environmental surveillance, selection, use, maintenance, and decontamination of PPE and its associated training.

The written PPE program should include policy statements, procedures, and guidelines. Copies should be made available to all employers, and a reference copy should be made available at the worksite. Technical data on equipment, maintenance manuals, relevant regulations, and other essential information should also be collected and maintained.

6. *Incident command system (ICS).* Paragraph 1910.120(q)(3)(ii) requires the implementation of an ICS. The ICS is an organized approach to effectively control and manage operations at an emergency incident. The individual in charge of the ICS is the senior official responding to the incident. The ICS is not much different than the "command post" approach used for many years by the fire service. During large complex fires involving several companies and many pieces of apparatus, a command post would be established. This enabled *one* individual to be in charge of managing the incident, rather than having several officers from different companies making separate, and sometimes conflicting, decisions. The individual in charge of the command post would delegate responsibility for performing various tasks to subordinate officers. Additionally, all commur.ications were routed through the commana post to reduce the number of radio transmissions and eliminate confusion.

However, strategy, tactics, and all decisions were made by one individual.

The ICS is a very similar system, except it is implemented for emergency response to all incidents, both large and small, that involve hazardous substances.

For a small incident, the individual in charge of the ICS may perform many tasks of the ICS. There may not be any, or little, delegation of tasks to subordinates. For example, in response to a small incident, the individual in charge of the ICS, in addition to normal command activities, may become the safety officer and may designate only one employee (with proper equipment) as a back-up to provide assistance if needed. OSHA does recommend, however, that at least two employees be designated as back-up personnel since the assistance needed may include rescue.

To illustrate the operation of the ICS, the following scenario might develop during a small incident, such as an overturned tank truck with a small leak of flammable liquid. The first responding senior officer would implement and take command of the ICS. That person would size-up the incident and determine if additional personnel and apparatus were necessary; would determine what actions to take to control the leak; and, determine the proper level of personal protective equipment. If additional assistance is not needed, the individual in charge of the ICS would implement actions to stop and control the leak using the fewest number of personnel that can effectively accomplish the tasks. The individual in charge of the ICS then would designate himself as the safety officer and two other employees as a back-up in case rescue may become necessary. In this scenario, decontamination procedures would not be necessary.

A large complex incident may require many employees and difficult, time-consuming efforts to control. In these situations, the individual in charge of the ICS will want to delegate different tasks to subordinates in order to maintain a span of control that will keep the number of subordinates, that are reporting, to a manageable level.

Delegation of task at large incidents may be by location, where the incident scene is divided into sectors, and subordinate officers coordinate activities within the sector that they have been assigned.

Delegation of tasks can also be by function. Some of the functions that the individual in charge of the ICS may want to delegate at a large incident are: medical services; evacuation; water supply; resources (equipment, apparatus); media relations; safety; and, site control (integrate activities with police for crowd and traffic control). Also for a large incident, the individual in charge of the ICS will designate several employees as back-up personnel; and a number of safety officers to monitor conditions and recommend safety precautions.

Therefore, no matter what size or complexity an incident may be, by implementing an ICS there will be *one individual in charge* who makes the decisions and gives directions; and, all actions, and communications are coordinated through one

central point of command. Such a system should reduce confusion, improve safety, organize and coordinate actions, and should facilitate effective management of the incident.

7. *Site Safety and Control Plans.* The safety and security of response personnel and others in the area of an emergency response incident site should be of primary concern to the incident commander. The use of a site safety and control plan could greatly assist those in charge of assuring the safety and health of employees on the site.

A comprehensive site safety and control plan should include the following: summary analysis of hazards on the site and a risk analysis of those hazards; site map or sketch; site work zones (clean zone, transition or decontamination zone, work or hot zone); use of the buddy system; site communications; command post or command center; standard operating procedures and safe work practices; medical assistance and triage area; hazard monitoring plan (air contaminate monitoring, etc.); decontamination procedures and area; and other relevant areas. This plan should be a part of the employer's emergency response plan or an extension of it to the specific site.

8. *Medical surveillance programs.* Workers handling hazardous substances may be exposed to toxic chemicals, safety hazards, biologic hazards, and radiation. Therefore, a medical surveillance program is essential to assess and monitor workers' health and fitness for employment in hazardous waste operations and during the course of work; to provide emergency and other treatment as needed; and to keep accurate records for future reference.

The *Occupational Safety and Health Guidance Manual for Hazardous Waste Site Activities* developed by the National Institute for Occupational Safety and Health (NIOSH), the Occupational Safety and Health Administration (OSHA), the U.S. Coast Guard (USCG), and the Environmental Protection Agency (EPA); October 1985 provides an excellent example of the types of medical testing that should be done as part of a medical surveillance program.

Appendix D—References

The following references may be consulted for further information on the subject of this standard:

1. OSHA Instruction DFO CPL 2.70— January 29, 1986, *Special Emphasis Program: Hazardous Waste Sites.*
2. OSHA Instruction DFO CPL 2–2.37A— January 29, 1986, *Technical Assistance and Guidelines for Superfund and Other Hazardous Waste Site Activities.*
3. OSHA Instruction DTS CPL 2.74— January 29, 1986, *Hazardous Waste Activity Form, OSHA 175.*
4. *Hazardous Waste ınspections Reference Manual,* U.S. Department of Labor, Occupational Safety and Health Administration, 1986.
5. Memorandum of Understan...ng Among the National Institute for Occupational Safety and Health, the Occupational Safety and Health Administration, the United States Coast Guard and the United States Environmental Protection Agency, *Guidance for Worker Protection During Hazardous*

9336 Federal Register / Vol. 54, No. 42 / Monday, March 6, 1989 / Rules and Regulations

Waste Site Investigations and Clean-up and Hazardous Substance Emergencies. December 18, 1980.

6. *National Priorities List,* 1st Edition, October 1984; U.S. Environmental Protection Agency, Revised periodically.

7. *The Decontamination of Response Personnel,* Field Standard Operating Procedures (F.S.O.P.) 7; U.S. Environmental Protection Agency, Office of Emergency and Remedial Response, Hazardous Response Support Division, December 1984.

8. *Preparation of a Site Safety Plan,* Field Standard Operating Procedures (F.S.O.P.) 9; U.S. Environmental Protection Agency, Office of Emergency and Remedial Response, Hazardous Response Support Division, April 1985.

9. *Standard Operating Safety Guidelines;* U.S. Environmental Protection Agency, Office of Emergency and Remedial Response, Hazardous Response Support Division, Environmental Response Team; November 1984.

10. *Occupational Safety and Health Guidance Manual for Hazardous Waste Site Activities,* National Institute for Occupational Safety and Health (NIOSH), Occupational Safety and Health Administration (OSHA), U.S. Coast Guard (USCG), and Environmental Protection Agency (EPA); October 1985.

11. *Protecting Health and Safety at Hazardous Waste Sites: An Overview,* U.S. Environmental Protection Agency, EPA/625/9–85/006; September 1985.

12. *Hazardous Waste Sites and Hazardous Substance Emergencies,* NIOSH Worker Bulletin, U.S. Department of Health and Human Services, Public Health Service, Centers for Disease Control, National Institute for Occupational Safety and Health; December 1982.

13. *Personal Protective Equipment for Hazardous Materials Incidents: A Selection Guide;* U.S. Department of Health and Human Services, Public Health Service, Centers for Disease Control, National Institute for Occupational Safety and Health; October 1984.

14. *Fire Service Emergency Management Handbook,* International Association of Fire Chiefs Foundation, 101 East Holly Avenue, Unit 10B, Sterling, VA 22170, January 1985.

15. *Emergency Response Guidebook,* U.S. Department of Transportation, Washington, DC, 1987.

16. *Report to the Congress on Hazardous Materials Training, Planning and Preparedness,* Federal Emergency Management Agency, Washington, DC, July 1986.

17. *Workbook for Fire Command,* Alan V. Brunacini and J. David Beageron, National

Fire Protection Association, Batterymarch Park, Quincy, MA 02269, 1985.

18. *Fire Command,* Alan V. Brunacini, National Fire Protection, Batterymarch Park, Quincy, MA 02269, 1985.

19. *Incident Command System,* Fire Protection Publications, Oklahoma State University, Stillwater, OK 74078, 1983.

20. *Site Emergency Response Planning,* Chemical Manufacturers Association, Washington, DC 20037, 1986.

21. *Hazardous Materials Emergency Planning Guide,* NRT–1, Environmental Protection Agency, Washington, DC, March 1987.

22. *Community Teamwork: Working Together to Promote Hazardous Materials Transportation Safety.* U.S. Department of Transportation, Washington, DC, May 1983.

23. *Disaster Planning Guide for Business and Industry,* Federal Emergency Management Agency, Publication No. FEMA 141, August 1987.

(The Office of Management and Budget has approved the information collection requirements in this section under control number 1218–0139)

[FR Doc. 89–4992 Filed 3–1–89; 11:54 am]

BILLING CODE 4510-26-M

APPENDIX K
TITLE III LIST OF LISTS

TITLE III LIST OF LISTS

Chemicals Subject to
Reporting Under Title III
of the Superfund Amendments
and Reauthorization Act
(SARA) of 1986

July 1987

Office of Toxic Substances
U.S. Environmental Protection Agency
Washington, DC 20460

SARA TITLE III
CONSOLIDATED CHEMICAL LIST

This consolidated chemical list includes chemicals subject to reporting requirements under Title III of the Superfund Amendments and Reauthorization Act of 1986 (SARA).* It has been prepared to help firms handling chemicals determine whether they need to submit reports under Sections 304 or 313 of Title III and, for a specific chemical, what reports need to be submitted.

The list includes chemicals referenced under four federal statutory provisions, as follows:

(1) SARA Section 302 Extremely Hazardous Substances, the presence of any of which, in sufficient quantities, requires certain emergency planning activities to be conducted. Releases of these substances are also subject to reporting under Section 304 of Title III. The final rule was published on April 22, 1987 (52 FR 13378).

(2) CERCLA Hazardous Substances ("RQ" Chemicals), releases of which are subject to reporting under the Comprehensive Environmental Response, Compensation, and Liability Act of 1980 (CERCLA, or "Superfund"). Such releases are also subject to reporting under Section 304 of Title III. CERCLA hazardous substances, and their reportable quantities, are listed in 40 CFR Part 302, Table 302.4.

(3) SARA Section 313 Toxic Chemicals, emissions or releases of which must be reported annually as part of SARA Title III's community right-to-know provisions. The proposed Section 313 rule containing these chemicals was published on June 4, 1987 (52 FR 21152).

(4) RCRA Hazardous Wastes from the P and U lists only (40 CFR 261.33), which consist of lists of specific chemicals. RCRA hazardous wastes consisting of waste streams on the F and K lists are not included here; such waste streams are also CERCLA hazardous substances. This listing is provided as an indicator to companies that they may already have data on a specific chemical that can be used for Title III reporting purposes.

There are four columns in the consolidated list corresponding to these four statutory provisions. An "X" in one of these four columns indicates that a chemical is subject to reporting requirements under the appropriate statutory provision. A fifth column, headed "State," is left blank, to be checked if state reporting requirements apply to a chemical. The heading "Section 304" over the Section 302 and CERCLA lists indicates that the

---

* Not specified in this consolidated list are all chemicals subject to the reporting requirements in Section 311 and 312 of SARA Title III. These hazardous chemicals, for which material safety data sheets (MSDS) must be developed under Occupational Safety and Health Act Hazard Communication Standards, are identified as such by broad criteria, rather than by enumeration. There are over 50,000 such substances that satisfy the criteria.

APPENDIX K 383

reporting requirements in Section 304 of SARA Title III apply to Section 302 extremely hazardous substances and CERCLA hazardous substances. As indicated, most chemicals on the consolidated list are subject to reporting requirements under more than one statutory provision.

The chemicals on this list are ordered by Chemical Abstracts Service (CAS) registry number. Categories of chemicals, which do not have CAS registry numbers, but which are cited under CERCLA and Section 313, are placed at the end of the list. The listed chemicals are grouped by fours to facilitate reading.

For additional copies of this document address requests to:

TSCA Assistance Office (TS-799)
Office of Toxic Substances
U.S. Environmental Protection Agency
401 M Street, SW
Washington DC 20460

Phone; (202) 554-1411

SARA
TITLE III
CONSOLIDATED CHEMICAL LIST
5/20/87

Page 1

		Section 304				
CAS Number	Chemical Name	§302	CERCLA	§313	RCRA	STATE
50-00-0	Formaldehyde	X	X	X	X	
50-07-7	Mitomycin C	X	X		X	
50-14-6	Ergocalciferol	X				
50-18-0	Cyclophosphamide		X		X	
50-29-3	DDT		X		X	
50-32-8	Benzo[a]pyrene		X		X	
50-55-5	Reserpine		X		X	
51-21-8	Fluorouracil	X				
51-28-5	2,4-Dinitrophenol		X	X	X	
51-43-4	1,2-Benzenediol,4-[1-hydroxy-2-(methylamino)ethyl]-		X		X	
51-75-2	Mechlorethamine	X		X		
51-79-6	Carbamic acid, ethyl ester		X	X	X	
51-83-2	Carbachol chloride	X				
52-68-6	Trichlorophon	X	X	X		
52-85-7	Famphur		X		X	
53-70-3	Dibenz[a,h]anthracene		X		X	
53-86-1	Indomethacin	X				
53-96-3	Acetamide, N-9H-fluoren-2-yl-		X	X	X	
54-11-5	Nicotine	X	X		X	
54-62-6	Aminopterin	X				
55-18-5	Ethanamine, N-ethyl-N-nitroso-		X	X	X	
55-21-0	Benzamide			X		
55-63-0	Nitroglycerine		X	X	X	
55-91-4	Isofluorphate	X	X		X	
56-04-2	Methylthiouracil		X		X	
56-23-5	Carbon tetrachloride		X	X	X	
56-25-7	Cantharidin	X				
56-38-2	Parathion	X	X	X	X	
56-49-5	Benz[j]aceanthrylene,1,2-dihydro-3-methyl-		X		X	
56-53-1	Diethylstilbestrol		X		X	
56-55-3	Benz[a]anthracene		X		X	
56-72-4	Coumaphos	X	X			
57-12-5	Cyanides (soluble cyanide salts)		X	X	X	
57-14-7	Dimethylhydrazine	X	X	X	X	
57-24-9	Strychnine	X	X		X	
57-47-6	Physostigmine	X				
57-57-8	Propiolactone, beta-	X		X		
57-64-7	Physostigmine, salicylate (1:1)	X				
57-74-9	Chlordane	X	X	X	X	

APPENDIX K   385

Page 2

CAS Number	Chemical Name	Section 304 §302	CERCLA	§313	RCRA	STATE
57-97-6	1,2-Benzanthracene, 7,12-dimethyl-		x		x	
58-36-6	Phenoxarsine, 10,10'-oxydi-	x				
58-89-9	Lindane	x	x	x	x	
58-90-2	Phenol, 2,3,4,6-tetrachloro-		x		x	
59-50-7	4-Chloro-m-cresol		x		x	
59-88-1	Phenylhydrazine hydrochloride	x				
59-89-2	N-Nitrosomorpholine			x		
60-00-4	Ethylenediamine tetraacetic acid (EDTA)		x			
60-09-3	4-Aminoazobenzene			x		
60-11-7	Benzenamine, N,N-dimethyl-4-phenylazo-		x	x	x	
60-29-7	Ethane, 1,1'-oxybis-		x		x	
60-34-4	Methylhydrazine	x	x	x	x	
60-35-5	Acetamide			x		
60-41-3	Strychnine, sulfate	x				
60-51-5	Dimethoate	x	x		x	
60-57-1	Dieldrin		x		x	
61-82-5	Amitrole		x		x	
62-38-4	Phenylmercury acetate	x	x		x	
62-44-2	Acetamide, N-(4-ethoxyphenyl)-		x		x	
62-50-0	Ethyl methanesulfonate		x		x	
62-53-3	Aniline	x	x	x	x	
62-55-5	Ethanethioamide		x	x	x	
62-56-6	Carbamide, thio-		x	x	x	
62-73-7	Dichlorvos	x	x	x		
62-74-8	Sodium fluoroacetate	x	x		x	
62-75-9	Nitrosodimethylamine	x	x	x	x	
63-25-2	Carbaryl		x	x		
64-00-6	Phenol, 3-(1-methylethyl)-, methylcarbamate	x				
64-18-8	Formic acid		x		x	
64-19-7	Acetic acid		x			
64-67-5	Diethyl sulfate			x		
64-86-8	Colchicine	x				
65-30-5	Nicotine sulfate	x				
65-85-0	Benzoic acid		x			
65-86-1	Orotic acid	x				
66-75-1	Uracil, 5-[bis(2-chloroethyl)amino]-		x		x	
66-81-9	Cycloheximide	x				
67-56-1	Methanol		x	x	x	
67-63-0	Isopropyl alcohol (mfg.-strong acid processes)			x		
67-64-1	Acetone		x	x	x	
67-66-3	Chloroform	x	x	x	x	
67-72-1	Ethane, 1,1,1,2,2,2-hexachloro-		x	x	x	
68-76-8	Triaziquone			x		

Page 3

CAS Number	Chemical Name	§302	Section 304 CERCLA · §313		RCRA	STATE
70-25-7	Guanidine, N-nitroso-N-methyl-N'-nitro-		X		X	
70-30-4	Hexachlorophene		X		X	
70-69-9	Propiophenone, 4'-amino-	X				
71-36-3	1-Butanol		X	X	X	
71-43-2	Benzene		X	X	X	
71-55-6	Methyl chloroform		X	X	X	
71-63-6	Digitoxin	X				
72-20-8	Endrin	X	X		X	
72-43-5	Ethane, 1,1,1-trichloro-2,2-bis(p-methoxyphenyl)-		X	X	X	
72-54-8	DDD		X		X	
72-55-9	DDE		X			
72-57-1	Trypan blue		X		X	
74-83-9	Methyl bromide	X	X	X	X	
74-85-1	Ethylene			X		
74-87-3	Methane, chloro		X	X	X	
74-88-4	Methane, iodo-		X	X	X	
74-89-5	Monomethylamine		X			
74-90-8	Hydrocyanic acid	X	X	X	X	
74-93-1	Methyl mercaptan	X	X		X	
74-95-3	Methane, dibromo-		X	X	X	
75-00-3	Chloroethane		X	X		
75-01-4	Ethene, chloro-		X	X	X	
75-04-7	Monoethylamine		X			
75-05-8	Acetonitrile		X	X	X	
75:07-0	Acetaldehyde		X	X	X	
75-09-2	Methane, dichloro-		X	X	X	
75-15-0	Carbon disulfide	X	X	X	X	
75-18-3	Dimethyl sulfide	X				
75-20-7	Calcium carbide		X			
75-21-8	Ethylene oxide	X	X	X	X	
75-25-2	Bromoform		X	X	X	
75-27-4	Dichlorobromomethane		X	X		
75-34-3	1,1-Dichloroethane		X		X	
75-35-4	1,1-Dichloroethylene		X	X		
75-36-5	Acetyl chloride		X		X	
75-44-5	Phosgene	X	X	X	X	
75-50-3	Trimethylamine		X			
75-55-8	Propyleneimine	X	X	X	X	
75-56-9	Propylene oxide	X	X	X		
75-60-5	Cacodylic acid		X		X	
75-64-9	tert-Butylamine		X			
75-65-0	tert-Butyl alcohol			X		
75-69-4	Methane, trichlorofluoro-		X		X	

CAS Number	Chemical Name	Section 304 §302	CERCLA	§313	RCRA	STATE
75-71-8	Dichlorodifluoromethane		x		x	
75-74-1	Tetramethyl lead	x				
75-77-4	Trimethylchlorosilane	x				
75-78-5	Dimethyldichlorosilane	x				
75-79-6	Methyltrichlorosilane	x				
75-86-5	Acetone cyanohydrin	x	x		x	
75-87-6	Acetaldehyde, trichloro-		x		x	
75-99-0	2,2-Dichloropropionic acid		x			
76-01-7	Pentachloroethane	x	x		x	
76-02-8	Trichloroacetyl chloride	x				
76-13-1	Chlorinated fluorocarbon (Freon 113)			x		
76-44-8	Heptachlc·		x	x	x	
77-47-4	Hexachlorocyclopentadiene	x	x	x	x	
77-78-1	Dimethyl sulfate	x	x	x	x	
77-81-6	Tabun	x				
78-00-2	Tetraethyllead	x	x		x	
78-34-2	Dioxathion	x				
78-53-5	Amiton	x				
78-59-1	Isophorone		x			
78-71-7	Oxetane, 3,3-bis(chloromethyl)-	x				
78-79-5	Isoprene		x			
78-81-9	iso-Butylamine		x			
78-82-0	Isobutyronitrile	x				
78-83-1	Isobutyl alcohol		x		x	
78-84-2	Isobutyra'dehyde			x		
78-87-5	1,2-Dichι opropane		x	x	x	
78-88-6	2,3-Dichloropropene		x			
78-92-2	sec-Butyl alcohol			x		
78-93-3	2-Butanone		x	x	x	
78-94-4	Methyl vinyl ketone	x				
78-97-7	Lactonitrile	x				
78-99-9	1,1-Dichloropropane		x			
79-00-5	Ethane, 1,1,2-trichloro-		x	x	x	
79-01-6	Trichloroethene		x	x	x	
79-06-1	Acrylamide	x	x	x	x	
79-09-4	Propionic acid		x			
79-10-7	Acrylic acid		x	x		
79-11-8	esistacetic acid	x		x		
79-19-6	Thiosemicarbazide	x	x		x	
79-21-0	Peracetic acid	x		x		
79-22-1	Methyl chloroformate	x	x		x	
79-31-2	iso-Butyric acid		x			
79-34-5	Ethane, 1,1,2,2-tetrachloro-		x	x	x	

Page 5

CAS Number	Chemical Name	§302	CERCLA	§313	RCRA	STATE
			Section 304			
79-44-7	Carbamoyl chloride, dimethyl·		X	X	X	
79-46-9	2-Nitropropane		X	X	X	
80-05-7	4,4'-Isopropylidenediphenol			X		
80-15-9	alpha,alpha-Dimethylbenzylhydroperoxide		X	X	X	
80-62-6	Methyl methacrylate		X	X	X	
80-63-7	Methyl 2-chloroacrylate	X				
81-07-2	1,2-Benzisothiazolin-3-one,1,1-dioxide, and salts		X	X	X	
81-81-2	Warfarin	X	X		X	
81-88-9	C.I. Food Red 15			X		
82-28-0	1-Amino-2-methylanthraquinone			X		
82-66-6	Diphacinone	X				
82-68-8	Benzene, pentachloronitro-		X	X	X	
83-32-9	Acenaphthene		X			
84-66-2	1,2-Benzenedicarboxylic acid, diethyl ester		X	X	X	
84-74-2	Dibutyl phthalate	X	X	X	X	
84-80-0	Phylloquinone	X				
85-00-7	Diquat		X			
85-01-8	Phenanthrene		X			
85-44-9	1,2-Benzenedicarboxylic acid anhydride		X	X	X	
85-68-7	Butyl benzyl phthalate		X	X		
86-30-6	N-Nitrosodiphenylamine		X	X		
86-50-0	Azinophos-methyl	X	X			
86-73-7	Fluorene		X			
86-78-3	1,3-Butadiene, 1,1,2,3,4,4-hexachloro-		X		X	
86-88-4	Antu	X	X		X	
87-62-7	2,6-Xylidine			X		
87-65-0	2,6-Dichlorophenol		X		X	
87-68-3	Hexachloro-1,3-butadiene			X		
87-86-5	Pentachlorophenol	X	X	X	X	
88-01-9	Aroclor		X			
88-05-1	Aniline, 2,4,6-trimethyl·	X				
88-06-2	Phenol,2,4,6-trichloro		X	X	X	
88-72-2	o-Nitrotoluene		X			
88-75-5	o-Nitrophenol		X	X		
88-85-7	Dinoseb	X	X		X	
88-89-1	Picric acid			X		
90-04-0	o-Anisidine			X		
90-43-7	2-Phenylphenol			X		
90-94-8	Michler's ketone			X		
91-08-7	Toluene 2,6-diisocyanate	X	X	X		
91-20-3	Naphthalene		X	X	X	
91-22-5	Quinoline		X	X		
91-58-7	beta-Chloronaphthalene		X		X	

APPENDIX K   389

Page 6

CAS Number	Chemical Name	Section 304 §302	CERCLA	§313	RCRA	STATE
91-59-8	2-Naphthylamine		x	x	x	
91-80-5	Methapyrilene		x		x	
91-94-1	(1,1'-Biphenyl)-4,4'diamine,3,3'dichloro-		x	x	x	
92-52-4	Biphenyl			x		
92-67-1	4-Aminobiphenyl			x		
92-87-5	Benzidine		x	x	x	
92-93-3	4-Nitrobiphenyl			x		
93-05-0	Diethyl-p-phenylenediamine	x				
93-72-1	Propionic acid, 2-(2,4,5-trichlorophenoxy)-		x		x	
93-76-5	2,4,5-T		x		x	
93-79-8	2,4,5-T esters		x			
94-11-1	2,4-D Esters		x			
94-36-0	Benz. peroxide			x		
94-58-6	Benzene, 1,2-methylenedioxy-4-propyl-		x		x	
94-59-7	Benzene, 1,2-methylenedioxy-4-allyl-		x	x	x	
94-75-7	2,4-D Acid		x	x	x	
94-79-1	2,4-D Esters		x			
94-80-4	2,4-D Esters		x			
95-47-6	Benzene, o-dimethyl-		x	x	x	
95-48-7	Cresylic acid	x	x	x	x	
95-50-1	Benzene, 1,2-dichloro-		x	x	x	
95-53-4	o-Toluidine			x		
95-57-8	2-Chlorophenol		x		x	
95-63-6	Pseudocumene	x		x		
95-80-7	Diaminotoluene		x	x	x	
95-94-3	Benzene, 1,2,4,5-tetrachloro-		x		x	
95-95-4	Phenol, 2,4,5-trichloro-		x	x	x	
96-09-3	Styrene oxide			x		
96-12-8	1,2-Dibromo-3-chloropropane		x	x	x	
96-33-3	Methyl acrylate			x		
96-45-7	Ethylenethiourea		x	x	x	
97-18-7	Phenol, 2,2'-thiobis(4,6-dichloro-	x				
97-56-3	C.I. Solvent Yellow 3		x			
97-63-2	Ethyl methacrylate		x		x	
98-01-1	2-Furancarboxaldehyde		x		x	
98-05-5	Benzenearsonic acid	x				
98-07-7	Benzotrichloride	x	x	x	x	
98-09-9	Benzenesulfonyl chloride	x	x		x	
98-13-5	Trichlorophenylsilane	x				
98-16-8	Benzenamine, 3-(trifluoromethyl)-	x				
98-82-8	Benzene, 1-methylethyl-		x	x	x	
98-86-2	Acetophenone		x		x	
98-87-3	Benzal chloride	x	x	x	x	

Page 7

Section 304

CAS Number	Chemical Name	§302	CERCLA	§313	RCRA	STATE
98-88-4	Benzoyl chloride		X	X		
98-95-3	Nitrobenzene	X	X	X	X	
99-08-1	m-Nitrotoluene		X			
99-35-4	Benzene, 1,3,5-trinitro-		X		X	
99-55-8	Benzenamine, 2-methyl-5-nitro-		X		X	
99-59-2	5-Nitro-o-anisidine			X		
99-65-0	m-Dinitrobenzene		X			
99-98-9	Dimethyl-p-phenylenediamine	X				
99-99-0	p-Nitrotoluene		X			
100-01-6	Benzenamine, 4-nitro-		X		X	
100-02-7	p-Nitrophenol		X	X	X	
100-14-1	Benzene, 1-(chloromethyl)-4-nitro-	X				
100-21-0	Terephthalic acid			X		
100-25-4	p-Dinitrobenzene		X			
100-41-4	Ethylbenzene		X	X		
100-42-5	Styrene		X	X		
100-44-7	Benzyl chloride	X	X	X	X	
100-47-0	Benzonitrile		X			
100-75-4	N-Nitrosopiperidine		X	X	X	
101-14-4	Benzenamine, 4,4'-methylenebis(2-chloro-		X	X	X	
101-55-3	Benzene, 1-bromo-4-phenoxy-		X		X	
101-61-1	4,4'-Methylene bis(N,N-dimethyl) benzenamine			X		
101-68-8	Methylene bis(phenylisocyanate) (MBI)			X		
101-77-9	4,4'-Methylene dianiline			X		
101-80-4	4,4'-Diaminodiphenyl ether			X		
102-36-3	Isocyanic acid, 3,4-dichlorophenyl ester	X				
103-23-1	Bis(2-ethylhexyl) adipate			X		
103-85-5	Phenylthiourea	X	X		X	
104-94-9	p-Anisidine			X		
105-46-4	sec-Butyl acetate		X			
105-67-9	2,4-Dimethylphenol		X	X	X	
106-42-3	Benzene, p-dimethyl-		X	X	X	
106-44-5	p-Cresol		X	X	X	
106-46-7	Benzene, 1,4-dichloro-		X	X	X	
106-47-8	Benzenamine, 4-chloro-		X		X	
106-50-3	p-Phenylenediamine			X		
106-51-4	p-Benzoquinone		X	X	X	
106-88-7	1,2-Butylene oxide			X		
106-89-8	Epichlorohydrin	X	X	X	X	
106-93-4	Ethane, 1,2-dibromo-		X	X	X	
106-96-7	Propargyl bromide	X				
106-99-0	Butadiene	X		X		
107-02-8	Acrolein	X	X	X	X	

CAS Number	Chemical Name	§302	Section 304 CERCLA	§313	RCRA	STATE
107-05-1	Allyl chloride		x	x		
107-06-2	1,2-Dichloroethane		x	x	x	
107-07-3	Chloroethanol	x				
107-10-8	1-Propanamine		x		x	
107-11-9	Allylamine	x				
107-12-0	Propionitrile	x	x		x	
107-13-1	Acrylonitrile	x	x	x	x	
107-15-3	Ethylenediamine	x	x			
107-16-4	Formaldehyde cyanohydrin	x				
107-18-6	Allyl alcohol	x	x		x	
107-19-7	Propargyl alcohol		x		x	
107-20-0	Chloroacetaldehyde	x	x		x	
107-21-1	Ethylene glycol			x		
107-30-2	Chloromethyl methyl ether	x	x	x	x	
107-44-8	Sarin	x				
107-49-3	Tepp	x	x		x	
107-92-6	Butyric acid		x			
108-05-4	Vinyl acetate monomer	x	x	x		
108-10-1	Methyl isobutyl ketone		x	x	x	
108-23-6	Isopropyl chloroformate	x				
108-24-7	Acetic anhydride		x			
108-31-6	2,5-Furandione		x	x	x	
108-38-3	Benzene, m-dimethyl-		x	x	x	
108-39-4	m-Cresol		x	x	x	
108-46-3	1,3-Benzenediol		x		x	
108-60-1	Bis(2-chloroisopropyl) ether		x	x	x	
108-67-8	Mesitylene	x				
108-78-1	Melamine			x		
108-88-3	Benzene, methyl-		x	x	x	
108-90-7	Benzene, chloro-		x	x	x	
108-91-8	Cyclohexylamine	x				
108-94-1	Cyclohexanone		x		x	
108-95-2	Phenol	x	x	x	x	
108-98-5	Thiophenol	x	x		x	
109-06-8	2-Picoline		x		x	
109-19-3	Butyl isovalerate	x				
109-61-5	Propyl chloroformate	x				
109-73-9	Butylamine		x			
109-77-3	Malononitrile	x	x		x	
109-86-4	2-Methoxyethanol			x		
109-89-7	Diethylamine		x			
109-99-9	Furan, tetrahydro-		x		x	
110-00-9	Furan	x	x		x	

Page 9

CAS Number	Chemical Name	§302	Section 304	§313	RCRA	STATE
110-16-7	Maleic acid			x		
110-17-8	Fumaric acid			x		
110-19-0	iso-Butyl acetate			x		
110-57-6	Trans-1,4-dichlorobutene	x				
110-75-8	2-Chloroethyl vinyl ether			x	x	
110-80-5	2-Ethoxyethanol			x		
110-82-7	Benzene, hexahydro-		x	x	x	
110-86-1	Pyridine		x	x	x	
110-89-4	Piperidine	x				
111-34-2	Butyl vinyl ether	x				
111-42-2	Diethanolamine			x		
111-44-4	Dichloroethyl ether	x	x	x	x	
111-54-6	1,2-Ethanediylbiscarbamodithioic acid			x	x	
111-69-3	Adiponitrile	x				
111-91-1	Bis(2-chloroethoxy) methane			x	x	
114-26-1	Propoxur			x		
115-02-6	Azaserine			x	x	
115-07-1	Propylene (Propene)			x		
115-21-9	Trichloroethylsilane	x				
115-26-4	Dimefox	x				
115-29-7	Endosulfan	x		x	x	
115-32-2	Kelthane			x	x	
115-90-2	Fensulfothion	x				
116-06-3	Aldicarb	x		x	x	
117-52-2	Coumafuryl	x				
117-79-3	2-Aminoanthraquinone			x		
117-80-6	Dichlone			x		
117-81-7	1,2-Benzenedicarboxylic acid,[bis(2-ethylhexyl)]ester		x	x	x	
117-84-0	Dioctyl phthalate	x	x	x	x	
118-74-1	Benzene, hexachloro		x	x	x	
119-38-0	Isopropylmethylpyrazolyl dimethylcarbamate	x				
119-90-4	(1,1'-Biphenyl)-4,4'diamine,3,3'dimethoxy-		x	x	x	
119-93-7	(1,1'Biphenyl)-4,4'-diamine,3,3'-dimethyl-		x	x	x	
120-12-7	Anthracene			x	x	
120-58-1	Benzene, 1,2-methylenedioxy-4-propenyl-			x	x	
120-71-8	p-Cresidine			x		
120-80-9	Catechol			x		
120-82-1	1,2,4-Trichlorobenzene			x	x	
120-83-2	2,4-Dichlorophenol		x	x	x	
121-14-2	Benzene, 1-methyl-2,4-dinitro-		x	x	x	
121-21-1	Pyrethrins			x		
121-29-9	Pyrethrins			x		
121-44-8	Triethylamine			x		

CAS Number	Chemical Name	§302	Section 304 CERCLA	§313	RCRA	STATE
121-69-7	N,N-Dimethylaniline			X		
121-75-5	Malathion		X			
122-09-8	alpha,alpha-Dimethylphenethylamine		X		X	
122-14-5	Fenitrothion	X				
122-66-7	1,2-Diphenylhydrazine		X	X	X	
123-31-9	Hydroquinone	X		X		
123-33-1	1,2-Dihydro-3,6-pyridazinedione		X		X	
123-38-6	Propionaldehyde			X		
123-62-6	Propionic anhydride		X			
123-63-7	Paraldehyde		X			
123-72-8	Butyraldehyde			X		
123-73-9	Crotonaldehyde	X	X		X	
123-86-4	Butyl acetate		X			
123-91-1	1,4-Diethylene dioxide		X	X	X	
123-92-2	iso-Amyl acetate		X			
124-04-9	Adipic acid		X			
124-40-3	Dimethylamine		X		X	
124-41-4	Sodium methylate		X			
124-48-1	Chlorodibromomethane		X			
124-65-2	Sodium cacodylate	X				
124-87-8	Picrotoxin	X				
126-72-7	1-Propanol, 2,3-dibromo-, phosphate (3:1)		X	X	X	
126-98-7	Methacrylonitrile	X	X		X	
126-99-8	Chloroprene			X		
127-18-4	Ethene, 1,1,2,2-tetrachloro-		X	X	X	
127-82-2	Zinc phenolsulfonate		X			
128-56-3	Sodium anthraquinone-1-sulfonate	X				
128-66-5	C.I. Vat Yellow 4			X		
129-00-0	Pyrene	X	X			
129-06-6	Warfarin sodium	X				
130-15-4	1,4-Naphthalenedione		X		X	
131-11-3	Dimethyl phthalate		X	X	X	
131-52-2	Sodium pentachlorophenate					
131-74-8	Ammonium picrate		X		X	
131-89-5	4,6-Dinitro-o-cyclohexylphenol		X		X	
132-64-9	Dibenzofuran			X		
133-06-2	Captan		X	X		
133-90-4	Chloramben			X		
134-29-2	o-Anisidine hydrochloride			X		
134-32-7	1-Naphthylamine		X	X	X	
135-20-6	Cupferron			X		
137-26-8	Bis(dimethylthiocarbamoyl)disulfide		X		X	
139-13-9	Nitrilotriacetic acid			X		

394 FUNDAMENTALS OF HAZARDOUS MATERIALS INCIDENTS

Page 11

		Section 304				
CAS Number	Chemical Name	§302	CERCLA	§313	RCRA	STATE
139-65-1	4,4'-Thiodianiline				x	
140-29-4	Benzyl cyanide	x				
140-76-1	Pyridine, 2-methyl-5-vinyl-	x				
140-88-5	Ethyl acrylate		x	x	x	
141-32-2	Butyl acrylate				x	
141-66-2	Dicrotophos	x				
141-78-6	Acetic acid, ethyl ester		x		x	
142-28-9	1,3-Dichloropropane		x			
142-71-2	Cupric acetate		x			
142-84-7	Dipropylamine		x		x	
143-33-9	Sodium cyanide (Na(CN))	x	x		x	
143-50-0	Kepone		x		x	
144-49-0	Fluoroacetic acid	x				
145-73-3	Endothall		x		x	
148-82-3	Alanine, 3-[p-bis(2-chloroethyl)amino]phenyl-,L-		x		x	
149-74-6	Dichloromethylphenylsilane	x				
151-38-2	Methoxyethylmercuric acetate	x				
151-50-8	Potassium cyanide	x	x		x	
151-56-4	Ethyleneimine	x	x	x	x	
152-16-9	Diphosphoramide, octamethyl-	x	x		x	
156-10-5	p-Nitrosodiphenylamine			x		
156-60-5	1,2-trans-Dichloroethylene		x		x	
156-62-7	Calcium cyanamide			x		
189-55-9	1,2:7,8-Dibenzopyrene		x		x	
191-24-2	Benzo[ghi]perylene		x			
193-39-5	Indeno(1,2,3-cd)pyrene		x		x	
205-99-2	Benzo[b]fluoranthene		x			
206-44-0	Benzo[j,k]fluorene		x		x	
207-08-9	Benzo[k]fluoranthene		x			
208-96-8	Acenaphthylene		x			
218-01-9	1,2-Benzphenanthrene		x		x	
225-51-4	Benz[c]acridine		x		x	
287-92-3	Cyclopentane	x				
297-78-9	Isobenzan	x				
297-97-2	Thionazin	x	x		x	
298-00-0	Parathion-methyl	x	x		x	
298-02-2	Phorate	x	x		x	
298-04-4	Disulfoton	x	x		x	
300-62-9	Amphetamine	x				
300-76-5	Naled		x			
301-04-2	Acetic acid, lead salt		x		x	
302-01-2	Hydrazine	x	x	x	x	
303-34-4	Lasiocarpine		x		x	

CAS Number	Chemical Name	§302	CERCLA	§313	RCRA	STATE
			Section 304			
305-03-3	Butanoic acid, 4-[bis(2-chloroethyl)amino] benzene-		x		x	
309-00-2	Aldrin	x	x	x	x	
311-45-5	Diethyl-p-nitrophenyl phosphate		x		x	
315-18-4	Mexacarbate	x	x			
316-42-7	Emetine, dihydrochloride	x				
319-84-6	alpha - BHC		x			
319-85-7	beta - BHC		x			
319-86-8	delta - BHC		x			
327-98-0	Trichloronate	x				
329-71-5	2,5-Dinitrophenol		x			
330-54-1	Diuron		x			
333-41-5	Diazinon		x			
334-88-3	Diazomethane			x		
353-42-4	Boron trifluoride compound with methyl ether (1:1)	x				
353-50-4	Carbon oxyfluoride		x		x	
357-57-3	Brucine		x		x	
359-06-8	Fluoroacetyl chloride	x				
371-62-0	Ethylene fluorohydrin	x				
379-79-3	Ergotamine tartrate	x				
460-19-5	Cyanogen		x		x	
463-58-1	Carbonyl sulfide			x		
465-73-6	Isodrin	x	x		x	
470-90-6	Chlorfenvinfos	x				
492-80-8	Auramine		x	x	x	
494-03-1	Chlornaphazine		x		x	
496-72-0	Diaminotoluene		x			
502-39-6	Methylmercuric dicyanamide	x				
504-24-5	Pyridine, 4-amino-	x	x		x	
504-60-9	1-Methylbutadiene		x		x	
505-60-2	Mustard gas	x		x		
506-61-6	Potassium silver cyanide	x	x		x	
506-64-9	Silver cyanide		x		x	
506-68-3	Cyanogen bromide	x	x		x	
506-77-4	Chlorine cyanide		x		x	
506-78-5	Cyanogen iodide	x				
506-87-6	Ammonium carbonate		x			
506-96-7	Acetyl bromide		x			
509-14-8	Tetranitromethane	x	x		x	
510-15-6	Ethyl 4,4'-c _nlorobenzilate		x	x	x	
513-49-5	sec-Butylamine		x			
514-73-8	Dithiazanine iodide	x				
528-29-0	o-Dinitrobenzene		x			
532-27-4	2-Chloroacetophenone				x	

Page 13

		Section 304				
CAS Number	Chemical Name	§302	CERCLA	§313	RCRA	STATE
534-07-6	Bis(chloromethyl) ketone	X				
534-52-1	Dinitrocresol	X	X	X	X	
535-89-7	Crimidine	X				
538-07-8	Ethylbis(2-chloroethyl)amine	X				
540-59-0	1,2-Dichloroethylene			X		
540-73-8	1,2-Dimethylhydrazine		X		X	
540-88-5	tert-Butyl acetate		X			
541-09-3	Uranyl acetate		X			
541-25-3	Lewisite	X				
541-41-3	Ethyl chloroformate			X		
541-53-7	Dithiobiuret	X	X		X	
541-73-1	Benzene, 1,3-dichloro-		X	X	X	
542-62-1	Barium cyanide		X		X	
542-75-6	1,3-Dichloropropene		X	X	X	
542-76-7	Propionitrile, 3-chloro-	X	X		X	
542-88-1	Chloromethyl ether	X	X	X	X	
542-90-5	Ethyl thiocyanate	X				
543-90-8	Cadmium acetate		X			
544-18-3	Cobaltous formate		X			
544-92-3	Copper cyanide		X		X	
554-84-7	m-Nitrophenol		X			
555-77-1	Tris(2-chloroethyl)amine	X				
556-61-6	Methyl isothiocyanate	X				
556-64-9	Methyl thiocyanate	X				
557-19-7	Nickel cyanide		X		X	
557-21-1	Zinc cyanide		X		X	
557-34-6	Zinc acetate		X			
557-41-5	Zinc formate		X			
558-25-8	Methanesulfonyl fluoride	X				
563-12-2	Ethion	X	X			
563-41-7	Semicarbazide hydrochloride	X				
563-68-8	Acetic acid, thallium(I) salt		X		X	
569-64-2	C.I. Basic Green 4			X		
573-56-8	2,6-Dinitrophenol		X			
584-84-9	Toluene 2,4-diisocyanate	X	X	X		
591-08-2	Acetamide, N-(aminothioxomethyl)-		X		X	
592-01-8	Calcium cyanide		X		X	
592-04-1	Mercuric cyanide		X			
592-85-8	Mercuric thiocyanate		X			
592-87-0	Lead thiocyanate		X			
593-60-2	Vinyl bromide			X		
594-42-3	Perchloromethylmercaptan	X	X		X	
597-64-8	Tetraethyltin	X				

CAS Number	Chemical Name	Section 304 §302	CERCLA	§313	RCRA	STATE
598-31-2	Bromoacetone		X		X	
606-20-2	Benzene, 1-methyl-2,6-dinitro-		X	X	X	
608-93-5	Benzene, pentachloro-		X		X	
609-19-8	3,4,5-Trichlorophenol		X			
610-39-9	3,4-Dinitrotoluene		X			
614-78-8	Thiourea, (2-methylphenyl)-	X				
615-05-4	2,4-Diaminoanisole			X		
615-53-2	Carbamic acid, methylnitroso-,ethyl ester		X		X	
621-64-7	Di-n-propylnitrosamine		X	X	X	
624-83-9	Methyl isocyanate	X	X	X	X	
624-92-0	Methyl disulfide	X				
625-16-1	tert-Amyl acetate		X			
625-55-8	Isopropyl formate	X				
626-38-0	sec-Amyl acetate		X			
627-11-2	Chloroethyl chloroformate	X				
628-63-7	Amyl acetate		X			
628-86-4	Fulminic acid, mercury(II)salt		X		X	
630-10-4	Carbamimidoselenoic acid		X		X	
630-20-6	Ethane, 1,1,1,2-tetrachloro-		X		X	
630-60-4	Ouabain	X				
631-61-8	Ammonium acetate		X			
633-03-4	C.I. Basic Green 1	X				
636-21-5	Benzenamine, 2-methyl-, hydrochloride		X	X	X	
639-58-7	Triphenyltin chloride	X				
640-15-3	Thiometon	X				
640-19-7	Fluoroacetamide	X	X		X	
644-64-4	Dimetilan	X				
646-06-0	Dioxolane	X				
675-14-9	Cyanuric fluoride	X				
676-97-1	Methyl phosphonic dichloride	X				
680-31-9	Hexamethylphosphoramide			X		
684-93-5	Carbamide, N-methyl-N-nitroso-		X	X	X	
692-42-2	Arsine, diethyl-		X		X	
696-28-6	Phenyl dichloroarsine	X	X		X	
732-11-6	Phosmet	X				
757-58-4	Hexaethyl tetraphosphate		X		X	
759-73-9	Carbamide, N-ethyl-N-nitroso-		X	X	X	
760-93-0	Methacrylic anhydride	X				
764-41-0	2-Butene, 1,4-dichloro-		X		X	
765-33-4	Glycidylaldehyde		X		X	
786-19-6	Carbophenothion	X				
814-49-3	Diethyl chlorophosphate	X				
814-68-6	Acrylyl chloride	X				

Page 15

CAS Number	Chemical Name	§302	Section 304 CERCLA	§313	RCRA	STATE
815-82-7	Cupric tartrate		X			
823-40-5	Diaminotoluene		X		X	
824-11-3	Trimethylolpropane phosphite	X				
842-07-9	C.I. Solvent Yellow 14			X		
900-95-8	Stannane, acetoxytriphenyl·	X				
919-86-8	Demeton·S·methyl	X				
920-46-7	Methacryloyl chloride	X				
924-16-3	1·Butanamine, N·butyl·N·nitroso·		X	X	X	
930-55-2	N·Nitrosopyrrolidine		X		X	
933-75-5	2,3,6·Trichlorophenol		X			
933-78-8	2,3,5·Trichlorophenol		X			
944-22-9	Fonofos	X				
947-02-4	Phosfolan	X				
950-10-7	Mephosfolan	X				
950-37-8	Methidathion	X				
959-98-8	alpha · Endosulfan		X			
961-11-5	Tetrachlorvinphos			X		
989-38-8	C.I. Basic Red 1			X		
991-42-4	Norbormide	X				
998-30-1	Triethoxysilane	X				
999-81-5	Chlormequat chloride	X				
1024-57-3	Heptachlor epoxide		X			
1031-07-8	Endosulfan sulfate		X			
1031-47-6	Triamiphos	X				
1066-30-4	Chromic acetate		X			
1066-33-7	Ammonium bicarbonate		X			
1066-45-1	Trimethyltin chloride	X				
1072-35-1	Lead stearate		X			
1111-78-0	Ammonium carbamate		X			
1116-54-7	Ethanol, 2,2'·(nitrosoimino)bis·		X		X	
1120-71-4	1,2·Oxathiolane, 2,2·dioxide		X	X	X	
1122-60-7	Nitrocyclohexane	X				
1124-33-0	Pyridine, 4·nitro·, 1·oxide	X				
1129-41-5	Metolcarb	X				
1163-19-5	Decabromodiphenyl oxide			X		
1185-57-5	Ferric ammonium citrate		X			
1194-65-6	Dichlobenil		X			
1300-71-6	Xylenol		X			
1303-28-2	Arsenic pentoxide	X	X		X	
1303-32-8	Arsenic disulfide		X			
1303-33-9	Arsenic trisulfide		X			
1306-19-0	Cadmium oxide	X				
1309-64-4	Antimony trioxide		X			

CAS Number	Chemical Name	§302	CERCLA	§313	RCRA	STATE
1310-58-3	Potassium hydroxide		x			
1310-73-2	Sodium hydroxide		x	x		
1313-27-5	Molybdenum trioxide			x		
1314-20-1	Thorium dioxide			x		
1314-32-5	Thallic oxide	x	x		x	
1314-56-3	Phosphorus pentoxide	x				
1314-62-1	Vanadium pentoxide	x	x		x	
1314-80-3	Phosphorus pentasulfide		x		x	
1314-84-7	Zinc phosphide	x	x		x	L
1314-87-0	Lead sulfide		x			
1314-96-1	Strontium sulfide		x		x	
1319-72-8	2,4,5-T amines		x			
1319-77-3	Cresol(s)		x	x	x	
1320-18-9	2,4-D Esters		x			
1321-12-6	Nitrotoluene		x			
1327-52-2	Arsenic acid		x		x	
1327-53-3	Arsenous oxide	x	x		x	
1330-20-7	Benzene, dimethyl-		x	x	x	
1331-17-5	Propylene glycol, allyl ether	x				
1332-07-6	Zinc borate		x			
1332-21-4	Asbestos		x	x		
1333-83-1	Sodium bifluoride		x			
1335-32-6	Lead subacetate		x		x	
1335-87-1	Hexachloronaphthalene	x		x		
1336-21-6	Ammonium hydroxide		x			
1336-36-3	POLYCHLORINATED BIPHENYLS (PCBs)		x	x		
1338-23-4	2-Butanone peroxide		x		x	
1338-24-5	Naphthenic acid		x			
1341-49-7	Ammonium bifluoride		x			
1344-28-1	Aluminum oxide			x		
1397-94-0	Antimycin A	x				
1405-87-4	Bacitracin	x				
1420-07-1	Dinoterb	x				
1464-53-5	Diepoxybutane	x	x	x	x	
1558-25-4	Trichloro(chloromethyl)silane	x				
1563-66-2	Carbofuran	x	x			
1582-09-8	Trifluralin			x		
1600-27-7	Mercuric acetate	x				
1615-80-1	N,N'-Diethylhydrazine		x		x	
1622-32-8	Ethanesulfonyl chloride, 2-chloro-	x				
1634-04-4	Methyl tert-butyl ether				x	
1642-54-2	Diethylcarbamazine citrate	x				
1746-01-6	2,3,7,8-Tetrachlorodibenzo-p-dioxin (TCDD)		x			

400 FUNDAMENTALS OF HAZARDOUS MATERIALS INCIDENTS

CAS Number	Chemical Name	§302	Section 304 CERCLA	§313	RCRA	STATE
1752-30-3	Acetone thiosemicarbazide	x				
1762-95-4	Ammonium thiocyanate		x			
1836-75-5	Nitrofen			x		
1863-63-4	Ammonium benzoate		x			
1888-71-7	Hexachloropropene		x		x	
1897-45-6	Chlorothalonil			x		
1910-42-5	Paraquat	x				
1918-00-9	Dicamba		x			
1928-38-7	2,4-D Esters		x			
1928-47-8	2,4,5-T esters		x			
1928-61-6	2,4-D Esters		x			
1929-73-3	2,4-D-Esters		x			
1937-37-7	Direct Black 38			x		
1982-47-4	Chloroxuron	x				
2001-95-8	Valinomycin	x				
2008-46-0	2,4,5-T amines		x			
2032-65-7	Methiocarb	x	x			
2074-50-2	Paraquat methosulfate	x				
2097-19-0	Phenylsilatrane	x				
2104-64-5	EPN	x				
2164-17-2	Fluometuron			x		
2223-93-0	Cadmium stearate	x				
2231-57-4	Thiocarbazide	x				
2234-13-1	Octachloronaphthalene			x		
2235-25-8	Ethylmercuric phosphate	x				
2238-07-5	Diglycidyl ether	x				
2244-16-8	Carvone	x				
2275-18-5	Prothoate	x				
2303-16-4	Diallate		x	x	x	
2312-35-8	Propargite		x			
2497-07-6	Oxydisulfoton	x				
2524-03-0	Dimethyl phosphorochloridothioate	x				
2540-82-1	Formothion	x				
2545-59-7	2,4,5-T esters		x			
2570-26-5	Pentadecylamine	x				
2587-90-8	Phosphorothioic acid, O,O-dimethyl·	x				
2602-46-2	Direct Blue 6			x		
2631-37-0	Promecarb	x				
2636-26-2	Cyanophos	x				
2642-71-9	Azinphos-ethyl	x				
2650-18-2	C.I. Acid Blue 9, diammonium salt			x		
2665-30-7	Phosphonothioic acid, methyl·,O·(4·nitrophenyl) O·phenyl est	x				
2703-13-1	Phosphonothioic acid, methyl·, O·ethyl O·(4·(methylthio)phen	x				

Page 18

		Section 304				
CAS Number	Chemical Name	§302	CERCLA	§313	RCRA	STATE
2757-18-8	Thallous malonate	X				
2763-96-4	Muscimol	X	X		X	
2764-72-9	Diquat			X		
2778-04-3	Endothion	X				
2832-40-8	C.I. Disperse Yellow 3			X		
2921-88-2	Chlorpyrifos		X			
2944-67-4	Ferric ammonium oxalate		X			
2971-38-2	2,4-D Esters		X			
3012-65-5	Ammonium citrate, dibasic		X			
3037-72-7	Silane, (4-aminobutyl)diethoxymethyl-	X				
3048-64-4	Vinylnorbornene	X				
3118-97-6	C.I. Solvent Orange 7			X		
3164-29-2	Ammonium tartrate		X			
3165-93-3	Benzenamine, 4-chloro-2-methyl-,hydrochloride		X		X	
3251-23-8	Cupric nitrate		X			
3254-63-5	Phosphoric acid, dimethyl 4-(methylthi:- phenyl ester	X				
3288-58-2	O,O-Diethyl S-methyl dithiophosphate		X		X	
3486-35-9	Zinc carbonate		X			
3569-57-1	Sulfoxide, 3-chloropropyl octyl	X				
3615-21-2	Benzimidazole, 4,5-dichloro-2-(trifluoromethyl)-	X				
3689-24-5	Sulfotep	X	X		X	
3691-35-8	Chlorophacinone	X				
3734-97-2	Amiton oxalate	X				
3735-23-7	Methyl phenkapton	X				
3761-53-3	C.I. Food Red 5			X		
3813-14-7	2,4,5-T amines		X			
3844-45-9	C.I. Acid Blue 9, disodium salt			X		
3878-19-1	Fuberidazole	X				
4044-65-9	Bitoscanate	X				
4098-71-9	Isophorone diisocyanate	X				
4104-14-7	Phosacetim	X				
4170-30-3	Crotonaldehyde	X	X		X	
4301-50-2	Fluenetil	X				
4418-66-0	Phenol, 2,2'-thiobis[4-chloro-6-methyl-	X				
4549-40-0	Ethenamine, N-methyl-N-nitroso-		X	X	X	
4680-78-8	C.I. Acid Green 3			X		
4835-11-4	Hexamethylenediamine, N,N'-dibutyl-	X				
5281-13-0	Piprotal	X				
5344-82-1	Thiourea, (2-chlorophenyl)-	X	X		X	
5836-29-3	Coumatetralyl	X				
5893-66-3	Cupric oxalate		X			
5972-73-6	Ammonium oxalate		X			
6009-70-7	Ammonium oxalate		X			

Page 19

CAS Number	Chemical Name	§302	Section 304 CERCLA	§313	RCRA	STATE
6369-96-6	2,4,5-T amines			x		
6369-97-7	2,4,5-T amines			x		
6484-52-2	Ammonium nitrate (solution)			x		
6533-73-9	Thallous carbonate	x	x		x	
6923-22-4	Monocrotophos	x				
7005-72-3	4-Chlorophenyl phenyl ether		x			
7421-93-4	Endrin aldehyde		x			
7428-48-0	Lead stearate		x			
7429-90-5	Aluminum (fume or dust)			x		
7439-92-1	Lead		x	x	x	
7439-96-5	Manganese and compounds			x	x	
7439-97-6	Mercury		x	x	x	
7440-02-0	Nickel	x	x	x		
7440-22-4	Silver		x	x	x	
7440-23-5	Sodium		x			
7440-28-0	Thallium		x	x		
7440-36-0	Antimony		x	x		
7440-38-2	Arsenic		x	x	x	
7440-39-3	Barium and compounds			x	x	
7440-41-7	Beryllium		x	x	x	
7440-43-9	Cadmium		x	x	x	
7440-47-3	Chromium		x	x	x	
7440-48-4	Cobalt	x		x		
7440-50-8	Copper		x	x		
7440-62-2	Vanadium (fume or dust)			x		
7440-66-6	Zinc		x	x		
7446-08-4	Selenium dioxide		x		x	
7446-09-5	Sulfur dioxide	x				
7446-11-9	Sulfur trioxide	x				
7446-14-2	Lead sulfate		x			
7446-18-6	Thallous sulfate	x	x		x	
7446-27-7	Lead phosphate		x		x	
7447-39-4	Cupric chloride		x			
7487-94-7	Mercuric chloride	x				
7488-56-4	Selenium disulfide		x		x	
7550-45-0	Titanium tetrachloride	x		x		
7558-79-4	Sodium phosphate, dibasic		x			
7580-67-8	Lithium hydride	x				
7601-54-9	Sodium phosphate, tribasic		x			
7631-89-2	Sodium arsenate	x	x			
7631-90-5	Sodium bisulfite		x			
7632-00-0	Sodium nitrite		x			
7637-07-2	Boron trifluoride	x				

Page 20

CAS Number	Chemical Name	Section 304 §302	CERCLA	§313	RCRA	STATE
7645-25-2	Lead arsenate		X			
7646-85-7	Zinc chloride		X			
7647-01-0	Hydrochloric acid	X	X	X		
7647-18-9	Antimony pentachloride		X			
7664-38-2	Phosphoric acid		X	X		
7664-39-3	Hydrogen fluoride	X	X	X	X	
7664-41-7	Ammonia	X	X	X		
7664-93-9	Sulfuric acid	X	X	X		
7681-49-4	Sodium fluoride		X			
7681-52-9	Sodium hypochlorite		X			
7697-37-2	Nitric acid	X	X	X		
7699-45-8	Zinc bromide		X			
7705-08-0	Ferric chloride		X			
7718-54-9	Nickel chloride		X			
7719-12-2	Phosphorous trichloride	X	X			
7720-78-7	Ferrous sulfate		X			
7722-64-7	Potassium permanganate		X			
7722-84-1	Hydrogen peroxide	X				
7723-14-0	Phosphorus	X	X	X		
7726-95-6	Bromine	X				
7733-02-0	Zinc sulfate		X			
7738-94-5	Chromic acid		X			
7757-82-6	Sodium sulfate (solution)			X		
7758-29-4	Sodium phosphate, tribasic		X			
7758-94-3	Ferrous chloride		X			
7758-95-4	Lead chloride		X			
7758-98-7	Cupric sulfate		X			
7761-88-8	Silver nitrate		X			
7773-06-0	Ammonium sulfamate		X			
7775-11-3	Sodium chromate		X			
7778-39-4	Arsenic acid		X		X	
7778-44-1	Calcium arsenate	X	X			
7778-50-9	Potassium bichromate		X			
7778-54-3	Calcium hypochlorite		X			
7779-86-4	Zinc hydrosulfite		X			
7779-88-6	Zinc nitrate		X			
7782-41-4	Fluorine	X	X		X	
7782-49-2	Selenium		X	X	X	
7782-50-5	Chlorine	X	X	X		
7782-63-0	Ferrous sulfate		X			
7782-82-3	Sodium selenite		X			
7782-86-7	Mercurous nitrate		X			
7783-00-8	Selenous acid	X	X		X	

404 FUNDAMENTALS OF HAZARDOUS MATERIALS INCIDENTS

		Section 304				
CAS Number	Chemical Name	§302	CERCLA	§313	RCRA	STATE
7783-06-4	Hydrogen sulfide	X	X		X	
7783-07-5	Hydrogen selenide	X				
7783-18-8	Ammonium thiosulfate		X			
7783-20-2	Ammonium sulfate (solution)			X		
7783-35-9	Mercuric sulfate		X			
7783-46-2	Lead fluoride		X			
7783-49-5	Zinc fluoride		X			
7783-50-8	Ferric fluoride		X			
7783-56-4	Antimony trifluoride		X			
7783-60-0	Sulfur tetrafluoride	X				
7783-70-2	Antimony pentafluoride	X				
7783-80-4	Tellurium hexafluoride	X				
7784-34-1	Arsenous trichloride	X	X			
7784-40-9	Lead arsenate		X			
7784-41-0	Potassium arsenate		X			
7784-42-1	Arsine	X				
7784-46-5	Sodium arsenite	X	X			
7785-84-4	Sodium phosphate, tribasic		X			
7786-34-7	Mevinphos	X	X		X	
7786-81-4	Nickel sulfate		X			
7787-47-5	Beryllium chloride		X			
7787-49-7	Beryllium fluoride		X			
7787-55-5	Beryllium nitrate		X			
7788-98-9	Ammonium chromate		X			
7789-00-6	Potassium chromate		X			
7789-06-2	Strontium chromate		X			
7789-09-5	Ammonium bichromate		X			
7789-42-6	Cadmium bromide		X			
7789-43-7	Cobaltous bromide		X			
7789-61-9	Antimony tribromide		X			
7790-94-5	Chlorosulfonic acid		X			
7791-12-0	Thallous chloride	X	X		X	
7791-23-3	Selenium oxychloride	X				
7803-51-2	Phosphine	X	X		X	
7803-55-6	Ammonium vanadate		X		X	
8001-35-2	Camphechlor	X	X	X	X	
8001-58-9	Creosote		X		X	
8003-19-8	Dichloropropane · Dichloropropene (mixture)		X			
8003-34-7	Pyrethrins		X			
8014-95-7	Sulfuric acid		X			
8023-53-8	Dichlorobenzalkonium chloride	X				
8065-48-3	Demeton	X				
9004-66-4	Ferric dextran		X		X	

		Section 304				
CAS Number	Chemical Name	§302	CERCLA	§313	RCRA	STATE
10022-70-5	Sodium hypochlorite		x			
10025-65-7	Platinous chloride	x				
10025-73-7	Chromic chloride	x				
10025-87-3	Phosphorus oxychloride	x	x			
10025-91-9	Antimony trichloride		x			
10025-97-5	Iridium tetrachloride	x				
10026-11-6	Zirconium tetrachloride		x			
10026-13-8	Phosphorus pentachloride	x				
10028-15-6	Ozone	x				
10028-22-5	Ferric sulfate		x			
10031-59-1	Thallous sulfate	x	x		x	
10034-93-2	Hydrazine sulfate			x		
10039-32-4	Sodium phosphate, dibasic		x			
10043-01-3	Aluminum sulfate		x			
10045-89-3	Ferrous ammonium sulfate		x			
10045-94-0	Mercuric nitrate		x			
10049-04-4	Chlorine dioxide			x		
10049-05-5	Chromous chloride		x			
10049-07-7	Rhodium trichloride	x				
10099-74-8	Lead nitrate		x			
10101-53-8	Chromic sulfate		x			
10101-63-0	Lead iodide		x			
10101-89-0	Sodium phosphate, tribasic		x			
10102-06-4	Uranyl nitrate		x			
10102-18-8	Sodium selenite	x	x			
10102-20-2	Sodium tellurite	x				
10102-43-9	Nitric oxide	x	x		x	
10102-44-0	Nitrogen dioxide	x	x		x	
10102-45-1	Thallium(I) nitrate		x		x	
10102-48-4	Lead arsenate		x			
10108-64-2	Cadmium chloride		x			
10124-50-2	Potassium arsenite	x	x			
10124-56-8	Sodium phosphate, tribasic		x			
10140-65-5	Sodium phosphate, dibasic		x			
10140-87-1	Ethanol, 1,2-dichloro-, acetate	x				
10192-30-0	Ammonium bisulfite		x			
10196-04-0	Ammonium sulfite		x			
10210-68-1	Cobalt carbonyl	x				
10265-92-6	Methamidophos	x				
10294-34-5	Boron trichloride	x				
10311-84-9	Dialifos	x				
10361-89-4	Sodium phosphate, tribasic		x			
10380-29-7	Cupric sulfate ammoniated		x			

Page 23

CAS Number	Chemical Name	§302	CERCLA	§313	RCRA	STATE
			Section 304			
10415-75-5	Mercurous nitrate		x			
10421-48-4	Ferric nitrate		x			
10476-95-6	Methacrolein diacetate	x				
10544-72-6	Nitrogen dioxide		x		x	
10588-01-9	Sodium bichromate		x			
11096-82-5	Aroclor 1260		x			
11097-69-1	Aroclor 1254		x			
11104-28-2	Aroclor 1221		x			
11115-74-5	Chromic acid		x			
11141-16-5	Aroclor 1232		x			
12002-03-8	Paris green	x	x			
12039-52-0	Thallium(I) selenide		x		x	
12054-48-7	Nickel hydroxide		x			
12108-13-3	Manganese, tricarbonyl methylcyclopentadienyl	x				
12122-67-7	Zineb			x		
12125-01-8	Ammonium fluoride		x			
12125-02-9	Ammonium chloride		x			
12135-76-1	Ammonium sulfide		x			
12427-38-2	Maneb			x		
12672-29-6	Aroclor 1248		x			
12674-11-2	Aroclor 1016		x			
12771-08-3	Sulfur monochloride		x			
13071-79-9	Terbufos	x				
13171-21-6	Phosphamidon	x				
13194-48-4	Ethoprophos	x				
13410-01-0	Sodium selenate	x				
13450-90-3	Gallium trichloride	x				
13454-96-1	Platinum tetrachloride	x				
13463-39-3	Nickel carbonyl	x	x		x	
13463-40-6	Iron, pentacarbonyl·	x				
13463-67-7	Titanium dioxide			x		
13494-80-9	Tellurium	x				
13560-99-1	2,4,5-T salts		x			
13597-99-4	Beryllium nitrate		x			
13746-89-9	Zirconium nitrate		x			
13765-19-0	Calcium chromate		x		x	
13814-96-5	Lead Fluoborate		x			
13826-83-0	Ammonium fluoborate		x			
13952-84-6	sec-Butylamine		x			
14017-41-5	Cobaltous sulfamate		x			
14167-18-1	Salcomine	x				
14216-75-2	Nickel nitrate		x			
14258-49-2	Ammonium oxalate		x			

Page 24

CAS Number	Chemical Name	§302	Section 304 CERCLA	§313	RCRA	STATE
14307·35·8	Lithium chromate		x			
14307·43·8	Ammonium tartrate		x			
14639·97·5	Zinc ammonium chloride		x			
14639·98·6	Zinc ammonium chloride		x			
14644·61·2	Zirconium sulfate		x			
15271·41·7	Bicyclo[2.2.1]heptane·2·ca: itrile, 5·chloro·6·((((methyla	x				
15699·18·0	Nickel ammonium sulfate		x			
15739·80·7	Lead sulfate		x			
15950·66·0	2,3,4·Trichlorophenol		x			
16071·86·6	Direct Brown 95			x		
16543·55·8	N·Nitrosonornicotine			x		
16721·80·5	Sodium hydrosulfide		x			
16752·77·5	Methomyl	x	x		x	
16871·71·9	Zinc silicofluoride		x			
16919·19·0	Ammonium silicofluoride		x			
16919·58·7	Ammonium chloroplatinate	x				
16923·95·8	Zirconium potassium fluoride		x			
17702·41·9	Decaborane(14)	x				
17702·57·7	Formparanate	x				
18883·66·4	D·Glucopyranose, 2·deoxy·2·(3·methyl·3·nitrosoureido)·		x		x	
19287·45·7	Diborane	x				
19624·22·7	Pentaborane	x				
20816·12·0	Osmium tetroxide	x	x	x	x	
20830·75·5	Digoxin	x				
20830·81·3	Daunomycin		x		x	
20859·73·8	Aluminum phosphide		x	x		x
21548·32·3	Fosthietan	x				
21564·17·0	Thiocyanic acid, 2·(benzothiazolylthio)methyl ester	x				
21609·90·5	Leptophos	x				
21908·53·2	Mercuric oxide	x				
21923·23·9	Chlorthiophos	x				
22224·92·6	Fenamiphos	x				
23135·22·0	Oxamyl	x				
23422·53·9	Formetanate	x				
23505·41·1	Pirimifos·ethyl	x				
23950·58·5	3,5·Dichloro·N·(1,1·dimethyl·2·propynyl)benzamide		x		x	
24017·47·8	Triazofos	x				
24934·91·6	Chlormephos	x				
25154·54·5	Dinitrobenzene (mixed)		x			
25154·55·6	Nitrophenol (mixed)		x			
25155·30·0	Sodium dodecylbenzene sulfonate		x			
25167·82·2	Trichlorophenol		x			
25168·15·4	2,4,5·T esters		x			

Page 25

CAS Number	Chemical Name	Section 304 §302	CERCLA	§313	RCRA	STATE
25168-26-7	2,4,-D Esters		X			
25321-14-6	Dinitrotoluene		X			
25321-22-6	Dichlorobenzene (mixed)		X	X		
25376-45-8	Diaminotoluene		X	X	X	
25550-58-7	Dinitrophenol		X			
26264-06-2	Calcium dodecylbenzene sulfonate		X			
26419-73-8	Carbamic acid, methyl-, O-(((2,4-dimethyl-1, 3-dithiolan-2-y	X				
26471-62-5	Benzene, 2,4-diisocyanatomethyl-		X		X	
26628-22-8	Sodium azide (Na(N3))	X	X		X	
26638-19-7	Dichloropropane		X			
26952-23-8	Dichloropropene		X			
27137-85-5	Trichloro(dichlorophenyl)silane	X				
27176-87-0	Dodecylbenzenesulfonic acid		X			
27323-41-7	Triethanolamine dodecylbenzene sulfonate		X			
27774-13-6	Vanadyl sulfate		X			
28300-74-5	Antimony potassium tartrate		X			
28347-13-9	Xylylene dichloride	X				
28772-56-7	Bromadiolone	X				
30525-89-4	Paraformaldehyde		X			
30674-80-7	Methacryloyloxyethyl isocyanate	X				
32534-95-5	2,4,5-TP acid esters		X			
33213-65-9	beta - Endosulfan		X			
36478-76-9	Uranyl nitrate		X			
37211-05-5	Nickel chloride		X			
39156-41-7	2,4-Diaminoanisole sulfate			X		
39196-18-4	Thiofanox	X	X		X	
42504-46-1	Isopropanolamine dodecylbenzene sulfonate		X			
50782-69-9	Phosphonothioic acid, methyl-, S-(2-(bis(1-methylethyl)amino	X				
52628-25-8	Zinc ammonium chloride		X			
52652-59-2	Lead stearate					
52740-16-6	Calcium arsenite		X			
53467-11-·	2,4-D Esters		X			
53469-21	Aroclor 1242		X			
53558-25-1	Pyriminil	X				
554?8-87-4	Ferric ammonium oxalate		X			
561c?-09-4	Lead stearate		X			
58270-08-9	Zinc, dichloro(4,4-dimethyl-5((((methylamino) carbonyl)oxy)i	X				
61792-07-2	2,4,5-T esters		X			
62207-76-5	Cobalt, ((2,2'-(1,2-ethanediylbis (nitrilomethylidyne))bis(6	X				

Page 26

Chemical Category	Section 304 §302	CERCLA	§313	RCRA	STATE
Organorhodium Complex	x				
Barium Compounds			x		
Cobalt Compounds			x		
Cyanide and Compounds		x	x		
Glycol Ethers			x		
Manganese Compounds			x		
Polybrominated Biphenyls (PBBs)			x		
Antimony and Compounds		x	x		
Arsenic and Compounds		x	x		
Beryllium and Compounds		x	x		
Cadmium and Compounds		x	x		
Chlordane (Technical Mixture and Metabolites)		x			
Chlorinated Benzenes		x			
Chlorinated Ethanes		x			
Chlorinated Naphthalene		x			
Chlorinated Phenols		x	x		
Chloroalkyl Ethers		x			
Chromium and Compounds		x	x		
Coke Oven Emissions		x			
Copper and Compounds		x	x		
DDT and Metabolites		x			
Dichlorobenzidine		x			
Diphenylhydrazine		x			
Endosulfan and Metabolites		x			
Endrin and Metabolites		x		x	
Haloethers		x			
Halomethanes		x			
Heptachlor and Metabolites		x			
Lead and Compounds		x	x		
Mercury and Compounds		x	x		
Nickel and Compounds		x	x		
Nitrophenols		x			
Nitrosamines		x			
Phthalate Esters		x			
Polynuclear Aromatic Hydrocarbons		x			
Radionuclides		x			
Selenium and Compounds		x	x		
Silver and Compounds		x	x		
Thallium and Compounds		x	x		
Zinc and Compounds		x	x		

GLOSSARY

The following glossary includes terms, abbreviations, and acronyms that are used in dealing with hazardous materials, especially in right-to-know laws and regulations, as well as in other documents that deal with hazardous materials.

AACT — American Academy of Clinical Toxicology.

AAPCC — American Association of Poison Control Centers.

ABIH — American Board of Industrial Hygiene, 510 Edgewood Drive, Collingswood, NJ 08108.

Accident — An unanticipated occurrence that does or could cause personal injury and/or property damage.

Accident mechanism — A series of events culminating in unexpected injury.

Accident site — An incident site. The location of an unexpected occurrence, failure, or loss, at a facility, or along a transportation route, or along an energy corridor, resulting in a release of hazardous materials.

ACGIH — American Conference of Governmental Industrial Hygienists, 6500 Glenway Avenue, Building D-7, Cincinnati, OH 45211-4438.

Acid — A chemical compound that can react with a base to form a salt. A number of acids, such as sulfuric, perchloric, and hydrofluoric acid, are very dangerous.

ACM — Asbestos-containing materials.

ACS — American Chemical Society, 1255 Sixteenth St., NW, Washington DC 20036.

Act — The Superfund Amendments and Reauthorization Act of 1986.

Acute

Of short duration. Acute health effects are those that occur during or immediately after exposure to hazardous chemicals.

Acutely toxic
chemicals

Chemicals that can cause severe short- and long-term health effects after a single, brief exposure. When ingested, inhaled, or absorbed through the skin, these chemicals can cause damage to living tissue, impairment of the central nervous system, severe illness, or even death.

Aerosol

A suspension of fine solid or liquid particles in gas. Smoke and fog are examples of aerosols.

AHERA

Asbestos Hazard Emergency Response Act.

AIChE

American Institute of Chemical Engineers.

AIHA

American Industrial Hygiene Association, 475 Wolf Ledges Parkway, Akron, OH 44311.

Airborne release

Release of a chemical into the atmosphere.

Alkali

A basic substance. See Base.

Ambient

Surrounding. Ambient temperatures are temperatures of the surrounding area (e.g., air or water).

ANDO

Ammonium nitrate and diesel oil combination. Used as an explosive.

Anhydrous

Dry, without water.

ANSI

American National Standards Institute, 1430 Broadway, New York, NY 10018

ASME

American Society of Mechanical Engineers.

Asphyxiant

Substance that interferes with oxygenation of tissues with the result that the individual may suffocate.

(1) *Simple asphyxiant.* Physiologically inert gas that acts by diluting atmospheric oxygen below that required for tissue respiration. Examples: Carbon dioxide, helium, methane, and hydrogen.

(2) *Chemical asphyxiant.* Chemical asphyxiants, through their direct chemical action, either prevent the uptake of oxygen by the blood or interfere with the transportation of oxygen from the lungs to the tissues or prevent normal oxygenation of tissues, even though the blood is well oxygenated. Examples:

Carbon monoxide, hydrogen cyanide, hydrogen sulfide.

ASSE
American Society of Safety Engineers, 850 Busse Highway, Park Ridge, IL 60068.

ASTM
American Society for Testing and Materials, 1916 Race St., Philadelphia, PA 19103.

ATSDR
Agency for Toxic Substances and Disease Registry (DHHS)

Auto-ignition temperature
The minimum temperature at which a substance will ignite in air when there in no ignition source. For liquids it is defined as the lowest temperature at which a drop of solvent will ignite spontaneously. (Measured as the temperature of a hot plate on which solvent ignites when added in drops.) The auto ignition temperature of toluene is 52°C and 125°F for carbon disulfide.

Base
A chemical compound that reacts with an acid to form a salt.

Blasting agent
An agent designed for blasting which has been tested and "found to be so insensitive that there is very little probability of accidental initiation to explosion or transition from deflagration or detonation."

BLEVE
Boiling Liquid Expanding Vapor Explosion.

Boiling point
The temperature at which the vapor pressure of a liquid equals the atmospheric pressure and boils. Boiling point is usually given at a standard pressure of 760 mmHg at sea level.

Burning
The rapid oxidation of a fuel.

By-product
Material other than principal product(s) produced or generated as a consequence of an industrial process. Usually differentiated from waste products.

CA
Chemical Abstracts (American Chemical Society).

CAER
See Community Awareness and Emergency Response Program.

Carcinogen
A substance which causes cancer.

CAS
Chemical Abstract Service.

CAS number
Chemical Abstract Service Registry Number.

Caustic
A substance that attacks tissue by chemical action.

Caution	A term used on a precautionary label to denote a lower degree of hazard rather than "danger" or "warning."
CDC	Centers for Disease Control, 1600 Clifton Road, Atlanta, GA 30333.
Ceiling concentration	The maximum concentration of toxic substance that is allowed at any time. Cannot be exceeded without causing serious harm.
CEPP	Chemical Emergency Preparedness Program (EPA).
CERCLA	Comprehensive Environmental Response, Compensation, and Liability Act of 1980 ("Superfund").
CFR	Code of Federal Regulations.
CGA	Compressed Gas Association, 500 Fiftieth Ave, New York, NY 10036.
Chemical Hazards Response Information System/Hazard Assessment Computer System (CHRIS/HACS)	Developed by the U.S. Coast Guard. HACS is a computerized model of the four CHRIS manuals that contain chemical-specific data. FOSCs uses HACS to find answers to specific questions during a chemical spill/ response.
Chemical Transportation Emergency Center	The AMC Chemical Transportation Emergency Center. This center provides immediate information on what to do in case of a spill, leak, fire, or exposure to hazardous chemicals. This center is manned 24 hours a day. (800) 424-9300.
CHEMTREC	See Chemical Transportation Emergency Center.
Chlorine Emergency Plan (CHLOREP)	A 24-hour mutual aid program operated by the Chlorine Institute. Response is activated by a CHEMTREC call to the designated CHLOREP contact, who then notifies the appropriate team leader, based upon CHLOREP's geographical sector assignments for teams. The team leader in turn calls the individual at the incident scene and determines what advice and assistance is necessary. The team leader then determines whether or not to dispatch his team to the scene.

CHRIS	See Chemical Hazards Response Information System.
Chronic	Of long duration or having frequent recurrence or exposures. Chronic health effects are those that become apparent or continue for some time after exposure to a hazardous materials.
CMA	Chemical Manufacturers Association, 1825 Connecticut Ave., Washington DC 20009.
CNS	Central nervous system.
Combustible liquid	Any liquid with a flash point from 100 to 200°F, as measured by the tests specified in Section 173.115, 49 CFR, except any mixture having one component or more with a flash point at 200°F or higher, that makes up at least 99% of the total volume of the mixture.
Combustion	A chemical oxidation reaction characterized by visible flame and heat evolution.
Combustion product	Material produced or generated during the burning or oxidation of a material.
Community Awareness and Emergency Response (CAER) Program	Program developed by the CMA, to assist chemical plant managers in taking the initiative in cooperating with local communities to develop integrated (community/industry) plans for responding to releases of hazardous materials.
Conditions	A particular state of being of a person, thing, or set of things.
Contingency plan	A document to identify and catalog the elements required to respond to an emergency, to define responsibilities and specific tasks, and to serve as a response guide.
Corrosive	A term used to describe a substance that actively attacks metal or tissue. Example: acids.
COSC	County on-scene coordinator.
Critical facilities	Facilities essential to emergency response, such as fire stations, police stations, hospitals, and communications centers.
Cryogenic	Relating to low temperatures. Danger of frostbite exists when working with cryogenic materials.
CVS	Cardiovascular system.

CWA — Clean Water Act.

Damage — Harm to an inanimate object which disrupts its intended functional continuity.

Danger — Risk of potential harm.

Decontamination — The removal of hazardous substances from employees and their equipment to the extent necessary to preclude the occurrence of foreseeable adverse health effects.

Dike — A barrier such as a low wall or embankment designed to prevent a spill from spreading or flooding during an emergency response. Can be sand, soil, hay, or whatever absorbent materials are available.

Dispersion mechanism — The process by which a material scatters in one or more directions.

Disposal — The removal of waste material to a controlled site or facility that is specifically designed and permitted to receive such waste.

Dissociation product — A substance produced as a result of the decomposition or dissolution of another substance. Usually used in reference to water pollution. Includes decomposition product and degradation product.

DOC — U.S. Department of Commerce.

DOD — U.S. Department of Defense.

DOE — U.S. Department of Energy.

DOI — U.S. Department of the Interior.

DOJ — U.S. Department of Justice.

DOL — U.S. Department of Labor.

DOT — U.S. Department of Transportation.

Dust — Airborne solid particles in the 0.1 to 25 microns size range.

EBS — See Emergency Broadcast System

EHS — See Extremely hazardous substance.

Emergency — A situation arising with or without warning, causing or threatening injury or death, disruption to normal life for numbers of people in excess of those which can be dealt with by the public services operating under normal conditions, and which requires prompt action, special mobilization, and organization of those services.

Emergency Broadcast
System   A volunteer program of the broadcasting industry that allows use of its facilities to transmit emergency information.

Emergency functions   Includes warning and communications services, relocation or evacuation of persons from stricken areas, temporary restoration of utilities, plant security, transportation, welfare, engineering, search, rescue, health, law enforcement, fire fighting and other activities necessary to protect the welfare of the community.

Emergency operations
simulations (EOS)   An exercise used primarily to headmistress the conduct of centralized operations in an Emergency Operations Center or interim EOS facility that simulates the use of available resources.

Emergency response   Action to ameliorate the consequences of an emergency situation.

Emergency services   Planning and preparation necessary to carry out emergency functions, other than functions for which military or other federal agencies are primarily responsible, to prevent or minimize the loss of life or property caused by disaster of any kind.

Endothermic   A process or chemical reaction that absorbs heat.

Environment   Includes water, air, land, and the interrelationship that exists among and between water, air, land, and all living things.

EOC   Emergency Operations Center: the facility at which Emergency Government operates during a major emergency.

EOS   See Emergency operations simulations.

EPA   U.S. Environmental Protection Agency.

EPCRA   The Emergency Planning and Community Right-To-Know Act. Title III of SARA.

ERP   Emergency response personnel to include; police, fire, and emergency medical services.

Etiologic agents   Viable microorganisms or their toxins that cause or may cause human disease.

Evaluation   The decision-making process resulting in a judgment or degree of risk.

Event — An occurrence, act of decision, or other action which is a discrete and identifiable occurrence in an ordered sequence of occurrences; connotes action, dynamic rather than static.

Exercise — A simulated accident or release set up to test emergency response methods and for use as a training tool. Evaluations are made of these simulated responses to determine levels of competencies.

Exothermic — A process or chemical reaction that emits heat.

Explosion-proof — An electrical apparatus so designed that an explosion of flammable gas inside the enclosure will not ignite flammable gases outside, such as instrumentation, flashlights, etc., that are declared "intrinsically safe".

Explosive — A substance that can decompose violently, releasing large quantities of gas and heat. Includes black powder, dynamite, TNT, priming explosives, nitroglycerine, detonators, blasting caps, munitions, chemical ammunition, rocket boosters, propellant explosives, low explosives, high explosives.

Explosive A — A detonating or otherwise maximum hazard.

Explosive B — A flammable explosive, including fireworks, rocket ammunition, smokeless powder, empty projectile ammunition, railway torpedoes, toy torpedoes.

Explosive C — An article containing small quantities of A and/or B explosives, including small arm ammunition, fuses, blasting cord, tracer fuses, electric delay igniters, jet starter cartridges, signal flares, trick matches, actuating cartridges.

Extremely hazardous substance (EHS) — Chemicals identified by the EPA and listed under Section 302 of EPCRA. The EPA list is based primarily upon the inherent acute toxicity of the substance, with occasional revisions that include criteria of chronic toxicity, reactivity, volatility, dispersability, combustibility, flammability, or other hazards of the material.

Facility Emergency Coordinator — Facility representative for each hazardous materials

	waste site or facility who participates in the emergency planning process and is the designated point of contact for communications
FDA	U.S. Food and Drug Administration.
FEMA	U.S. Federal Emergency Management Agency.
Fenceline	Outermost perimeter of a hazardous waste site or designated facility.
Flammable limits	The minimum (lower) and maximum (upper) concentration (%v/v) of a fuel in air that would explode or burn. The term explosive limits is also used. The flammable limit of toluene in air is 1.27 to 7.0%.
Flammable liquid	Any liquid with a flash point less than 100°F as measured by the tests specified in Section 173.115, 49 CFR, with the following exceptions:

(1) A flammable liquid with a vapor pressure greater than 40 psi at 100°F, as defined in Section 173.300.

(2) Any mixture having one component or more with a flash point of 100°F or higher that makes up at least 99% of the total volume of the mixture.

(3) A water-alcohol solution containing 24% or less alcohol by volume if the remainder of the solution does not meet the definition of a hazardous material.

Flash point	The lowest temperature at which a material will ignite when an ignition source is introduced.
FOSC	Federal or First On-Scene Coordinator/Commander.
GIT	Gastrointestinal tract.
HACS	See Chemical Hazards Response Information System.
Hazard	Any condition that has the potential for causing damage to life, property, and/or the environment, or which may interrupt or interfere with the expected, orderly progress of an activity
Hazard analysis	The procedure for identifying potential sources of a hazardous material release, determining the vulnerability of an area to a hazardous materials release, and comparing hazards to determine risks to a community.

Hazard category

A system of classifications of hazards to life, health, or property.

(1) "Immediate (acute) health hazard" includes "highly toxic," "toxic," "irritant," "sensitizer," or "corrosive," and other hazardous materials that cause an adverse effect to a target organism and which effect usually occurs rapidly as a result of short-term exposure.

(2) "Delayed (chronic) health hazard" includes "carcinogens," "fetotoxins," "mutagens," "teratogens," and other hazardous materials that cause an adverse effect to a target organism after the period of initial exposure.

(3) "Fire hazard" includes "flammables," "combustibles," "pyrophoric," "oxidizers," and other hazardous materials.

(4) "Sudden release of pressure hazard" includes hazardous materials that are "explosive" or "compressed."

(5) "Reactive hazard" includes "unstable reactives," "organic peroxides," "water reactives," and other hazardous materials.

Hazard Communication
Standard

29 CFR 1910.1200 or the regulation giving workers the right to know about the risks and proper handling of hazardous chemicals in the work place.

Hazard identification

Provides information on which facilities have hazardous material, what those materials are, how much there is at each facility, how the chemicals are stored and handled, and whether or not they are used at high temperatures.

Hazardous chemical

A kind of hazardous material defined by federal law in OSHA 29 CFR as any chemical which presents a health or physical hazard, except as follows:

(1) Any food, food additive, color additive, drug, or cosmetic regulated by the Food and Drug Administration (FDA);

(2) Any substance present as a solid in any manufactured item to the extent exposure

does not occur during normal conditions of use;

(3) Any substance used for personal, family, or household purposes or present in the same form and concentration as a product packaged for distribution and used by the general public;

(4) Any substance used in a research laboratory or a hospital or other medical facility under the direct supervision of a technically qualified individual;

(5) Any substance used in routine agricultural operations or as a fertilizer held for sale by a retailer for sale to the ultimate customer.

Hazardous material   Any substance or material in any quantity or form that may be harmful to humans, animals, crops, water systems, or other elements of the environment. Used in the context of this book, the term may include hazardous chemical, hazardous substance, toxic chemical, toxic substance, explosive gases (compressed, liquefied, or dissolved), flammable and combustible liquids, flammable solids or substances, oxidizing substances, poisonous solids and infectious substances, radioactive materials, or corrosives.

Hazardous substance   As designated under CERCLA, incorporated substances listed under the Clean Water Act, the Clean Air Act, RCRA, and TSCA Section 7.

Hazmat   See Hazardous material.

HCA   Hazard Communications Act.

HCS   Hazard Communication Standard.

HEW   U.S. Department of Health, Education, and Welfare, now called Health and Human Services.

HHS   U.S. Department of Health and Human Services, 200 Independence Ave., S.W., Washington DC 20202.

HSDB   Hazardous Substance Databank of the National Library of Medicine.

IARC   International Agency for Research on Cancer.

IDLH   Immediately Dangerous to Life or Health.

Ignition temperature   See Auto-ignition temperature.

IMCO   International Maritime Consultation Organization.

Incident

An actual or potential spill or other release of a hazardous material requiring emergency response.

Incident Command Post

Facility located a safe distance upwind from an accident site, where the on-scene coordinator or Incident Commander, responders, and technical representatives can make response decisions, deploy manpower and equipment, maintain liaison with media, and handle communications.

Incident Commander

The ranking on-scene official of the jurisdiction within which a hazardous incident occurs.

Insidious

Acting in imperceptible steps; having a gradual cumulative effect.

Irritant

A substance that causes a local inflammatory reaction upon contact. Does not usually include Poison A substances.

LC_{50}

The concentration of a toxic material which, when administered to test animals, kills half of the population in a specified period of time.

LD_{50}

The dose of a toxic material which, when administered to test animals, kills half the population in a specified period of time.

LEL

Lower explosive limit. See Flammable limits.

Level of concern (LOC)

The concentration of hazardous substance above which there may be serious irreversible health effects or death as a result of a single exposure.

Local emergency

The existence of conditions of disaster or of extreme peril to the safety of persons or property within the territorial limits of a county, city, or town which conditions are or are likely to be beyond the control of the services, personnel, equipment, and facilities of such political subdivisions as determined by its governing body and which require the combined efforts of other political subdivisions.

Material Safety Data Sheet (MSDS)

A compilation of information required under EPCRA and the OSHA Hazard Communication Standard on the identity of hazardous chemicals, health and physical hazards, exposure limits, fire-fighting requirements, disposal requirements, and

	other precautions when storing and using the chemical.
Melting point	The temperature at which a solid is at equilibrium with its liquid phase.
mg/m^3	Milligrams per cubic meter.
Mixture	A heterogeneous association of substances where the various individual substances retain their identities and can usually be separated by mechanical means.
mppcf	Million particles per cubic foot (of air).
MSDS	See Material Safety Data Sheet.
Mutagen	A substance which can alter the genetic structure.
NACA	National Agricultural Chemicals Association.
National Fire Academy (NFA)	A component of FEMA's National Emergency Training Center located in Emmettsburg, MD. It provides courses in fire prevention and control training for the fire service and allied services. Courses on campus are offered in technical, management, and prevention subject areas. An off-campus delivery system is operated in conjunction with state fire training programs.
National Response Center (NRC)	A communications center for activities related to response actions, located at Coast Guard Headquarters, Washington DC. The National Response Center receives and relays notices of discharges or releases to the appropriate OSC, disseminates OSC and RRT reports to the NRT when appropriate, and provides facilities for the NRT to use in coordinating a national response action when required. The toll-free number (800-424-8802) is operational 24 hours a day for reporting actual or potential pollution accidents.
NCRIC	National Chemical Response and Information Center (CMA).
NEIC	EPA's National Enforcement and Investigation Center, Denver, CO 80202.
Neoplastic effect	A new or abnormal growth, as in a tumor.
NETC	National Emergency Training Center.
Neurological	Relating to the nervous system.

NFA   See National Fire Academy.
NFPA   National Fire Protection Association.
ng   The abbreviation for the nanogram, which is one billionth of a gram, sometimes represented in the following mathematical notation: 10^{-9} g.
NIOSH   National Institute of Occupational Safety and Health, 4676 Columbia Parkway, Cincinnati, OH 45226.
NIOSHTIC   Electronic, on-line database for occupational health information.
NOAA   National Oceanic and Atmospheric Administration.
NRC   Nuclear Regulatory Commission; National Response Center (see).
NRT   National Response Team. Federal team charged under EPCRA with planning for and responding to hazardous material emergencies. Team consists of representatives of 14 different federal agencies: DOD, DOI, DOT/RSPA, DOT/USCG, EPA, DOC, FEMA, DOS, USDA, DOJ, HHS, DOL, NRC, and DOE.
NRT-1   Hazardous Materials Emergency Planning Guide, prepared by the National Response Team.
NSC   National Safety Council, 425 North Michigan Ave., Chicago, IL 60611.
NUREG 0654/FEMA-
  REP-1   Criteria for preparation and evaluation of radiological emergency response plans and preparedness in support of nuclear power plants, prepared by NRC and FEMA.
OHM-TADS   The oil and hazardous materials technical assistance data system developed by the EPA.
On-scene   The pre-designated local, state, or federal official responsible for the coordination of a hazardous materials response action, as outlined in the pertinent Emergency Response Plan or Guide.
Operational field
  exercises   Any exercise designed primarily for the purpose of testing and evaluations of the integrated emergency preparedness capability of community/county and/or state jurisdictions in an operational environment. Includes mobilization of personnel and resources sufficient to demonstrate on-site coordination and support.

Organic peroxide	Organic peroxide is an organic compound containing the bivalent –O–O– structure and which may be considered a derivative of hydrogen peroxide where one or more of the hydrogen atoms have been replaced by organic radicals. One or more of the oxygen atoms can be readily released. Examples: Chlorate, permanganate, inorganic peroxide.
OSC	On-scene coordinator.
OSHA	Occupational Safety and Health Administration, United States Department of Labor, Washington DC 20210.
OSHRC	Occupational Safety and Health Review Commission, 1925 K Street, NW, Washington DC 20006.
Oxidizer	A substance that readily reacts with oxidizable materials (organics, solvents, etc.) to release oxygen and stimulate combustion.
Pesticide Safety Team Network (PSTN)	Operated by the National Agriculture Chemicals Association to minimize environmental damage and injury arising from accidental pesticide spills or leaks.
pg	Picogram, one trillionth of a gram, 10^{-12} g.
pH	Means of expressing degree of acidity or alkalinity of a solution on a scale of 0 to 14. Acids have a pH lower than 7. A neutral solution, such as water, has a pH of 7.0. Alkaline solutions have a pH higher than 7.
Placard	A sign posted for notification that hazardous materials are being used in that work area, being held in storage, or are on board the vehicle.
Plume	Effluent cloud resulting from a continuous source release.
Poison	A substance that is toxic at low levels. The general definition is a substance with LD_{50} of 50 mg/kg body weight or less (as determined by animal studies).
Poison A	Extremely dangerous poisons; poisonous gases or liquids of such nature that a very small amount of the gas, or vapor of the liquid, mixed with air is dangerous to life. Examples: Hydrocyanic acid, arsine, phosgene.

Poison B	Less dangerous poisons; substances, liquids, or solids, including pastes and semi-solids, other than Class A or irritating materials, which are know to be so toxic to man as to afford a hazard to health during transportation, or which in the absence of adequate data on human toxicity are presumed to be toxic to man. Examples: Aniline, arsenic.
PPM	Parts per million.
PSTN	See Pesticide Safety Team Network.
Pulmonary	Relating to lungs and lung functions.
Radioactive material	Any material, or combination of materials, that spontaneously emits ionizing radiation and has a specific activity greater than 0.002 microcuries per gram.
Radioactivity	The property of some elements to emit gamma rays or alpha or beta particles by the disintegration of the nucleus of the atom.
RCRA	Resource Conservation and Recovery Act.
Reaction	(1) Chemical: The combining of two or more materials into another, the chemical breakdown of a chemical material, or the combining of a chemical into a polymer; includes oxidation.
	(2) Physical: The response to a force applied.
Reactive	A class of materials capable of chemical reactions.
Recovery	Short-term activities necessary to return vital systems and facilities to minimum operating standard and long-term activities required to return life to normal improved levels.
Reducing agent	Any substance which reduces another. Incompatible with oxidizers.
Regional response team (RRT)	The federal response team established under EPCRA to respond to hazardous materials emergencies of a magnitude such that the incident cannot be handled at the state level.
Regulated substance	A legal term referring to hazardous materials, when defined in various federal rules and regulations.
Release	Any spilling, leaking, pumping, pouring, emitting, emptying, discharging, injecting, escaping, leaching, dumping, or disposing into the environment

	(including the abandonment or discarding of barrels, containers, and other closed receptacles) of any hazardous material.
REP	Roentgen equivalent physical. The dose of ionizing radiation that produces energy absorption of 93 ergs per gram of body tissue.
Reportable quantity (RQ)	The quantity of a hazardous substance that triggers reporting under CERCLA; if a substance is released in a quantity that exceeds its RQ, the release must be reported to the NRC, as well as to the AERC and the community emergency coordinator, for an area likely to be affected by the release.
Response	The efforts to minimize the risks created in an emergency by protecting the people, the environment, and property, and the efforts to return the scene to normal pre-emergency conditions.
Risk	(1) Exposure to a hazard or danger.
	(2) A measure of the probability and severity of damage to life, property, and/or the environment that may occur if a hazard manifests itself.
Risk analysis	Assessment of the probable damage that may be caused to the community by a hazardous material release.
RQ	See Reportable quantity.
RSPA	Research and Special Programs Administration (DOT).
SARA	See Superfund Amendments and Reauthorization Act of 1986.
SCBA	Self-contained breathing apparatus.
Sensitizer	A substance which can cause an allergic-like reaction.
Site Safety and Health Officer	The individual located at an incident or hazardous waste site who is responsible to the employer or commander and has the authority and knowledge necessary to implement the site safety and health plan and verify compliance with applicable safety and health requirements.
SPCC	Spill prevention control and countermeasures.
Special populations	Groups of people that may be more susceptible than

the general population due to preexisting health conditions (e.g., asthmatics) or age (e.g., infants and the elderly) to the toxic effects of an accidental release.

Specific gravity
The ratio of density of a substance to that of water under specified conditions. The specific gravity of toluene at 20°C (68°F compared to water at 4°C [39°F]) is 0.8669.

State of emergency
The duly proclaimed existence of conditions of disaster or extreme peril to the safety of persons or property or environment caused by air pollution, fire, flood or floodwater, storm, epidemic, riot, earthquake, or other causes, except those resulting in a state of war emergency, which are or likely to be beyond the control of services, personnel, equipment, and facilities of any single county, city, or town and which require the combined efforts of jurisdictions.

Superfund
Another name for CERCLA and the trust fund established under CERCLA to provide money the OSC can use during cleanup of hazardous materials incidents or waste sites.

Superfund Amendments and Reauthorization Act of 1986 (SARA)
The Act Title III.

Systemic effects
These affect an entire organ or the entire body.

TCC
Tagliabue closed cup — a test procedure to determine flammability/combustibility.

Teratogen
A substance which may cause physical defects in the developing embryo.

Threshold Planning Quantity (TPQ)
A quantity designated for each chemical on the list of extremely hazardous substances (EHSs) that triggers notification by facilities of the AERC that such facilities are subject to emergency planning under EPCRA.

Title III
See EPCRA.

TLV
Threshold limit value — an estimate of the average concentration of a toxic substance that can safely be tolerated over an 8-hour workday for a 40-hour workweek throughout a working lifetime.

TOC

Toxic Chemical Release Form

Toxic cloud

Toxic substance

Toxicity

Toxicity class

TOXNET

TPQ

TWA

UEL

Unstable

USCG

USDA

USGS

USNRC

Vapor density

Tagliabue open cup — test procedure.

Information form required to be submitted by facilities that manufacture, process, or use (in quantities above a specified amount) chemicals listed in Section 313 of EPCRA.

Airborne mass of gases, vapors, fumes, or aerosols of toxic materials.

A substance which can produce injury or illness through ingestion, inhalation, or absorption.

The ability of a substance to cause damage to living tissue, impairment of central nervous system function, illness, or death when ingested, inhaled, or absorbed by the skin.

An EPA classification of pesticides and chemicals according to acute toxicity.

Electronic database network of the National Library of Medicine.

See Threshold Planning Quantity.

Time-weighted average — the sum of the products of the toxicant concentration and the exposure duration divided by the total exposure time. Normally based on an 8-hour exposure, but by the following example can be any time: A 2-hour exposure of 150 ppm, 2 hours at 75 ppm, 4 hours at 50 ppm, and 2 hours at 10 ppm. $(2 \times 150) + (2 \times 75) + (4 \times 50) + (2 \times 10) = 670$. If the TWA were 100 ppm, this would be an acceptable level of exposure, but if the TWA were 50 ppm then this would be an example of overexposure and out of compliance with the regulations.

Upper explosive limit. See Flammable limits.

The readiness of a material to react or otherwise behave unexpectedly or erratically.

U.S. Coast Guard.

U.S. Department of Agriculture.

U.S. Geological Survey.

U.S. Nuclear Regulatory Commission.

The relative density of a vapor as compared to air at the same temperature. The vapor density of

	ether, for example, is 2.55 and that of toluene is 3.14.
Vapor dispersion	The movement of vapor clouds or plumes in air due to wind, gravity spreading, and mixing.
Vapor pressure	The pressure exerted by the vapor above the liquid surface. The vapor pressure of toluene is 21.8 mmHg at 20°C (68°F).
Viscosity	The internal resistance to change exhibited by a liquid or a gas. The viscosity of toluene is 0.587 cP at 20°C (68°F) and 21 cP for ethylene glycol at 20°C (68°F).
Warning	A term indicating an intermediate degree of hazard in precautionary labeling — between danger and caution.

INDEX

431

calibration
  field instruments 169
  oxygen-monitoring equipment 176
cancer
  carcinogens 29–31
  definitions 20
  fear of 31
  history of 29
  liver function 16
  radiation and 163
carbamates 143, 144
carbon dioxide
  blood circulatory system and 12
  oxygen-monitoring equipment and 176
  respiration, anatomy and physiology 4, 5
carbon monoxide
  incident site hazard categories 172
  oxygen-monitoring equipment and 176
carcinogens
  defined 20
  toxicology 29–31
cardiopulmonary resuscitation 144
cardiovascular system *See* blood circulatory system
cartridges, respirators 190, 192
ceiling, TLV 59 *See also* Threshold Limit Values
central nervous system 8
  carcinogens 30
  classes of toxins 21, 22
  symptoms of pesticide poisoning 143
CERCLA *See* Comprehensive Environmental Response, Compensation, and Liability Act
cerebellum 9, 42
cerebrum 9, 42
certification, respirators 197
CFR *See* Code of Federal Regulations, specific titles
CGI *See* combustible-gas indicator

chemical agents, defined 19
chemical names, Material Safety Data Sheet information 83
chemical-resistant clothing, water operations 129
chemicals *See also* pesticide incidents; toxicology
  atmospheric *See also* airborne contaminants
  chemistry 36–39
  classes of toxins 21, 22
  in cleanup operations 78
  defined 19
  exposure effects *See* exposure effects
  exposure standards 51
  hazardous, defined 35
  incident/site hazard categories 172
  incompatible 35, 38–39
  liver function 16
  Material Safety Data Sheet information 83
  personal protection equipment 129–131
  routes of entry 40
  safety 40
Chemical Transportation Emergency Center 115–116
chemistry
  defined 35
  Material Safety Data Sheet information 83
CHEMTREC 115–116
chlorophenols 143–144
chromosomes
  anatomy and physiology of reproductive system 9
  mutagens, defined 20
  sex cells 10
chronic exposure 21
  defined 19
  pesticides 141–142
chronic toxins, classes of 27–28
cilia 4

incident command checklist 121
Incident Commander responsibilities 121
safety plan for response team 104
explosions
  chemical safety 40
  combustibility monitoring equipment 176–178
  gas detectors 184
  instrumentation and 172–183
explosives
  classification of hazardous materials 106–108
  international warning systems 65
  Material Safety Data Sheet information 83
  personal protective equipment 127
  placarding 114–115
  United Nations hazard classification 112–113
exposure, definitions 3, 19
exposure effects
  chemical safety 40
  chemistry 36–39
  circulatory system 43–44
  definitions 35
  digestive system 41–42
  endocrine system 43
  eye 42–43
  incompatible chemicals 38–39
  industrial hygiene 44–46
  liver 44
  nervous system 42
  reproductive system 43
  respiratory system 41
  routes of entry 40
  skin 40–41
  urinary system 44
exposure limits 179–180 *See also* Permissible Exposure Limits
"extremely toxic" category 22, 23
eyes
  anatomy and physiology 9, 10
  chronic exposures 21
  exposure effects 42–43

pesticide exposures 142–144
radiation and 159
selection of protective equipment 135, 136

federal regulation *See* environmental protection; NIOSH; OSHA; OSHA standards; worker protection
female sex cells 10–11
female urinary system 14–15
fertilization 11
fetal development
  hazardous chemical effects 43
  teratogens 20
FIDs *See* flame ionization detectors
field decontamination 149–150
field instruments 168
  air monitoring 168
  construction specifications 169–170
  guidelines for use 171–172
  safety considerations 172–174
field maintenance 152
filters, decontamination of equipment 150–151
fire fighters
  personal protection equipment 126–127, 130, 133
  personnel hazards 101
  pesticides, smoke from 142
  self-contained breathing apparatus requirements 192
  worker protection regulations of National Fire Protection Association 63, 64, 67
fire hazard
  classification of hazardous materials 106–109
  combustibility monitoring equipment 176–177
  gas detectors 184
  incident/site hazard categories 172
  instrumentation and 172–183
  Material Safety Data Sheet information 83

OSHA standards 81–82
program elements 90–91
hazardous chemicals *See also*
chemicals
defined 35
routes of entry 40
safety 40
hazardous materials *See also* toxicology
classification of 106 *See also*
classification of materials
exposure effects *See also* exposure
effects
incident command checklist 121
warning system *See* warning
systems
Hazardous Materials Transportation
Act 55
hazardous waste
classification of hazardous
materials 106
defined 71
generators 72, 74
identification of 71–72
OSHA standards for site construction 52–53
pesticide poisoning incidents 144
storage, treatment, and disposal
facilities 76–77
health hazards 20 *See also* toxicology
Material Safety Data Sheet
information 83
worker protection *See* worker
protection
heart 12, 42
heart rate
pesticide poisoning, first aid in 144
symptoms of pesticide poisoning
143
heat 7, 185 *See also* temperature
heavy metals
carcinogen classification 30–31
chronic exposures 21
in cleanup sites 78
as teratogen 27
hematopoietic toxins 21, 22

hepatotoxins 21, 22
hormonal carcinogens 30
hormones
hazardous chemical effects 43
transport in blood 13
hygiene *See* personal hygiene
hyperthermia 98, 185
hypothermia 98–99, 185

identification of hazards, determination
of levels of protection 133–137
IDLH *See* immediately dangerous to
life and health conditions
ignitability 72 *See also* fire hazards
ileum 7
immediately dangerous to life and
health (IDLH) conditions
oxygen deficiency and 184
respirator use 189, 191
immunosuppressive carcinogens 30, 31
Incident Commander 101, 119–122
incident command system 101
cessation of response 121–122
command checklist 121
responsibilities of Incident
Commander 120–121
incidents
equipment repair during 151
personnel *See* personnel hazards
pesticide *See* pesticide incidents
response safety plan 103–104
site monitoring *See* site monitoring
incompatible chemicals 38–39
chemistry 35–39
defined 35
safety 40
individual differences in biological
response 25–26
industrial hygiene 44–46
defined 35
history of 45–46
infectious agents 66
inflammable materials *See* fire hazards
information on health hazards *See*
hazard communication
infrared radiation 156, 158